Ross

Vehicle Safety and Future Mobility

Hans-Leo Ross

Vehicle Safety and Future Mobility

Social, Legal and Safety Solutions

With contributions from Ole Hans and Krystian Radlak

HANSER

About the author:
Hans-Leo Ross, Lorsch

Print-ISBN: 978-1-56990-267-7
E-Book-ISBN: 978-1-56990-323-0

Bibliographic information of the German National Library:
The German National Library lists this publication in the German National Bibliography; detailed bibliographic data are available on the Internet at http://dnb.d-nb.de.

Vilshofener Straße 10 | 81679 Munich | info@hanser.de
www.hanserpublications.com
www.hanser-fachbuch.de
Editor: Julia Stepp
Production Management: le-tex publishing services GmbH, Leipzig
Cover concept: Marc Müller-Bremer, www.rebranding.de, München
Cover design: Max Kostopoulos
Cover picture: © Max Kostopoulos, with the use of chatgpt.com
Typesetting: Eberl & Koesel Studio, Kempten

Contents

Forewords

In this book, Hans-Leo Ross embarks on a critical exploration of contemporary transportation systems, which have historically been designed for human operation. This poses significant challenges as we consider the role of Artificial Intelligence (AI) in automating these systems, given that current solutions and regulations are deeply rooted in human-driven paradigms.

Hans-Leo Ross raises the essential question of whether AI should replicate human behavior—the behavior of the driver it aims to replace—or if new methodologies can be developed to create safer and more efficient mobility solutions. He suggests that it may be through innovative automation concepts that genuine progress can be achieved.

This book seeks to illuminate a path toward a compromise that is acceptable to society. Hans-Leo Ross aims to address pressing concerns about the integration of AI into transportation systems, emphasizing the importance of prioritizing safety, efficiency, and societal acceptance.

This book marks an exciting and important journey as Hans-Leo Ross envisions a future where technology enhances our mobility experience, all while safeguarding the values and safety we hold dear.

Prof. Dr.-Ing. habil. Mario Trapp, June 2025
Executive Director of the Fraunhofer Institute for Cognitive Systems IKS
Chair for Engineering Resilient Cognitive Systems, Technical University of Munich

Hans-Leo Ross brings decades of hands-on experience in the field of autonomous driving to this book. He explores current developments such as risk and safety management, the role of the driver in autonomous and teleoperated vehicles, and the application of artificial intelligence methods.

From the perspective of a seasoned engineer, he presents not only well-founded insights but also practical approaches that are highly relevant for those involved in the development of autonomous or semi-autonomous vehicles.

I therefore highly recommend this book to all engineers working in the automotive sector. Moreover, professionals in aerospace and robotics will also find valuable and transferable knowledge within these pages.

Prof. Dr.-Ing. Jürgen Adamy, June 2025
Department of Control Methods and Intelligent Systems, Technical University of Darmstadt

Preface

In the ever-accelerating trajectory of mobility—particularly within the domain of automotive innovation—the prevailing discourse has long been captivated by technological triumphs and the sheer brilliance of engineering. While the critical importance of these advancements is beyond dispute, this book advocates for a necessary recalibration: placing the human being unequivocally at the center of our pursuit of safer and more sustainable mobility. Previous analyses may have inadvertently marginalized the profound and intricate ways in which human beings are shaped by, and in turn shape, the dynamic landscape of transportation. We contend that a truly comprehensive understanding of safe mobility can only be attained by acknowledging the multifaceted roles individuals inhabit. They are not merely users and operators, drivers and passengers, but also—crucially—vulnerable road users: cyclists, pedestrians, and countless others who navigate our increasingly complex thoroughfares.

To grasp this central human element, however, we must delve deeper than superficial categorizations. A human being is not reducible to software, nor solely the product of an intelligent brain. Rather, the human capacity to navigate the world, interact with complex systems, and ensure safety has evolved through intricate learning loops—loops forged through interpersonal connection, the experience of diverse environments, the richness of cultural variation, and the sophisticated use of language. Human competence in the physical domain is inextricably linked to the mastery of bodily coordination. This coordination transcends basic neural networks; it involves the purposeful deployment of limbs and body segments, often mediated via the spinal cord, through energy and information transfer along what appear to be deterministic pathways to actuators such as muscles. This embodied intelligence, refined over millennia, forms the bedrock of safe interaction with mobility systems.

The spectrum of potential endangerment is far from uniform; it varies dramatically across different modes of transport, each presenting unique challenges and necessitating bespoke safety paradigms. Simultaneously, the societal lens through which we view mobility is broadening. Issues once relegated to the periphery—such as noise pollution and environmental degradation—are now rightly assuming prominence in public discourse and regulatory frameworks. This necessitates a more inclusive approach, one that accords greater attention to the safety, comfort, and inherent rights of non-motorized road users and the communities profoundly affected by transportation systems. Safety, in its most authentic form, must be examined through the dual prisms of societal expectation and individual perception. It is an undeniable truth that comparable risks are often perceived and evaluated differently across demographic groups and cultural contexts. This subjectivity is a pivotal factor in the design and implementation of safety measures that are both acceptable and effective.

We find ourselves at a transformative juncture, where intelligent solutions—powered by sophisticated digital systems and Artificial Intelligence—are reshaping the very fabric of mobility. These technologies offer enticing possibilities: heightened efficiency, unprecedented speeds, and novel business paradigms. Yet, this technological ascent casts long shadows. While certain demographics may reap substantial benefits, others express legitimate concerns, perceiving these technologies as intrusive, disruptive, or even threatening to their well-being and livelihoods.

A focal point of contemporary debate is the emergence of autonomous systems—machines engineered to learn and, ostensibly, to make independent decisions. This raises a critical ethical and societal question: Can we, with integrity, delegate life-critical decision-making authority to machines? Especially in scenarios where human lives, animal welfare, or environmental integrity may be jeopardized in pursuit of a transport objective?

Should our collective focus not instead pivot towards automation solutions that demonstrably offer clear, transparent benefits and are meticulously managed by accountable organizations? Is it not imperative that human beings retain ultimate oversight—actively guiding development and diligently monitoring deployment—to ensure these advanced mobility solutions serve humanity responsibly? Within this framework, Artificial Intelligence is not excluded; rather, it is precisely positioned as a powerful tool fulfilling a clearly defined purpose. Those who benefit from AI-driven solutions must also bear full responsibility for the products and services they underpin. Accountability for their application—especially in cases of misuse or liability—must reside unequivocally with human agents.

It is within this context that a further, indispensable truth must be acknowledged: a mobility solution suitable for a mature and responsible society must always be underpinned by a stable and accountable organizational structure. Such an organization must act with integrity, operating within the bounds of legal frameworks while responding conscientiously to the evolving needs of society. It must strive to develop mobility systems that are not only technically sound but also socially acceptable, environmentally sustainable, and purposefully aligned with the public good. These are not merely technical puzzles demanding an engineering solution; they are profound societal challenges demanding collective deliberation and action.

Over the past four years, my esteemed colleagues—Dr. Krystian Radlak and doctoral candidate Ole Hans—and I, Hans-Leo Ross, have been deeply engaged in intensive research on remote driving technologies. Throughout this endeavor, safety has remained the immutable core of our focus. Our efforts culminated in a significant milestone in 2023, when we successfully introduced the first vehicles operating without a physically present human driver onto public roads in Hamburg. It is noteworthy that commercial services based on similar technologies are already operational in cities such as Las Vegas.

This book addresses not only engineers and technical experts, but also speaks directly to investors, managers, consultants, and policymakers. It is intended as a resource for jurists and economists alike—those whose decisions shape the regulatory, financial, and ethical contours of our mobility future. Our profound aspiration is to illuminate the vast panoply of opportunities presented by this exhilarating new digital epoch. By doing so, we aim to proactively chart a course towards a new, unequivocally responsible horizon for the development and deployment of safe and secure future mobility solutions. We hold a steadfast conviction that the future of mobility must not merely be "smart;" it must be just, inclusive, and profoundly human-centered.

In the compilation of this book, we have judiciously employed Artificial Intelligence as a translational and structuring tool. Nevertheless, we have relied unwaveringly on our accumulated knowledge, professional judgment, and a deeply ingrained sense of responsibility for the accuracy and implications of the ideas presented herein. The foundational concepts, guiding values, and ultimate responsibility for this work remain entirely human.

Hans-Leo Ross, June 2025

1 Human Mobility

Human mobility, encompassing migration, daily movement, and travel, has been a constant feature throughout history. It shapes societies, cultures, and economies across time and space. The desire to explore, seek better opportunities, and connect with distant places is deeply ingrained in human nature. Whether driven by curiosity, necessity, or ambition, mobility is a fundamental need. People move due to a combination of pressures at home (push factors) and attractions elsewhere (pull factors). These factors include economic opportunities, safety, cultural exchange, and personal aspirations. Throughout history, mobility has been influenced by technological advancements, societal changes, and geopolitical events. Trade, commerce, and exchange of goods and ideas thrive through mobility. The movement of people facilitates economic growth by connecting markets, fostering innovation, and enabling specialization. Mobility allows for the exchange of languages, traditions, art, and knowledge. It enriches societies by promoting diversity and cross-cultural understanding. Scholars, scientists, and artists have historically travelled to share knowledge, collaborate, and learn from different cultures. Think of the Silk Road or the Renaissance. Mobility enables access to healthcare, education, and resources. It can improve quality of life and well-being.

Around 1.75 million years ago, Homo erectus began their migration out of Africa, spreading across Eurasia. Modern humans (Homo sapiens) evolved from their early hominid predecessors between 200,000 and 300,000 years ago. They developed language capacity around 50,000 years ago. Homo sapiens colonized all of Africa about 150,000 years ago. Some members of this species moved out of Africa 70,000 years ago (or even earlier, around 125,000 years ago) into Asia and Europe.[1] [2] By 40,000 BCE, modern humans had spread across Australia, Asia, and Europe. They adapted to

1 *https://en.wikipedia.org/wiki/History_of_human_migration*

2 *https://bing.com/search?q=human+migration+history*

diverse environments, following animal herds and river courses. Approximately 20,000 to 15,000 years ago, humans migrated to the Americas. They crossed the Bering Strait land bridge, eventually reaching the southern tip of South America. In the Americas, they encountered indigenous populations, possibly including those from Australia or Oceania.

The pace and extent of migration depended on climate conditions, population pressures, and technological innovations (such as boats). The transition from hunting and gathering to agriculture (Neolithic Revolution) led to population growth and further migrations. Indo-European migrations shaped Europe and parts of Asia. Even in ancient China, the Silk Road facilitated the movement of goods, technologies, and information. The trade in silk, spices, and ideas shaped civilizations. Military conquests, territorial expansion, and strategic interests have driven mobility. Empires sought dominance through conquest and control of trade routes. Today, we witness large-scale migration due to conflict, persecution, and environmental factors. People risk dangerous journeys, often relying on traditional means of travel. Despite modern innovations, traditional modes of movement (like walking) persist. Not everyone can access advanced transport technologies. Human mobility is both a timeless phenomenon and a dynamic force that shapes our world. It reflects our aspirations, challenges, and the intricate interplay of historical, social, and technological factors.

1.1 Motivation

Since the early 20th century, the concepts of individual mobility and mass mobility have undergone significant transformations. These changes reflect shifts in technology, societal norms, and political landscapes. Henry Ford revolutionized the automotive industry by introducing the Model T and pioneering assembly line production. This democratized car ownership, making automobiles accessible to a broader population. The assembly line not only streamlined production but also led to the establishment of a robust supplier industry. Ferdinand Porsche built upon Ford's ideas. His development of the Volkswagen Beetle aimed to provide an affordable and reliable vehicle for the masses. The Beetle became an iconic symbol of accessibility and practicality. After World War II, the motor vehicle symbolized economic prosperity and progress. It stood as a testament to the "economic miracle." Commercial transportation shifted from ships and railways to trucks and buses, playing a crucial role in global trade and logistics.

Aviation also experienced growth during this period. However, environmental concerns were often ignored. Despite its expansion, aviation faced criticism for its impact on the environment.

Individual mobility evolved from a luxury to a necessity, while mass mobility transformed the way goods and people moved across the globe.

The economic growth did not consider resources, environmental sustainability, or social needs. Instead, it focused on economic success, expansion, performance, and patriarchal corporate cultures, which were socially recognized and even envied by society.

The automobile became an integral part of everyday life. As we continue to explore new modes of transportation, balancing progress with environmental responsibility remains crucial. However, the automotive industry has faced numerous challenges over the years. The "Dieselgate" scandal exposed the lengths to which manufacturers would go to circumvent environmental regulations, leading to a loss of trust among consumers. The COVID-19 pandemic and subsequent waves brought about a reevaluation of public transportation and the necessity of personal vehicles, as people sought safer travel options.

Environmental concerns have come to the forefront, with discussions on the impact of vehicles on urban spaces and the consumption of land for infrastructure. Attempts at electrification and a shift towards sustainable energy sources have been met with mixed success, often hindered by economic and political crises, such as the conflicts in Europe and Gaza. The rise of right-wing populism, exemplified by figures like Donald Trump, has influenced nationalist and self-centered behaviors, affecting cooperation and progress on global issues.

In Germany, expectations for public transportation improvements have not been met, partly due to the neglect of digital infrastructure. Renewable energy sources and alternative mobility solutions have yet to show stable trends for the future. Advanced concepts like the Hyperloop and maglev trains remain far from becoming mainstream modes of transport.

Although Electric Vehicles (EVs) are a step towards sustainability, are expensive and quickly become outdated due to rapid innovation cycles. Used EVs with aging batteries suffer from significant depreciation. The dream of a fully autonomous and software-driven future has been tempered by reality, as incidents involving autonomous vehicles from companies like Cruise and Waymo have shown the limitations of current technology.

Remote mobility solutions may offer alternatives for urban environments, while autonomous driving could be more suited for highways. China's push for autonomous driving infrastructure and the export of Battery Electric Vehicles (BEVs) to Europe raises questions about maintenance, service, and repair, especially in the event of accidents. Workshops struggle with the complexity of software-defined vehicles, and over-the-air updates cannot solve all issues.

The question of vehicle ownership is also evolving. Leasing may address concerns about depreciation, but if a vehicle's value is defined by its software, does more software equate to higher value? Or is software merely a tool for functionality, usability, and quick problem-solving?

The journey of individual mobility from the days of Ford and Porsche to the present has been marked by innovation and setbacks. As we look to the future, the industry must navigate technological advancements, societal shifts, and environmental imperatives to redefine what personal transportation means in a rapidly changing world. The path forward will likely be a blend of traditional concepts and emerging technologies, striving for a balance between individual needs and collective responsibilities.

1.2 Development of Human Mobility

Human mobility is the ability and willingness of humans to move from one place to another, either within or across geographic boundaries. It has been a fundamental aspect of human history, culture, and society, and has shaped the development of civilizations, innovations, and identities. The development of humanity has been closely intertwined with the need for mobility. Not only did early humans follow herds or watercourses to secure food, but they also engaged in transportation tasks, by moving tools, shelter, and food to storage or distribution points. The technological evolution began in Mesopotamia with the invention of the wheel, and who knows, perhaps it will continue for centuries, culminating in something as futuristic as the "hyperwarp".

The concept of mobility extends beyond physical movement—it encompasses social, economic, and cultural dimensions. Here are some key points:

- **Historical mobility:** Early humans migrated to find better resources, escape harsh conditions, or explore new territories. This mobility was essential for survival and adaptation. The invention of the wheel in ancient Mesopotamia revolutionized transportation, enabling the movement of goods and people more efficiently.
- **Technological advances and mobility:** Throughout history, technological innovations (such as ships, trains, and automobiles) have enhanced mobility. These developments facilitated trade, communication, and cultural exchange. The ongoing quest for faster, more efficient modes of transportation continues until today, with concepts like the "hyperwarp" capturing our imagination.
- **Urbanization and mobility:** Urbanization—concentration of people in cities—has increased the need for efficient transportation systems. Cities rely on mobility for economic growth, social interaction, and access to services. Sustainable urban development considers mobility as a critical factor, aiming to balance accessibility, environmental impact, and quality of life.[3]
- **Human development and mobility:** Mobility is integral to human development. It enables education, employment, and social participation.

[3] *https://odi.org/en/publications/urbanisation-and-human-mobility-fact-sheets*

However, the impact of mobility isn't always straightforward. It interacts with social structures, affecting both individuals and societies. Understanding this complexity is crucial for informed policymaking.[4] [5]

In summary, mobility has shaped our past, continues to shape our present, and will likely play a significant role in our future. Whether it's the humble wheel or the futuristic "hyperwarp," our journey toward progress remains interconnected with our ability to move and adapt.

The wheel is one of the oldest and most influential inventions in human history. It is estimated that the wheel was invented around 3500 BCE in Mesopotamia, and was used for pottery making, transportation, and warfare. The wheel enabled humans to move faster, farther, and more efficiently, and facilitated the expansion of trade, communication, and empires. The wheel also inspired other inventions, such as the chariot, the cart, the bicycle, and the car.

The Romans demonstrated an innovative approach to the wheel by constructing an extensive and systematic road network. These roads, designed for the efficient movement of their wheel-based chariots and carts, played a crucial role in the expansion and administration of the Roman Empire. The roads facilitated trade, military movements, and cultural exchange, thereby solidifying the Romans' dominance in their era.

In ancient China, the wheel was integral to various aspects of life and technology. From wheelbarrows and chariots to water wheels, the usage of the wheel was diverse. Notably, the development of the wheel significantly influenced the evolution of transportation and agriculture in ancient China.

In contrast, in pre-Columbian America, the wheel was known but not widely used in practical applications. Despite their sophisticated civilizations, neither the Mayans, Aztecs, nor the Incas used wheels for transportation. The wheel was primarily used in toys and ceremonial objects, possibly due to the challenging terrain and lack of domesticated large animals for pulling wheeled vehicles.

The transition from wheel to horse coaches to steam locomotives was a significant development in human mobility and transportation. It involved the invention, improvement, and replacement of different technologies and modes of travel over time. Horse coaches were vehicles that used horses to pull passengers or goods along roads or rails. Horses have been used for transporting goods and people for thousands of years, since they were first domesticated around 3500 BCE. Horses have many advantages for transportation, such as speed, endurance, adaptability, and intelligence. Here are some examples of how horses have been used for transportation throughout history:

---

[4] *https://link.springer.com/chapter/10.1007/978-3-031-42798-5_1*

[5] *https://hdr.undp.org/content/mobility-and-human-development*

- **Trade and transportation:** Horses have been used to carry people and their possessions across long distances, facilitating the expansion of trade, communication, and empires. Horses have also been used to pull carts, chariots, wagons, and carriages, increasing the capacity and comfort of transportation. Furthermore, horses have been used for mail delivery, stagecoach services, and public transportation.
- **War:** Horses have been used for warfare, giving soldiers a new advantage of mobility and power. They have been used to pull heavy artillery and equipment, as well as to carry soldiers into battle. Horses have also been used for reconnaissance, communication, and cavalry charges. The use of horses in warfare led to the development of the saddle, stirrups, and horse collar.
- **Sport and recreation:** Horses have been used for sport and recreation, providing humans with entertainment and enjoyment. They have been used for racing, jumping, showing, polo, hunting, and rodeo. Horses have also been used for pleasure riding, tourism, and therapy.

Horses are still used for transportation in some parts of the world, especially in less economically developed regions where they are more affordable and accessible than motorized vehicles. However, horses are mostly used for riding and sport in the western world, where they are valued for their companionship and beauty.

Here are some examples of other animals that have been used for transportation throughout history:

- **Mules in the Himalayas:** Mules and other pack animals are commonly used in the Himalayan region due to the rugged and steep terrain. Their sure-footedness, strength, and endurance make them ideal for carrying loads over long distances in high-altitude areas. They are used for transporting goods, supplies, and sometimes people across narrow mountain paths.
- **Elephants in India:** Elephants have been used for transportation in India for centuries, particularly in the logging industry, where their strength and dexterity allow them to move large logs in dense forests. They are also used for riding in certain tourist areas and for ceremonial purposes. Elephants are well-suited to the Indian climate and terrain, and their use helps in navigating areas that are inaccessible by vehicles.
- **Camels in Africa:** Known as the "ships of the desert," camels are perfectly adapted to the harsh desert conditions in Africa. They can carry heavy loads for long distances without food or water, making them ideal for transportation across arid landscapes. They are used for carrying goods, people, and even for pulling ploughs in agricultural fields.
- **Llamas in South America:** In the Andean region of South America, llamas have been used as pack animals for thousands of years. They are well-adapted to high altitudes and rocky terrain, making them ideal for transporting goods across the Andes mountains. Their wool is also used for making clothing and blankets.

There are many further examples where animals are used for transportation based on different geographical conditions. The animals were also domesticated by people so that they were optimally adapted to the special requirements and the given infrastructure. Many animals have even been optimized for different tasks through breeding. Transportation systems and road networks have been intricately designed and optimized over centuries, often tailored to the specific needs of the animals that were integral to these networks. This symbiotic relationship between humans, animals, and the environment has shaped the course of history and continues to influence modern infrastructure.

In North America, the Pony Express stands as a testament to this optimization. This mail service relied on horse riders who could change mounts at relay stations, allowing mail to be delivered over vast distances with unprecedented speed. The routes and relay stations were strategically located considering the endurance and speed of the horses, demonstrating a deep understanding of the animals' capabilities.

Similarly, the caravan routes across Africa and from Europe to Asia, including the famous Silk Road, were designed considering the needs of the pack animals, primarily camels. These routes traversed harsh and diverse terrains, from scorching deserts to towering mountains. The success of these trade routes hinged on the remarkable resilience of camels to harsh conditions, and the infrastructure was developed to support their needs for rest and sustenance.

In essence, the development of transportation and road networks has been a dynamic process, adapting to the strengths and limitations of various animals. This has resulted in a rich tapestry of interconnected routes that have not only facilitated trade and communication but also cultural exchange and the spread of ideas across continents.

Horse coaches were widely used in Europe and America from the 16th to the 19th century and offered a relatively comfortable and convenient way of travelling. They were also used for mail delivery, stagecoach services, and public transportation. However, horse coaches had some limitations, such as speed, capacity, cost, and dependence on animal welfare.

Non-animal modes of transportation are ways of moving people or goods that do not rely on animals, such as horses, camels, or dogs. There are many types of non-animal modes of transportation, but they can be broadly classified into four categories: road, rail, air, and water.

- Road transportation uses vehicles that run on roads, such as cars, buses, trucks, motorcycles, and bicycles. Road transportation is the most common and flexible mode of transportation, as it can reach almost any destination and carry various types of cargo. However, road transportation also has some disadvantages, such as traffic congestion, pollution, accidents, and high maintenance costs.

- Rail transportation uses vehicles that run on rails, such as trains, trams, and subways. Rail transportation is faster, more efficient, and more reliable than road transportation, as it can carry large volumes of passengers or goods at high speeds and avoid road obstacles. However, rail transportation also has some limitations, such as high initial investment, fixed routes, and dependence on electricity.
- Air transportation uses vehicles that fly in the air, such as airplanes, helicopters, and balloons. Air transportation is the fastest and most convenient mode of transportation, as it can reach any destination in the world and overcome geographical barriers. However, air transportation also has some drawbacks, such as high cost, noise, environmental impact, and security risks.
- Water transportation uses vehicles that float on water, such as ships, boats, and ferries. Water transportation is the oldest and most economical mode of transportation, as it can carry huge amounts of cargo over long distances and use natural waterways. However, water transportation also has some challenges, such as low speed, weather dependence, and piracy.[6]

Non-animal modes of transportation have evolved over time, from the invention of the wheel to the development of the hyperwarp, a hypothetical device that could warp space and time and allow faster-than-light travel. Non-animal modes of transportation have also diversified, from the use of conventional vehicles to the use of alternative vehicles, such as electric cars, solar planes, and hoverboards. Non-animal modes of transportation have shaped human history, culture, and society and will continue to do so in the future.

The development of machines for transportation of goods and people is a fascinating topic that covers the history, science, and engineering of various modes of travel. Machines are devices that use energy to perform work, and transportation is the movement of people or goods from one place to another. Together, machines and transportation have shaped human civilization, culture, and society and have enabled humans to explore, communicate, and trade across the world.

According to Britannica, the growth of the ability and the need to transport large quantities of goods or numbers of people over long distances at high speeds in comfort and safety has been an index of civilization and in particular of technological progress.[7] Some of the earliest machines for transportation were the wheel, the sail, and the oar, which were used to move carts, boats, and ships. These machines relied on human or animal power, or natural forces such as wind and water currents.

With the advent of the Industrial Revolution, new machines for transportation emerged, such as the steam engine, the locomotive, and the bicycle. These machines used steam, coal, or human power to move trains, cars, and bikes. They increased the

[6] *https://transportgeography.org/contents/chapter5/transportation-modes-modal-competition-modal-shift*

[7] *https://www.britannica.com/technology/transportation-technology*

speed, efficiency, and convenience of transportation, and stimulated the development of other technologies, such as roads, railways, bridges, and tunnels.

In the twentieth century, machines for transportation reached new heights, literally and figuratively, with the invention of the airplane, the rocket, and the helicopter. These machines used gasoline, jet fuel, or electricity to fly in the air, or even in space. These machines revolutionized human mobility and transportation, as they could reach any destination in the world and overcome geographical barriers.

In the twenty-first century, machines for transportation are becoming more intelligent, autonomous, and sustainable, with the development of the electric car, the maglev train, and the drone. These machines use batteries, magnets, or solar power to move on the ground, on rails, or in the air. These machines also use sensors, computers, and artificial intelligence to navigate, communicate, and optimize their performance.

The development of machines for transportation of goods and people is a continuous and dynamic process that involves various factors, such as technology, environment, economy, politics, and culture. It also involves various challenges, such as competition, resistance, adaptation, and transformation. The development of machines for transportation of goods and people is a key element of human history, and a potential driver of human future.

Steam locomotives were machines that used steam power to move trains along rails. They were invented in the early 19th century in Britain and soon spread to other parts of the world. Steam locomotives revolutionized human mobility and transportation, as they could carry large numbers of people and goods at high speeds over long distances. Steam locomotives also stimulated the development of other technologies, such as railways, bridges, tunnels, and signals.

The transition from wheel to horse coaches to steam locomotives was not a linear or smooth process, but rather a complex and dynamic one. It involved various factors, such as technology, environment, economy, politics, and culture. It also involved various challenges, such as competition, resistance, adaptation, and transformation. The transition from wheel to horse coaches to steam locomotives was a key element of the Industrial Revolution, and a major milestone in human history.

The car is a modern invention that revolutionized human mobility in the twentieth century. The first practical car was invented by Karl Benz in 1885 in Germany and was powered by an Internal Combustion Engine (ICE). The car offered humans a new mode of personal, individual, and private transportation, and increased their mobility, autonomy, and convenience. The car also stimulated the development of other technologies, such as roads, highways, bridges, and traffic systems.

The hyperwarp is a hypothetical invention that could enable human mobility in the future. It is a concept of a device that could warp space and time and allow humans to travel faster than light. The hyperwarp could offer humans a new mode of interstellar and intergalactic transportation, and expand their mobility, exploration, and discovery. The hyperwarp could also challenge the current understanding of physics, cosmology, and reality.

Human mobility is a dynamic and complex phenomenon that reflects the interaction of human creativity, curiosity, and necessity. It has been influenced by various factors, such as technology, environment, economy, politics, and culture. It has also influenced various outcomes, such as innovation, diffusion, adaptation, and transformation. Human mobility is a key element of human history, and a potential driver of human future.

The development of transportation systems and the dependency on energy are closely related topics that reflect the evolution of human mobility, technology, and society. Transportation systems are the networks and modes that enable the movement of people and goods from one place to another, such as roads, railways, airways, and waterways. Energy is the capacity to perform work, and it can be derived from various sources, such as fossil fuels, electricity, biofuels, and hydrogen. Here is a brief overview of how transportation systems and energy dependency have changed over time:

- **Pre-industrial era:** Before the Industrial Revolution, most transportation systems relied on human or animal power, or natural forces such as wind and water currents. The main modes of transportation were walking, riding horses, sailing boats, and using carts or carriages. The energy sources were mainly biomass (wood, crops, and animal feed) and muscle power. The speed, efficiency, and convenience of transportation were low, and the mobility of people and goods was limited by distance, terrain, and weather.
- **Industrial era:** The Industrial Revolution, which started in the late 18th century, brought about major changes in transportation systems and energy dependency. New machines and modes of transportation emerged, such as the steam engine, the locomotive, the bicycle, the car, and the airplane. These modes used steam, coal, oil, or gasoline as energy sources, and they increased the speed, efficiency, and convenience of transportation. The mobility of people and goods expanded significantly, and new infrastructures, such as roads, railways, bridges, and tunnels, were built to support the transportation networks.
- **Post-industrial era:** The post-industrial era, which began in the late 20th century, witnessed further developments in transportation systems and energy dependency. New technologies and modes of transportation emerged, such as the electric car, the maglev train, the drone, and the hyperloop. These modes used electricity, magnets, solar power, or batteries as energy sources, and they improved the speed, efficiency, and convenience of transportation. The mobility of people and goods became more diverse, flexible, and intelligent, and new challenges, such as congestion, pollution, and climate change, emerged as a result of the transportation activities.

The development of transportation systems has undergone several stages, each characterized by a shift in energy sources and consumption patterns. Here's a brief overview:

- **Industrialization phase:** The Industrial Revolution marked the transition from reliance on human and animal power to mechanized forms of transportation. The invention of the steam engine revolutionized transportation, powering trains and ships, and paving the way for the development of road vehicles.
- **Combustion engines:** The advent of the Internal Combustion Engine led to the development of automobiles and trucks, transforming road transportation. This period also saw the use of steam and combustion engines in rail and water transport.
- **Wind energy for ships:** In the early phase, wind energy played a significant role in maritime transport. Sailing ships harnessed the power of the wind for propulsion, a renewable and efficient energy source.
- **Electrification:** The introduction of electric engines marked a significant shift in rail and road transportation. Electric trams, trains, and trolleybuses became common in many cities, offering a cleaner alternative to combustion engines. Even for road vehicles during the 19th century several solutions were developed based on electric propulsion systems. In Europe during the 1960s, the main railroads had been electrified. In the 1920s, not only electric trams were introduced in many cities, but also electric overhead line buses, thus establishing a city-wide electric overhead line system over the main roads.
- **Air transportation:** The 20th century saw the development of air transportation systems. It started with gliders, followed by airplanes powered by combustion engines, and eventually led to the development of rockets for space travel.
- **Efficiency approach:** With the growth of urban centers and long-distance travel, there was a need for mass transportation systems that could move large numbers of people efficiently and quickly. This led to the development of high-speed trains, larger aircraft, and more efficient ships.
- **Energy crisis and efficiency:** The oil crisis in the 1970s triggered a global focus on energy efficiency. This led to advancements in fuel-efficient engines, hybrid vehicles, and eventually electric vehicles.
- **Digitalization and decarbonization:** In the current era, there is a strong emphasis on reducing carbon emissions and improving efficiency through digital solutions. This includes optimizing logistics and control systems, transitioning to renewable energy sources, and developing electric and hydrogen-powered vehicles.
- **Artificial Intelligence (AI):** AI is now being used to optimize routes, improve safety, and increase efficiency in the transportation of people and goods. Autonomous vehicles and drones are examples of AI applications in transportation.

The evolution of transportation systems reflects our growing understanding of energy sources and technology, as well as our changing societal needs and environmental concerns. The future of transportation will likely continue this trend, with a focus on sustainability, efficiency, and intelligent systems.

1.3 Energy Base for Transportation

Transportation systems have evolved over time, and so have the energy sources used to power them. From the early days of wind, water, and horse-drawn carriages to the modern era of gasoline, diesel, fuel cells, and hydrogen combustion engines, the transportation sector has come a long way.

The steam engine was the first successful heat engine, and it was followed by the Internal Combustion Engine in the 19th century as an improvement over the steam engine for many applications. The Internal Combustion Engine cannot be attributed to any single inventor, but it was developed by many inventors and engineers over several decades.

The first successful gas engine was built in France in 1860 by Étienne Lenoir, and it operated on illuminating gas. The Lenoir engine was essentially a converted double-acting steam engine with slide valves for admitting gas and air and for discharging exhaust products. Although the Lenoir engine developed little power and utilized only about four percent of the energy in the fuel, hundreds of these devices were in use in France and Britain within five years.

The first gasoline-fueled engine was built by German engineer Karl Friedrich Michael Vaillant Benz in 1879. The gasoline engine was a significant improvement over the steam engine for many applications because it was more efficient, lighter, and more compact. The gasoline engine was also more convenient to operate because it was self-contained and did not require a separate furnace for burning fuel.

Hydrogen has been used as a fuel for transportation since the early days of the zeppelin airships. The long-distance transportation of hydrogen is a major challenge, but it may be possible to decarbonize green hydrogen transportation by envisioning hydrogen-filled rigid or frameless airships (blimps) which work on the buoyancy of hydrogen to lift and transport.

Hydrogen fuel cells are also being used in transportation. Fuel cells are electrochemical devices that convert the chemical energy of a fuel (hydrogen) and an oxidant (usually oxygen) directly into electrical energy. Fuel cells are more efficient than Internal Combustion Engines and produce no harmful emissions. Fuel cell vehicles are powered by electric motors, which are driven by the electricity generated by the fuel cells.

Google co-founder Sergey Brin is reportedly working on a relief aid zeppelin that will use the world's largest mobile hydrogen fuel cell. The H2 Clipper is another example of a hydrogen-filled airship that uses hydrogen fuel cells coupled with electric engines to enable it to fly distances of over 6000 miles at a speed of 175 miles per hour.

According to the International Energy Agency (IEA), the transportation sector is responsible for 28% of annual energy use in the United States, with on-road vehicles accounting for more than 82% of all transportation energy use. The IEA also reports that CO_2 emissions from the transport sector grew at an annual average rate of 1.7% from 1990 to 2022, faster than any other end-use sector except for industry.

To get on track with the Net Zero Emissions (NZE) by 2050 Scenario, CO_2 emissions from the transport sector must fall by more than 3% per year to 2030. Strong regulations and fiscal incentives, as well as considerable investment in infrastructure to enable low- and zero-emission vehicle operations, will be needed to achieve these emissions reductions.

1.4 History of Vehicle Technology

The general design of motor vehicles was always driven by safety considerations. One of the most design feature of today's motor vehicles is shown on a plate in the Karl Benz Museum in Ladenburg (Figure 1.1). The text could be translated as follow:

"Why did automobiles used to have the steering wheel on the right side? This question is often asked by our museum visitors. We want to give you the answer. Even with the first horse-drawn carriages, the driver sat on the right side because he had the task of helping when alighting and opening the door. This allowed the nobility, whom he was allowed to chauffeur, to comfortably step onto the sidewalk. Moreover, the edges of the roads outside of cities were still unpaved. There were densely grown avenue trees and quite deep roadside ditches. All these obstacles could be better observed by the coachman from his seat on the right side. This tradition initially continued with automobiles. It was not until the mid-1920s of the last century that it was realized that the actual danger came from oncoming traffic. There were many serious accidents because the drivers of the motor vehicles did not correctly estimate the distances. As a result, from about 1925 onwards, the steering wheels of automobiles were fundamentally arranged on the left."

Figure 1.1 Right or left-steering (source: information board at the Karl Benz Museum, Ladenburg, photographed 2023)

The text above and in Figure 1.1 is remarkable because it highlights a critical shift in automotive design philosophy that prioritizes the driver's ability to perceive risks from oncoming traffic. This change was driven by the realization that the position of the steering wheel could significantly impact safety. With the steering wheel on the left side, drivers have a better view angle at intersections, allowing for earlier and clearer visibility of oncoming vehicles. Additionally, this positioning facilitates an earlier perception of road traffic participants on the right side, such as cyclists and pedestrians, which is crucial for preventing accidents. These advantages underscore the importance of design considerations in vehicle safety and the ongoing evolution of automotive standards to enhance protection for all road users.

Another significant and earlier aspect of automotive safety evolution coming from the experiences of Karl Benz, the inventor of what is widely regarded as the first practical modern automobile. His first car, the Benz Patent Motor Car, was a three-wheeled vehicle.[8] This design choice was initially due to the simplicity and efficiency of construction. However, it soon became apparent that three-wheeled vehicles had a higher risk of tipping over, especially during turns or when navigating uneven terrain. The tipping risk associated with three-wheeled vehicles led to several accidents, prompting the auto industry to focus on four-wheeled designs. Four wheels provided a more stable base, significantly reducing the risk of rollovers and improving overall vehicle safety. This shift was part of the industry's broader realization that vehicle stability was paramount for safe operation, especially as cars began to travel at higher speeds and on a variety of road surfaces.

---

[8] *https://group.mercedes-benz.com/company/tradition/company-history/1885-1886.html?source=post_page-----1d8f10cad8e5--------------------?r=dai*

An additional aspect of Karl Benz's experiences that is noteworthy involves his wife, Bertha Benz, who was the first person to drive his automobile over a long distance. During her historic journey, she had to stop at a pharmacy to purchase ligroin (a petroleum solvent) to refuel the car, which was the only available fuel for automobiles at the time. This event is significant as it demonstrated a great deal of trust in automotive technology, showing that even a woman—who at the time was rarely involved with technology—was capable of refueling the vehicle.

This issue underscores the importance of Human-Machine Interface (HMI) in the automotive industry, highlighting that refueling and charging should be straightforward and transparent processes. It's not just a matter of convenience but also of safety. The ease with which Bertha Benz refueled the car reflects the need for clear and user-friendly HMI designs that ensure all drivers can operate vehicles and their features safely and efficiently, a principle that remains just as relevant in today's era of advanced automotive technology.

Over the years, the automotive industry has made significant strides in safety technology. These advancements have helped to reduce the risk of accidents and mitigate the consequences of collisions.

Here are some of the key milestones:

- **1902—Bosch's high voltage ignition:** Bosch's high voltage ignition allowed for better control over voltage surges, leading to more reliable engine performance and reducing the risk of engine malfunctions.
- **1903—windshield wipers by Mary Anderson:** Mary Anderson invented the windshield wiper, which improved visibility in bad weather, thus reducing the risk of accidents due to poor visibility conditions.
- **1908—4-wheel brake system by Giustino Cattaneo/Isotta Fraschini Company:** This system enhanced braking performance and stability, leading to shorter stopping distances and a reduced risk of rear-end collisions.
- **1915—electronic components by Ford:** By installing headlights in automobiles, Ford improved night vision and reduced the risk of accidents in the dark.
- **1919/1921—hydraulic brakes by Loughead and Duesenberg Model A:** The introduction of hydraulic brakes provided better braking force and reliability, increasing safety during emergency stops.
- **1924—dual-circuit brake system in Chrysler 70:** The first production car with a dual-circuit brake system offered an additional level of safety by maintaining braking ability even if one circuit failed.
- **1929—electrically operated windshield wiper by Baron Rothschild:** The electrification of the windshield wiper ensured consistent and reliable visibility improvement.

- **1934—dual-circuit brake by Bugatti/Lockheed:** The introduction of the dual-circuit brake with a tandem master cylinder provided a redundant safety measure to ensure the availability of the braking system.
- **1950—electric lights as turn indicators:** Switching from manual semaphores to electric lights improved the visibility of turn signals and reduced the risk of misunderstandings on the road.
- **1952—crumple zones and safety steering column by Béla Barényi:** The invention of crumple zones and the safety steering column helped to better protect occupants in a collision.
- **1951—airbag patent by Walter Linderer:** The airbag was developed to cushion occupants in a collision and reduce the risk of injury.
- **1958—Anti-Lock Braking System (ABS) by Dunlop:** ABS prevented the wheels from locking during braking, leading to shorter stopping distances and a reduced risk of tire blowouts.
- **1966—three-point seatbelt by Nils Bohlin:** The three-point seatbelt was designed to keep occupants inside the vehicle during a collision, reducing the risk of severe injuries.
- **1967—electronic fuel injection:** This technology reduced fuel consumption and emissions, contributing to a more environmentally friendly and economical driving experience.
- **1978—electronic ABS by Bosch:** The electronic ABS further improved braking performance and stability, helping to reduce the risk of accidents due to brake failure.
- **1980—airbag in Mercedes-Benz W126 (S-Class):** The deployment of airbags in vehicles provided additional protection for occupants, reducing the risk of severe head and upper body injuries.
- **1991—Bosch patent for vehicle control system:** The system with a main and backup computer offered an additional level of safety for vehicle control.
- **1993—sensory parking aid by Ford and ESP by Bosch and Mercedes-Benz:** The sensory parking aid reduced the risk of collisions while parking, and ESP improved vehicle stability and reduced the risk of skidding and rollovers.

These achievements demonstrate how the automotive industry has made driving safer through innovations in safety technology. Each development has compensated for specific risks and contributed to reducing the number of traffic accidents and their severity.

1.5 Accidents Change Technology

George Stephenson was a British civil engineer and inventor who is credited with developing the first steam locomotive. The development of the steam engine was a significant milestone in the history of transportation, but it was not without its challenges. One of the most significant problems with early steam engines was the risk of boiler explosions. Stephenson's solution to this problem was to develop a safety valve that would release excess steam from the boiler if the pressure became too high. The safety valve was a simple but effective device that helped to prevent boiler explosions and made steam engines much safer to operate.

The Hindenburg disaster is one of the most infamous airship accidents in history. The disaster occurred on May 6, 1937, when the German airship Hindenburg caught fire and crashed while attempting to dock at the Lakehurst Naval Air Station in New Jersey, USA. The cause of the disaster is still a matter of debate, but it is widely believed that the explosion was caused by a spark that ignited the hydrogen gas used to lift the airship. The Hindenburg was filled with hydrogen gas, which is highly flammable and can ignite easily. There are several theories about what caused the spark that ignited the hydrogen gas. Some people believe that the spark was caused by static electricity, while others think that it was caused by a lightning strike. It is also possible that the spark was caused by a backfiring engine or a leak in one of the fuel lines. Despite the many theories, the exact cause of the Hindenburg disaster remains unknown. However, it is clear that the disaster had a significant impact on the future of airship travel.

The use of lead in gasoline dates back to the 1920s when tetraethyl lead was added to petrol to reduce engine knocking. Lead was an "antiknock agent" that improved the efficiency of vehicles and the performance of the engine. However, lead proved to be a toxic pollutant, particularly for children, and using it in gasoline was polluting the air in towns and cities across the world, with large impacts on human health. There had been concerns about the health impacts of leaded gasoline throughout most of the first half of the 20th century, but these worries were quickly dismissed due to a lack of empirical evidence. It wasn't until 1969 that the first clinical studies showed that leaded gasoline had toxic impacts on humans. From there, the evidence of the extent of lead poisoning and its potential link with poor cognitive performance in children continued to grow. Many countries began to phase out leaded gasoline in the 1970s, and the US Environment Protection Agency issued guidelines to reduce lead content in 1983. However, it wasn't until 1986 that Japan became the first country to ban leaded gasoline completely. In 2021, Algeria became the last country to ban leaded gasoline.

1.6 Role of Humans

The role of humans in product development is a topic that has been explored by various researchers and practitioners in different fields. Humans are involved in product development both as customers and as developers. As customers, humans have needs, preferences, and expectations that influence their perception of product value and quality. As developers, humans have skills, creativity, and knowledge that enable them to design, test, and launch new or improved products. The human plays a special role in product development and also in the use of the product. Although we have always considered this from a marketing technical perspective, it is rarely considered from a purely technical standpoint.

The use of AI is a significant topic of discussion, but the discussion extends to questions such as: What degree of automation does the human being need? Which level of automation benefits humans? And which functions may benefit a few people but hinder and disturb many others in their needs and sense of safety and security?

Regarding the topic of automated driving, some new aspects have been accounted. The human being is not completely out of the loop in many transport systems. Even legislation for public road traffic considers a human supervisor and teleoperation goes even a step further: A human driver still drives the car but not from inside the vehicle. Teledriving or teleoperation means driving a vehicle in public or private ground from a protected indoor place. These developments drive the need to investigate how humans drive cars, perceive situations, risks, etc., and how they react to various events in a timely and appropriate manner.

Autonomous driving technology has made significant strides in recent years, but the transition to fully autonomous vehicles in urban environments remains a complex and ongoing journey. While there are promising advancements, several critical challenges persist. In this book, we'll explore why technical systems are not yet capable of mastering the risks associated with urban driving and why a complete replacement of human drivers is still a distant goal.

Urban environments are dynamic and unpredictable. Roads are filled with pedestrians, cyclists, construction zones, and other vehicles. While sensors and cameras on autonomous vehicles can detect objects, understanding their intentions and predicting their behavior accurately is challenging. Human drivers rely on intuition, context, and social cues to navigate these situations effectively, which remains a challenge for AI-driven systems. Driving involves split-second decisions influenced by a multitude of factors: traffic rules, weather conditions, road signs, and social norms. Autonomous systems follow predefined algorithms, but they struggle with nuanced decision-making. For instance, should a self-driving car prioritize avoiding a collision with a pedestrian or protect its passengers? These ethical dilemmas are difficult to encode into rigid rules. Edge cases—rare and unexpected scenarios—are common in urban driving. A child chasing a ball onto the road, a sudden construction detour, or a malfunctioning traffic light—all require adaptive responses. While machine learning models improve over time, they may not encounter these edge cases frequently enough to learn robust solutions.

Human drivers possess intuition and common sense, honed through years of experience. They can anticipate the behavior of other road users, interpret non-verbal cues, and adjust their driving accordingly. These intangible qualities are hard to replicate in code. Driving involves empathy, patience, and negotiation. Humans communicate through eye contact, hand gestures, and subtle movements. Autonomous vehicles lack emotional intelligence, leading to potential misunderstandings and conflicts on the road.

Rather than an abrupt shift, a hybrid approach is more practical. Semi-autonomous features (e.g., adaptive cruise control, lane-keeping assistance) gradually introduce automation while retaining human oversight. This approach allows for iterative improvements and safety validation. Legal and regulatory frameworks must adapt to accommodate autonomous vehicles. Liability, insurance, and accountability need careful consideration. Policymakers recognize the need for a phased transition, balancing innovation with public safety.

While the vision of fully autonomous urban driving is compelling, we must acknowledge the current limitations. Technical systems are advancing, but they are not yet ready to replace human drivers entirely. The road ahead involves collaboration, innovation, and a patient evolution towards safer and more efficient transportation.

1.6.1 Role of Humans in Product Development

One of the key challenges in product development is to understand the needs of markets and customers. This requires conducting research and analysis to identify the problems, opportunities, and solutions that can create value for the customers. Some of the methods and tools that can help with this process include surveys, interviews, focus groups, personas, user stories, and prototyping. These methods allow the developers to gather feedback and insights from the customers and to test their assumptions and hypotheses. By involving the customers in the product development process, the developers can ensure that the products are functional, appealing, and profitable.

Another important aspect of human factors in product development is the design of the product itself. This involves creating the product concept, features, specifications, and aesthetics that can satisfy the customer needs and wants. The design process also considers the technical and economic feasibility of the product, as well as the environmental and social impacts. Some of the skills and tools that can help with the design process include creativity, problem-solving, attention to detail, and communication. The developers also need to collaborate with other stakeholders, such as engineers, marketers, manufacturers, and suppliers, to ensure the quality and efficiency of the product development process.

A third aspect of human factors in product development is the management of the product lifecycle. This involves planning, organizing, and controlling the activities and resources that are required to develop, launch, and maintain the product in the market. The product lifecycle may also include improving the manufacturing or distribution processes, updating the product features, or discontinuing the product when it becomes obsolete. Some of the methods and tools that can help with the product lifecycle management include project management, quality management, risk management, and product portfolio management. These methods allow the developers to monitor and evaluate the performance and profitability of the product, as well as to respond to changing customer needs and market conditions.

Human factors play a vital role in product development, both as customers and as developers. By understanding and integrating human factors in the product development process, the developers can create better products that can deliver value to the customers and the company.

1.6.2 Example for Technology Decisions

The battle between Westinghouse and Tesla about Direct Current (DC) or Alternating Current (AC) was a historical conflict that took place in the late 19th century over the best way to distribute electricity. It is also known as the War of the Currents or the Battle of Currents.

Thomas Edison, the inventor of the first practical light bulb, supported Direct Current, which is a constant flow of electricity in one direction. He had built the first power plant in New York City in 1882 using DC generators. However, DC had a major drawback: It was difficult to convert to different voltages, which limited its transmission range and efficiency.

Nikola Tesla, a brilliant engineer from Croatia, worked for Edison for a short time, but left after Edison refused to fund his research on Alternating Current, which is a periodic reversal of the direction of electricity. Tesla believed that AC was superior to DC, as it could be easily transformed to different voltages using transformers and could travel longer distances with less power loss. Tesla sold his patents on AC technology to George Westinghouse, an industrialist who had founded the Westinghouse Electric Company in 1886. Westinghouse was a fierce competitor of Edison, and he saw the potential of AC to challenge Edison's monopoly on electricity distribution.[9]

[9] *https://www.history.com/news/what-was-the-war-of-the-currents*

Edison, feeling threatened by the rise of AC, launched a propaganda campaign to discredit AC and convince the public that it was dangerous. He staged demonstrations of animals being electrocuted by AC, and even advocated for the use of AC in the electric chair, hoping to associate it with death[10] He also tried to prevent Westinghouse from using his patents on light bulbs, which were essential for AC lighting.[11] Westinghouse and Tesla fought back by showcasing the advantages of AC in various events and projects. They won the contract to supply electricity to the 1893 World's Fair in Chicago, which became a dazzling display of AC-powered lights and devices. They also built the first hydroelectric power plant at Niagara Falls in 1896, which transmitted AC power to Buffalo, New York, 26 miles away, proving its feasibility and efficiency.[12]

The War of the Currents ended in the late 1890s, when Edison's company merged with another rival, Thomson-Houston, to form General Electric. GE eventually adopted AC as the standard for electricity distribution, as it became clear that it was more economical and practical than DC. Today, most of the electricity we use is still generated and transmitted by AC, but some devices, such as computers, LEDs, solar cells, and electric vehicles, run on DC. There are also methods for converting DC to higher and lower voltages, making it more versatile and stable.[13] The War of the Currents was a pivotal moment in the history of electricity, and the legacy of Edison, Tesla and Westinghouse lives on in the modern power grid.

1.6.3 Horse or Steam Locomotive

A race between a horse and a locomotive was a historical event that took place in 1830 in the United States. It was a contest between the old and the new modes of transportation, and it attracted a lot of attention and curiosity. The race was between Peter Cooper's Tom Thumb, the first American-built steam locomotive, and a horse-drawn railcar on the Baltimore and Ohio Railroad. Tom Thumb was a small and experimental engine that Cooper had built to demonstrate the feasibility of steam power on the railroad.[14] The horse-drawn railcar was a common sight on the tracks, as the B&O had not yet fully switched to steam engines.[15] The race was arranged by the B&O directors, who wanted to impress their investors and the public with the potential of steam power. The race took place on August 28, 1830, on a 13-mile stretch of track from Baltimore to Ellicott's Mills. A large crowd gathered to witness the spectacle, and

[10] *https://www.history.com/news/what-was-the-war-of-the-currents*

[11] *https://www.nikolateslalegend.com/wiki/tesla/war-of-currents-ac-vs-dc*

[12] *https://www.history.com/news/what-was-the-war-of-the-currents*

[13] *https://www.energy.gov/articles/war-currents-ac-vs-dc-power*

[14] *https://www.fhwa.dot.gov/rakeman/1830.htm*

[15] *https://www.history.com/news/industrial-revolution-horse-train-railroads*

some even rode on the railcars to follow the race.[16] The race started with Tom Thumb pulling ahead of the horse, showing its superior speed and power. However, Tom Thumb suffered a mechanical failure when a belt slipped off the blower, causing the engine to lose steam pressure. The horse took advantage of the situation and passed Tom Thumb, reaching the finish line first.

The horse may have won the race, but the Tom Thumb had proved its point. The race demonstrated that steam locomotives were faster, stronger, and could be more reliable than horses, and that they could handle the terrain and the curves of the railroad. The race also sparked the public interest and enthusiasm for steam power and paved the way for the rapid development and expansion of the railroad industry in the United States.

The race between the horse and the locomotive was a landmark moment in the history of transportation and technology, and it marked the beginning of the end of the era of horse power. Today, horsepower versus kilowatt is still the battle between ICE-driven motor vehicles and vehicles driven by an electric motor.

1.6.4 Role of Humans as Users

The world of transportation and mobility involves a diverse cast of individuals, each playing a crucial role in keeping our societies moving. From drivers and passengers to logistics planners and software developers, there are multifaceted roles that contribute to efficient and seamless mobility. Whether it's the thrill of driving, the convenience of being a passenger, or the behind-the-scenes coordination, everyone has a part to play.

In the following, various roles of individuals in the context of mobility solutions are described.

Driver

The drivers are responsible for operating the vehicle. They control the steering, acceleration, and braking. These are their motivation and benefits:

- **Personal transportation:** Drivers use vehicles to reach their desired destinations efficiently.
- **Driving pleasure:** Some individuals enjoy the act of driving itself.
- **Service providers:** Taxi drivers, rideshare drivers, and delivery drivers offer transportation services to others.
- **Status and image:** Owning and driving a specific car model can convey prestige or personal style.

[16] *https://www.fhwa.dot.gov/rakeman/1830.htm*

Passenger

Passengers are individuals being transported. They rely on drivers or automated systems for mobility. These are their motivation and benefits:

- **Convenience:** Passengers can relax during the journey without the responsibility of driving.
- **Efficient transportation:** Passengers reach their destinations without the effort of operating a vehicle.

Vehicle Owner

Vehicle owners possess and utilize the vehicle. They may own personal cars, company fleets, or rental vehicles. These are their Motivation and benefits:

- **Personal mobility:** Owners have access to transportation whenever needed.
- **Symbol of status:** Owning a high-end or luxury vehicle reflects social standing.

Logistics Planner

Logistic planners are professionals who are responsible for planning and coordinating transportation tasks. These are their motivation and benefits:

- **Efficient organization:** Logistics planners optimize routes, schedules, and resource allocation.
- **Streamlined operations:** They ensure timely deliveries and minimize costs.

Operator (e. g., Teledriver)

Operators remotely control vehicles, especially in automated or remote driving scenarios. These are their motivation and benefits:

- **Providing services:** Teledrivers offer ride services or manage autonomous vehicles.
- **Ensuring safety:** Operators monitor and intervene when needed.

Car Rental Service Staff

Those are individuals who are involved in car rental services, including rental agents and maintenance staff. These are their motivation and benefits:

- **Customer service:** Rental agents facilitate vehicle rentals and address customer needs.
- **Vehicle maintenance:** Staff ensure that rental cars are clean, fueled, and well-maintained.

Pickup and Delivery Service Staff

Those are people who are responsible for transporting goods or packages. These are their motivation and benefits:

- **Reliable delivery:** Staff transport items to specific locations, meeting customer expectations.
- **Efficient logistics:** They contribute to supply chains and e-commerce.

Charging Station Operator

Those are individuals who manage electric vehicle charging stations. These are their motivation and benefits:

- **Supporting electric mobility:** Operators maintain charging infrastructure.
- **Eco-friendliness:** They are enabling eco-friendly transportation.

Fuel Station Attendant

Those are the staff who work at traditional fuel stations. These are their motivation and benefits:

- **Assisting with refueling:** Attendants ensure smooth fueling processes.
- **Safety and service:** They provide assistance and maintain fuel station operations.

Software Programmer (e. g., App Developer)

Those are developers creating mobility-related software and apps. These are their motivation and benefits:

- **User-friendly experiences:** Programmers design apps for ride-hailing, navigation, or charging services.
- **Enhancing mobility solutions:** Their work improves accessibility and convenience.

Preceding, the diverse roles that contribute to a well-functioning mobility system, ensuring efficient and safe movement of people and goods, were highlighted.

While the software developer is a prominent player within the development of a vehicle or any other mobility solution, they are just one piece of the puzzle. In subsequent chapters, we'll explore these roles in greater detail, examining their strengths, weaknesses, abilities, and potential pitfalls.

1.7 Development of Road Traffic Regulations

With the invention of the automobile, the need for traffic rules had to be further developed. In continental Europe, right-hand traffic had already been established for horse-drawn carriages, and in England, left-hand traffic was the norm. These local principles have persisted to this day.

Initially, rules were established in cities and in Germany on the basis of regional authorities. However, by the end of the 19th century, more and more national laws were being defined. These laws varied greatly, depending on the need and necessity for regulations. With the further internationalization of traffic, particularly after the Second World War, efforts were made at UN level to harmonize and internationalize these rules. This led to the development of comprehensive traffic laws and regulations that are recognized and enforced globally. These efforts underscore the importance of international cooperation in ensuring road safety and order. The Geneva and Vienna World Agreements are two significant pieces of legislation that have shaped global road traffic laws:

- **Geneva Convention on Road Traffic (1949):** This agreement, signed predominantly by American nations as well as African and Asian states, has had a significant influence on road traffic laws in these regions. It includes provisions for the issuance of international driver's licenses, which are recognized among the signatory states.
- **Vienna Convention on Road Traffic (1968):** This convention, to which Germany is a signatory, has largely influenced European road traffic laws. It also includes provisions for international driver's licenses, which are valid throughout the EU.

While both conventions aim to regulate international road traffic and ensure safety and order, they differ in their scope and the specifics of the regulations they impose. For instance, the driver's licenses issued under the Vienna Convention are different from those issued under the Geneva Convention. This difference is a key factor contributing to the variations in road traffic acts between North America and Europe.

In addition to these conventions, the Geneva and Vienna World Agreements form the basis for homologation and all road traffic law. They are the basis for the approval regulations of UNECE, GB in China, and the NHTSA in North America or the respective associated countries, the driver's license or driving license laws, the road traffic rules, and the resulting jurisprudence. These principles are broken down worldwide at the UN level, in Europe at the EU level to country law and rights of municipalities (for example for stationary traffic in Germany). Thus, the Vienna World Agreement forms the basis for all road traffic laws worldwide.

In 2016, the Vienna World Agreement was expanded. This expansion allows for systems, in addition to the driver in the car or a guide for wagons, to lead a vehicle in public road traffic.

This comprehensive influence of the Geneva and Vienna World Agreements underscores their importance in shaping road traffic laws and regulations globally. Their impact extends from international organizations to local municipalities, truly forming the backbone of global road traffic laws.

1.8 Evolutionary Development of Road Traffic and Means of Transportation

The development of road traffic rules and driving license regulations has not followed an evolutionary process. The need and necessity for mobility have existed since time immemorial, leading to the establishment of rules in early cultures about how roads or trade routes should be created, what they should look like, and how they should be managed and maintained by stakeholders.

Financing has long been done through tolls, but martial aspects have also shaped the design of traffic networks. The means of transport themselves have developed out of the capabilities of various cultures, leading to the creation of different means of transport depending on transport needs, the capabilities of different cultures, and the purpose of the respective means of transport based on affordable infrastructure.

With the increase in traffic density, further regulation became necessary. The respective political instances balanced the benefits for the population (travel needs, food transport, etc.), economic interests including military strategies, and a corresponding need for regulation for means of transport and traffic infrastructure emerged.

Humans, as participants in traffic, had to adapt to the circumstances. For example, the driver had to learn the appropriate skills. For the guide for wagons or coachmen, it was sufficient to learn how to handle the draft animals and to know on which side of the road they can move best. However, the driver of automobiles also had to learn to deal with the technical conditions of the vehicle.

This historical perspective underscores the complexity and diversity of traffic rules and regulations across different cultures and societies. It also highlights the importance of understanding these differences in the context of global mobility and transportation.

The availability of energy has significantly shaped transport routes and the types of transport means. In the days of the Pony Express or post coach stations, considerations had to be made not only for the travelers and their comfort, leading to the establishment of hotels along the roads. The animals and their guides also had to be provided for, and protection against robbers and theft had to be ensured.

As vehicles began to populate roads, a new infrastructure need arose: fuel stations. These establishments emerged as critical waypoints on our roadways, serving as the lifeblood for the burgeoning number of automobiles. Over time, the handling of vehicle refueling at these stations has been continually developed and refined, evolving in tandem with advancements in automotive technology. This narrative explores the fascinating journey of how fuel stations came into existence, their evolution, and the ongoing innovations in vehicle refueling processes.

The evolution of railways has been a testament to human ingenuity and technological advancement. One of the most significant milestones in this journey has been the electrification of railways. This transformative shift ensured that electrical energy was always available for the locomotive, marking a departure from the era of steam and coal. The impact of this change was profound and multifaceted. It not only enhanced the efficiency and reliability of railway transport but also led to a substantial reduction in the necessary human resources. The roles of the stoker on the steam locomotive and personnel for coaling and maintenance were significantly diminished, if not entirely eliminated. This narrative delves into the fascinating story of railway electrification, exploring its origins, its implications, and its enduring influence on the railway industry.

With the advent of automobiles, drivers had to learn a new set of skills, one of which was refueling their vehicles. This task required understanding the peculiarities of both the vehicle and the fueling system. Over time, as drivers became more adept at this task, the role of the gas station attendant began to fade. In the early days, gas station attendants, also known as pump attendants, played a crucial role. They would pump gas, clean windshields, and often check the vehicle's oil and tire pressure. However, as self-service pumps were introduced and became more prevalent, the need for these attendants decreased. Today, many gas stations are unattended, operating entirely on self-service. Drivers pull up to the pump, refuel their vehicles, and pay with a credit card, all without the need for another person to be involved in the process. This shift towards self-service has made the process of refueling more efficient and convenient for drivers, marking a significant milestone in the evolution of automobile usage and fueling practices. It's a testament to how advancements in technology and infrastructure can transform everyday tasks and industries.

As we transition towards a more sustainable future, Electric Vehicles (EVs) have emerged as a key player in transforming our transportation landscape. However, this shift brings with it a new set of challenges, particularly in developing a robust and efficient charging infrastructure for EVs. Today, it is evident how complex this task is, requiring a rethinking of roles and activities to adapt to these new needs. From ensuring adequate coverage of charging stations to managing peak demand times, the task of building an effective EV charging infrastructure is multifaceted. It involves not just the physical installation of charging stations, but also the development of smart grid technologies, pricing strategies, and user-friendly interfaces for EV drivers.

Parallel to these developments, we find ourselves in a new digital world where automation is increasingly becoming the norm. From self-driving cars to smart homes, we are continually looking for ways to automate processes to make them more efficient and resource-saving. In the context of EV charging infrastructure, automation could play a pivotal role in streamlining operations, improving user experience, and optimizing energy use. This book explores these two intertwined aspects: the challenges with EV charging infrastructure and the role of automation in addressing these challenges. We delve into this exciting and rapidly evolving field, shedding light on the complexities, opportunities, and the future of EV charging infrastructure and automation.

The evolution of transportation infrastructure and regulations has been driven by a multitude of factors. While the pursuit of efficiency and positive advancements has certainly played a significant role, it's important to acknowledge that accidents and even disasters have also been instrumental in shaping these developments.

The need for efficiency has always been a primary driver of innovation in transportation. From the development of faster and more fuel-efficient vehicles to the optimization of traffic flow through intelligent transport systems, the quest for efficiency has led to significant advancements in both infrastructure and regulations.

However, as stated above, it's not just the positive aspects that have driven these developments. Accidents, and in some cases, major disasters, have highlighted the need for robust safety regulations and the development of safer, more resilient infrastructure. These unfortunate incidents have served as clear reminders of the potential risks associated with transportation, prompting authorities to implement stringent safety standards and invest in the development of safer transport systems. Furthermore, the development of the vehicles had been significantly influenced by more efficient production means, the development of vehicle safety features, and improving the handling of processes from fueling and charging up to driving the vehicle. In response to these incidents, regulations have been tightened and infrastructure has been optimized to mitigate risks. This includes everything from improved vehicle safety standards and stricter driver licensing requirements to the development of safer road infrastructure and more efficient emergency response systems.

While the pursuit of efficiency and positive advancements has been a key driver of progress in transportation, it's the lessons learned from accidents and disasters that have truly shaped the landscape of transportation infrastructure and regulations. These incidents have underscored the importance of safety and resilience, leading to the development of safer, more reliable transport systems.

The interfaces between different modes of transport present a complex challenge that cannot be solved by software improvements alone. Consider the issue of parking when switching from a car to a train or airplane. Where do vehicles get parked? How is security ensured? These are logistical challenges that require physical infrastructure and planning. Then there's the question of the "last mile"—the final leg of a journey from a transportation hub to a final destination. Buses, bicycles, and scooters all play a role here, but their integration is not always seamless. Is the bus accepted for the last mile? How do bicycles and scooters fit into the picture?

The coexistence of different modes of transport on the road is another aspect. On highways, it's relatively straightforward—pedestrians and vehicles operating under 60 km/h are not allowed. In urban and suburban areas, though, the interaction between cars, buses, bicycles, scooters, and pedestrians can be complex and requires careful regulation and infrastructure design.

Looking ahead, we must also consider whether owning a personal vehicle will continue to be accepted, given the manufacturing costs of a car, energy costs, availability of energy, and land use for parking spaces and charging stations, not to mention wind farms and photovoltaic parks for energy generation.

1.9 Challenges for Future Mobility Solutions

The development of future mobility solutions is a complex and multifaceted process. It involves the evolution of traffic rules, transportation infrastructure, and interfaces between various modes of transport. However, these elements have developed at different paces and in different ways across the globe.

All people on Earth have become accustomed to different roles and tasks in relation to transportation. These roles and tasks have not only influenced the characteristics of vehicles and traffic rules but have also permeated all levels of human culture.

Today, there is a growing motivation, both politically and environmentally, to prioritize public transport over individual car ownership. This shift demands a change in roles within this network. The charging infrastructure for Electric Vehicles must grow closer to the vehicle itself. This cannot be solved solely by software on an app but requires measures within the car and necessitates the driver's appropriate handling of information. Even the charging plug must fit the respective vehicle. The traditional concept of a fuel station every 20 km is no longer sufficient. A completely different charging infrastructure must be established, leading to a shift in the dominant players in the transport industry. The oil multinationals no longer dominate; instead, energy providers are becoming increasingly important.

The development of future mobility solutions is a dynamic and ongoing process. It involves not only technological advancements but also changes in societal norms, regulations, and infrastructure. As we move forward, it is crucial to continue exploring and implementing innovative solutions that meet our evolving transportation needs while also considering environmental sustainability and efficiency. The development of future mobility solutions requires a holistic approach that considers not just technological advancements, but also infrastructure development, regulatory changes, and shifts in societal attitudes towards transportation. It's a complex task, but one that holds the promise of more efficient, sustainable, and inclusive mobility solutions for all.

2 Social Acceptance

In the ever-evolving world of technology, the development of new products is often heralded as the pinnacle of human achievement. However, the successful integration of these innovations into society hinges not only on their technical merits but also on their social acceptance. This chapter explores the critical importance of social acceptance in the development of future mobility solutions, highlighting its significance and the interplay between legislative frameworks and societal readiness.

As we venture into the era of automated driving and electromobility, social acceptance emerges as a linchpin for the widespread adoption and success of these innovations. Legislators play a pivotal role in establishing the baseline for social acceptance through regulations that ensure safety, environmental sustainability, and ethical considerations. These laws act as the foundation upon which societal acceptance is built, providing a sense of security and trust in the new technologies.

However, the responsibility of fostering social acceptance extends beyond the legislative framework. Society itself must be prepared to embrace these changes, adapting to new paradigms in mobility. This readiness involves not only acknowledging the functional benefits of these innovations but also recognizing their potential to inspire and excite. One of the challenges facing the adoption of Electric Vehicles (EVs) is the need to create enthusiasm and excitement among consumers. Unlike the roaring engines of a V12 Ferrari or Lamborghini, the silent operation of EVs lacks the auditory thrill that many car enthusiasts crave. Moreover, the rolling noise of electric cars is sometimes perceived as a nuisance rather than an advantage.

To overcome these obstacles, it is imperative to engage in marketing efforts that highlight the unique attributes of EVs. Emphasizing their environmental benefits, cutting-edge technology, and the sheer joy of driving a vehicle that represents the future can create a sense of excitement. Without such efforts, the appeal to early adopters diminishes, as they seek to showcase their innovation and forward-thinking by being the first to own these impressive electric vehicles.

While the benefits of automated driving are clear—improved safety, reduced traffic congestion, and increased accessibility—there is a lingering question about its appeal to passionate drivers. Many automotive enthusiasts derive immense satisfaction from the tactile experience of controlling a vehicle, reveling in the thrill of the drive. The notion of relinquishing control to automation can seem antithetical to this desire.

To address this, it is crucial to communicate the specific advantages of automation. For instance, highway assistants designed for long, monotonous stretches can relieve the burden of tedious driving at constant speeds, allowing drivers to appreciate the journey without sacrificing their love for the open road. Furthermore, showcasing scenarios where automation enhances the driving experience, such as navigating through traffic or providing advanced safety features, can help bridge the gap between traditional driving enthusiasts and the new wave of automated solutions.

Social acceptance is as vital as product innovation in the successful integration of future mobility solutions. While legislators set the foundation, society must be prepared to embrace these changes and find excitement in the possibilities they offer. By highlighting both the practical and emotional benefits of innovations like electromobility and automated driving, we can foster a sense of enthusiasm and readiness, ensuring that these technologies are not only accepted but celebrated.

2.1 Social Acceptance of Mobility Solutions

Mobility solutions are an integral part of our daily lives, shaping how we commute, travel, and interact with our environment. The social acceptance of these solutions is crucial for their successful implementation and widespread use. This book explores the needs of individuals, groups, national economies, and economic communities in relation to mobility solutions.

Individual Needs

At the individual level, mobility solutions must cater to a variety of needs. These include convenience, affordability, accessibility, and safety. Individuals require reliable and efficient transportation that fits their lifestyle and budget. Furthermore, solutions must be inclusive, catering to people of all ages, abilities, and circumstances.

Group Needs

When considering groups, such as families or organizations, additional factors come into play. Shared mobility solutions, like carpooling or ridesharing, can offer cost-effective and environmentally friendly alternatives. However, these require careful coordination, trust among users, and a shift in traditional mindsets about vehicle ownership and privacy.

National Economies

On a larger scale, national economies need mobility solutions that support economic growth and sustainability. Infrastructure development, job creation, and the promotion of local industries are all tied to the transportation sector. Governments must balance these economic considerations with environmental and social impacts, promoting solutions that reduce congestion, pollution, and inequality.

Economic Communities

At the level of economic communities, such as the European Union, mobility solutions contribute to the integration and cohesion of member states. Cross-border transportation networks facilitate trade, tourism, and cultural exchange. However, these must be managed in a way that respects national sovereignty, regional disparities, and local customs.

The social acceptance of mobility solutions is a complex issue that involves a wide range of human needs and societal considerations. It requires a holistic approach that considers the individual, group, national, and international perspectives. By addressing these diverse needs, we can promote mobility solutions that are not only efficient and sustainable, but also equitable and widely accepted.

2.1.1 Stakeholder Requirements

Mobility solutions encompass a wide range of transportation methods, including road, air, rail, and water. Each of these solutions has a diverse set of stakeholders, each with their own requirements, benefits, and burdens. These stakeholders and their respective roles and interests are listed below.

Road Mobility Solutions

Stakeholders in road mobility solutions include drivers, passengers, vehicle manufacturers, infrastructure developers, and government regulators. Drivers and passengers seek safety, efficiency, and comfort. Manufacturers aim for profitability and innovation, while infrastructure developers focus on sustainable and efficient design. Regulators balance public safety, environmental impact, and economic growth.

Air Mobility Solutions

Air mobility stakeholders include airlines, passengers, aircraft manufacturers, airports, and aviation authorities. Airlines strive for operational efficiency and customer satisfaction. Passengers prioritize safety, comfort, and punctuality. Manufacturers aim for technological advancement and profitability. Airports seek efficient operations, while aviation authorities ensure safety and regulatory compliance.

Rail Mobility Solutions

Stakeholders in rail mobility include train operators, passengers, train manufacturers, infrastructure providers, and regulatory bodies. Operators and passengers prioritize safety, reliability, and speed. Manufacturers aim for innovation and cost-effectiveness. Infrastructure providers focus on sustainable and efficient design, while regulatory bodies ensure safety and fair competition.

Water Mobility Solutions

Water mobility stakeholders include ship operators, passengers, shipbuilders, port authorities, and maritime regulators. Operators and passengers value safety, reliability, and comfort. Shipbuilders strive for technological advancement and durability. Port authorities aim for efficient operations, and maritime regulators ensure safety and environmental protection.

The diverse stakeholders in mobility solutions each have their own unique requirements and bear different burdens. Balancing these interests is a complex task that requires careful consideration and collaboration. By understanding these dynamics, we can work towards mobility solutions that meet the needs of all stakeholders and contribute to a sustainable and efficient transportation system.

2.1.1.1 Typical Legal Regulations for Various Modes of Transport

Legislators play a crucial role as stakeholders in mobility solutions. They are responsible for creating and enforcing laws that govern the use of various modes of transport. These laws must strike a balance between societal acceptance, economic benefits, and the potential impacts on individuals or groups.

Balancing Economic Benefits and Societal Impacts

Laws must be designed to foster societal acceptance of mobility solutions. This involves ensuring that the economic benefits of these solutions, such as job creation and infrastructure development, are balanced against potential negative impacts on individuals or groups. For example, while a new highway may stimulate economic growth, it may also lead to increased noise and air pollution in nearby residential areas.

The Role of Interest Groups in Democracies

In democratic societies, interest groups can band together to advocate for their shared interests. This can involve organizing demonstrations, submitting petitions to parliaments, or lobbying politicians. Politicians and political groups, in turn, are dependent on their voter base and must take these interests into account when making decisions.

The Influence of Legal Structures and Historical Context

Legal structures and the democratic foundations of different nations have often evolved historically, leading to a focus on different aspects of law. For instance, some nations may prioritize civil law aspects (typical of civil law jurisdictions), while others may prioritize common law aspects (typical of common law jurisdictions). Depending on these priorities, laws may be prioritized differently.

Economic Interests as Drivers of Legal Balancing

Often, economic interests are the driving force behind the balancing of different laws. For instance, a law that promotes the use of electric vehicles may be driven by the economic interest in promoting the domestic electric vehicle industry, even as it also serves environmental and public health interests.

In the different means of transportation, the legislators have different basis for the relevant regulations on the different territorial areas. Some examples are listed below.

Road transport:

- **National regulations:** In many countries, road transport is regulated by national laws and regulations that cover aspects such as vehicle standards, driver licensing, traffic rules, and safety inspections.
- **European regulations:** The European Union has harmonized legal provisions for road transport across its member states. This includes technical specifications, taxation, social provisions, and environmental measures.
- **Global/UN regulations:** While there isn't a global governing body for road transport, the United Nations has developed conventions and agreements that many countries adhere to, such as the Vienna Convention on Road Traffic.

Air transport:

- **National regulations:** Each country has its own aviation authority that regulates air transport within its jurisdiction, including aircraft certification, air traffic control, and airline operations.
- **European regulations:** The European Union has established a single aviation market, with common rules for licensing, safety, air traffic management, and passenger rights.
- **Global/UN regulations:** The International Civil Aviation Organization (ICAO), a UN agency, sets international standards and regulations for aviation safety, security, efficiency, and environmental protection.

Water transport:

- **National regulations:** Countries regulate their waterways through national maritime laws, covering areas like vessel registration, safety standards, and environmental regulations.
- **European regulations:** The EU has regulations for maritime and inland waterway transport, including public service obligations, safety requirements, and market access conditions.
- **Global/UN regulations:** The International Maritime Organization (IMO), a UN agency, develops international regulations for shipping, including safety, environmental impact, and legal matters.

Rail transport:

- **National regulations:** National rail transport is governed by each country's specific laws and regulations, which cover areas like safety standards, network access, and operator licensing.
- **European regulations:** The EU has established a single European railway area, with harmonized rules for market access, safety, interoperability, and passenger rights.
- **Global/UN regulations:** While there isn't a global governing body for rail transport, the Intergovernmental Organisation for International Carriage by Rail (OTIF) and the Organisation for Cooperation between Railways (OSJD) develop uniform technical and legal standards that many countries follow.

Regulations are in any of the means of transportation a matrix with different historical backgrounds. This makes even coexistence in the same traffic environment very difficult.

2.1.1.2 Different Roles of Governmental Authority

Legislative Instances and Separation of Powers

The principle of separation of powers divides the institutions of government into three branches: legislative, executive, and judicial. Each has separate and independent powers and areas of responsibility so that the powers of one branch are not in conflict with those of the other branches. This book will explore these instances and their roles in the context of road traffic law.

Legislative Branch

The legislative branch is responsible for making laws. In the context of road traffic law, this could involve creating laws related to speed limits, seat belt use, or drunk driving. The legislative process often involves consultation with experts, public input, and careful consideration of the potential impacts of the law.

Executive Branch

The executive branch is responsible for implementing and enforcing laws. In the context of road traffic law, this could involve agencies that oversee driver licensing, vehicle registration, road maintenance, and traffic enforcement. The executive branch ensures that the laws passed by the legislative branch are carried out effectively and fairly.

Judicial Branch

The judicial branch is responsible for interpreting laws and adjudicating disputes. In the context of road traffic law, this could involve courts that handle traffic violations, accidents, or disputes between drivers. The judicial branch ensures that the laws are interpreted correctly and that disputes are resolved in accordance with the law.

Examples of Legislative, Executive, and Judicial Instances in Road Traffic Law

Legislative instances

Legislative instances are responsible for creating laws. In the context of road traffic law, these could include the following:

- **Parliaments:** National parliaments, such as the Bundestag in Germany or the House of Commons in the UK, pass laws related to road safety, vehicle standards, and driver licensing.
- **EU Commission:** The European Commission proposes legislation for road transport that applies across the European Union, such as regulations on vehicle emissions and cross-border traffic rules.
- **Councils:** Local councils or city councils, like the Stadtrat in Munich or the London Assembly, can set local traffic regulations, such as speed limits in residential areas or parking restrictions.
- **Regional governments:** Regional governments in Germany have legislative powers in certain areas of transport policy, such different authorities on different types of roads like country roads or urban roads.

In the context of traffic management, authority is often divided among various governmental bodies depending on the type of traffic and the geographical area. For instance, cities or municipalities typically have jurisdiction over stationary traffic, which includes aspects such as parking regulations, parking enforcement, and the designation of loading zones. This is because these aspects of traffic management often have direct impacts on local residents and businesses, and decisions about them can be best made at the local level.

On the other hand, regional governments often have jurisdiction over moving traffic on public roads. This includes aspects such as speed limits, traffic signal timings, and the enforcement of traffic rules. These aspects of traffic management require a broader perspective that takes into account regional traffic patterns and infrastructure.

This division of authority allows for more effective and responsive traffic management, as each level of government can focus on the aspects of traffic that are most relevant to its jurisdiction. However, coordination and cooperation among different levels of government are essential to ensure a smooth and efficient transportation system.

Executive instances

Executive instances are responsible for enforcing laws. In the context of road traffic law, these could include the following:

- **Police:** The police enforce traffic laws, for example by issuing fines for speeding or drunk driving.
- **Prosecutors:** Prosecutors bring cases against individuals who have committed traffic offenses.
- **Ministry of Interior:** The Ministry of Interior organizes internal security, including traffic safety. It may oversee the police and other law enforcement agencies.

Judicial instances

Judicial instances are responsible for interpreting laws and adjudicating disputes. In the context of road traffic law, these could include the following:

- **Courts:** Courts adjudicate disputes and determine penalties for traffic offenses. This could range from local traffic courts to higher regional courts.
- **Ministry of Justice:** The Ministry of Justice organizes the judicial system, including the courts that handle traffic cases.

Each of these instances plays a crucial role in the governance of road traffic. They work together to create, enforce, and interpret road traffic laws, ensuring the safety and efficiency of road transport. Since they have different regional priorities, different groups of interest, lobbyists and different historical background, the dependencies and basics of their regulations are often very heterogeneous.

2.2 Establishment of Fundamental Rules for Road Traffic

The need to create a standardized set of regulations for road traffic became increasingly apparent in the mid-20th century due to the rapid growth of motor vehicle use and international travel. The United Nations recognized this necessity and sought to establish comprehensive frameworks to ensure road safety, facilitate international transportation, and harmonize traffic regulations. This led to the development of the Vienna Convention on Road Traffic and the Geneva Conventions on Road Traffic.

Geneva Conventions on Road Traffic

Before the Vienna Convention, the Geneva Conventions on Road Traffic, adopted in 1949, served as an initial effort to establish common traffic rules. The goals of these conventions included the following:

- **facilitating international movement:** creating a framework that allows vehicles to move seamlessly between countries by recognizing each nation's driving permits and vehicle registrations
- **enhancing road safety:** introducing basic traffic regulations and safety measures to protect drivers, passengers, and pedestrians
- **harmonization of laws:** encouraging countries to adopt similar traffic laws to reduce the risk of accidents and misunderstandings among international drivers

Vienna Convention on Road Traffic

The Vienna Convention on Road Traffic, concluded in 1968, aimed to promote the development of international road traffic and to increase road safety through uniform traffic rules. Its key objectives included the following:

- **standardization of traffic signs and signals:** establishing consistent and recognizable traffic signs, signals, and road markings to prevent confusion and enhance safety for international drivers
- **driver and vehicle requirements:** setting minimum standards for driver competence and vehicle safety, including provisions for driving licenses and vehicle registration
- **rules of the road:** harmonizing road traffic rules, such as right-of-way, speed limits, and overtaking regulations, to ensure that drivers can operate vehicles safely across different countries

The Vienna and Geneva Conventions on Road Traffic were pivotal in establishing foundational rules for international road safety and traffic regulation. By standardizing traffic signs, signals, and rules, and setting minimum standards for drivers and vehicles, these conventions have significantly contributed to safer and more efficient international road travel. As global mobility continues to evolve, the principles established by these conventions remain crucial to maintaining road safety and facilitating cross-border transportation.

2.3 Societal versus Individual Risk

Individual Risk in the context of public road traffic encompasses various factors and affects different stakeholders. Understanding both societal and individual risks is crucial to developing effective safety measures and policies. This section delves into these risks, examining the roles of various road users and highlighting the diverse influences on risk tolerance and acceptance.

Societal risk refers to the collective risk to the community or population as a whole. This type of risk is often managed through regulations, infrastructure improvements, and public policies aimed at enhancing road safety for everyone. Societal risk involves the following:

- **Other road users:** The safety of all road users, including motorists, cyclists, pedestrians, and vulnerable groups like children, is a communal concern. Measures such as traffic regulations, speed limits, and safe crossings aim to mitigate risks.
- **Public perception and policy:** Societal risk is heavily influenced by cultural, political, and social factors. For instance, the promotion of environmentally friendly transportation methods can shift public tolerance and acceptance of certain risks.
- **Public transport vs. private vehicles:** The choice between using personal cars, rented vehicles, taxis, robotaxis, or public transport affects the overall risk landscape. Each mode has different levels of risk associated with it, influencing public policy and individual preferences.

Individual risk pertains to the risk experienced by a single person. This type of risk is subjective and varies greatly based on personal circumstances and perceptions. Key factors include the following:

- **Role of different road users:** Drivers, passengers, cyclists, and pedestrians each face unique risks. Vulnerable road users, like children, are at higher risk and need special protections.
- **Personal situations and stress:** Personal circumstances, such as stress levels and whether one is experiencing eustress (positive stress) or distress (negative stress), significantly impact tolerance for risk. Different roles and situations can lead to varied risk acceptance thresholds.
- **Cultural and political influences:** The cultural background and political stance of individuals shape their attitudes towards risk. For instance, someone from a culture that prioritizes public transport may have a lower tolerance for the risks associated with private vehicle use.
- **Influence of politics and public events:** Public and political discourse can significantly affect societal risk tolerance and acceptance. Policies advocating for greener transport options or significant road safety campaigns can alter public perception and drive changes in individual behaviors and societal norms.

Understanding the nuances between societal and individual risk in public road traffic is essential for developing effective safety measures. Cultural, political, and personal factors all play a role in shaping risk perceptions and tolerances. By addressing both collective and individual concerns, we can create safer roads for all users.

2.4 Approach of Tolerable Risk

The concept of tolerable risk is a fundamental aspect of risk management, particularly in safety-critical and safety-involved systems. Tolerable risk is a risk that is accepted in view of the benefits obtained by accepting it. The cost in inconvenience or in money is balanced against the scale of risk and a compromise is accepted.

One of the key tools used in the assessment of tolerable risk is the Farmer's curve, also known as the F-N curve. This graphical representation of societal risk generated by an activity or project provides a visual means of assessing the frequency and severity of potential accidents. It allows for a comprehensive understanding of the risk landscape and aids in making informed decisions about risk tolerability.

The Farmer's curve, also known as the F-N curve, is a tool that was historically developed in the context of nuclear power plants to manage and quantify risk. It plots the cumulative frequency (F) of events causing N or more fatalities, providing a visual representation of societal risk.

The curve is divided into three regions, each representing a different level of risk:

- **Unacceptable risk:** This region represents risks that are too high to be tolerated under any circumstances. Any risk falling into this area requires immediate action for risk reduction.
- **Tolerable risk:** This region represents risks that are accepted under certain conditions, such as the benefits outweighing the risks or the impracticality of further risk reduction measures. Risks in this area are deemed tolerable but should be reduced as much as possible.
- **Negligible risk:** This region represents risks that are so low that they are generally accepted without further mitigation.

The Farmer's curve helps in defining these areas of risk and provides a quantitative measure for effective risk reduction. It allows us to see where a particular risk falls on the curve and thus determine whether it is unacceptable, tolerable, or negligible.

Moreover, the Farmer's curve also guides the implementation of qualitative measures for effective risk reduction. For instance, if a risk falls into the unacceptable region, immediate and substantial measures must be taken to reduce it. If it falls into the tolerable region, continuous efforts should be made to lower the risk as much as possible, even if it's already accepted.

The Farmer's curve, derived from the nuclear power industry, provides a robust framework for managing and reducing risk, defining areas of tolerable and unacceptable risk, and guiding the implementation of effective risk reduction measures. It is a powerful tool for answering the question "How safe is safe enough?" in a quantitative and understandable way. It allows for a balance between the benefits of risk-taking and the need for safety, ultimately contributing to more effective and responsible decision-making.

The approach of tolerable risk, supported by tools like the Farmer's curve provides a robust framework for managing risk in a wide range of contexts. It allows for a balance between the benefits of risk-taking and the need for safety, ultimately contributing to more effective and responsible decision-making.

The Farmer's curve or F-N curve plots the cumulative frequency (F) of events causing N or more fatalities. A wider view could allow a better adaptation of the approach to mobility solutions:

- On the x-axis, we could have the severity of damages. This could be measured in terms of the number of fatalities, financial loss, or any other measure of harm. The aspects of financial losses are a possible view for insurance assumptions.
- On the y-axis, we have the frequency, probability, or duration of these events. This could be the number of times the event occurs in a given time period, the likelihood of the event occurring, or the length of time the event lasts.

The curve is divided into three regions by two lines (Figure 2.1): the "intolerable line" and the "negligible line:"

- **Intolerable risk:** This region is above the "intolerable line." Any risk falling into this area is considered too high to be tolerated under any circumstances. Immediate action is required to reduce the risk. This could involve implementing safety measures, changing procedures, or even stopping the activity altogether.
- **Tolerable risk (ALARP):** This region is between the "intolerable line" and the "negligible line." Risks in this area are deemed tolerable but should be reduced as much as possible. This is also known as the ALARP region (As Low As Reasonably Practicable). In this region, quantitative risk reduction measures are applied to lower the risk as much as reasonably practicable. This could involve implementing additional safety measures, improving existing measures, or changing procedures to reduce the likelihood or severity of the event.
- **Negligible risk:** This region is below the "negligible line." Risks in this area are so low they are generally accepted without further mitigation. No additional measures are required to reduce the risk.

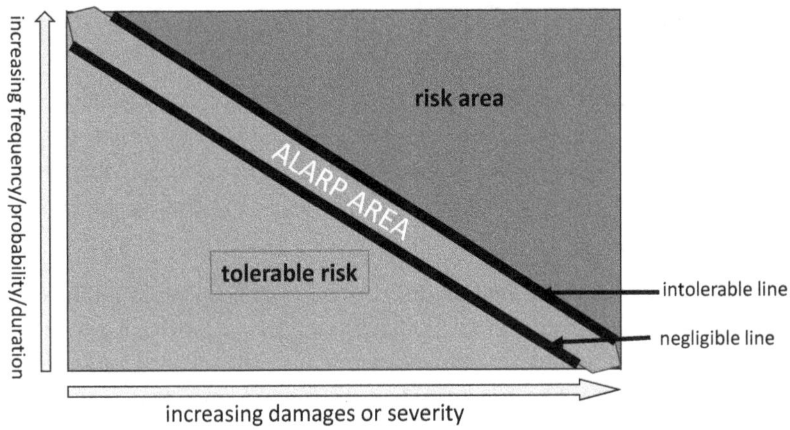

Figure 2.1 Basic interpretation of the farmer curve

By plotting the risks on the Farmer's curve, we can see where they fall in these regions and make informed decisions about whether the level of safety is sufficient or whether further risk reduction measures are needed. This makes the Farmer's curve a powerful tool for managing risk in a wide range of contexts. It allows for a balance between the benefits of risk-taking and the need for safety, ultimately contributing to more effective and responsible decision-making. It is a powerful tool for answering the question "How safe is safe enough?" in a quantitative and understandable way. It allows for a balance between the benefits of risk-taking and the need for safety, ultimately contributing to more effective and responsible decision-making.

2.4.1 Quantitative Risk Management

Risk management, particularly its quantitative aspect, plays a pivotal role in decision-making processes across various domains. Methods for quantitative risk management provide a structured, quantifiable approach to assess and mitigate risks, striking a balance between the benefits of risk-taking and the imperative need for safety.

Quantification in risk management is of paramount importance as it lends transparency and traceability to the decision-making process. It allows for a clear, objective assessment of risks, thereby facilitating informed decision-making. Quantitative measures provide a common language that enables clear communication of risks and their potential impacts. This transparency aids in the understanding and acceptance of decisions, fostering trust among all stakeholders.

In addition to these methods, this book also explores various strategies for managing traffic risks, such as enforcement of traffic laws, road safety campaigns, infrastructure improvements, vehicle safety features, and driver education and training. These strategies, backed by quantitative data, have proven effective in reducing traffic accidents and fatalities.

The effectiveness of these methods, however, can vary depending on various factors such as the specific context, the level of enforcement, and public compliance. By providing a comprehensive overview of these methods, this book aims to contribute to more effective, transparent, and responsible decision-making in risk management.

Here some typical examples:

- **ALARP (As Low As Reasonably Practicable):** This principle is based on the concept of "acceptable" or "tolerable" risks. It allows analysts and decision-makers to define boundaries to combine probability-consequence scales. The ALARP level is reached when the time, trouble, and cost of further reduction measures become unreasonably disproportionate to the additional risk reduction obtained.
- **MEM (Minimum Endogenous Mortality):** MEM is a method to obtain absolute values for the acceptance of risks based on the natural mortality rate of society. Used widely in Germany as a methodology for risk acceptance, it is considered an acceptable risk if it is lower than the mortality of human individuals in the general population.
- **Risk-benefit analysis:** This method seeks to quantify the risk and benefits and hence their ratio. It involves analyzing the expected or actual failure modes of a product and identifying actions to reduce or mitigate their effect. The analysis is often complex as it requires evaluation of a large amount of data and involves weighing risks and benefits, which depend on attitudes and preferences about which reasonable people disagree.
- **Positive risk balance:** This approach is about ensuring that the benefits of sufficiently mitigating residual risk outweigh the potential harm. For example, in the context of automated vehicles, this includes the expectation that automated vehicles cause fewer crashes on average compared to those made by drivers.

Depending on the application and the context, the methods could be calibrated in various ways. Such methods could be used to analysis risks in road traffic in the following areas:

- **Enforcement of traffic laws:** Strict enforcement of traffic laws, including speed limits, drunk driving laws, and seatbelt usage, has been shown to reduce traffic accidents and fatalities.
- **Road safety campaigns:** These campaigns aim to raise awareness among drivers about the dangers of risky behaviors such as speeding, drunk driving, and distracted driving.
- **Infrastructure improvements:** This includes measures such as better road design, installation of traffic calming measures, and improved signage and lighting.

- **Vehicle safety features:** Modern vehicles come equipped with a variety of safety features such as seat belts, airbags, Anti-Lock Braking Systems (ABS), and Electronic Stability Control (ESC) that can significantly reduce the risk of accidents and injuries. It is the basis for today's safety standards such as ISO 26262.
- **Driver education and training:** Providing drivers with proper education and training can equip them with the skills and knowledge needed to drive safely and responsibly.

Remember, the effectiveness of these methods can vary depending on various factors such as the specific context, the level of enforcement, and public compliance. However, the use of these methods should not lead to complacency or a false sense of safety and security. Risks that are quantified and deemed "acceptable" or "tolerable" should not be ignored or neglected. They still require continuous monitoring and management to ensure that they remain within acceptable levels. Any change in circumstances or new information could alter the risk profile, potentially moving a previously "acceptable" risk into the "unacceptable" category.

Moreover, these quantitative methods serve to establish traceability and comparability in the decision-making process. They allow for clear, objective assessment of risks and their potential impacts, facilitating transparency in decision-making. This transparency aids in the understanding and acceptance of decisions, fostering trust among all stakeholders.

While quantitative risk analysis methods are invaluable tools in risk management, they should not be used as an excuse to ignore or neglect risks. Instead, they should be used as part of a comprehensive risk management strategy that includes continuous monitoring and management of all identified risks. This approach ensures that risks are not only quantified but also effectively managed, contributing to safer and more responsible, traceable, and transparent decision-making.

2.5 Risk Management Systems (RMS)

The concept of tolerable risk fits neatly into general management approaches, such as the 7-step approach. This approach, which includes steps like defining the problem, disaggregating the problem, prioritizing features, creating a work plan, analyzing each piece, synthesizing the information, and communicating findings, can be effectively used to manage tolerable risk. By defining the risk, breaking it down into manageable parts, and systematically analyzing each part, managers can make informed decisions about what level of risk is tolerable.

Risk Management Systems (RMSs) focus on identifying, assessing, and prioritizing risks to minimize the negative impact on an organization. This process involves several key steps:

- **risk identification:** spotting potential risks that could affect the organization, such as financial uncertainties, legal liabilities, or strategic management errors
- **risk assessment:** evaluating the likelihood and potential impact of identified risks
- **risk mitigation:** developing strategies to manage and mitigate these risks, such as risk avoidance, reduction, sharing, or retention
- **monitoring and review:** continuously monitoring risks and the effectiveness of mitigation strategies and adjusting them as needed

Safety Management Systems (SMSs) are more specific, focusing on maintaining and improving safety standards to prevent accidents and incidents. The core elements of an SMS include the following:

- **Safety policy:** establishing a clear safety policy, outlining the organization's commitment to safety and its objectives
- **Safety Risk Management:** identifying, analyzing, and controlling safety risks to prevent accidents and incidents
- **Safety assurance:** ensuring that safety measures are effective and being properly implemented through continuous monitoring and audits
- **Safety promotion:** promoting a safety culture within the organization through training, communication, and continuous improvement

The key differences between an RMS and an SMS could be summarized as follows:

- **Scope:** The RMS encompasses a broader range of risks, including financial, operational, strategic, and compliance risks, while the SMS specifically targets safety-related risks.
- **Focus:** The RMS aims to protect the organization from various potential threats, whereas SMS is dedicated to preventing accidents and ensuring the safety of personnel and operations.
- **Implementation:** The RMS involves a wide range of stakeholders, including risk managers, executives, and department heads, whereas the SMS primarily involves safety officers, employees, and management teams focused on safety.

The RMS is often project-specific, addressing risks tied to particular goals or initiatives, while the SMS is a broader, systemic approach to safety management that permeates all levels of an organization:

- **Risk Management System (RMS):** The RMS is typically applied within the scope of a specific project, system, or initiative. It's about identifying and managing potential risks that could impact the success and performance of the particular undertaking. For instance, in a new product launch, the RMS would assess risks like market acceptance, supply chain issues, or regulatory compliance. It's focused and project-centric, ensuring that potential threats to the project's objectives are identified, assessed, and mitigated.

- **Safety Management System (SMS):** On the other hand, the SMS is more of an overarching control mechanism for organizations. It is designed to manage and enhance safety across all organizational activities. The SMS integrates safety into the daily operations and culture of the entire organization, from executive decision-making down to routine operational procedures. It's about ensuring ongoing adherence to safety standards, continuous improvement in safety practices, and fostering a culture of safety throughout the organization.

The automotive industry is increasingly focusing on Safety Management Systems to establish an organizational framework for various safety tasks within a company. This shift emphasizes integrating safety into every aspect of the organization's operations and culture. In this book, the following chapters will delve deeper into the implementation and benefits of SMS in the automotive sector.

2.5.1 Types of Risk Management Systems

There are several types of Risk Management Systems tailored to different industries and organizational needs. Each type focuses on specific aspects of risk management and employs unique methodologies. Here's a brief overview of some of the main systems:

Enterprise Risk Management (ERM)

Description: Enterprise Risk Management (ERM) is a holistic approach that focuses on identifying, assessing, and managing risks across an entire organization. It aims to align risk management with the organization's strategic objectives.

Example: A multinational corporation uses ERM to address risks in finance, operations, compliance, and reputation.

Project Risk Management (PRM)

Description: Project Risk Management (PRM) targets risks associated with specific projects. It focuses on identifying potential project-related risks, assessing their impact, and developing mitigation strategies.

Example: A construction company uses PRM to manage risks such as delays, cost overruns, and safety hazards on a building project.

Financial Risk Management (FRM)

Description: Financial Risk Management (FRM)deals with managing financial risks, including market risk, credit risk, liquidity risk, and operational risk. It uses financial instruments and techniques to mitigate these risks.

Example: A bank employs FRM to manage its exposure to interest rate fluctuations, credit defaults, and market volatility.

Operational Risk Management (ORM)

Description: Operational Risk Management (ORM) focuses on risks arising from an organization's day-to-day operations. It includes risks related to processes, people, systems, and external events.

Example: A manufacturing company implements ORM to address risks such as equipment failure, supply chain disruptions, and human errors.

Strategic Risk Management (SRM)

Description: Strategic Risk Management (SRM) is concerned with risks that can impact an organization's long-term strategic goals. It involves assessing strategic initiatives and external factors that could affect the organization's direction.

Example: A technology firm uses SRM to evaluate risks associated with entering new markets or launching innovative products.

Compliance Risk Management (CRM)

Description: Compliance Risk Management (CRM) focuses on ensuring that an organization adheres to laws, regulations, and internal policies. It aims to prevent legal penalties, financial losses, and reputational damage.

Example: A healthcare provider uses CRM to comply with healthcare regulations and patient data protection laws.

Health, Safety, and Environment Risk Management (HSE)

Description: Health, Safety, and Environment Risk Management (HSE) addresses risks related to health, safety, and environmental impact. It aims to create a safe workplace and minimize environmental harm.

Example: An oil and gas company uses HSE to manage risks such as workplace accidents, environmental spills, and regulatory compliance.

Information Security Risk Management (ISRM)

Description: Information Security Risk Management (ISRM) focuses on protecting an organization's information assets from cyber threats, data breaches, and other information security risks.

Example: An IT company implements ISRM to safeguard its data against hacking, malware, and unauthorized access.

Summary

Each of the aforementioned systems employs specific processes and tools to manage risks effectively, ensuring that organizations can mitigate potential threats and capitalize on opportunities.

These are the overall goals of risk management:

- **minimize negative impacts:** reduce the likelihood and severity of potential risks
- **ensure business continuity:** maintain smooth operations and achieve strategic objectives despite uncertainties
- **enhance decision-making:** provide a structured approach to identifying and managing risks, supporting informed decisions
- **protect resources:** safeguard financial, human, and physical resources from potential threats
- **compliance and governance:** ensure adherence to regulatory requirements and internal policies, fostering a culture of risk awareness

By systematically addressing each step, organizations can effectively manage risks and create a safer, more resilient operational environment.

2.5.2 Risk Management in 7 Steps

The 7-Step Risk Management System is a structured approach to identifying, evaluating, and mitigating risks within an organization. The overarching goal of risk management is to minimize the potential negative impacts of risks, ensuring business continuity and achieving organizational objectives.

Here's a detailed explanation of each step and its specific goals:

(1) Risk Identification

Goal: to uncover all potential risks that could affect the project or organization

Explanation: This step involves systematically identifying risks using various techniques such as brainstorming, expert interviews, and risk checklists. The aim is to compile a comprehensive list of risks, no matter how unlikely or insignificant they may seem.

(2) Risk Analysis

Goal: to understand the nature of identified risks and their potential impacts

Explanation: This step involves assessing the likelihood and consequences of each risk. By analyzing risks, organizations can determine which risks are most significant and prioritize them accordingly.

(3) Risk Evaluation

Goal: to prioritize risks based on their assessed impact and likelihood

Explanation: After analyzing the risks, this step involves comparing them against risk criteria to decide which risks require immediate attention and which can be monitored. This prioritization helps focus resources on the most critical risks.

(4) Risk Treatment

Goal: to develop and implement strategies to manage risks

Explanation: In this step, organizations determine how to address each prioritized risk. Strategies can include avoiding the risk, reducing its impact, transferring the risk to another party, or accepting it. The goal is to mitigate risks to an acceptable level.

(5) Risk Monitoring and Review

Goal: to ensure that risk management strategies are effective and updated as needed

Explanation: This step involves continuous monitoring of risks and the effectiveness of risk treatment measures. Regular reviews help ensure that new risks are identified, and existing risks are managed appropriately.

(6) Risk Communication and Consultation

Goal: to keep stakeholders informed and engaged in the risk management process

Explanation: Effective communication ensures that all relevant parties understand the risks and the measures being taken. Consultation with stakeholders helps gather insights and ensures buy-in for risk management strategies.

(7) Risk Documentation and Reporting

Goal: to maintain a record of risk management activities and provide transparency

Explanation: This step involves documenting all aspects of the risk management process, including identified risks, analysis results, treatment plans, and monitoring activities. Reporting to stakeholders ensures accountability and transparency.

The 7-Step Risk Management System is not explicitly applied in automotive industries, but the different steps could be recognized in several fields of applications.

Here are some examples of the different steps:

Risk Identification

Example: In an automotive company, identify potential risks such as software bugs in automated driving systems, supply chain disruptions, or regulatory changes. This first step is about listing all possible risks.

Risk Analysis

Example: Assess the likelihood and impact of each identified risk. For instance, analyze the probability of software bugs occurring and the potential impact on vehicle safety and company reputation.

Risk Evaluation

Example: Compare the identified and analyzed risks against risk criteria to prioritize them. In the case of the automotive company, determine which risks require immediate action and which can be monitored over time.

Risk Treatment

Example: Develop strategies to manage the prioritized risks. For example, implement regular software updates and rigorous testing protocols to mitigate the risk of software bugs.

Risk Monitoring and Review

Example: Continuously monitor the risks and review the effectiveness of risk treatment plans. In our automotive scenario, regularly review the results of software tests and update risk management strategies accordingly.

Risk Communication and Consultation

Example: Keep all stakeholders informed about the risks and the measures being taken. This could involve regular updates to the management team about software testing results and ongoing risk management efforts.

Risk Documentation and Reporting

Example: Document all risk management activities and report to relevant stakeholders. This might include detailed reports on the risk assessments, treatment plans, and their outcomes, ensuring transparency and accountability in the automotive company's risk management process.

By following these steps, organizations can systematically identify, assess, and manage risks to minimize negative impacts and enhance decision-making.

2.6 Area of Tolerable Risk

Tolerable Risk: Balancing Interests and Societal Well-Being

In the automotive industry, managing risk is critical to ensuring safety and advancing technology responsibly. It is essential to apply the approach of tolerable risk to develop an understanding of the interplay between tolerable risk, legal regulations, and the obligations of those introducing new technologies.

These are some examples from the automotive sector to illustrate the concepts:

- **Legal regulations and risk management:** Legal frameworks are designed to protect both individual rights and the collective welfare. They establish rules and standards that prevent undue harm while allowing technological progress. In the automotive sector, regulations ensure that risks, such as those posed by autonomous vehicles, are manageable and acceptable.
- **Individual freedoms versus societal well-being:** While individual freedoms are essential, they must not override societal interests. Regulations limit risk-taking to protect everyone. For example, speed limits on roads balance personal freedom with public safety, ensuring that drivers do not endanger others.
- **Preventing excessiveness:** Risk Regulations set boundaries to prevent reckless behavior, curbing actions that could harm people, property, or the environment. For instance, stringent emission standards for vehicles aim to reduce environmental pollution, preventing excessive risk accumulation.
- **Acceptable risk levels**: Regulations define what level of risk is acceptable, ensuring that activities benefit society without endangering it. Building codes for automotive manufacturing facilities, for example, ensure safe construction practices that protect workers and the surrounding community.

Product Liability and Producers' Obligations

- **Introduction of technologies:** Those interested in introducing new automotive technologies must consider their societal impact. Innovations such as Electric Vehicles (EVs) should enhance well-being by reducing emissions, not harm it.
- **Compensation for damages:** If an automotive technology causes harm, such as faulty airbags leading to injuries, the responsible party must compensate for damages. Product liability laws hold manufacturers accountable, ensuring they address and rectify any issues promptly.
- **Producers' rights and responsibilities:** Producers have the right to innovate but also bear significant responsibilities. They must ensure safety, quality, and adherence to regulations. For example, automotive companies must conduct rigorous testing to confirm that self-driving cars are safe before they hit the roads.

- **Social contract:** The basis of this system is a social contract. Innovators benefit from society, and in return, they must prioritize safety and mitigate risks. This ensures that technological advancements do not come at the expense of public safety.

New technology inherently carries unknown risks and uncertainties regarding its potential impact. Start-ups and developers face significant pressure from investors and their own ambition to bring new technology to market. They must demonstrate the business case for the technology to investors, which creates immense pressure to introduce their ideas to the market quickly, even at higher risks, to prove the concept can eventually be self-sustaining.

Balancing technological development, growth, proof of concept, and business scalability—essentially predicting from individual success to a mass-market solution—is incredibly challenging. Authorities also face the dual pressure of promoting innovation while simultaneously protecting society from the risks of new technologies. Finding the balance between trust, acceptance, and risk tolerance is a significant challenge for society, government, administration, the judiciary, and the willingness to foster innovation:

- **Balancing interests:** Legal regulations aim to strike a balance between individual freedoms and societal well-being. They set boundaries to prevent excessive risk-taking that could harm people, property, or the environment. By defining acceptable risk levels, regulations ensure that activities benefit society without endangering it.
- **Public safety:** Regulations exist to protect public safety and health. They establish standards for various domains, such as workplace safety, transportation, and food production, to minimize risks. In the automotive sector, compliance with regulations ensures that risks are managed responsibly, from manufacturing to road safety.
- **Risk perception and trust:** Society's perception of risk is influenced by legal frameworks. When activities adhere to regulations, they gain public trust. Conversely, non-compliance erodes trust and may lead to social rejection. For example, strict adherence to safety standards in the automotive industry enhances consumer confidence in new vehicle technologies.
- **Accountability and liability:** Legal requirements assign responsibility and liability. Entities, including individuals, companies, and institutions, must follow rules or face consequences. This accountability fosters trust and acceptance. Automotive recalls due to safety issues, for instance, hold manufacturers accountable and reinforce their commitment to public safety.
- **Consistency and predictability:** Regulations provide consistency across sectors. Predictable rules allow stakeholders to plan, invest, and innovate confidently. This

stability contributes to societal acceptance. For example, consistent safety standards in vehicle manufacturing create a predictable environment that encourages technological advancement while ensuring public safety.

Legal regulations define the boundaries of acceptable risk, ensuring that activities align with societal values and safety. Those introducing new automotive technologies must act responsibly, compensating for any harm caused. This balanced approach ensures progress without sacrificing safety, fostering a trusted and innovative automotive industry.

3 Safety Management Systems (SMSs)

Ole Hans

3.1 Introduction

The ongoing automation and digitalization in the automotive industry presents new challenges for the safety of vehicles and their systems. To address these challenges, the concept of Safety Management Systems (SMSs) is gaining increasing importance in the automotive industry especially for highly automated vehicle fleets. Originally developed and implemented in the aviation industry to meet the high safety standards of that sector, SMSs are now being increasingly adopted in the automotive industry. The relevance of structured safety management for ensuring safe mobility is undeniable, particularly in light of the growing complexity of modern vehicles. This evolution highlights the necessity of transferring and adapting proven safety concepts from aviation to the automotive industry in order to continually enhance safety standards across both sectors. In this context, leveraging the experiences and best practices from aviation plays a critical role in the development of a comprehensive and effective SMS for the automotive sector.

Furthermore, today's systems are developed by various roles within organizations, necessitating that all involved parties adhere to the same rules and structures. If these standards are not uniformly applied, the resulting components may not integrate properly within the final product, leading to incompatibility in the derived systems. An effective SMS also consolidates various safety domains, such as functional safety, Safety of the Intended Functionality (SOTIF), security, and product safety. By harmonizing these domains under a unified framework, the SMS facilitates a more holistic approach to safety management, ensuring that all safety aspects are considered and addressed throughout the development and operational lifecycle of automotive systems.

3.1.1 Introduction of Safety Management Systems in the Automotive Industry

An SMS is a comprehensive and structured approach to safety management that was introduced to systematically identify, assess, and mitigate safety risks, thereby reducing the likelihood of accidents and incidents. An SMS is based on a continuous improvement process consisting of four main components: safety policy, Safety Risk Management (SRM), safety assurance, and safety promotion.

SMSs are becoming increasingly important in the automotive industry. With the increasing complexity of modern vehicles, the introduction of driver assistance systems and the development towards autonomous vehicles, the demands on vehicle safety, vehicle production, and vehicle fleet management are growing. Traditional approaches to safety management, which are limited to technical inspections and selective safety measures, are no longer sufficient. Instead, the modern automotive industry requires integrated and holistic safety solutions that consider all aspects of vehicle development, production, and use.

Adopting and adapting SMS principles from aviation provides the automotive industry with a proven framework for overcoming these challenges. In particular, the systematic approach to identifying and assessing risks and the continuous monitoring and improvement of safety measures are key elements that can be implemented in the automotive industry. Given the increasing complexity of modern vehicles and the introduction of new technologies such as autonomous driving, a robust SMS is becoming increasingly important.

3.1.2 Relevance for Safe Mobility in the Automotive Sector

This book addresses the current and future challenges of road safety. The inclusion of a chapter on SMSs is of great importance in this context, as it provides a comprehensive and proactive approach to improving vehicle safety, vehicle production, and vehicle fleet management. By integrating SMS into the automotive industry, manufacturers, suppliers, and operators can not only improve the technical safety of their products, but also establish organizational and procedural safety standards. The implementation of SMSs in the automotive industry offers a holistic perspective on safety management, encompassing both preventive and reactive measures.

A key aspect of SMSs and a major part of this chapter is the seamless integration of various disciplines and their work products, such as risk management, architecture, design, verification, validation, integration, and release processes. Each of these elements plays a critical role in the overall safety lifecycle. For instance, risk management serves as the foundation by providing detailed risk analyses that inform the development of appropriate safety measures across all phases from system architecture to final product release. These risk analyses help ensure that safety requirements are consistently addressed and verified through the entire development process, enabling a structured approach to safety that is both proactive and responsive to potential hazards.

A central concern of the book is to promote a culture of safety in which all stakeholders—from managers to engineers to end users—actively contribute to safety. This chapter shows how such a culture can be established and maintained in the long term. It makes it clear that safety management is not just a technical challenge, but also an organizational and cultural one. This chapter offers valuable insights into the strategic dimension of vehicle safety and shows how the automotive industry can benefit from the experience of the aviation industry.

The chapter also provides valuable insights into the benefits of systematic safety management, which includes preventive and reactive measures. It explains how safety can be continuously improved through ongoing monitoring and analysis of safety data as well as regular audits and training. This is particularly important in view of the increasing digitalization and networking of vehicles, as new technologies also bring new safety risks.

3.1.3 Bridging the Gap between Aviation and the Automotive Industry

The aim of this section is to transfer the principles and methods of SMSs from the aviation industry to the automotive industry. It will show which specific adaptations are necessary to meet the special requirements and challenges of the automotive sector. The section will present practical examples of how these components can be implemented in the automotive industry. Successes as well as challenges and solutions will be discussed. A particular focus will be on adapting the SMS principles to the specific requirements and circumstances of the automotive industry.

Another important aspect is the promotion of a safety culture within automotive companies. The section will explain how a culture of safety can be created and sustainably promoted through targeted measures such as regular training, safety campaigns, and the involvement of employees in the safety process.

Ultimately, the section should help to deepen the understanding of the advantages and benefits of systematic safety management and provide practical impetus for the implementation of SMSs in the automotive industry. By building a bridge from aviation to the automotive industry, valuable insights and best practices can be used to increase road safety and thus make an important contribution to Vision Zero—zero road deaths and serious injuries.

3.2 Fundamentals of Safety Management Systems

This section provides a comprehensive introduction to the basics of safety management by examining the term "safety" in its historical development and their different concepts. Over time, the understanding of safety has evolved, which is reflected in different approaches known as Safety I, Safety II, and Safety III. These concepts illustrate how safety management has adapted in response to changing requirements and findings.

The section begins with the evolution of the concept of safety (Section 3.2.1). It explains how the definition and understanding of safety has changed over time. Originally, safety was primarily understood as the absence of accidents or damage. However, this understanding has evolved with technological advances and the increasing complexity of systems. The traditional safety paradigm, which focuses on the prevention of errors and accidents, is presented in Section 3.2.1.1 on Safety I. Building on this, a more modern approach is described in Section 3.2.1.2 on Safety II, which emphasizes the capability of a system. This approach recognizes that not all risks can be completely eliminated, and that flexibility and resilience are important factors for safety. With Safety III (Section 3.2.1.3), another development step is introduced that deepens the systemic perspective on safety. Safety III integrates approaches from Safety I and Safety II and expands the understanding of safety by considering the interactions between the various system elements and the context in which a system is operated. In Section 3.2.1.4, the three safety concepts are compared once again, and their respective strengths and weaknesses are analyzed. This summary helps to understand the evolutionary development of safety approaches and recognize their relevance for modern safety management systems.

Section 3.2.2 provides a precise definition of what constitutes an SMS. It describes the basic components and principles required to establish an effective system for ensuring safety in organizations. Finally, Section 3.2.3 traces the development of these systems. It shows how they have evolved from earlier, less formal safety practices to today's highly structured and systematic approaches that are used in numerous industries worldwide. This introduction lays the foundation for an in-depth understanding of the complex and dynamic nature of safety management and prepares the reader to contextualize the subsequent content of the chapter.

3.2.1 The Evolution of the Term "Safety"

In general, safety is often defined as the absence of accidents and incidents. However, this definition falls short as it fails to capture important aspects. For example, if someone is moving on thin ice and there are currently no accidents, they would be considered safe according to this definition. However, in the next moment, the ice could break, and they could fall into the water, demonstrating that this definition of safety is incomplete.

Safety scientists (e.g., Bristow et al. 2012) have pointed out the inadequacy of this definition, and it is now often supplemented by another definition: "freedom from unacceptable risks" (ISO/IEC Guide 51:2014). However, these two definitions do not express the same idea. Referring to the concept of risk introduces uncertainties, and assessing the extent and significance of the risk becomes necessary. Safety in this sense relies on an assessment of the risk associated with a potential accident. This raises many questions about what risk is, how the extent of the risk is determined, and how to conclude what constitutes an unacceptable risk. These questions are crucial for assessing the soundness and validity of this interpretation of the safety concept (Pich et al. 2002, Bristow et al. 2012), yet they are rarely discussed in the safety science literature.

Safety is generally defined as the quality of a system that is necessary and sufficient to ensure that the number of events that could be harmful to workers, the public, or the environment is acceptably low. For example, the WHO defines patient safety as "the prevention of errors and adverse effects on patients associated with healthcare."

According to Hollnagel et al. (Hollnagel et al. 2015), safety management begins either with the occurrence of accidents (actual negative outcomes) or the perception of risks (potential negative outcomes). Negative outcomes have usually been explained by the existence of a cause—and the solution has been to eliminate or control that cause.

In recent decades, as highlighted by Leveson, the engineering world has undergone a major technological revolution, while safety approaches have been unable to keep pace with these changes and deliver the expected benefits in terms of increased safety and reduced accidents (Leveson 2012). Furthermore, complexity—defined as the inability to assess the impact of actions due to too many system variables interacting (Pich et al. 2002)—is increasing dramatically, forcing risk managers and analysts to examine less obvious systems (Bristow et al. 2012).

Safety is a state characterized by protection from harm or other dangers. The term "state" refers to the situation at a specific point in time. A common definition of safety describes it as a condition where the number of adverse outcomes (such as accidents, incidents, or near misses) is minimized (Hollnagel 2014). This implies that safety is not just the absence of negative events but also involves the control of recognized hazards to achieve an acceptable level of risk.

An intriguing concept in the field of safety is the idea of "safety as a non-event" (Hollnagel 2014). Weick originally spoke of "reliability as a dynamic non-event," but the parallels to safety are clear (Weick 2001). The notion of a "non-event" suggests that safety exists when no negative events occur, meaning when nothing goes wrong. However, the dynamic nature of this state requires active and continuous vigilance. It is not sufficient to rely on passive measures and to simply layer defenses and protections. Instead, safety must be maintained through ongoing attention and proactive action. This perspective highlights the fact that safety is not static but a living, dynamic process that must be continuously nurtured. It requires active participation and an awareness that the absence of problems should not be taken for granted. Safety is a continuous challenge, demanding adaptability and constant improvement.

3.2.1.1 Safety I

Safety I is a traditional and reactive safety concept that focuses on preventing accidents and minimizing damage after they occur (Hollnagel 2018). This concept was particularly widespread in safety-critical industries such as nuclear power and aviation between the 1960s and 1980s (Hollnagel et al. 2015) and is still applied today in many industries, such as oil and gas (Aven 2022).

The reactive approach of Safety I aims to identify and eliminate the causes of high-risk events, such as accidents or incidents, as soon as they occur. Safety I is based on simple cause-and-effect relationships and provides a problem-solving approach by identifying, eliminating, or mitigating hazards, as well as reducing the consequences of dangerous events (Hollnagel et al. 2015). The causes of accidents are considered to be failures and malfunctions of system components (Hollnagel 2018). In this context, traditional engineering risk analysis methods such as Fault Tree Analysis (FTA) or Event Tree Analysis (ETA) are used to determine the root cause of problems (Hollnagel 2016). Safety I assume that risk assessments using these methods serve to estimate the operational risk as accurately as possible based on identifiable errors or malfunctions in system components. This is done by identifying relevant scenarios and calculating probabilities to control and largely avoid risk events. The risk assessment then provides a statement on whether the accident risk is acceptable.

Safety I is particularly applicable to simple, linear, comprehensible, and complex systems (Aven 2022). According to Hollnagel (Hollnagel 2018), a comprehensible system is one whose operating principles are known, can be easily described, and do not change during the description. A comprehensible complex system has many components, where, despite the complexity, it is possible to understand the interactions and functions of the system (Aven 2022).

An example of Safety I from the automotive industry is the introduction of seat belts and airbags in vehicles. These passive safety systems are intended to minimize injuries in the event of an accident by protecting the vehicle occupants from harm. However, they are only activated when an accident actually occurs.

While Safety I is an important aspect of overall safety management, it is considered to have limited ability to predict and prevent accidents before they happen. Moreover, it is not sufficient to blame a single factor for an accident (Hollnagel 2018, Hollnagel 2016), as according to Safety II and explained in Section 3.2.1.2, a combination of variables, consisting of states, conditions, and events, must be used to explain an observed accident (Aven 2022).

A notable consequence of the Safety I approach is that we analyze situations and try to learn from them where, by definition, there was a lack of safety. Another interesting consequence is that the safety level is inversely proportional to the number of negative outcomes. When many things go wrong, the safety level is considered low; when few things go wrong, the safety level is considered high. In other words, the more negative outcomes there are, the lower the safety, and vice versa. A perfect safety level means there are no negative outcomes and therefore nothing to measure. Unfortunately, this makes it very difficult, if not impossible, to prove that efforts to improve safety have been successful, thus making it difficult to argue for the allocation of further resources (Hollnagel et al. 2015).

3.2.1.2 Safety II

The traditional approach to safety management, described as Safety I, is based on the assumption that accidents and incidents are caused by clearly identifiable errors and malfunctions in specific components. These components can be technological, procedural, human, or organizational in nature. The idea is simple: If something goes wrong, there is a clear cause that must be found and eliminated to prevent future incidents. This perspective is deeply rooted in our history and offers a seemingly logical and straightforward method for addressing safety issues.

However, over time, it has become evident that this approach is increasingly inadequate, especially in highly complex, dynamic systems that are more common today. A central problem with Safety I is that it focuses almost exclusively on negative events, which means on the few instances where something has gone wrong. This perspective, however, neglects the far more frequent situations where everything goes right and ignores the processes and mechanisms that ensure a system typically functions successfully. This leads to an imbalance: We know a lot about what goes wrong, but very little about what works well.

Another issue is that Safety I operates retrospectively. It tries to determine the causes of an incident after it has occurred and takes action to prevent it from happening again in the future. This approach does not explain, though, how and why errors arise and persist in the first place. It assumes that human error or technical defects are the primary causes of accidents, overlooking the fact that people and systems constantly adapt to changing conditions in order to succeed. These adaptations are often not recognized as part of normal, successful system operations, but are only noticed when they fail. This retrospective focus of Safety I makes it difficult to evaluate the success of safety measures. At best, a successful safety measure results in nothing happening but this "nothing" is hard to measure or prove. Additionally, the "find and fix" model on which Safety I is based is increasingly unsuitable for complex socio-technical systems. These systems are so intertwined and dynamic that simple cause-and-effect relationships are often insufficient to explain or prevent undesirable events.

The development of the Swiss cheese model in the 1970s was an attempt to better capture this complexity (Figure 3.1). It demonstrated that undesirable events are often caused by a combination of active failures and latent conditions (Hollnagel et al. 2015). However, even this model is reaching its limits in the face of the ever-increasing complexity of modern systems (Hollnagel 2018, Reason et al. 2006). Today's systems are so dynamic and multifaceted that it is almost impossible to identify and eliminate all potential sources of failure before they become a problem.

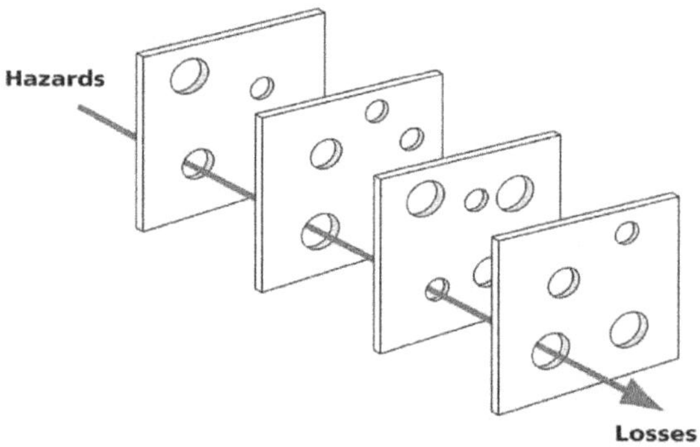

Figure 3.1
Swiss cheese model (source: *https://commons.wikimedia.org/wiki/File:Swiss_cheese_model_of_accident_causation.png*, author: Davidmack)

Thus, Safety I is no longer effective because it focuses too much on errors and problem elimination, rather than understanding how and why systems are generally successful (Figure 3.2). This focus on the negative leaves little room for analyzing and learning from positive outcomes. In an increasingly complex world, however, we need precisely this understanding to make systems safer.

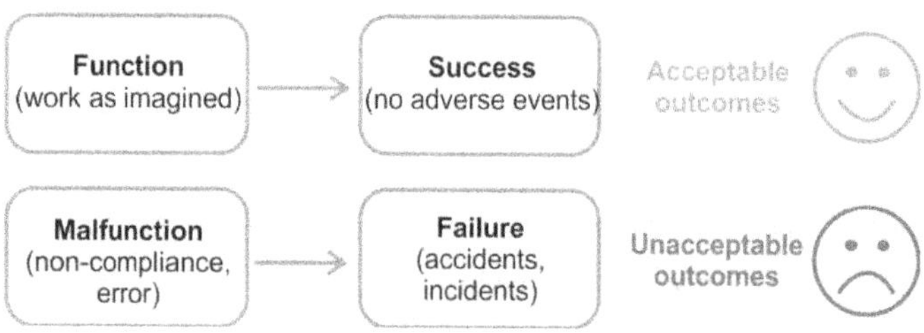

Figure 3.2 Safety I assumes that things that go right and things that go wrong happen in different ways. In Safety I, the starting point for safety management is that something has gone wrong (source: Hollnagel et al. 2015).

Safety II represents a major paradigm shift in the approach to safety by reversing and expanding the traditional view. While Safety I focuses mainly on avoiding errors and preventing accidents by examining the causes and contributing factors of undesirable events, Safety II goes one step further. It focuses on the question of why systems work successfully in the vast majority of cases and how these successes can be maintained and promoted. Safety II is therefore a modern, proactive approach to safety that emphasizes the ability of a system to function as intended under variable conditions while achieving as many intended and acceptable outcomes as possible. In contrast to Safety I, which is reactive to errors and failures, Safety II focuses on understanding and promoting what is going right in the system. This perspective makes it possible not only to identify potential sources of error, but also to understand the underlying mechanisms that ensure the success of a system.

A key feature of Safety II is the recognition of human adaptability as a critical factor in system performance. People constantly adapt their behavior to changing working conditions to achieve the desired results. This adaptability is becoming increasingly important, particularly in complex systems, as it enables uncertainties, ambiguities, and conflicting goals to be managed. Safety II views acceptable and undesirable outcomes as two sides of the same coin, both resulting from the same day-to-day performance adjustments.

Safety II offers a more comprehensive approach to safety, as it aims not only to prevent errors, but also to maximize system performance under variable conditions. Safety I has undoubtedly contributed to improving safety by focusing on analyzing accidents and identifying sources of error. However, in an increasingly complex and dynamic working environment, these reactive measures alone are no longer sufficient. Safety II offers a further development here by analyzing everyday processes and their success factors. It asks how systems usually work successfully and what adjustments are necessary to achieve consistently positive results under variable conditions. This approach makes it possible to better anticipate developments and future events and proactively manage risks. Safety II's proactive focus on the variability of working conditions and the promotion of positive outcomes therefore provides a more holistic approach to improving safety. By examining not only failures but also success, it creates a deeper understanding of the dynamics of complex systems and provides a more robust foundation for continuous Safety Improvement.

In the automotive industry, for example, Safety II can be implemented by continuously monitoring and analyzing daily operating data. By using modern technologies such big data analytics, data on vehicle performance can be collected and analyzed in real time. This data can then be used to identify patterns and adjust improve vehicle safety and reliability. One concrete example is the further development of driver assistance systems. Instead of focusing solely on preventing accidents, Safety II analyzes how these systems are used in everyday life and what adjustments can be made to optimize their performance. This can help to identify potential problems at an early stage and rectify them before they lead to accidents due to human error, for example.

In summary, while Safety I focuses on responding to negative events, Safety II promotes a deeper understanding of day-to-day work processes and the factors that lead to positive outcomes (Table 3.1). Compared to Safety I, the Safety II perspective requires a different set of methods and techniques to manage performance variability.

By analyzing what is going right, systems can be made more resilient and safer to prevent future accidents. In the automotive industry, this approach can be implemented using advanced technologies and data analytics to continuously improve vehicle safety and reliability. Safety II emphasizes learning from what works well and adapting to different conditions to create a safe and resilient system that can anticipate and prevent accidents before they happen. In line with Safety II thinking, day-to-day performance variability provides the adaptations needed to respond to different conditions and is therefore the reason things go well. Safety II emphasizes learning from what goes well, along the lines that it is important to learn from what happens every day, e. g., from performance fluctuations and performance adjustments, because this is the reason why things sometimes go wrong and because it is the most effective way to improve performance (Hollnagel et al. 2015). In general, it is more useful to base learning on small but frequent events than on rare events with serious consequences.

Table 3.1 Criteria for comparing Safety I and Safety II according to Peñaloza et al. (based on Peñaloza et al. 2019)

	Safety I	Safety II
understanding of Safety	reducing what goes wrong	increasing what goes right
learning basis	unwanted events	everyday work
practices and tools	reactive and proactive in terms of addressing failures, accidents and unacceptable risks	reactive and proactive in terms of understanding successes and surprises in everyday work
strategy to cope with performance variability	Harm should be prevented, e. g., by barriers, standardized processes, and compliance with procedures.	Variability needs to be monitored and managed, e. g., reconciling work-as-imagined and work-as-done.

However, the way forward is not to replace Safety I with Safety II, but to combine both ways of thinking. It is still the case that most adverse events are relatively simple—or can be treated as relatively simple without having serious consequences—and that they can therefore be treated with the known methods. However, there are a growing number of cases where this approach does not work. In these cases, it is necessary to adopt a Safety II.

Provan et al. define safety as the ability of a system to fulfill its intended purpose while preventing harm to people (Provan et al. 2020). They combine Hollnagel's Safety II definition with its emphasis on success and the Safety I definition's focus on accidents and incidents (harm, see Figure 3.3). The Safety II system allows people to dynamically align the pursuit of safety and effectiveness, as there are always multiple competing goals, limited resources and pressure to achieve more. This comprehensive approach of Safety II helps to create a safe and resilient system that is constantly improving and is able to anticipate and prevent accidents before they happen.

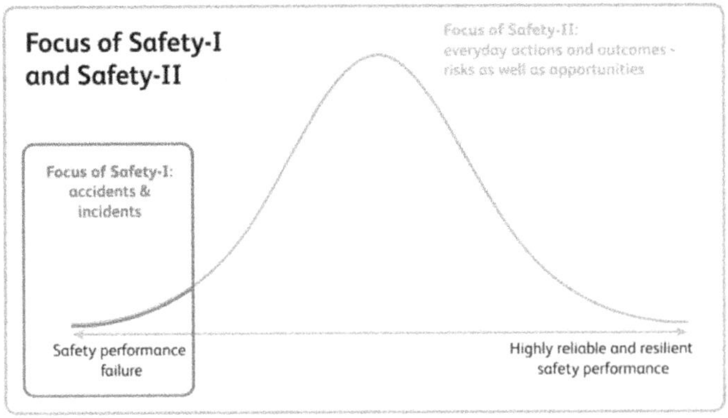

Figure 3.3 Focus on Safety I and Safety II (source: McCarthy 2020)

3.2.1.3 Safety III

It is important to emphasize that Safety I and Safety II are two complementary perspectives on safety and should not be seen as conflicting or contradictory approaches. Many existing practices can still be applied, albeit possibly with a different focus (Hollnagel et al. 2015). In recent decades, as Nancy Leveson points out, technological development has undergone a tremendous transformation (Leveson 2012). However, traditional safety approaches have not been able to keep pace with these changes and achieve the hoped-for improvements in terms of increasing safety and reducing accidents. The increasing complexity of modern systems, characterized by the interaction of numerous system variables, makes it difficult to accurately assess the impact of measures (Pich et al. 2002). This forces safety experts and analysts to look at less obvious and often hidden systems (Bristow et al. 2012).

With this in mind, this section introduces the concept of Safety III. This concept offers a more holistic and resilience-oriented approach to safety by integrating and further developing the principles of Safety I and Safety II to meet the challenges of modern, complex, and dynamic systems (Table 3.2).

Table 3.2 A comparison of Safety I, II, and II (based on Simon Tulloch 2023)

	Safety I	Safety II	Safety III
Definition of Safety	as few things as possible go wrong	as many things as possible go right	Safety is defined as freedom from unacceptable losses as identified by the system stakeholders. The goal is to estimate, mitigate, or control hazards, which are the states that can lead to these losses.

Safety III, a relatively new concept, is coming to the fore and offers a more proactive and systemic approach to safety. Traditionally, Safety I was based on the assumption that accidents could be prevented by identifying and eliminating causes. These causes, often referred to as "root causes," are determined by tracing events that lead to an accident. But Safety III challenges this approach by emphasizing that the choice of a particular event as a root cause is often arbitrary. In a chain of events, any event could theoretically be labeled a "root cause," depending on how far back in the chain you go. This perspective leads to the realization that it makes little sense to consider individual causes in isolation, as they are often only part of a larger, systemic problem.

"It's important to learn from what happens every day—from performance fluctuations and performance adjustments—because that's why things sometimes go wrong, and because it's the most effective way to improve performance." (Hollnagel et al. 2015)

This approach aims to draw conclusions about potential improvements by analyzing successful processes. Safety II emphasizes the importance of learning from what works well to increase the number of successful processes and thereby improve overall safety. However, even this approach has its limitations, especially when it comes to fully grasping the complexity of modern systems and proactively responding to potential hazards.

Safety III, on the other hand, is based on systems theory and goes far beyond traditional approaches. It looks not only at individual components or events, but at the entire system and the complex interactions that take place within this system. In this context, accidents are no longer seen as the result of isolated errors or omissions, but as the consequence of inadequate management and control within a socio-technical system (Leveson 2022). Systems theory emphasizes that it is not sufficient to analyze only the direct causes of an accident. Rather, the entire network of relationships and dependencies within the system must be considered to gain a deep understanding of the underlying dynamics.

A central aspect of Safety III is the rejection of linear causality, as is often assumed in Safety I. Safety III recognizes that accidents are often caused by complex interactions between different system components that cannot be explained by simple cause-and-effect relationships. This perspective leads to a more comprehensive understanding of causality that considers both feedback loops and systemic factors that can jointly lead to an accident. In this sense, causality is understood as a network of conditions that interact to produce certain outcomes.

"The cause is the sum of the positive and negative conditions which, taken together, constitute the totality of the conditions of each description, from the fulfillment of which the consequence inevitably follows." (Mill 1873)

Safety III also emphasizes the need for proactive safety management. While Safety I and II are largely reactive and based on the analysis of past accidents, Safety III aims to predict future risks and prevent them through targeted measures. This proactive approach is supported by the integration of modern technologies and data analysis tools, which make it possible to identify potential hazards at an early stage and take appropriate measures. Instead of just reacting to accidents before they happen, Safety III aims to identify hazards in advance and minimize them through preventive measures. Safety III goes one step further by emphasizing the importance of resilient systems. Resilience in this context means that a system not only relies on the performance of human operators, but also prevents hazards as far as possible through careful design. Human operators are integrated into the development of these systems as a critical component. The system itself will be designed to be able to cope with unforeseen events and operators will be equipped with the necessary tools to act effectively in emergencies.

Another key aspect of Safety III is recognizing the inevitability and necessity of change within a system. While planned changes must be carefully analyzed to ensure that they do not introduce new hazards, unplanned changes, whether due to external circumstances or assumptions that turn out to be incorrect, must also be taken into account. Safety III therefore requires not only the development of sophisticated SMSs, but also a safety culture that is flexible and adaptable to change.

A strong safety culture promotes continuous safety improvement and ensures that everyone involved understands the importance of safety processes and actively contributes to them. In her book *Engineering a Safer World*, Nancy Leveson describes in detail how such an SMS can be set up. She emphasizes the need for a clearly defined safety culture and the implementation of management structures and information systems that enable effective monitoring and control of safety processes.

An example of Safety III in the automotive industry would be the ability of highly Automated Vehicle (AV) fleets to learn and adapt in real-time to unexpected situations in complex environments. When a vehicle successfully responds to a new traffic obstacle, like a construction site, this knowledge is immediately shared with the entire fleet through a central network. This enables all vehicles to flexibly respond to changing conditions and continuously enhance their safety mechanisms without requiring all possible scenarios to be predefined.

In summary, Safety III represents a significant paradigm shift in safety research. It offers a more comprehensive and proactive approach to accident prevention that recognizes the complexity of modern socio-technical systems and aims to identify and control hazards in advance. Safety III thus goes far beyond traditional approaches and offers a forward-looking perspective on safety that ensures both the protection of human lives and the sustainable success of organizations in hazardous industries. In the long term, safety-oriented technology and design is the key to success and profit. In the automotive industry, for example, Safety III could include the use of data from sensors and cameras in vehicles to predict and prevent accidents. This could include using data from cameras and radar systems to detect potential collisions and automatically apply the brakes to avoid an accident. In addition, data from sensors could be used to monitor the health and well-being of drivers and passengers and intervene if necessary in order to prevent accidents.

3.2.1.4 Summary

In the following, the concepts of Safety I, II and III are summarized and compared based on their models, addressees and risk management.

3.2.1.4.1 Safety I, Safety II and Safety III Models

The concepts of Safety I, II and III represent different approaches to the consideration and management of risks and accidents in complex systems:

- **Safety I** is based on the assumption that it is possible to develop models that accurately represent the actual system and the associated activities that could lead to an accident. By creating such models, risk events can be identified, controlled and largely avoided. The basic idea is that the risk of accidents can be reduced to an acceptable level by accurately identifying and eliminating the underlying causes.
- In contrast, **Safety II** rejects this assumption. Proponents of this approach argue that it is simply impossible to create such detailed models for many real, complex systems, which are often referred to as "intractable systems". These systems are so complex and dynamic that they cannot be fully captured by a single model. Instead, the focus is on how the system functions under normal conditions and how it behaves under different circumstances to better understand and manage risks.
- **Safety III** builds on the findings of Safety II and goes one step further. While recognizing the importance of causality models, this approach criticizes the traditional linear models that are often used to analyze accidents. Such models, which view negative outcomes because of errors and latent conditions, are considered insufficient to capture the complexity of modern systems. Safety III therefore advocates an alternative methodology based on systems theory, in particular the System-Theoretic Accident Model and Processes (STAMP) model developed by Nancy Leveson. This model considers not only linear causal relationships, but also the dynamic interactions within a system to enable a more comprehensive understanding of accidents and their prevention.

3.2.1.4.2 Possible Addresses

The different safety approaches, Safety I, II and III, are aimed at different types of systems, which vary according to their complexity and predictability:

- **Safety I** is ideal for simple, linear and manageable as well as complicated systems. A straightforward system is characterized by the fact that the principles of its functioning are well known. It can be described simply and with few details, without being subject to changes during the description process. Even if a system is complicated, it can still be comprehensible if we have a good understanding of the relationships between its many components and know how the system works as a whole. In such cases, it is possible to create a detailed and precise description of the system.

- **Safety II**, on the other hand, is aimed at socio-technical systems that are difficult to solve. The term "socio-technical" refers to the complex interactions between the social and technical components of a system. This type of system is unsolvable if the principles of its functioning are only partially known or even completely unknown. The description of such systems is often fraught with great uncertainties and ambiguities, and the system can change before the description has even been completed. In complex, unmanageable systems, it is impossible to predict all possible scenarios, as the performance of the system cannot be determined solely by understanding the individual components and their states. The scenario that occurs can be unexpected and surprising.
- **Safety III** attempts to integrate both approaches and considers both linear and complex socio-technical systems. This approach recognizes that while traditional linear models are appropriate for simpler systems, they are not sufficient to fully capture the dynamics and complexity of modern socio-technical systems. Safety III therefore provides an extended framework that considers both predictability in linear systems and uncertainty and surprise in complex systems.

3.2.1.4.3 Risk Assessment

The different safety approaches, Safety I, II and III, are based on different methods for analyzing and assessing risks in technical systems:

- **Safety I** uses traditional technical risk analysis methods such as Fault Tree Analysis (FTA) and Event Tree Analysis (ETA). The main aim of these approaches is to precisely assess the risk of activities. This is done by identifying relevant scenarios and calculating the associated probabilities. Based on the results of such risk assessments, it can then be judged whether the identified risk can be classified as acceptable or tolerable.
- **Safety II** criticizes the limitations of traditional risk assessments, especially in the context of complex, unpredictable systems. These approaches, which are based on Safety I, are criticized for not sufficiently taking into account important safety aspects such as variability, human and organizational factors, and dependencies between system elements. In difficult, complex systems, traditional risk assessments are often unable to provide accurate estimates and predictions, which severely limits their usefulness.
- **Safety III** emphasizes the importance of a comprehensive hazard analysis to gain a deeper understanding of the system and identify potential hazards. It emphasizes the limitations of traditional risk assessment, especially when analyzing complex systems. Safety III recognizes that traditional methods are often inade-

quate to fully capture the dynamics and complexity of modern systems and therefore argues for an expansion of the analytical framework to better address these challenges.

3.2.2 Definition of a Safety Management System

Safety management and **Safety Management Systems (SMSs)** are two closely related concepts, each playing a critical role in maintaining safety within organizations, but they differ in scope, structure, and application.

Safety management refers to the broad process of managing safety measures across an organization. It involves all activities aimed at reducing risks, preventing accidents, and ensuring safe working conditions. This approach is inherently flexible, allowing it to be tailored to the specific needs of different industries, organizations, or individual projects. Safety management encompasses various tasks, such as conducting risk assessments, developing safety policies, providing training, monitoring safety practices, and continuously improving safety protocols. The primary goal of safety management is to create a proactive culture where safety is a fundamental aspect of everyday operations, and potential hazards are identified and addressed before they can lead to incidents.

Safety management is a proactive way to mitigate safety risks and improve safety performance. When effectively implemented, it can lead to a documented, process-based approach to safety. This proactive stance ensures that safety is not merely a reactive response to incidents, but a fundamental and continuous effort embedded in the organization's operations. There are several approaches within safety management that are not mutually exclusive but can serve as guidelines for choosing the focus and objectives in safety research and practice. These include the following:

- **Best practice approach:** This approach focuses on identifying and applying the most effective methods and strategies that have been proven to work in safety management. It involves learning from successful examples within the industry or other sectors and adapting these practices to fit the specific context of the organization. The best practice approach encourages continuous improvement by benchmarking against high standards and striving to replicate success.
- **Systems approach:** The systems approach to safety management views an organization as a complex system of interrelated components. It emphasizes understanding the interactions between these components and how they contribute to safety outcomes. By analyzing the system as a whole, rather than in isolated parts, organizations can identify potential weaknesses and design more effective safety measures that address root causes rather than symptoms. This holistic perspective is crucial for managing safety in complex environments where multiple factors interact to create risks.

- **Cultural aspects:** The cultural aspects approach recognizes that safety is deeply influenced by the organizational culture (the shared values, beliefs, and behaviors that shape how safety is perceived and practiced within the organization). A strong safety culture is one where safety is prioritized at all levels, from leadership to front-line workers. This approach involves fostering a culture where safety is ingrained in the daily routines, decision-making processes, and communication practices of the organization. By promoting a positive safety culture, organizations can enhance employee engagement in safety practices and reduce the likelihood of accidents.

On the other hand, an SMS is a formal, structured framework designed to systematically manage safety. It is a set of organized processes and procedures that are documented and implemented within an organization to achieve and maintain high safety standards. Unlike the more general concept of safety management, an SMS is typically governed by specific standards and regulations, such as ISO 45001 (ISO 45001:2018), which provides a framework for occupational health and SMSs, or the International Civil Aviation Organization (ICAO) standards, which outline safety management requirements for the aviation industry.

It is intended to improve safety performance through the continuous identification of hazards, the collection and analysis of safety data and information, and the ongoing assessment of safety risks (Li/Guldenmund 2018). An SMS represents a business-like approach to managing safety risks, focusing on systematic control and continuous improvement. Generally defined as the management procedures, elements, and activities aimed at enhancing an organization's safety performance, an SMS integrates all safety management activities into a coherent system.

The primary goal of an SMS is to protect human beings, the environment, equipment, and property from unacceptable risk. To achieve this, the SMS is built around key components, which are described more in detail in Section 3.3, that ensure a consistent and methodical approach to safety. These tools structure the systematic control of safety, ensuring that all aspects of safety management are addressed in a coordinated manner:

1. **Safety policy and objectives:** establishing a clear safety policy that outlines the organization's commitment to safety and setting measurable objectives to guide safety initiatives
2. **Risk management:** systematically identifying hazards, assessing risks, and implementing controls to mitigate those risks
3. **Safety assurance:** monitoring and measuring the performance of safety processes to ensure they are effective and making adjustments as necessary
4. **Safety promotion:** fostering a safety culture through training, communication, and continuous learning to ensure that safety is a shared responsibility among all employees

An SMS is the formal, top-down business approach to managing safety risk, incorporating a systemic approach that includes the necessary organizational structures, accountability, policies, and procedures. The ultimate goal of an SMS is to minimize the number of accidents and serious incidents while maximizing safety (Boustras/Guldenmund 2017). In high-risk industries like aviation, safety management requires continuous vigilance. New hazards and risks are constantly emerging and must be controlled. An SMS helps organizations identify potential hazards, assess their severity, and manage risks in an ongoing, dynamic process. As Professor James Reason[1] insightfully noted: *"I compare managing safety to fighting a guerrilla war in which there are no final victories; it is a never-ending struggle to identify and eliminate or control hazards."* This underscores the importance of viewing safety management as a continuous process, rather than a one-time effort.

An effective SMS focuses on human and organizational factors, recognizing that these elements are crucial in controlling risks across all methods of operation. Ultimately, an SMS is a tool to ensure safe operations and reduce the likelihood of low-frequency, high-consequence incidents. It provides a structured and systematic approach to safety, but its success relies on the commitment of the entire organization to maintain and improve safety standards continuously. The structured nature of an SMS ensures that safety is not managed in a reactive or ad-hoc manner as explained in the Safety I concept, but through a deliberate and coordinated effort that involves all levels of the organization. Regular audits, reviews, and updates are integral to an SMS, ensuring that it evolves over time to address new challenges and incorporate lessons learned from past experiences.

In essence, while safety management represents the overall approach to managing safety within an organization, an SMS is the specific, standardized method of implementing that approach in a systematic, organized way. An SMS provides the framework for ensuring that safety practices are consistently applied, monitored, and improved, thereby offering a higher level of assurance that safety objectives will be met. By adopting an SMS, organizations can move beyond basic compliance and create a sustainable safety culture that prioritizes the well-being of employees, customers, and the broader community.

When breaking down the concept of SMS into its components, Li and Guldenmund suggest that an SMS can be understood either as a management system focused on safety or as a system designed to control safety (Li/Guldenmund 2018). An SMS, when deconstructed, intersects three key aspects: Safety, Management, and System. The evolution of an SMS over time is influenced by the individual advancements in these three areas. The concept of safety has already been thoroughly discussed in previous sections, so it will not be further examined here. The terms "Management" and "System," however, are broad in scope. Generally, management encompasses functions

---

1 Former professor of psychology at the University of Manchester and responsible for the introduction of the Swiss cheese model

such as planning, organizing, leading, and controlling (Robbins/Judge 2012). The fundamental principle of a system involves the input-process-output model (Hale et al. 1997, Hammer 1971, Waring 1996).

3.2.3 Historical Development of Safety Management Systems

The evolution of SMSs represents a pivotal shift in how industries, particularly aviation, approach safety. Historically, safety measures were predominantly reactive, focusing on the analysis of accidents and the implementation of corrective actions to prevent similar incidents in the future. This section explores the historical development of SMSs, highlighting the transition from reactive to proactive safety management and the integration of various factors such as technical, human, and organizational that have shaped modern safety practices.

The origins of safety management can be traced back to the early 1900s, particularly within the aviation industry, where safety efforts were initially centered around technical factors. During this period, the primary concern was the reliability and performance of machinery and equipment. The focus was on identifying and rectifying technological failures that were the leading causes of accidents. Engineers and safety professionals concentrated on improving the design, construction, and maintenance of aircraft to enhance their safety performance. This era saw significant advancements in aviation technology, which contributed to a gradual reduction in accident rates. Throughout the first half of the 20th century, the aviation industry experienced rapid technological advancements. These developments were driven by the need to improve the reliability and safety of aircraft, which were becoming increasingly complex. Innovations in materials, propulsion systems, and aerodynamics played a crucial role in reducing the incidence of technical failures. However, despite these advancements, accidents still occurred, often due to unforeseen technical issues or inadequate maintenance practices. As a result, safety management during this period remained largely reactive, with lessons learned from accidents being used to inform design improvements and maintenance protocols.

By the late 1960s, the focus of safety management began to shift. As technological advancements continued to improve the safety and reliability of aircraft, it became apparent that human error was a significant contributor to accidents. This realization marked the beginning of a new era in safety management, where the emphasis moved from purely technical factors to the role of human factors in safety. The concept of human factors in aviation gained prominence in the 1970s, driven by studies that revealed the critical role of human performance in the safety of flight operations. Human factors encompass a wide range of elements, including the cognitive, physical, and psychological aspects of human performance, as well as the interactions between individuals and the systems they operate. During this period, the aviation industry began to recognize that pilot errors, specification errors, instruction mistakes, and other human-related issues were major contributors to accidents. To address these challenges, new training programs and procedures were developed to enhance the performance of flight crews and other personnel involved in aviation operations. These initiatives were based on the understanding that human performance is influenced by a variety of factors, including workload, stress, and fatigue, and that addressing these factors could significantly reduce the likelihood of accidents.

The evolution of safety management took another significant step forward in the mid-1990s with the recognition of organizational factors as a critical component of safety. While technical and human factors had been the primary focus of safety efforts up until this point, it became increasingly clear that the organizational environment in which individuals operate plays a crucial role in shaping safety outcomes.

This shift in perspective was driven by a growing body of research that highlighted the impact of organizational culture, policies, and management practices on safety performance. The concept of organizational factors encompasses a wide range of elements, including leadership, communication, resource allocation, and the overall safety culture within an organization. The realization that these factors could either support or undermine safety efforts led to the development of new approaches to safety management that emphasized the importance of creating a positive safety culture and ensuring that safety was integrated into all aspects of an organization's operations.

The ICAO played a key role in promoting the integration of organizational factors into safety management. In its 2013 Safety Management Manual, the ICAO highlighted the importance of considering organizational factors in the design and implementation of SMSs (ICAO 2013a). The manual emphasized that effective safety management requires not only addressing technical and human factors but also understanding and managing the organizational context in which these factors operate. This approach recognizes that safety is a systemic issue that cannot be effectively managed without considering the broader organizational environment.

The historical development of SMSs reflects the evolution of safety practices from a focus on technical factors to a more holistic approach that considers human and organizational factors. This evolution has been driven by the recognition that safety is a complex, systemic issue that requires a proactive and integrated approach. The introduction of the SMS framework represents the culmination of these developments, providing a comprehensive and effective means of managing safety in modern organizations. As industries continue to evolve and new challenges emerge, the principles of SMS will remain central to ensuring the safety and well-being of individuals and communities around the world. The shift from reactive to proactive safety management marks a significant milestone in the history of safety practices. With the implementation of SMSs, organizations and mobility concepts are better equipped to manage safety risks and continuously improve their safety performance, ultimately contributing to a safer and more resilient industry.

3.3 Basic Principles of Safety Management Systems

SMSs form the backbone of a proactive approach to ensuring safety in various industries, including aviation, manufacturing, healthcare, and more. The adoption of an SMS is pivotal in managing risks, preventing accidents, and maintaining a culture of safety within organizations. At the heart of any SMS are four fundamental pillars: **Safety Policy**, **Safety Risk Management**, **Safety Assurance**, and **Safety Promotion** (Figure 3.4). These pillars provide a structured approach to managing safety and are interconnected, each supporting the others to create a robust and effective SMS. This section explores these four pillars in detail, elucidating their importance, components, and how they interrelate to form a robust safety management framework.

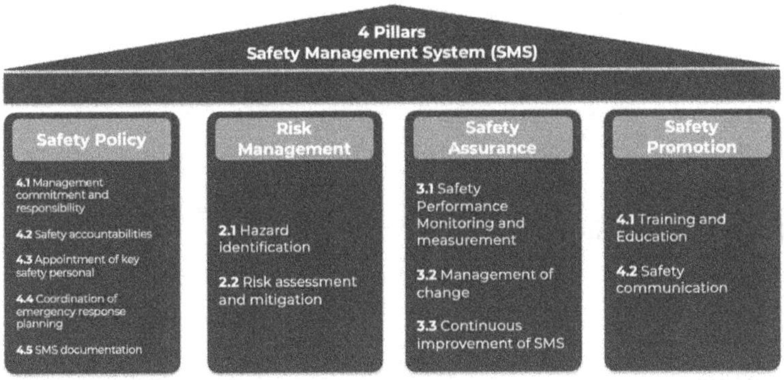

Figure 3.4 Overview about the 4 pillars of Safety Management Systems (SMS)

3.3.1 Safety Policy

The safety policy is the foundation of an SMS and represents the organization's commitment to safety. It sets the tone for the overall safety culture and creates a framework within which safety objectives are defined, responsibilities assigned, and safety-related activities managed. The safety policy must be a clear, concise and unambiguous document that is effectively communicated at all levels of the organization. A comprehensive safety policy includes the following key elements:

- **Commitment from top management:** The safety policy should explicitly demonstrate the commitment of the organization's leadership to safety. This includes a declaration of the importance of safety, the allocation of necessary resources, and a commitment to continuous improvement.
- **Safety objectives and goals:** The safety policy must outline specific safety objectives that the organization aims to achieve. These objectives should be measurable, realistic, and aligned with the overall strategic goals of the organization.

- **Responsibilities and accountabilities:** The safety policy must clearly define the roles and responsibilities of all personnel within the organization concerning safety. This includes the designation of a safety manager or officer who is responsible for overseeing the SMS.
- **Compliance with legal and regulatory requirements:** The safety policy should state the organization's commitment to complying with all applicable safety laws, regulations, and standards. This ensures that the organization operates within the legal framework and meets industry best practices.
- **Non-punitive reporting environment:** To encourage open reporting of safety concerns, the safety policy should include a commitment to a non-punitive reporting culture. Employees should feel secure in reporting safety hazards or incidents without fear of retribution.
- **Communication and documentation:** The safety policy should emphasize the importance of clear communication and thorough documentation of all safety-related activities. This ensures that safety information is accessible, accurate, and up to date.

3.3.2 Safety Risk Management (SRM)

Safety Risk Management (SRM) is the second pillar of an SMS and is critical for preventing accidents and incidents by addressing hazards before they escalate into serious safety issues within the organization. SRM is a proactive process that aims to anticipate potential hazards and implement measures to reduce or eliminate their impact before they result in accidents or incidents. It allows organizations to make informed decisions about risk and implement appropriate controls to maintain a safe working environment. SRM consists of several key steps:

- **Hazard identification:** The first step in SRM is to identify potential hazards that could pose a risk to safety. This can be done through various methods such as safety audits, employee reports, inspections, and analysis of past incidents. Hazard identification is an ongoing process and should be revisited regularly as new hazards may emerge.
- **Risk assessment:** Once hazards are identified, they must be assessed to determine the level of risk they pose. This involves evaluating the likelihood of the hazard occurring and the severity of its potential consequences. Risk assessment helps prioritize hazards based on their significance, allowing the organization to focus its resources on the most critical risks.
- **Risk mitigation:** After assessing the risks, the next step is to develop and implement measures to mitigate them. This can involve engineering controls, adminis-

trative controls, changes to procedures, training, and more. The goal is to reduce the risk to an acceptable level while maintaining operational efficiency.

- **Risk acceptance:** Not all risks can be completely eliminated. Some residual risk may remain even after mitigation efforts. The organization must decide whether the remaining risk is acceptable based on its risk tolerance and the potential impact on safety. If the risk is deemed unacceptable, further mitigation measures must be explored.
- **Monitoring and review:** SRM is a dynamic process that requires continuous monitoring and review. This ensures that the implemented risk controls are effective and that any changes in the operational environment are accounted for. Regular reviews also provide an opportunity to identify new hazards and reassess existing risks.

3.3.3 Safety Assurance

Safety assurance involves the ongoing monitoring and evaluation of the safety performance of the organization. It ensures that the safety management processes are functioning as intended and that safety objectives are being met. Safety assurance is essential for maintaining the integrity of the SMS and ensuring that it adapts to changes in the operational environment. It provides the mechanisms needed to verify that safety risks are being managed effectively and that the organization is meeting its safety objectives. The safety assurance key components are the following:

- **Performance monitoring:** Safety performance is continuously monitored through various indicators such as incident reports, safety audits, and safety data analysis. Performance monitoring helps identify trends, areas of concern, and opportunities for improvement.
- **Internal audits:** Regular internal audits (e.g., according to ISO 9001:2015) are conducted to assess the effectiveness of the SMS and ensure compliance with safety policies, procedures, and regulatory requirements. Audits provide an objective evaluation of the organization's safety management processes and highlight areas that may require corrective action.
- **Corrective and Preventive Actions (CAPA):** When safety issues are identified, corrective and preventive actions are implemented to address the root causes and prevent recurrence. CAPA is an essential part of the continuous improvement process, ensuring that safety concerns are resolved effectively and that lessons learned are applied to prevent future incidents.
- **Safety Performance Indicators (SPIs):** SPIs are metrics used to measure the organization's safety performance. They can be both leading indicators, e.g. predic-

tive, such as safety training completion rates and lagging indicators, e.g. reactive, such as the number of incidents. SPIs provide a quantifiable means of assessing the effectiveness of the SMS and identifying areas for improvement.

- **Management review:** Periodic management reviews are conducted to evaluate the overall performance of the SMS. During these reviews, management assesses safety performance data, audit results, and the effectiveness of corrective actions. The review process ensures that top management remains engaged in the SMS and provides strategic direction for future safety initiatives.

3.3.4 Safety Promotion

Safety promotion is essential for embedding safety into the organizational culture and ensuring that safety is not just a set of procedures, but a core value that influences every aspect of the organization. By promoting safety, organizations can create a work environment where safety is everyone's responsibility, and continuous improvement in safety performance is the norm. Safety promotion is about building awareness, providing training, and encouraging a commitment to safety at all levels of the organization. The main elements of safety promotion include the following:

- **Safety culture:** A strong safety culture is characterized by shared values, beliefs, and behaviors that prioritize safety. Safety culture is shaped by leadership, communication, and the attitudes of all employees toward safety. Promoting a positive safety culture involves engaging employees, encouraging reporting of safety issues, and recognizing and rewarding safe behaviors.
- **Training and education:** Continuous training and education are vital for maintaining a competent and safety-conscious workforce. Safety training should be provided to all employees, covering topics such as hazard recognition, emergency procedures, and specific safety protocols related to their roles. Regular refresher courses ensure that safety knowledge remains current and relevant.
- **Safety communication:** Effective communication is crucial for disseminating safety information and keeping employees informed about safety issues. This can be achieved through safety briefings, safety bulletins, and digital platforms. Clear and consistent communication helps reinforce the importance of safety and keeps safety at the forefront of employees' minds.
- **Employee involvement:** Engaging employees in safety activities is key to promoting a safety culture. Employees should be encouraged to participate in safety committees, hazard identification, and the development of safety procedures. Involving employees in safety decision-making processes fosters a sense of ownership and responsibility for safety.

- **Safety leadership:** Leadership plays a critical role in safety promotion. Leaders must lead by example, demonstrating their commitment to safety through their actions and decisions. They should actively participate in safety initiatives, support safety programs, and create an environment where safety is a priority.

3.4 Importance of Safety Management Systems for Organizations and Companies

SMSs have emerged as an indispensable tool for companies and organizations operating in both safety-critical industries and less obviously vulnerable sectors. Implementing an SMS is far more than a regulatory obligation. It is a strategic imperative that has a profound impact on the short, medium, and long-term success of an organization. This section discusses in detail the importance of SMSs and how they can have a lasting impact on the efficiency, reputation, and financial health of organizations. While implementation can be challenging in the short term, the medium-term and long-term benefits far outweigh the costs. A well-established SMS not only improves workplace safety, but also leads to greater efficiency, a stronger competitive position and long-term financial stability. Companies that invest in an effective SMS lay the foundation for sustainable success and secure a decisive advantage in an increasingly safety- and value-oriented world.

3.4.1 Short-Term Effects of SMSs

The implementation of an SMS brings with it some short-term challenges that companies often perceive as a burden. However, these negative effects are usually temporary and can be minimized through careful planning and implementation. Short-term challenges include the initial resource requirements. Setting up an SMS requires a significant investment in time, money, and human resources. Organizations may need to hire new professionals, conduct training and implement technology to meet the needs of an effective SMS. These investments can significantly increase operating costs in the initial phase and lead to a temporary decrease in profitability.

In addition, existing workflows and processes may be disrupted by the introduction of an SMS. Employees have to get used to new policies and procedures, which can lead to a temporary decrease in productivity. These changes may also be met with resistance from employees, especially if they feel that their previous ways of working are being challenged. Despite these challenges, SMS also offer tangible benefits in the short term. Introducing a structured system to identify and mitigate risks can quickly lead to a reduction in the number and severity of accidents and incidents. This not only improves workplace safety, but also reduces direct costs associated with workplace accidents, such as lost production, compensation claims, and insurance premiums.

However, if an SMS is implemented improperly, there can be an increase in errors. Insufficient training, poor communication, or a lack of clarity in new procedures may cause confusion and mistakes, potentially undermining the system's intended safety improvements. Therefore, a well-planned and carefully managed introduction is crucial to prevent such negative outcomes.

3.4.2 Mid-Term Effects of SMSs

Once the initial hurdles have been overcome, an SMS begins to have positive effects in the medium term. A well-implemented SMS contributes to the creation of a safety culture that has a lasting impact on employee behavior. Continuous training and sensitization to safety issues leads to greater awareness and a sense of responsibility in daily work processes. At this stage, the SMS begins to promote proactive safety measures. Companies can identify and eliminate potential hazards before they become serious problems. This not only leads to a reduction in accidents, but also to greater operational efficiency. Production interruptions and machine downtime due to safety incidents become less frequent, which reduces operating costs and increases productivity.

Another medium-term benefit of an SMS is the improvement in communication and collaboration within the company. Safety management requires collaboration between different departments and levels, which promotes information sharing and enables better coordination of company activities. This improved internal communication can also lead to innovation as employees develop new ideas and solutions through their involvement in safety processes.

Additionally, management will learn to leverage the SMS to monitor and influence operational activities. The structured approach of an SMS gives management better control over processes, enabling them to steer company operations more effectively. This oversight helps ensure compliance with safety standards while optimizing workflows, ultimately aligning safety with broader organizational goals.

3.4.3 Long-Term Effects of SMSs

In the long term, an SMS delivers its most comprehensive and sustainable benefits. Companies that have established a strong safety culture benefit from a stable and loyal workforce. Employees feel valued in a safe working environment and are more motivated to give their best. This leads to lower turnover and reduces the cost of recruiting and training new employees.

In addition, an effective SMS strengthens the company's reputation. In a world where social responsibility and sustainable business practices are becoming increasingly important, compliance with high safety standards is becoming a decisive competitive factor. Companies with a strong safety profile not only attract talented employees, but also customers and business partners who value safety and reliability. This can lead to a higher market position and better business relationships.

An SMS also contributes to financial stability. By reducing accident and damage costs, avoiding legal disputes, and lowering insurance premiums, companies can make considerable savings. At the same time, the continuous improvement of safety standards increases operational efficiency, leading to higher profit margins.

Another important aspect is regulatory compliance. Companies that invest in a comprehensive SMS at an early stage are better able to meet legal requirements and respond to changes in legislation. This protects the company from legal risks and possible penalties that could result from non-compliance with safety regulations.

Moreover, an SMS helps reduce product recalls and improves the handling of field events, particularly the communication with authorities and the public. A proactive approach to managing safety issues not only mitigates potential harm but also reinforces the company's credibility and trustworthiness. This ensures smoother resolution of incidents and enhances public relations in the long run. Safety is therefore also a crucial image factor for any company. If a business is publicly criticized due to recalls or safety deficiencies—whether by authorities or the media—it faces not only direct financial losses but also a long-term erosion of trust. This can damage brand reputation and customer loyalty. In this context, the balance between product safety, ergonomics, and usability becomes vital. A well-integrated SMS ensures that safety measures enhance rather than compromise the user experience, supporting both consumer trust and product functionality.

In the long term, an SMS also enables continuous improvement and adaptation to changing conditions. Through regular reviews and audits, companies can adapt and optimize their safety strategies, leading to a continuous development of the safety culture and overall company performance.

3.5 Legal and Normative Basis for Safety Management Systems

In a globalized world where technological innovations and industrial developments are advancing rapidly, safety plays a crucial role in various sectors. SMS are therefore essential to minimize risks, prevent accidents and ensure the protection of people, the environment and property. The legal and normative basis for SMSs varies considerably between different industries, but reflects a common goal: the identification, assessment and control of safety risks to prevent accidents and incidents. In aviation, these systems are strongly influenced by international organizations such as the ICAO and European regulators such as the European Aviation Safety Agency (EASA). In the automotive industry, SMSs are becoming increasingly important in the operation of highly automated vehicle fleets. In other safety-critical industries such as the nuclear industry, the requirements are particularly stringent and comprehensively regulated by international and national regulations. Regardless of the industry, SMSs are a central component of the organizational framework for ensuring the safety of people and the environment.

This section examines the legal and normative foundations of safety management systems in various industries, including aviation, the automotive industry, and other high-risk sectors such as the nuclear industry.

3.5.1 International Standards and Norms for SMSs in Aviation

Aviation is one of the most heavily regulated industries in the world, partly due to the high risks associated with aviation. The legal and normative foundations for SMS in aviation are based on international standards defined by organizations such as the ICAO, EASA and the Federal Aviation Administration (FAA) in the USA. These standards are crucial to ensure the safety of passengers, crew members, and aircraft. In this section, we will take a look at some of the key international organizations and standards in the field of SMS in aviation:

- **ICAO standards and guidelines:** The ICAO is a specialized agency of the United Nations (UN) responsible for coordinating and regulating aviation on an international level. ICAO plays a crucial role in developing standards and best practices for aviation safety, including setting requirements for SMSs. The ICAO created the legal basis for international civil aviation within the framework of the Chicago Convention. The ICAO has defined specific requirements for SMSs in Annex 19 of its convention (ICAO 2013b). Annex 19 obliges the contracting states to promote the introduction of SMSs in their national aviation sectors.
- **The International Air Transport Association (IATA):** IATA is a global airline trade association that develops standards for many aspects of aviation, including safety. IATA publishes the SMS Framework, which is used by airlines worldwide to structure and improve their safety processes. The IATA SMS Framework is based on ICAO requirements and provides detailed guidelines for implementing an effective SMS. It includes elements such as risk assessment, safety objectives and performance, safety reporting and analysis, and safety culture and awareness.
- **Federal Aviation Administration (FAA):** In the US, the FAA regulates the requirements for SMSs. Since 2015, the FAA has required air carriers to develop and im-

plement an SMS in accordance with 14 CFR Part 5 (FAA 2015). The FAA guidelines emphasize the need for a structured approach to risk management that includes continuous monitoring and improvement of safety performance.

- **European Aviation Safety Agency (EASA):** EASA is an agency of the EU responsible for promoting and monitoring aviation safety in Europe. EASA works closely with ICAO and has developed its own regulations and standards for SMSs that apply in Europe. EASA sets specific requirements for air carriers in Europe to ensure that they implement and maintain adequate safety management systems. These requirements include the training and education of personnel, the reporting of safety incidents and occurrences and regular safety audits and reviews. Regulation (EU) No. 965/2012 (European Union 2012) and Regulation (EU) No. 376/2014 (European Union 2014) are two of the key regulations that define the obligation to introduce an SMS for air carriers, maintenance organizations, aerodromes and air navigation services.
- **Regulation (EU) No. 965/2012:** This regulation defines the technical requirements and administrative procedures relating to flight operations. It obliges operators to implement an SMS that meets the requirements of ICAO Annex 19 (ICAO 2013b).
- **Regulation (EU) No. 376/2014:** This ordinance regulates the reporting system for incidents in civil aviation. It strengthens the safety culture by enabling the confidential and transparent reporting and analysis of incidents and risks.

3.5.2 Current Regulations and Applications in the Automotive Industry

At European level, **Regulation (EU) 2018/858** (European Union 2018) has created a legal framework for the type approval of vehicles that includes requirements for functional safety and the introduction of SMS. This regulation is intended to ensure that manufacturers manage safety risks appropriately throughout the entire life cycle of a vehicle.

In addition, the **UN ECE R157** regulation places significant emphasis on the necessity of implementing a robust SMS to ensure the safety and compliance of Automated Lane Keeping Systems (ALKS) throughout the vehicle's lifecycle. The regulation mandates that manufacturers demonstrate to type-approval authorities that an SMS is not only established but actively maintained, with processes, methodologies, and tools in place to manage safety effectively. This requirement reflects the complexity of automated driving systems and the need for a structured approach to managing risks. The SMS must cover all phases of the vehicle's development, from design and testing to failure tracking, remedy, and release. By integrating the SMS into the development lifecycle, manufacturers ensure that safety is prioritized at every step, from the initial design to post-production monitoring and field updates. Moreover, the regulation requires manufacturers to maintain processes for monitoring safety-relevant incidents, such as accidents caused by ALKS. This post-market surveillance is crucial for identifying and addressing potential safety gaps after vehicles have been released, ensuring continuous improvement and safety management even after production. UN ECE R157 also emphasizes the need for strong collaboration between manufacturers and suppliers. The SMS framework must extend to suppliers, ensuring that their safety management processes align with the manufacturer's standards. This includes establishing clear contractual agreements and interfaces to guarantee that all components meet the required safety criteria.

Part 2 Annex B of **ISO 26262:2018** deals with safety culture in relation to SMSs and defines safety culture as part of SMSs as a targeted process for creating and maintaining an effective safety culture. This process includes the clear definition and reinforcement of values, the establishment of expected behaviors and mindsets, the targeted investment of resources, the sanctioning of undesirable behaviors and the rewarding of desirable behaviors, and continuous improvement through methods such as PDCA cycle.

In addition to the current manageable regulations for the automotive sector, providers of highly automated vehicle fleets in the US are going one step further. These providers are inspired by the aviation industry and have already integrated a safety management system into their companies, according to their own statements.

The Aurora SMS is designed to proactively manage safety risks by ensuring the right information reaches the right people at the right time, promoting accountability and transparency across the company (Aurora Innovation Inc. 2022). Key components include the following:

- a safety concern reporting system which allows all employees to report safety concerns
- a safety review board which is a governance committee overseeing safety issues
- risk assessment and issue management which includes processes for identifying and managing safety risks
- a safety communications program which is a dynamic approach to keeping safety top-of-mind
- an independent safety organization which ensures impartial oversight of safety matters
- safety culture surveys which should gauge and improve the safety culture
- formal safety education which provides ongoing safety training for employees
- assurance programs with various programs to ensure compliance and safety

Additionally, Aurora has adopted other safety programs, such as the operational safety program, focusing on human factors, fatigue management, and incident response within the operations department. A safety culture program encourages all employees to actively participate in safety initiatives, including the "grounding policy," which empowers any employee to halt operations if they perceive a serious safety risk. This culture of shared responsibility is supported by the ability to report safety concerns without fear of reprisal, contributing to a strong and positive safety culture at Aurora.

In comparison to the Aurora SMS, Cruise SMS is a structured approach designed to ensure responsible safety decisions are made during product development and driverless operations (Cruise LLC 2022). This system is inspired by industries like aviation, nuclear energy, and oil and gas, assigns safety-related decisions to those accountable for the outcomes, ensuring they manage the necessary resources. At Cruise, the SMS involves everyone from the chief executive to individual contributors, fostering a companywide safety culture. The SMS team provides guidance to maintain consistent risk management and resolution, embedding safety into daily activities across the organization.

3.5.3 Other Industries

In safety-critical industries such as the nuclear industry, the requirements for SMSs are particularly strict, as the risks of malfunctions or accidents can have catastrophic consequences. Here, the legal and normative foundations are often country-specific, but are also based on international standards.

The International Atomic Energy Agency (IAEA) has developed comprehensive safety standards to support member states in the implementation of SMSs. The IAEA Safety Standards Series contains basic safety principles that apply to the operation of nuclear power plants, the handling of radioactive materials and other nuclear activities. These standards require member states to implement a robust SMS that includes the following elements:

- **Safety culture:** a strong safety culture that involves all employees and emphasizes the importance of safety
- **Risk management:** the identification, analysis and control of risks associated with the operation of nuclear facilities
- **Emergency planning:** the development and implementation of emergency plans that can be activated in the event of an accident or incident
- **Monitoring and improvement:** the continuous monitoring of safety performance and the implementation of improvement measures

Many countries have strict national laws and regulations that govern safety requirements for the nuclear industry. For example, the German Atomic Energy Act (AtG) requires operators of nuclear power plants to implement comprehensive safety measures that include an SMS. In the US, the Nuclear Regulatory Commission (NRC) also specifies requirements for safety management systems, which are regulated in 10 CFR Part 50 (U. S. Nuclear Regulatory Commission 2024).

In other safety-critical areas such as the chemical industry, mining and the offshore oil and gas industry, comprehensive legal and normative requirements for SMSs have also been established. These requirements aim to ensure the safety of people, the environment and facilities and often include similar principles to those in the aviation or nuclear industries, such as risk management, safety culture and emergency planning.

3.6 Necessity of Safety Management Systems in the Automotive Industry

In an era where the automotive industry is undergoing rapid transformation, the importance of SMSs cannot be overstated. As the industry continues to evolve with increasing automation and connectivity, the importance of a robust SMS cannot be overstated. By adopting a comprehensive approach to safety management, similar to that used in the aviation industry, the automotive sector can ensure the safe and efficient operation of automated vehicles, paving the way for a safer future on our roads. As the automotive world continues to push the boundaries of innovation, SMSs will be the cornerstone that ensures these advances are achieved without compromising safety.

The automotive industry is at the forefront of technological innovation, with significant advances in vehicle automation, electrification, and connectivity. However, these innovations bring new safety challenges. A typical modern mid-range vehicle today contains up to 100 control units, sourced from various suppliers. Unlike the harmonized standards of the machinery directive, the regulatory landscape in the automotive industry is more fragmented, with each party tailoring regulations and standards to meet their specific goals. This creates a massive potential for systematic errors at various organizational, functional, logical, and technical interfaces. Changes during development, variations in vehicle platform configurations, and field events such as recalls are difficult to coordinate because organizations are increasingly unable to manage this complex web of information. Every corrective action can itself introduce new risks.

This is exactly the challenge the automotive industry is currently facing: the inability to rely on established safety concepts in the face of growing complexity. As vehicles become more advanced and interconnected, managing these layers of complexity becomes crucial to ensure that safety is not compromised. An SMS provides the necessary framework to navigate these challenges, allowing for better coordination, risk mitigation, and a holistic approach to managing the safety of highly complex systems. With the increasing complexity of modern vehicles, there is a growing need for a structured approach to safety that goes beyond traditional quality assurance and risk management practices. This is where SMS comes into play.

3.6.1 Potential of SMSs in the Automotive Industry

One of the most critical aspects of an effective SMS is its seamless integration into the daily operations of a company. For an SMS to be successful, it should not be seen as an additional burden or a box-ticking exercise. Instead, it should be embedded in the company's culture, influencing every aspect of decision-making, from the design and manufacturing processes to the maintenance and operation of vehicles. A well-integrated SMS encourages a proactive approach to safety, where potential risks are identified and addressed before they lead to incidents. This requires the active participation of all employees, from top management to frontline workers. When everyone in the organization is committed to safety, it becomes a natural part of the work environment rather than an afterthought. In this way, SMS not only enhances safety but also contributes to the overall efficiency and effectiveness of the company.

Investing in SMS can lead to significant financial benefits for automotive companies. While the initial implementation of an SMS may require resources in terms of time, money, and training, the long-term benefits far outweigh these costs, as outlined in Section 3.4. One of the most apparent advantages is the potential to prevent accidents, which can have devastating consequences both in terms of human life and financial loss. Accidents and safety incidents in the automotive industry can result in costly recalls, legal liabilities, and damage to a company's reputation. By implementing an SMS, companies can identify and mitigate risks before they escalate into serious problems. This not only protects the company's bottom line but also builds trust with customers and regulators, which is crucial in today's competitive market. Moreover, a strong safety record can lead to lower insurance premiums and reduced regulatory scrutiny, further contributing to the financial health of the company. In essence, an effective SMS is not just a safety measure, it is a sound business investment that pays off by preventing costly accidents and ensuring the smooth operation of the company.

Operational safety is a critical concern in the automotive industry, particularly as vehicles become more complex and automated. SMSs plays a crucial role in ensuring that safety is maintained throughout the entire lifecycle of a vehicle, from design and manufacturing to operation and maintenance. In the design phase, an SMS helps engineers and designers identify potential safety hazards and incorporate safety features that mitigate these risks. During manufacturing, an SMS ensures that safety standards are consistently met and that any deviations are quickly identified and corrected. In the operation phase, an SMS helps monitor vehicle performance and detect any emerging safety issues, allowing for timely intervention before accidents occur.

Furthermore, SMS is essential in the maintenance of vehicles, where it ensures that all safety-critical components are regularly inspected and maintained. This is particularly important in automated and connected vehicles, where software updates and cybersecurity measures are crucial to maintaining safety. Cybersecurity is not only a standalone concern but also presents a significant risk to safety, especially in the context of advanced functionalities like remote driving, remote maintenance, and over-the-air (OTA) updates. As vehicles become increasingly interconnected, integrating cybersecurity into the overall maintenance strategy for the entire vehicle lifecycle becomes paramount. This requires a comprehensive approach that considers how cybersecurity threats can impact safety and reduce the range of potential solutions available for vehicle operations. As a result, the vehicle lifecycle comes into sharper focus, emphasizing the distinction between the System of Interest and the System Under Consideration (SUC), as defined by ISO 26262. This holistic view allows for a more effective integration of safety and cybersecurity practices, ensuring that every stage of the vehicle's life, from design and development to maintenance and updates, addresses both safety and cybersecurity concerns. By establishing a robust SMS that incorporates these elements, organizations can better manage risks and ensure the reliability and safety of automated vehicles throughout their operational lifespan.

By ensuring that all aspects of vehicle operation are covered by a robust SMS, companies can significantly reduce the risk of accidents and enhance the overall safety of their products. The development of automated vehicle fleets presents new challenges and opportunities for the automotive industry. As vehicles become increasingly automated, the responsibility for safety shifts from the driver to the vehicle itself, and by extension, to the manufacturer and operator. This shift requires a new approach to safety management, one that is comprehensive, proactive, and capable of addressing the unique risks associated with automated vehicles. In this context, SMS becomes even more critical. An effective SMS for automated vehicle fleets must encompass not only the safety of individual vehicles but also the safety of the entire fleet as a system. This includes monitoring and managing the interactions between vehicles, the infrastructure, and other road users. It also requires the ability to quickly identify and respond to emerging risks, such as software failures or cybersecurity threats. To achieve this, the automotive industry can learn from the aviation sector, where SMS has been successfully implemented for decades. In aviation, SMSs are used not only by manufacturers but also by operators and service providers, creating a comprehensive safety network that covers all aspects of flight operations. This model can be adapted to the automotive industry, where manufacturers, fleet operators, and service providers work together to ensure the safety of automated vehicle fleets. For example, manufacturers can implement SMSs to manage the safety of the design and production processes, ensuring that all vehicles meet the highest safety standards. Fleet operators can use SMSs to monitor the performance and safety of their vehicles in real-time, allowing them to quickly address any issues that arise. Service providers, such as maintenance and repair shops, can implement SMSs to ensure that all safety-critical systems are properly maintained and updated. By adopting a similar approach, the automotive industry can create a comprehensive safety network that ensures the safe operation of automated vehicle fleets, protecting both passengers and other road users.

3.6.2 SMSs for Highly Automated Vehicle Fleets

The safety of highly automated vehicle fleets is a complex challenge that can only be mastered through the cooperation of all stakeholders. The establishment and coordination of SMSs, based on the proven methods of aviation, provides a solid foundation for the safe operation of these new forms of mobility. Only through a clear distribution of responsibilities, regular coordination and an unconditional commitment to safety can a future be realized in which highly automated vehicles enrich our everyday lives safely and reliably.

At the heart of every SMS is the manufacturer of the highly automated vehicles. As in aviation, where companies such as Boeing or Airbus are responsible for designing safe airplanes, manufacturers such as Tesla, Waymo or others bear the main responsibility for the safety of their vehicles. Their SMS must ensure that every vehicle produced not only complies with applicable safety standards but is also robust against the challenges of real-world operation. Manufacturers must ensure that all safety-related information is made transparent and accessible to the other stakeholders involved—especially fleet operators. This is the only way to ensure that the vehicles can be operated safely in all conceivable scenarios.

The operator, comparable to an airline in aviation, assumes responsibility for the safe operation of the vehicle fleet. This requires a separate SMS tailored to the operational processes. The operator must ensure that the fleet is used safely in accordance with the manufacturer's specifications and taking into account the specific operating conditions. This includes planning routes, managing operating times, monitoring vehicle functions in real time and responding to any problems that arise. An effective SMS for the operator also takes into account the continuous training of personnel. Although the vehicles are highly automated, human intervention remains essential in certain situations, for example through remote operation concepts such as remote driving or remote assistance (Hans/Walter 2024). Staff must therefore be regularly trained in vehicle handling, emergency procedures, remote operation processes (Hans et al. 2024) and how to deal with unforeseen events. This is the only way to ensure the safety of the fleet even in critical situations.

Similar to airports in aviation, highly automated vehicle fleets also require specific hubs where vehicles can be loaded, serviced or rescheduled (Figure 3.5). These transportation hubs or specialized parking areas play a central role in safety management. Their SMS must be designed to ensure the safe operation of all processes at these locations. This applies not only to the physical safety of the vehicles and infrastructure, but also to digital safety, as these hubs are often central hubs for communication and fleet management. One safety-critical aspect of these hubs is the management of loading processes and the avoidance of collisions when several vehicles arrive or depart at the same time. Precise coordination and clear protocols are essential here in order to avoid accidents and guarantee a smooth process.

The maintenance and servicing of vehicles is another decisive factor for operational safety. Similar to aviation, highly automated vehicles require regular inspections and maintenance to ensure their functionality and safety. An SMS for workshops defines clear processes and responsibilities to ensure that all maintenance work is carried out to the highest standards. Software maintenance deserves particular attention. As highly automated vehicles are heavily dependent on their software, regular updates and safety checks must be carried out. Workshops must therefore work closely with manufacturers to ensure that all systems are up to date and that security gaps can be closed immediately.

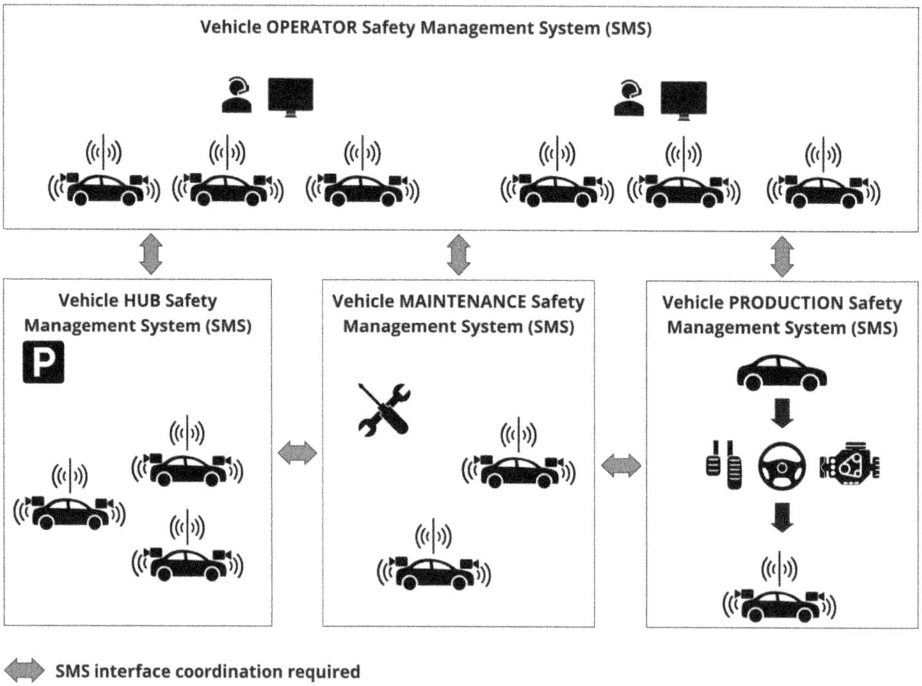

Figure 3.5 Simplified proposal for a Safety Management System (SMS) structure for highly automated vehicle fleets

One of the biggest challenges when implementing SMS for highly automated vehicle fleets is coordinating the various systems with each other. As manufacturers, operators, hubs, and workshops are often different companies, there is a risk of safety gaps arising at the interfaces. These interface errors can have fatal consequences if information is lost, or processes are not optimally coordinated. The establishment of clear communication channels and common standards is therefore of crucial importance. All stakeholders involved must work closely together and organize their SMS in such a way that they complement and support each other. This can be achieved through regular coordination meetings, joint training, and the establishment of central safety databases. In the early phases of the market for highly automated vehicles, it is conceivable that manufacturers will act as both operator and workshop. This can be particularly advantageous as it reduces the complexity of coordination processes and facilitates the establishment of a standardized SMS. If a company has complete control over the production, operation, and maintenance of the vehicles, safety processes can be implemented directly and without detours. However, even in such cases, care must be taken to ensure that safety standards are maintained at the highest level. The challenges may be different, but the responsibility for safety remains. As fleets grow and more players come into play, the careful separation and coordination of the various SMSs will become unavoidable.

3.7 Proactive Safety Culture in Safe Mobility

In today's fast-paced world, the demand for efficient and safe mobility systems has never been greater. The rise of smart cities, automated vehicles, and interconnected transportation networks has ushered in a new era of mobility. In the rapidly evolving landscape of global mobility, where technological advancements are reshaping the way we move and interact with our environment, ensuring safety has become more crucial than ever. The complexities of modern transportation systems, ranging from highly automated vehicles to interconnected public transport networks, necessitate a robust approach to safety management. A key component of this approach is the establishment and maintenance of a proactive safety culture within organizations responsible for mobility systems.

While the SMS provides essential structure and governance, the true power of safety management lies in the development of a safety culture, a shared, intuitive, and self-motivated commitment to safety at all levels of the organization. When a strong safety culture is present, employees don't just follow SMS requirements because they are mandated. They inherently understand and value the importance of safety in their daily tasks and decision-making. This culture leads to intuitive, proactive behavior where safety considerations are second nature, allowing for swift identification and resolution of potential risks before they escalate.

However, without this deep-seated safety culture, the implementation of SMSs can become merely a formality. When employees and managers alike lack intrinsic motivation or belief in the importance of safety, SMS guidelines may be reluctantly accepted, followed only to the extent necessary to meet regulatory requirements. This reluctance can create gaps in the effectiveness of safety management, with the SMS perceived as a burden rather than a tool for continuous improvement. In such environments, compliance-driven actions may replace genuine commitment, leading to minimal engagement with safety processes and an increase in potential risks.

A robust safety culture complements the SMS by embedding safety into the organizational DNA. Employees actively take ownership of safety issues, collaborate to address concerns, and consistently prioritize safety in both routine tasks and complex operations. In such environments, the SMS serves as a foundation, but the culture elevates it by fostering initiative, creativity, and collective responsibility. Safety becomes a shared value rather than just a set of rules to follow, creating an environment where individuals are motivated to improve safety not just for compliance but because it is a core part of their identity and purpose within the organization (Figure 3.6).

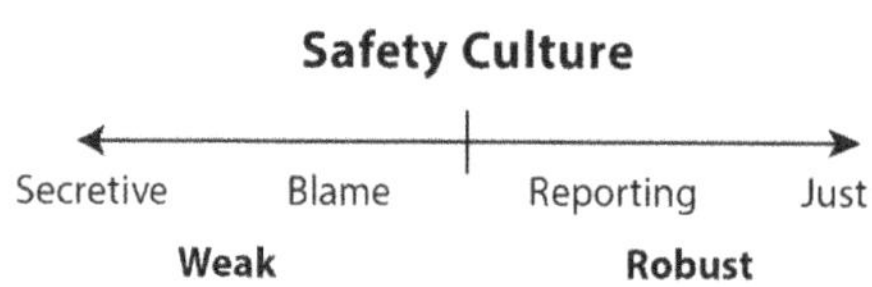

Figure 3.6 The safety culture perspective (source: Patankar/Duncan 2018)

In the context of modern mobility systems where innovation and complexity are rapidly increasing, the synergy between a well-implemented SMS and a strong safety culture is essential. As technological advancements introduce new challenges and risks, a proactive safety culture ensures that organizations can adapt to these changes effectively. Without this cultural foundation, even the most sophisticated SMS will struggle to manage the dynamic and interconnected risks of modern transportation systems.

The mobility sector has witnessed unprecedented changes in recent years. Autonomous vehicles, electric scooters, drones, and high-speed rail systems are just a few examples of innovations that are transforming how people and goods move. While these advancements offer tremendous benefits, they also introduce new risks and challenges. The interconnected nature of modern mobility systems means that a failure in one component can have cascading effects, potentially leading to severe consequences.

Safe mobility refers to the development and maintenance of transportation systems that prioritize the safety of all users whether they are pedestrians, cyclists, drivers, or passengers. It encompasses the entire ecosystem of transportation, including vehicles, infrastructure, regulations, and human behavior. In this context, the stakes are high. The consequences of failure can be catastrophic, resulting in loss of life, environmental damage, and severe economic repercussions. The challenges to safe mobility are multifaceted. On one hand, there is the ever-present risk of human error, which remains a leading cause of accidents across all modes of transportation. On the other hand, technological advancements, while beneficial, introduce new risks and uncertainties. For instance, highly automated vehicles promise to reduce human error, but they also raise concerns about software reliability, cybersecurity, and ethical decision-making. Additionally, the increasing interconnectivity of transportation systems means that a failure in one area can quickly propagate and cause widespread disruption.

In such a complex and rapidly evolving landscape, the establishment of both an SMS and a deeply ingrained safety culture is no longer optional, it is an essential strategy for any organization involved in mobility systems. Only by fostering a culture where safety is an intuitive, self-motivated priority can the full potential of an SMS be realized, ensuring that technological innovation and safety progress hand in hand.

Given these challenges, it is clear that traditional reactive approaches to safety, such as Safety I, those that focus on responding to incidents after they occur, are no longer sufficient. Instead, there is a pressing need for proactive measures that anticipate and mitigate risks before they manifest. This is where SMSs and a proactive safety culture come into play.

3.7.1 The Importance of a Proactive Safety Culture

While the structural elements of an SMS are crucial, they are only as effective as the culture that underpins them. A proactive safety culture is characterized by an organization-wide commitment to safety that goes beyond mere compliance with regulations. It is about embedding safety into the very fabric of the organization, influencing decision-making at all levels, and fostering an environment where safety is prioritized over other considerations, such as cost or efficiency. There are several reasons why a proactive safety culture is essential for safe mobility:

- **Prevention over reaction:** The primary benefit of a proactive safety culture is the prevention of incidents. By anticipating potential hazards and addressing them before they cause harm, organizations can significantly reduce the likelihood of accidents. This is particularly important in the context of modern mobility, where the complexity of systems and the speed of technological change can create new and unforeseen risks. A proactive safety culture ensures that these risks are identified and mitigated early, preventing incidents that could have far-reaching consequences.
- **Empowerment of employees:** In a proactive safety culture, all employees are encouraged to take ownership of safety. This means they are not only responsible for

following safety procedures but are also empowered to identify and report potential hazards, suggest improvements, and participate in safety initiatives. This bottom-up approach ensures that safety is everyone's responsibility, not just that of the safety department.

- **Continuous improvement:** A proactive safety culture fosters a mindset of continuous improvement. Rather than resting on past achievements, organizations with this culture are always looking for ways to enhance safety. This might involve adopting new technologies, refining processes, or learning from incidents and near-misses. Continuous improvement is essential in the mobility sector, where the rapid pace of change means that safety practices must evolve to keep up with new developments.
- **Adaptability to change:** In the fast-paced world of mobility, change is constant. New technologies, regulatory requirements, and market dynamics are continually shaping the industry. A proactive safety culture enables organizations to adapt to these changes effectively. By embedding safety into every aspect of their operations, organizations can respond quickly to new challenges and opportunities. For example, as highly automated vehicles become more prevalent, organizations with a proactive safety culture will be better equipped to address the unique safety challenges posed by these technologies.
- **Building trust with the public:** Safety is a key concern for customers, regulators, and other stakeholders in the mobility sector. Organizations that demonstrate a proactive commitment to safety are more likely to earn the trust and confidence of these stakeholders. This can translate into a competitive advantage, as customers are more likely to choose services and products from companies that prioritize safety. Furthermore, a strong safety record can enhance an organization's reputation, making it easier to attract and retain customers, employees, and investors. This is particularly important in the adoption of new technologies, such as highly automated vehicles, where public skepticism can be a significant barrier to acceptance.
- **Regulatory compliance and beyond:** While regulatory compliance is a baseline requirement, a proactive safety culture goes beyond mere compliance. Organizations with such a culture are not content with meeting the minimum safety standards; instead, they strive to exceed them. This proactive approach can help organizations stay ahead of regulatory changes and avoid penalties or legal issues. Moreover, by setting higher safety standards, these organizations can influence industry norms and contribute to the overall improvement of safety in the mobility sector.

The importance of a proactive safety culture is not just theoretical. It is supported by real-world evidence. Consider the aviation industry, which has long been a leader in safety management. The industry's adoption of SMSs and a proactive safety culture has been instrumental in reducing accident rates to historically low levels. Airlines and regulators have embraced a proactive approach, focusing on hazard identification, risk assessment, and continuous improvement. This has not only enhanced safety but has also contributed to the industry's strong safety reputation. In the field of rail transportation, organizations that have adopted a proactive safety culture have seen significant improvements in safety performance. By prioritizing safety and fostering a culture where employees are encouraged to report near-misses and potential hazards, these organizations have been able to identify and mitigate risks before they lead to accidents. In contrast, industries or organizations that have failed to embrace a proactive safety culture have often suffered the consequences. High-profile accidents, such as the Deepwater Horizon oil spill or the Fukushima nuclear disaster, have been attributed in part to a lack of a proactive safety culture. In these cases, warnings were ignored, safety procedures were bypassed, and the focus was on short-term gains rather than long-term safety.

3.7.2 Challenges in Implementing a Proactive Safety Culture

While the benefits of a proactive safety culture are clear, implementing and maintaining such a culture is not without challenges. One of the main challenges is ensuring that safety is consistently prioritized across all levels of the organization. This requires strong leadership and a commitment to safety that is communicated clearly and consistently. Leaders must set the tone for the organization, demonstrating through their actions that safety is a top priority.

In addition, leadership must be fully aware of its responsibility for the SMS and the impact it has on the organizational culture. The role of leadership is not only to implement the SMS but also to actively shape a culture where safety becomes an integral part of daily operations. Leaders must understand that their commitment to safety goes beyond compliance. It profoundly influences how employees perceive and engage with safety practices. When leadership demonstrates a genuine dedication to safety, it inspires employees to internalize these values, making safety a shared responsibility rather than just an obligation.

Another challenge is overcoming resistance to change. In some organizations, there may be a culture of complacency, where employees are reluctant to adopt new safety practices or technologies. This can be addressed through education and training, helping employees understand the importance of a proactive approach to safety and how it benefits them personally.

Resource constraints can also be a barrier to implementing a proactive safety culture. Developing and maintaining an SMS, conducting regular safety training, and investing in new safety technologies require time and money. Organizations must be willing to allocate the necessary resources to safety initiatives and recognize that these investments will pay off in the long run through reduced incidents and enhanced operational efficiency.

Finally, there is the challenge of maintaining a proactive safety culture over time. It is not enough to implement an SMS and assume that safety will take care of itself. Organizations must continually reinforce their commitment to safety, regularly review and update their safety practices, and ensure that all employees remain engaged in the safety process.

3.7.3 The Future of Safe Mobility and the Role of a Proactive Safety Culture

As we look to the future of mobility, the importance of a proactive safety culture will only grow. The increasing complexity of transportation systems, the rise of new technologies, and the growing interconnectedness of global transportation networks will all contribute to a more challenging safety environment. To navigate this environment, organizations will need to adopt not just the tools and processes of SMS but also the mindset of a proactive safety culture.

In this future, a proactive safety culture will be essential for several reasons. First, it will enable organizations to stay ahead of emerging risks, ensuring that new technologies and practices are implemented safely. Second, it will help build public trust, which will be crucial for the adoption of new transportation modes, such as highly automated vehicles and shared mobility services. Finally, it will support the ongoing effort to reduce accidents and incidents, contributing to the overall goal of zero fatalities in transportation.

The case for a proactive safety culture in safe mobility is compelling. It is not just about compliance with regulations or the avoidance of accidents; it is about creating a culture that prioritizes safety at every level of the organization. By adopting such a culture, organizations can not only enhance safety but also build trust, foster innovation, and ensure the long-term success of their operations. As the mobility landscape continues to evolve, the importance of a proactive safety culture will become increasingly clear, making it an indispensable element of any comprehensive approach to safe mobility.

3.8 Structural Integration of Safety Management Systems in the Automotive Industry

The increasing complexity and networking in the automotive industry requires careful and comprehensive implementation of SMSs to ensure the highest safety standards. This section examines the structural integration of SMSs in the automotive industry, analyzing key aspects that are essential for successful implementation. First, it examines how organizational changes and new responsibilities need to be created to ensure the effective application of SMS.

In modern automotive safety management, individual persons can no longer ensure an overview of all safety activities. As vehicles and transportation systems become more complex and safety responsibilities more distributed, it is essential that the responsibilities of different organizational units are clearly defined. This is particularly important because not only one organization is responsible for vehicle safety; there are multiple stakeholders involved, including vehicle owners, operators, holders, and drivers, each with distinct safety obligations. Furthermore, clear links must be established between the responsibilities of these stakeholders and those of the manufacturers and distributors. Ensuring that these roles are properly regulated is critical for the cohesive functioning of the SMS and for avoiding gaps in safety management across the entire vehicle lifecycle.

The section then looks at the integration of SMS into existing management systems to exploit synergies and avoid redundancies. Finally, the importance of training and raising employee awareness is emphasized, as this is crucial for the understanding and acceptance of safety measures in day-to-day business. This systematic approach ensures a seamless and sustainable anchoring of SMS in the automotive industry.

3.8.1 Organizational Changes and Responsibilities

The introduction of an SMS in a company is a far-reaching and complex process that goes far beyond the mere implementation of technical measures. An SMS can only be successful if it is firmly anchored at management level, as this is where it receives the necessary authority and support to be enforced throughout the company. This anchoring at management level inevitably leads to far-reaching organizational changes and a comprehensive redefinition of responsibilities. Raising awareness at the highest management level is therefore a key element of SMS integration. Management must not only develop a fundamental understanding of the importance of safety, but also make a commitment to actively promote it. This begins with the recognition that safety must not be seen as an isolated area, but as an integral part of the overall corporate strategy. This strategic integration of safety aspects is crucial to establishing a sustainable safety culture within the company. With the decision to introduce an SMS, management must undertake a comprehensive reassessment of existing organizational structures. It is important that clear and unambiguous responsibilities and roles are defined. This can include, for example, the creation of a specific position for a safety officer who is responsible for implementing and monitoring the SMS. Ideally, this role should report directly to senior management to ensure that safety issues are heard at the highest level and given the necessary priority. In addition, management must ensure that all stakeholders within the organization have the necessary resources and authority to effectively carry out their safety responsibilities. This may require the creation of new departments or the restructuring of existing teams to ensure that safety responsibilities are not just placed on the shoulders of one person or department but are distributed throughout the organization.

Moreover, every organizational restructuring or reorganization introduces additional risks. Without an SMS in place, these risks are often not transparent, leaving the organization vulnerable to oversight or mismanagement of critical safety functions. A well-implemented SMS helps identify, mitigate, and manage the risks associated with organizational changes, ensuring that safety is continuously maintained even as the company evolves. This underscores the importance of integrating safety management into the core structure of the organization, so that changes do not inadvertently compromise safety standards.

Another critical aspect of integrating an SMS is the adaptation and optimization of communication structures. Safety-relevant information must flow smoothly and quickly within the company to ensure that risks are identified at an early stage and appropriate measures can be taken as quickly as possible (e. g., the grounding of vehicle transportation fleets due to technical defects). This means that management must review and, if necessary, redesign existing communication channels to ensure that no safety-relevant information is lost or delayed in the hierarchy. Open and transparent communication is key to fostering a culture of trust and collaboration, which is essential to the effectiveness of an SMS.

The decision-making processes within the company must also be adapted to meet the requirements of an SMS. Traditionally, many decisions in companies are made on the basis of economic considerations, but when integrating an SMS, safety aspects must be given equal weight in the decision-making process. This means that management must ensure that decisions that have a potential impact on safety are not considered in isolation, but in the context of the company's overall safety objectives. This includes carrying out risk analyses and systematically assessing the potential consequences of decisions on safety.

Another key element of integrating an SMS is creating a culture of continuous improvement. An SMS should not be seen as a static system that is implemented once and then not changed again. Rather, it is a dynamic system that must be constantly reviewed, adapted, and further developed to meet changing requirements and challenges. Management must therefore establish mechanisms that enable the system to be regularly reviewed, such as internal and external audits, feedback loops or lessons learned processes. These mechanisms are necessary to ensure that the SMS remains up-to-date and continuously contributes to the improvement of safety standards. The commitment and active involvement of the entire workforce is also critical to the success of an SMS. Management must create an environment in which all employees at all levels of the organization understand the importance of safety and feel personally responsible for it. This requires targeted training and awareness-raising of employees to ensure that they not only know the principles of SMS, but also apply them in their daily actions. Training programs should aim to raise awareness of safety risks and provide employees with the tools to manage these risks effectively. In addition, management must act as role models and exemplify safety-conscious behavior to promote a culture of safety throughout the company.

3.8.2 Integration of SMSs into Existing Management Systems

Companies in safety-critical industries have traditionally implemented various management systems to monitor and control specific aspects of their operations. Among the most prominent are Operational Safety Management for, e. g., vehicle fleets, Functional Safety Management (ISO 26262-2:2018) and Quality Management System (QMS) (ISO 9001:2015). These systems were developed to address specific requirements and risks associated with operational management, the functional safety of technical systems and ensuring the consistent quality of products and services. However, the fragmented implementation of these systems is often not sufficient in itself to ensure a comprehensive level of safety within an organization:

- **Operational Safety Management** focuses on safety in operational business. It comprises measures and procedures aimed at minimizing risks in day-to-day business. Although it is an essential component of corporate safety, Operational Safety Management alone cannot cover all safety requirements. It often lacks the strategic overview that an SMS provides, especially when it comes to looking at operational risks in the context of the entire company.
- **Functional Safety Management** deals with the safety of systems based on electrical, electronic and programmable electronic devices. It plays a crucial role in industries where technical systems are critical to safety, such as the automotive industry, aviation or the process industry. Functional Safety Management addresses specific technical risks, but often without considering the overarching organizational risks and processes.
- **Quality Management Systems (QMSs)** aim to ensure product and service quality, which can also indirectly contribute to safety. There are overlaps between quality

and safety management, particularly in areas such as error prevention and process control. However, here too, isolated quality management is not sufficient to ensure the overall safety of the company.

It is also important to differentiate these management systems from **Business Management Systems (BMSs)**. While BMSs are primarily concerned with optimizing and automating business processes to improve efficiency and effectiveness, its focus is not on safety but on the overall performance and operational success of the business. BMSs center around streamlining workflows, reducing costs, and increasing productivity by aligning processes with business goals.

In addition, product recalls have a significant business impact, especially in industries like automotive, where safety defects can lead to large-scale recalls. Recalls can have far-reaching financial consequences, including direct costs from repairs and replacements, regulatory fines, and lawsuits. Moreover, they can severely damage a company's brand reputation, leading to customer dissatisfaction, lost market share, and long-term erosion of customer trust. Effective recall management requires seamless coordination between the manufacturer, suppliers, and regulatory bodies, which is facilitated by an integrated SMS.

A robust SMS allows businesses to manage recall processes proactively, minimizing financial and reputational risks by ensuring early identification of potential safety issues. Through continuous monitoring of field data and incidents, companies can detect and address defects before they necessitate widespread recalls. Furthermore, efficient recall handling, supported by an SMS, can help mitigate negative business impacts by fostering trust and transparency with customers and regulatory authorities. Thus, while the primary function of an SMS is safety, its influence extends deeply into business operations, particularly in mitigating the financial and reputational risks associated with recalls.

In contrast, BMSs focus more on process optimization and cost reduction, and while it may improve operational efficiency, it is not equipped to handle the risk management and regulatory compliance that are essential in safety-critical industries. Therefore, while BMSs and safety-related management systems may overlap in areas such as process improvement, SMSs play a more crucial role in addressing the business risks and consequences of safety failures, including recalls.

Although these different systems each offer a high degree of specialization and expertise, their isolated application can lead to incomplete risk coverage. There is a risk of information silos, inconsistent safety practices and inefficient use of resources as each system operates in isolation and does not always adequately consider the interactions with other relevant areas. It is therefore crucial to develop a holistic strategy that integrates or controls these management systems in a meaningful way to maximize safety throughout the company.

3.8.2.1 The SMS Approach as an Overarching Management System

An SMS offers the approach of anchoring safety as an integral and consistent principle across all company activities. As an overarching management system, the SMS has the task of promoting the safety culture in the company, defining safety objectives and coordinating the various individual management systems to ensure that safety is implemented consistently in all areas of the company. This can be useful for several reasons:

- **Holistic perspective:** As a higher-level system, the SMS offers a holistic perspective that ensures that safety considerations are incorporated into all decision-making processes and operational procedures. This prevents safety-critical aspects from being lost or neglected in the various management systems. An SMS can identify safety gaps that could arise from isolated or uncoordinated management practices and promotes synergy between different systems.
- **Avoiding information silos:** A key problem in companies with multiple management systems is the creation of information silos. These occur when different departments or teams develop and implement their own safety strategies without coordinating them with other parts of the organization. The SMS acts as a bridge here, ensuring that safety information and practices are shared and aligned across all departments. This leads to a coherent safety strategy based on common goals and methods.
- **Standardization and consistency:** The introduction of an SMS as an overarching management system enables standardization of safety practices across the entire company. This helps to ensure that all departments and teams work to the same principles and procedures, increasing the consistency and effectiveness of safety measures. Uniform safety standards reduce the risk of misinterpretation or misapplication of security policies.
- **Cultural integration:** A key element of an overarching SMS is the creation and promotion of a safety culture within the company. While individual management systems such as quality management or Functional Safety Management focus strongly on technical standards and processes, an SMS aims to anchor safety as a fundamental value and integral part of the corporate philosophy. This requires continuous communication and training of all employees as well as clear leadership and support from senior management.

3.8.2.2 Integration of SMSs into Existing Management Systems

An alternative approach to introducing an SMS is to integrate the SMS into existing management systems rather than introducing it as an overarching system. This approach has the advantage of using existing structures and processes to integrate safety management practices into daily operations. This approach has the following advantages:

- **Resource efficiency:** Integration into existing management systems can lead to a more efficient use of resources, as no separate structures need to be created for the SMS. Safety requirements and practices can be embedded into existing processes, minimizing the need for additional resources and reducing implementation costs.
- **Avoidance of redundancies:** In companies that already operate well-developed management systems such as quality management or environmental management, integrating the SMS can help to avoid redundancies. Safety processes can be seamlessly integrated into existing workflows, avoiding duplication and inefficient parallel structures. This can be particularly beneficial if similar auditing and compliance requirements already exist, and the SMS can simply build on these.
- **Flexibility and customization:** By integrating into existing systems, the company remains more flexible and can adapt the SMS to the specific needs and circumstances of the company. Instead of implementing a rigid, overarching system, the SMS can be designed to fill specific gaps in the existing systems and adapt flexibly to changes. This also facilitates the continuous improvement of the system, as adjustments can be made more easily within the existing management systems.

Despite the advantages mentioned, there are also challenges when integrating an SMS into existing management systems. One of the biggest challenges is ensuring that the integrated SMS is given the same visibility and priority as other management areas. There is a risk that safety aspects lose importance or are not given sufficient consideration due to integration into another system. It is therefore crucial that the integrated SMS has clear reporting and control mechanisms, and that top management clearly communicates and supports the importance of safety management.

Looking at specific management systems within an organization, the integration of Operational Safety Management into an SMS ensures that operational safety measures are not only assessed locally but in the overall context of the organization. An SMS allows operational risks to be aligned with strategic objectives and ensures that safety measures are applied consistently across the organization. This integration promotes a more comprehensive view of risk and improves the organization's ability to respond effectively to safety incidents.

By integrating Functional Safety Management into an SMS, technical safety can be placed in a broader safety strategy context. The SMS enables a link between technical safety requirements and the company's operational and strategic objectives. This ensures that the technical systems are not only functionally safe, but also in line with the company's overall safety policies and strategies. This integration improves the coherence and effectiveness of safety measures across technical and organizational boundaries.

An SMS can also integrate and extend quality management processes to ensure that safety aspects are systematically incorporated into quality management activities.

Linking safety and quality management ensures that safety considerations are integrated into all phases of the product lifecycle, from development to production and maintenance. This helps to improve both product quality and safety, as potential risks can be identified and addressed at an early stage.

3.8.3 Training and Awareness of Employees for Safety Management Systems

In today's fast-paced and increasingly complex work environments, SMSs have become essential tools for ensuring the safety and well-being of employees and the operational efficiency of organizations. The successful implementation of an SMS depends not only on the framework itself but also on the commitment and active participation of every employee within the organization.

Training and awareness of employees are the bedrock of an effective SMS. Without the active participation of a well-informed workforce, even the most well-designed SMS will fail to achieve its objectives. Training provides employees with the knowledge and skills they need to fulfill their safety responsibilities, while awareness initiatives ensure that safety remains a priority in their daily activities. Investing in training and awareness programs is not just a regulatory requirement or a best practice. It is a strategic imperative for any organization that values safety and operational excellence. By empowering employees with the tools and knowledge they need to contribute to the SMS, organizations can cultivate a culture of safety, enhance communication and reporting, and promote continuous improvement. Ultimately, this leads to safer work environments, more efficient operations, and a stronger, more resilient organization.

However, while training is fundamental, it is not enough on its own. Leadership and the active demonstration of the desired safety culture are far more critical to embedding safety into the fabric of the organization. Employees are far more likely to embrace and internalize safety values when they see those values consistently modeled by their leaders. When leadership visibly prioritizes safety, holds themselves accountable, and integrates safety into their decision-making, it sends a powerful message that safety is more than just a set of rules. It becomes and is a core value. Leaders who live and breathe safety inspire employees to do the same, creating an environment where safety becomes an intuitive, collective responsibility. Therefore, leadership's role in setting the tone and leading by example is essential to making an SMS truly effective and sustainable over time.

This section delves into the critical importance of training and awareness programs for employees concerning SMSs, elucidating why such initiatives are indispensable for achieving safety excellence and fostering a culture of safety throughout the organization.

3.8.3.1 The Necessity of Employee and Leadership Training for SMSs

An SMS is designed to be proactive, identifying hazards before they manifest into incidents, and it promotes a culture of continuous improvement. For an SMS to be effective, every employee must have a fundamental understanding of its principles, objectives, and components. Without proper training, employees may view SMSs as merely a set of bureaucratic procedures or, worse, as an unnecessary burden that detracts from their primary duties. This misconception can lead to a lack of engagement, where employees fail to recognize their roles in the safety management process, potentially compromising the entire system. Training is thus crucial to ensure that employees understand the purpose of SMSs, how they apply to their daily work, and why their active participation is essential.

One of the core elements of an SMS is hazard identification and risk management. Employees are often the first to encounter potential hazards in their work environment. However, the ability to identify and report these hazards effectively depends on their knowledge and understanding of what constitutes a hazard and the processes in place for managing risks. Training programs equip employees with the necessary skills to recognize hazards and assess risks accurately. They learn to use tools such as risk matrices, understand the hierarchy of controls, and apply these concepts to mitigate risks in real-time. Without this knowledge, employees might overlook significant risks, leading to incidents that could have been prevented. Therefore, training is not just about imparting knowledge but also about empowering employees to take an active role in maintaining safety.

In addition to that, regulations often include specific requirements for employee training and awareness programs. Compliance with these requirements is not optional; it is a legal obligation that, if neglected, can result in severe penalties, including fines, sanctions, or even the shutdown of operations. Training programs ensure that employees are aware of the regulatory landscape and understand their responsibilities within it. This is particularly important in industries such as aviation, healthcare, and manufacturing, where safety standards are stringent, and the consequences of non-compliance can be catastrophic. By investing in training, organizations not only meet their legal obligations but also protect themselves from the financial and reputational damage that can result from regulatory violations.

However, training for leaders is even more critical. Leaders play a pivotal role in shaping the organizational culture and setting the tone for safety within the workplace. When leaders are well-trained in SMS principles and practices, they are better equipped to model the behaviors they want to see in their teams. They can effectively communicate the importance of safety, provide guidance, and inspire their employees to prioritize safety in their daily activities. Additionally, well-trained leaders are more adept at recognizing potential risks and intervening proactively, which is essential for preventing incidents before they occur.

Therefore, it is essential to invest in leadership training as part of the SMS implementation process. This training should focus not only on the technical aspects of SMSs but also on leadership skills such as communication, team dynamics, and conflict resolution. By ensuring that leaders understand the significance of their role in promoting safety, organizations can create an environment where safety is truly valued and embedded in every aspect of operations. This holistic approach ensures that safety culture thrives at all levels, ultimately leading to more effective and resilient SMSs.

3.8.3.2 The Importance of Employee Awareness in SMSs

A robust SMS is underpinned by a strong safety culture, where safety is ingrained in the organizational values, beliefs, and practices. However, such a culture cannot be established through top-down mandates alone. It requires the active participation of all employees, from frontline workers to senior management. Awareness programs play a pivotal role in fostering this culture by continually reinforcing the importance of safety and encouraging employees to prioritize it in their daily activities.

Awareness initiatives, such as safety briefings, workshops, and campaigns, help to keep safety at the forefront of employees' minds. They remind employees of the importance of safety procedures, the potential consequences of neglecting them, and the role each individual plays in maintaining a safe workplace. Over time, these efforts contribute to the development of a shared commitment to safety, where every employee feels responsible for their own safety and that of their colleagues.

Therefore, effective communication is a cornerstone of a successful SMS. Employees need to feel comfortable reporting safety concerns, incidents, and near-misses without fear of retribution. Awareness programs can help to establish an open communication environment by educating employees about the importance of reporting and the processes in place to support it. When employees are aware that their reports are valued and lead to tangible improvements, they are more likely to participate in the SMS actively. This, in turn, enhances the organization's ability to identify and address safety issues promptly, preventing minor issues from escalating into major incidents. Awareness programs also highlight the organization's commitment to transparency and continuous improvement, further encouraging employee engagement in safety initiatives.

Finally, the SMS is not a static system; it requires ongoing assessment and improvement to remain effective in a changing environment. Employees play a critical role in this process by providing feedback, identifying areas for improvement, and contributing to the development of new safety strategies. Awareness programs can help employees understand the importance of their contributions and how they can positively impact the SMS. Through regular training and awareness initiatives, employees are kept informed about changes to the SMS, new safety procedures, and lessons learned from past incidents. This ongoing education ensures that the SMS evolves with the organization and that employees are always equipped with the latest knowledge and skills to maintain a safe working environment.

3.9 Technological Developments and Innovations

The automotive sector is in a phase of profound change, driven by rapid technological advances and an increasing focus on sustainability, safety and efficiency. These developments require SMSs to adapt and evolve to meet the new challenges and opportunities created by these innovations. The rapid development of new technologies in the automotive sector, particularly in the areas of automated driving, poses new challenges for SMSs. SMSs, traditionally designed to ensure the safety of drivers and passengers in largely manually operated vehicles, must now integrate increasingly complex, technological systems that require a high degree of automation, human-machine interaction and remote control. These developments require not only an expansion and adaptation of existing safety protocols, but also a profound realignment of the entire safety philosophy in both the transportation sector in general and the automotive sector.

The introduction of highly automated driving systems, in particular Level 3 (L3) and Level 4 (L4) according to the classification of the Society of Automotive Engineers (SAE), has profound effects on the design and implementation of SMS (SAE International 2021). In L3 systems, where the vehicle drives autonomously in certain situations but may need to be handed over to the driver at any time, the requirements change fundamentally. The need to ensure that the driver can take control of the vehicle safely and in a timely manner has led to the development of new protocols. These protocols include, in particular, monitoring driver alertness and readiness and providing clear and timely instructions for the handover. One of the key technological developments in this context is the Human-Machine Interface (HMI). A robust HMI is crucial to ensure clear communication between vehicle and driver, thus increasing safety during the transition process. In addition, drivers must either be fully trained to use these systems effectively or it must be clearly understandable how to use it, although the definition of clearly understandable can be questioned. This training and the HMI itself are integral parts of the SMS to ensure that human error is minimized, and transitions can be made as safe as possible.

With the further development of L4 systems, in which the vehicle can drive fully automatically in defined environments or under certain conditions, the focus of safety requirements is shifting considerably once again. While the driver still plays a central role in L3 systems, this is largely pushed into the background in L4. Nevertheless, safety is still of crucial importance, especially regarding new concepts such as **remote operation**, which are becoming increasingly important in L4 systems. Here, technologies such as remote assistance, remote monitoring and remote driving, as visualized in Figure 3.7, are integrated into the SMS to maintain safety and availability, especially in the operational area of vehicle fleets (Hans/Walter 2024), even when no driver is present in the vehicle, instead in a remote-control room. These developments bring new safety considerations, especially with regard to the human factor.

The concept of remote operation encompasses various approaches in which a vehicle is no longer controlled exclusively by a driver sitting in the vehicle but by a remote operator. This includes remote monitoring, in which the condition of the vehicle and its surroundings are continuously monitored, remote assistance, in which an operator can intervene to provide support if necessary, and remote driving, in which the vehicle is controlled entirely from a distance. These technologies offer significant benefits, particularly in situations where it is not practical or safe to have a human driver in the vehicle. However, they also bring new challenges for SMS, particularly in terms of the human factor and the associated risks (Hans et al. 2024).

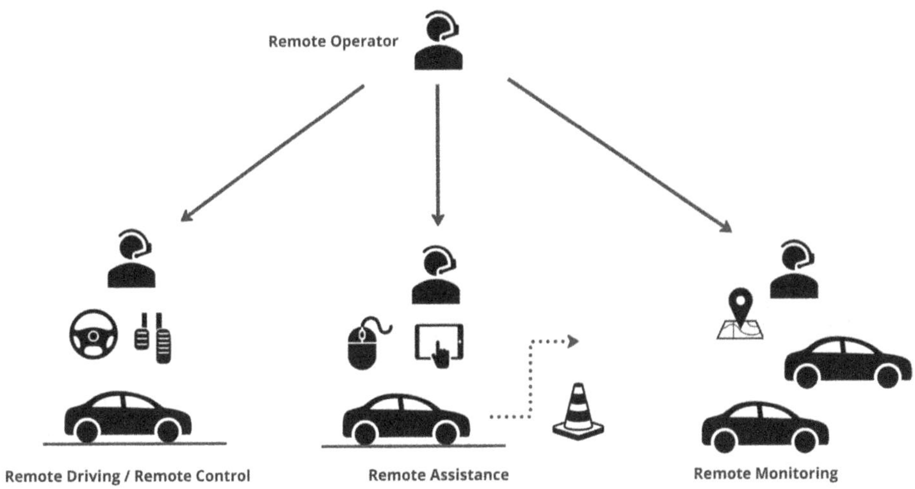

Figure 3.7 Remote operation concepts

The increasing implementation of remote-operation technologies such as remote driving, where vehicles are controlled by a remote operator, poses new challenges for SMSs. The biggest concern here is the human factor, specifically how the remote operator's behavior, attention and decision making can affect the safety of the vehicle (Schwindt-Drews et al. 2024). Unlike traditional driving, where the driver is physically present in the vehicle, remote driving creates potential delays in communication and control that can affect situational awareness and responsiveness. To minimize these risks, SMSs must include protocols that ensure thorough selection and training of remote operators. These operators must not only be technically proficient, but also able to make quick and accurate decisions under stressful conditions. In addition to the human factors, SMS must also take into account the specific environmental conditions under which remote operation takes place. These can differ significantly from the conditions under which a human driver operates in the vehicle. For example, in poor weather conditions, in confusing terrain or in urban environments, there may be significant challenges that affect the remote operator's ability to safely control the vehicle (Hans et al. 2023). The integration of these technologies into SMSs therefore requires the development of specific protocols for the selection, training, and monitoring of remote operators. These protocols must ensure that operators have the necessary skills and knowledge to meet the high safety requirements.

The evolution of the automotive sector offers both opportunities and challenges for the further development of SMSs. On the one hand, new technologies and innovations offer the potential to significantly improve safety by reducing human error and enabling more proactive risk management. On the other hand, these technologies also bring with them new complexities and risks that SMSs must address. One of the key challenges for the further development of SMSs is to keep up with the rapid pace of technological change. As new systems and technologies are introduced, SMSs need to be continually updated to ensure their effectiveness. This requires continued investment in research and development as well as collaboration between automotive manufacturers, regulators and other stakeholders to establish industry-wide safety standards. Another challenge is finding the balance between safety, cost, and user experience. While advanced safety features can prevent accidents, they can also increase the complexity and cost of vehicles, potentially making them less accessible to consumers. SMSs must therefore strike a balance between maximizing safety and ensuring that vehicles remain affordable and user-friendly. Finally, the challenge is to ensure that SMSs are flexible and adaptable enough to cope with the diverse range of vehicles, technologies, and use cases in the modern automotive sector. This includes considering different driving environments, from urban centers to rural areas, as well as addressing the specific needs of different user groups, such as elderly or disabled drivers.

3.10 Conclusion and Summary

The development of the concept of safety has changed significantly over time and today comprises the concepts of **Safety I**, **Safety II**, and **Safety III**. Safety I focuses on the prevention of errors and accidents by analyzing past incidents to minimize future risks. Safety II, on the other hand, focuses on the ability of a system to function successfully under varying conditions. Safety III expands this view by integrating resilience to address not only known, but also unknown and unexpected hazards. Safety III also emphasizes the importance of organizational and technical control in managing risks. It recognizes that safety cannot solely be viewed from a technical perspective. It must also account for the dynamics of human behavior and organizational structures.

Safety III is fundamentally based on the risk of loss of control at the organizational level, which subsequently influences the technical level, as losses are often closely linked to the loss of control over transportation systems. Traditionally, the driver was responsible for vehicle control in traffic, but as automation and connectivity increase, the roles and responsibilities are shifting. The lines between operators and manufacturers are becoming blurred, necessitating a reevaluation of accountability and control measures. In this new landscape, it is crucial for organizations to develop a comprehensive understanding of the risks associated with both the operational and technical aspects of safety. This involves not only implementing robust SMSs but also fostering collaboration between various stakeholders, including vehicle manufacturers, operators, and regulatory bodies. By doing so, organizations can better manage risks and enhance safety outcomes, ensuring that both human and technological elements work seamlessly together to mitigate potential hazards.

The SMS based on Safety III is a structured framework that helps organizations achieve their safety goals. It is based on four fundamental principles: **safety policy**, **Safety Risk Management**, **safety assurance**, and **safety promotion**. These principles originate from the aviation industry, where they have proven to be particularly effective. The safety policy defines the safety objectives and obligations of an organization. Safety Risk Management covers the identification, assessment, and control of safety risks. Safety assurance ensures that the system is continuously monitored and improved, while safety promotion aims to promote a strong safety culture within the organization.

The historical development of SMSs shows that these concepts first emerged in highly regulated industries such as aviation. In recent decades, however, other industries, including the automotive industry, especially in the operation of highly automated vehicle fleets, have also recognized the need for such systems. An SMS is of great importance to organizations and companies as it allows them to systematically manage safety risks while promoting a proactive safety culture. In the short term, implementing an SMS can be costly and time-consuming, but the investment pays off in the long term by reducing the risk of accidents and incidents, thereby not only protecting lives but also reducing costs and enhancing the organization's reputation.

The legal basis for SMSs varies by industry and region, with the aviation and automotive industries having particularly strict requirements. In aviation, SMS has been mandatory for years, while the automotive industry is still in the early stages of implementation, particularly with regard to automated vehicle fleets. Nevertheless, the potential of SMS in the automotive industry is enormous, as the advancing automation and networking of vehicles brings with it new safety challenges.

The need for an SMS extends across various stakeholders within the automotive industry, including manufacturers, operators, and workshops. It is essential for all stakeholders to implement an SMS to ensure safety and meet regulatory requirements. A proactive safety culture in mobility is becoming increasingly important, especially at a time when technologies such as highly automated or remote-controlled vehicles are on the rise.

The structural integration of SMSs into the automotive industry requires organizational changes and a clear allocation of responsibilities. It is important that the SMS is combined or integrated with existing management systems such as the Quality Management System (QMS) or functional safety systems in a sensible hierarchical manner in order to create synergies and avoid redundancies. In addition, training and sensitizing employees to the SMS is of central importance to ensure that the system functions effectively.

After all, SMSs are critical in a number of industries and their implementation brings significant benefits in both the short and long term. While implementing an SMS can be costly and resource intensive initially, the long-term benefits outweigh the costs in terms of increased safety, reduced costs and enhanced reputation. Especially in the automotive industry, where the challenges of new technologies and automation are increasing, a strong and well-integrated SMS is essential. Organizations that successfully implement an SMS are better equipped to manage safety risks and promote a proactive safety culture that ultimately protects lives and ensures the long-term success of the business.

References

Aurora Innovation Inc.: Aurora's 2022 Safety Report. 2022. *https://blog.aurora.tech/safety/auroras-2022-safety-report* (accessed August 26, 2024)

Aven, T.: A risk science perspective on the discussion concerning Safety I, Safety II and Safety III. 2022

Boustras, G./Guldenmund, F. W. (eds.): Safety management in small and medium sized enterprises (SMEs). CRC Press 2017

Bristow, M./Fang, L./Hipe, K. W.: System of systems engineering and risk management of extreme events: concepts and case study. In: Risk Anal. 2012, 32 (11), pp. 1935–1955 (DOI: 10.1111/j.1539-6924.2012.01867.x)

Cruise LLC: Cruise Safety Report 2022. Washington, D. C., USA 2022. *https://assets.ctfassets.net/95kuvdv8zn1v/zKJHD7X22fNzpAJztpd5K/ac6cd2419f2665000e4eac3b7d16ad1c/Cruise_Safety_Report_2022_sm-optimized.pdf* (accessed August 26, 2024)

European Union: Commission Regulation (EU) No 965/2012 of 5 October 2012 laying down technical requirements and administrative procedures related to air operations pursuant to Regulation (EC) No 216/2008 of the European Parliament and of the Council. Official Journal of the European Union, L 296, October 25, 2012

European Union: Regulation (EU) No. 376/2014 of the European Parliament and of the Council of 3 April 2014 on the reporting, analysis, and follow-up of occurrences in civil aviation, amending Regulation (EU) No. 996/2010 and repealing Directive 2003/42/EC. Official Journal of the European Union, L 122, April 24, 2014

European Union: Regulation (EU) 2018/858 of the European Parliament and of the Council of 30 May 2018 on the approval and market surveillance of motor vehicles and their trailers, and of systems, components, and separate technical units intended for such vehicles, amending Regulations (EC) No. 715/2007 and (EC) No. 595/2009 and repealing Directive 2007/46/EC. Official Journal of the European Union, L 151, June 14, 2018

FAA – Federal Aviation Administration: 14 CFR Part 5 – Safety Management Systems. Federal Register, Vol. 80, No. 9, January 8, 2015

Hale, A. R./Heming, B. H. J./Carthey, J./Kirwan, B.: Modelling of safety management systems. In: Saf. Sci. 26 (1–2), 1997, pp. 121–140. *https://doi.org/10.1016/S0925-7535(97)00034-9*

Hammer, W.: Why system safety programs can fail. Paper. System Safety Conference, Greenbelt, MD, USA, 26–28 May 1971

Hans, O./Avezum, M./ Borysov, S./Ross, H.-L./Adamy, J.: Operational Design Domain Qualification Framework for Remotely Driven Vehicles in Urban Environment. In: 2023 IEEE International Automated Vehicle Validation Conference (IAVVC), 2023, pp. 1–6

Hans, O./Walter, B.: ODD design for automated and remote driving systems: A path to remotely backed autonomy. In: 2024 IEEE the 9th International Conference on Intelligent Transportation Engineering (ICITE). 2024

Hans, O./Radlak, K./Adamy, J.: Human-Factor focused Application of STPA to Remotely driven Vehicles. In: 2024 IEEE World Forum on Public Safety Technology (WFPST), 2024, pp. 120–125

Hollnagel, E.: Barriers and Accident Prevention. Routledge, London 2016

Hollnagel, E.: Is safety a subject for science? In: Safety Science, Volume 67, August 2014, pp. 21–24

Hollnagel, E.: Safety-I and Safety-II. The past and future of safety management. CRC Press 2018

Hollnagel, E./Wears, R. L./Braithwaite, J.: From Safety-I to Safety-II. A white paper. The Resilient Health Care Net: Published simultaneously by the University of Southern Denmark, University of Florida, USA, and Macquarie University, Australia. 2015

ICAO – International Civil Aviation Organization: Annex 19 – Safety Management. 1st Edition. 2013b (or relevant edition depending on updates)

ICAO – International Civil Aviation Organization: Safety Management Manual (SMM). 3rd Edition. Montreal 2013a. *https://www.icao.int/SAM/Documents/2017-SSP-GUY/Doc%209859%20SMM%20Third%20edition%20en.pdf*

ISO 9001:2015: Quality management systems – Requirements. International Organization for Standardization, Geneva 2015

ISO 26262:2018: Road vehicles – Functional safety. Annex B. 2018

ISO 26262-2:2018: Road vehicles – Functional safety – Part 2: Management of functional safety. International Organization for Standardization, Geneva 2018

ISO 45001:2018: Occupational health and safety management systems – Requirements with guidance for use. International Organization for Standardization, Geneva 2018

ISO/IEC Guide 51:2014: Safety aspects – Guidelines for their inclusion in standards. International Organization for Standardization, Geneva 2014

Leveson, N. G.: Engineering a safer world. Systems thinking applied to safety. MIT Press, Boston, MA, USA 2012

Leveson, N.: Safety III: A Systems Approach to Safety and Resilience. 2022

Li, Y./Guldenmund, F. W.: Safety management systems. A broad overview of the literature In: Safety science 103 (2018), pp. 94–123

McCarthy, P.: The application of Safety II in commercial aviation – the operational learning review (OLR). In: *Harris, D./Li, W. C. (eds.):* Engineering Psychology and Cognitive Ergonomics. Cognition and Design. 17th International Conference, EPCE 2020, Held as Part of the 22nd HCI International Conference, HCII 2020, Copenhagen, Denmark, July 19–24, 2020, Proceedings, Part II 22. Springer International Publishing 2020, pp. 368–383

Mill, J. S.: Autobiography. Longmans, Green, Reader, and Dyer, London 1873

Patankar, M./Duncan, J.: Safety Culture. Image Wisely, July 2018. *https://www.imagewisely.org/Imaging-Modalities/General-Radiation-Safety/Safety-Culture*

Peñaloza, G. A./Wasilkiewicz, K./Saurin, T. A./Herrera, I. A./Formoso, C. T.: Safety-I and Safety-II. Opportunities for an integrated approach in the construction industry. In: REA Symposium on Resilience Engineering Embracing Resilience 2019

Pich, M. T./Loch, C. H./Meyer, A. de: On uncertainty, ambiguity, and complexity in project management. In: Manage Sci. 2002, 48 (8), pp. 1008–1023 (DOI: 10.1287/mnsc.48.8.1008.163)

Provan, D. J./Woods, D. D./Dekker, S. W. A./Rae, A. J: Safety II professionals: How resilience engineering can transform safety practice. In: Reliability Engineering & System Safety, 2020, 195, 106740

Reason, J./Hollnagel, E./Paries, J.: Revisiting the Swiss cheese model of accidents. In: Journal of Clinical Engineering, 27(4), 2006, pp. 110–115

Robbins, S. P./Judge, T. A.: Organizational Behavior. 15th Edition. Prentice Hall, Upper Saddle River, NJ, USA 2012

SAE International: Taxonomy and Definitions for Terms Related to Driving Automation Systems for On-Road Motor Vehicles. SAE Standard J3016_202104. Warrendale, PA, USA 2021

Schwindt-Drews, S./Frenzel, J./Hans, O./Abendroth, B./Adamy, J.: Competencies and technological support for remote operator of automated vehicles. A literature review and perspectives. In: AHFE 2024 International Conference Proceedings. 2024

Tulloch, S.: Safety III – More Prequel than Sequel. LinkedIn Pulse 2023. *https://www.linkedin.com/pulse/safety-iii-more-prequel-than-sequel-simon-tulloch* (accessed October 16, 2024)

U.S. Nuclear Regulatory Commission: Code of Federal Regulations. Title 10: Energy. Part 50 – Domestic Licensing of Production and Utilization Facilities. 2024. *https://www.ecfr.gov/current/title-10/chapter-I/part-50* (accessed August 26, 2024)

Waring, A.: Safety Management Systems. Chapman and Hall, London 1996

Weick, K. E.: Making Sense of the Organization. Blackwell Publishing, Oxford, UK 2001

4 Holistic Safety

Holistic Safety refers to a comprehensive approach to safety that considers all aspects of a system or organization. Instead of focusing solely on individual safety measures, it takes into account the entire environment, including physical, technical, and human factors. The goal is to create a harmonious and secure overall system.

The term can be applied in many other facets:

- **Holistic security:** This encompasses not only physical security but also information security, cybersecurity, and protection against internal threats such as unauthorized data access and manipulation. The aim is to identify and close all potential security gaps.
- **Sociological security concepts:** This perspective considers safety from a societal viewpoint. It involves the well-being and security of communities and individuals, including social justice, economic security, and psychological safety.

There are also other risks under consideration of cybersecurity which do not lead directly to harm of people but cause intolerable risks to society and individuals:

- **Privacy violations:** Unauthorized access to personal data stored in vehicle systems can lead to privacy breaches, exposing sensitive information about drivers and passengers. This can result in identity theft, financial loss, and a loss of trust in transportation systems.
- **Intellectual Property (IP) Theft:** Cyberattacks targeting the proprietary technology and software of automotive manufacturers can lead to the theft of intellectual property. This not only affects the competitiveness of companies but also poses risks to innovation and economic stability.
- **Data manipulation:** Hackers can manipulate data within vehicle systems, leading to misinformation about vehicle performance, maintenance needs, or navigation. This can cause confusion and potentially dangerous situations for drivers.

By adopting this holistic approach, all potential risks and threats are considered and addressed to create a safe and stable environment.

When considering safety as freedom from intolerable risks, the sociological aspect and cybersecurity are also part of safety. Security also involves the protection of intellectual property and privacy, as their breaches can lead to risks that society deems unacceptable. In the transportation sector, safety is often associated with accidents or intolerable disruptions or impairments affecting individuals or entire groups within society.

Examples in the transportation sector:

- **Physical safety:** Ensuring that vehicles are equipped with advanced safety features such as airbags, Anti-Lock Braking Systems (ABS), and Electronic Stability Control (ESC) to prevent accidents and protect passengers.
- **Technical safety:** Implementing robust cybersecurity measures to protect against hacking of vehicle systems, which could lead to unauthorized control or data breaches.
- **Human factors:** Training drivers and promoting safe driving behaviors to reduce the risk of accidents caused by human error. This includes campaigns against drunk driving, distracted driving, and encouraging the use of seat belts.
- **Sociological safety:** Addressing the safety needs of vulnerable road users such as pedestrians and cyclists by designing safer road infrastructure, including dedicated bike lanes and pedestrian crossings.

Considering all the risks in public road traffic that could impact individuals and society, a holistic safety approach should be adopted to address all kinds of risks.

4.1 Why Holistic Safety?

In modern motor vehicles and other means of transportation, the characteristics of functions and features dominate rather than characteristics of their mechanical design or the mechanical design of their propulsion system. Also, safety could be expressed by the presence of safety belts, crash zones, and other visible characteristics of a vehicle. However, features and functions are system properties that also manifest their safe and secure properties through the way they were created in the development process. Therefore, we need to think about a holistic safety approach that can demonstrably ensure the necessary safety aspects for a means of transport.

With the increasing proportion of software (SW-)based systems, the sources for the requirements are also increasing. SW not only has advantages in terms of flexibility but also disadvantages compared to a function that was implemented by a mechanical component. Especially in the field of automation, there are many activities where humans interact directly with mechanics, like a steering wheel or a brake pedal. Not only the Human-Machine Interface (HMI) but also the information and energy transfer to ensure the physical action is often replaced by electronics and software.

Software-based functions have several disadvantages when compared to mechanical functional solutions. For instance, software-based functions are more prone to security vulnerabilities and can be harder to debug and troubleshoot. They are also less reliable and can be more fragile than mechanical solutions. Additionally, software-based functions can be more complex and require more resources to operate than mechanical solutions.

On the other hand, mechanical functional solutions are often less flexible and can be harder to modify than software-based functions. They are also less scalable and can be more expensive than software-based solutions, especially when it comes to maintenance and upgrades. Mechanical solutions are also more deterministic and provide a more robust basis for energy transport (also physical energy transport), haptic feedback, real-time behavior, deterministic behavior, maintenance, immunity versus noise, and stress.

However, the choice between software-based and mechanical functional solutions ultimately depends on the specific use case and requirements. In some cases, a hybrid approach that combines both software and mechanical solutions may be the best option.

Energy transfer: Mechanical functional solutions are often more efficient than software-based functions when it comes to energy transfer. This is because mechanical solutions can directly convert energy from one form to another, whereas software-based functions require additional hardware to perform the same task. As a result, software-based functions can consume more energy and generate more heat than mechanical solutions.

Haptic feedback: Haptic feedback is an area where mechanical functional solutions excel. Mechanical solutions can provide more realistic and precise haptic feedback than software-based functions. This is because mechanical solutions can directly interact with the user's sense of touch, whereas software-based functions rely on sensors and other hardware to simulate haptic feedback. As a result, software-based functions can be less responsive and less accurate than mechanical solutions.

Real-time behavior: Real-time performance is another area where mechanical functional solutions have an advantage over software-based functions. Mechanical solutions can respond to input signals almost instantaneously, whereas software-based functions require additional processing time to interpret input signals and generate output signals. As a result, software-based functions can be less responsive and less accurate than mechanical solutions in real-time applications.

Deterministic behavior: Mechanical functional solutions are often more deterministic than software-based functions. This means that mechanical solutions can produce the same output for a given input every time, whereas software-based functions can produce different outputs for the same input due to factors such as system load, memory allocation, and other variables.

Maintenance: Mechanical functional solutions are often easier to maintain than software-based functions. This is because mechanical solutions are often simpler and require less specialized knowledge to repair and maintain. In contrast, software-based functions can be more complex and require specialized knowledge to maintain and upgrade.

Immunity versus noise: Mechanical functional solutions are often more immune to noise and other environmental factors than software-based functions. This is because mechanical solutions can be designed to filter out unwanted signals and interference, whereas software-based functions can be more susceptible to noise and other environmental factors.

Stress: Mechanical functional solutions are often more robust and can handle higher levels of stress than software-based functions. This is because mechanical solutions can be designed to withstand high levels of stress and strain, whereas software-based functions can be more fragile and prone to failure under high levels of stress.

Humans quickly adapt to the deterministic behavior of machines. Especially at the Human-Machine Interface (HMI), they perceive the situation with all their senses. They observe the control unit (e. g., display) and compare the machine's behavior, incorporating their tactile information into how they operate the machine. Temperature is felt and replaced by sensors in automation. Overheating or even fire can be detected by appropriate sensors.

It is therefore necessary to translate physical processes, which humans intuitively and systematically learn or even inherit, into machine actions that behave in a manner humans are accustomed to. The more mature the implementation, the stiffer and deterministic the coupling between HMI and actuator must be. The more experience a person has with a specific relationship between operation and system response, or the more frequently a person is accustomed to a particular action, the more they expect such fixed deterministic behavior between HMI and machine behavior.

The following conclusion can be drawn: If mechanical behavior can be replicated, then the human understanding of the process is also comprehensible, because they can imagine the interaction of the mechanical parts.

This fixed, stable, and rigid behavior is very difficult to replicate in systems of electronics and software due to ergonomic reasons and variances in human tactile perception. A mechanical solution typically offers a higher functional density compared to an electronic or software-based implementation of a function. However, an electronic implementation is always more flexible for adaptation and adjustment, and it has the ability to produce different functional variants under various external influences. This flexibility is especially valued in software-intensive systems, where the ability to reconfigure the function during runtime is a significant advantage.

In a microcontroller or even a microprocessor, you will never achieve the same functional density as in a custom ASIC (Application-Specific Integrated Circuit). However, you lose the flexibility and the ability to make changes during runtime. For example, a mechanical steering column is much less maintenance-intensive over its lifetime compared to a steer-by-wire solution. The steering column also transmits the desired steering forces and ensures deterministic haptics and the energy transfer from the driver's hand to the steering knuckles is deterministic.

Considering safety only for an electrical and electronic (EE) function without relating it to the real and physical world is very questionable. Sensors are used to convert certain physical effects into electronic signals so that they can be processed by a system. Actuators have the task of converting electronic information generated by a system into physical movements or actions. Therefore, it is questionable to consider EE or software errors in isolation without knowing the physical function.

In a dynamic system like a car, evaluating safety in terms of nominal performance must also be questioned. On the other hand, the threat of hacker attacks is different. A hacker cannot threaten a mechanical steering column with their means. However, sawing through the steering knuckles would be a massive act of sabotage, but there are other rules and best practices for this.

It is certainly sensible to evaluate a highly dynamic system from different perspectives and in terms of safety. However, analyzing and evaluating risks based on the potential extent of damage over an entire lifecycle using fundamental methods should be questioned.

4.2 What Is Holistic Safety?

According to the Australian Radiation Protection and Nuclear Safety Agency (ARPANSA), holistic safety is a best practice approach to safety management that focuses on managing the safety of technological, human, and organizational aspects. It aims to sustainably balance and optimize the health of people, animals, and ecosystems by recognizing that the health of humans, domestic and wild animals, plants, and the wider environment are closely linked and interdependent (ARPANSA 2024).

Holistic safety is a way of more thoroughly and comprehensively managing operational safety. Organizational safety performance is analyzed by examining technological, human, and organizational aspects including the complex interrelationships and interdependencies between these three aspects. It is about ensuring that the interaction between these three aspects is appropriately addressed, such as how people safely use technology, and how the organization provides training to people and ensures the technology is safe to use.[1)]

Holistic safety is important because it is a way of helping prevent incidents and accidents, as well as other benefits such as safer and more secure operations, reduced risks through more comprehensive understanding of operational risks, and capturing key aspects of safety that might otherwise be missed.

Holistic safety refers to an approach to managing safety that considers the interactions between different components of a system and how they contribute to safety outcomes. This approach emphasizes the importance of analyzing the system as a whole, rather than in isolated parts, to identify potential weaknesses and design more effective safety measures that address root causes rather than symptoms.

The cultural aspects approach to safety recognizes that safety is deeply influenced by the organizational culture, including shared values, beliefs, and behaviors. A strong safety culture is one where safety is prioritized at all levels, from leadership to front-line workers. This approach involves fostering a culture where safety is ingrained in the daily routines, decision-making processes, and communication practices of the organization. By promoting a positive safety culture, organizations can enhance employee engagement in safety practices and reduce the likelihood of accidents.

In a system architecture, design decisions and allocations, including safety requirements, must be verified to ensure that they fulfil the requirements. This involves consolidating, verifying, and validating the decisions at each level of the system.

Holistic safety and a safety culture are related concepts, but they are not the same thing. Holistic safety is a proactive approach to safety management that involves a comprehensive understanding of the system and its components, as well as the ability to anticipate and mitigate risks before they become problems. A safety culture, on the other hand, is a cultural norm that values safety and encourages employees to report near-misses and potential hazards.

While holistic safety is a specific approach to safety management, a safety culture is a broader concept that can be applied to any organization. A safety culture can be seen as an extension of holistic safety, as it provides the foundation for a proactive approach to safety management. In other words, holistic safety is a specific set of practices and techniques that are used to implement a safety culture.

The themes of safety management, safety culture, and holistic safety are interconnected in the sense that they all contribute to the overall safety of a system or organization. Safety management involves the implementation of policies, procedures, and systems to prevent and mitigate risks, while safety culture refers to the shared values, beliefs, and behaviors that promote a safety-focused mindset within an organization. Holistic safety, on the other hand, takes a comprehensive approach to safety by con-

---

[1] *https://www.arpansa.gov.au/regulation-and-licensing/safety-security-transport/holistic-safety/learn-about-holistic-safety*

sidering not only the technical aspects but also the organizational and cultural factors that can impact safety. By integrating these three themes, organizations can create a culture of safety that is proactive, preventive, and reactive, and that considers the interdependence of different components and factors to achieve a holistic and sustainable approach to safety management.

In order to bring the different safety areas such as usability, legal safety requirements (e. g., homologation, road traffic regulations), SOTIF (Safety of the Intended Functionality), functional safety, and cybersecurity into a holistic safety perspective, it is important to consider the interdependencies and interactions between these areas. For example, usability and legal safety requirements are closely related, as a product must meet both technical and legal requirements to be considered safe. Similarly, functional safety and cybersecurity are interconnected, as a system‘s technical safety can be compromised by cyber threats.

To address these risks, it is important to consider the potential hazards and risks associated with each safety area. For example, in the area of usability, risks may include user error or misuse, while in the area of cybersecurity, risks may include hacking or malware attacks. In the area of functional safety, risks may include system failures or malfunctions, while in the area of SOTIF, risks may include physical damage to the system or injury to users.

The term “operational safety” can be used as a general framework for addressing all behavioral risks in the public and individual transportation sectors, as well as the automation tasks involved in these sectors. However, it is important to note that operational safety is just one aspect of overall safety, and other factors such as environmental and social safety must also be considered.

Risks that can lead to vehicle misbehavior in public road traffic can be attributed to various causes. Here are some examples:

- **Safety in Use—driver misbehavior:** This can stem from driver inattention, distraction, or deliberate violations of regulations. For instance, a driver using their phone while driving is distracted and may not notice a pedestrian crossing the street.
- **Safety in Use—abuse or incorrect use of functions:** Drivers might misuse certain functions or use them in inappropriate situations. For example, using cruise control in heavy traffic can lead to accidents due to sudden stops and starts.
- **Cybersecurity—vehicle manipulation for performance:** Drivers can modify their cars to get more performance or use functions not intended for their type of driving. An example is tampering with the engine to increase speed, which can lead to loss of vehicle control.
- **Cybersecurity—sabotage and active manipulation:** This can involve various groups such as drivers, owners, illegal tuners, or hackers, as well as organizations

aiming to enable theft or terrorism. An example is hackers remotely controlling a vehicle's systems to cause a crash.

- **Safety of the Intended Functionality (SOTIF)—design flaws and inadequate measures:** Poor vehicle design, insufficient technical measures, or overly weak measures like inadequate sensors or undersized brake systems can lead to accidents without driver fault. For instance, weak brakes might not stop a car in time to avoid a collision.
- **Functional safety—technical failures:** These can occur in mechanical parts, electronics, and software, leading to potential vehicle misbehavior. An example is a software glitch that prevents a car's Anti-Lock Braking System from functioning correctly.
- **Operational safety (behavioral risks)—dangerous situations with multiple causes:** Situations leading to injuries can have many causes, often found in different safety domains and their risks. For instance, a poorly designed traffic intersection combined with driver distraction and vehicle malfunction can result in a serious accident.

Each of these points highlights the complexity and variety of factors that can contribute to vehicle-related risks in traffic systems.

Vehicle misbehavior in public road traffic can arise from numerous causes, each presenting unique challenges to safety. Driver misbehavior, such as inattention, distraction, or deliberate violations of traffic regulations, plays a significant role. For example, a driver using their phone while driving is distracted and may fail to notice a pedestrian crossing the street. Additionally, the misuse or incorrect use of vehicle functions, such as engaging cruise control in heavy traffic, can lead to accidents due to abrupt stops and starts.

Cybersecurity concerns also contribute to vehicle misbehavior. Drivers may manipulate their cars to gain more performance or use functions not intended for their type of driving, such as tampering with the engine to increase speed, which can result in loss of control. Moreover, vehicles can be targeted by various groups, including drivers, owners, illegal tuners, and hackers, or by organizations aiming to enable theft or terrorism. For instance, hackers might remotely control a vehicle's systems to cause a crash.

Design flaws and inadequate measures are additional risks. Poor vehicle design, insufficient technical measures, or overly weak components, such as inadequate sensors or undersized brake systems, can lead to accidents without any fault on the driver's part. For example, weak brakes might not stop a car in time to avoid collision.

Technical failures in mechanical parts, electronics, and software can also cause vehicle misbehavior. A software glitch that prevents a car's Anti-Lock Braking System from functioning correctly is an example. Dangerous situations that lead to injuries often have multiple causes, found in various safety domains and their associated risks. For instance, a poorly designed traffic intersection combined with driver distraction and vehicle malfunction can result in a serious accident.

These points underscore the complexity and variety of factors contributing to vehicle-related risks in traffic systems, highlighting the need for a comprehensive approach to identifying and mitigating these risks.

4.3 Functional Safety and Security

As vehicles evolve with increased connectivity, functional safety and cybersecurity must go hand-in-hand. Enter ISO 26262 and ISO 21434—the twin pillars of modern automotive standards! But what sets them apart, and where do they interact?

As automotive systems become more complex, ensuring safety and security is crucial part of engineering (Figure 4.1). Here's how ISO 26262 and ISO 21434 tackle these areas.

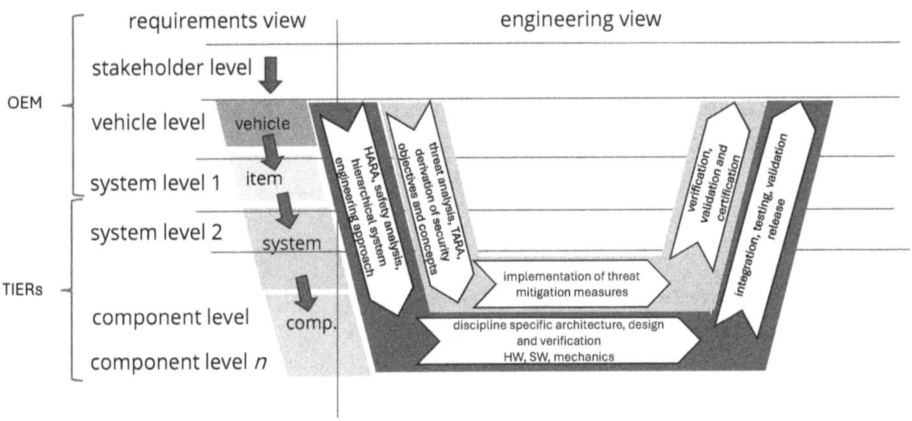

Figure 4.1 Safety and security in requirements and engineering's view

ISO 26262—Functional Safety (extended by ISO 21448; SOTIF, Safety of the Intended Functionality)

Objective: prevent hazards from hardware/software malfunctions that could cause accidents

Wider view including SOTIF: assure correct functioning of system function in motor vehicles

Risk approach: avoiding systematic failure, based on a structure system engineering process; uses Automotive Safety Integrity Levels (ASIL) to classify risks based on severity, exposure, and controllability; extended for SOTIF: risk due to functional insufficiencies or inadequacies, uncertainties and lack of nominal performance

Lifecycle coverage: encompasses the vehicle's entire safety lifecycle—from concept to decommissioning; lifecycle needs to be enhanced from item level to vehicle level within the dedicated operational context

Key practices: applies HARA (Hazard Analysis and Risk Assessment) and safety analysis such as Failure Mode Effects Analysis (FMEA), Fault Tree Analysis (FTA), and rigorous testing to ensure system reliability; for SOTIF the design assurance and validation process needs to be extended; interfacing to safety in use based on useability methods both standards claiming to cover misuse, misbehavior etc.

Management systems: Quality Management Systems are extended by Functional Safety Management Systems. Typical requirements are specified in ISO 26262:2018, Part 2. In current homologation regulations from UNECE further requirements for Safety Management Systems are defined.

ISO 21434—Automotive Cybersecurity

Objective: protect systems from cyberattacks that could compromise safety, privacy, or functionality

Risk approach: follows a Threat Analysis and Risk Assessment (TARA) process to identify potential threats and their impacts

Life cycle coverage: spans the cybersecurity lifecycle, covering secure design, implementation, and incident response

Key practices: uses vulnerability assessments, penetration testing, and planning for incident response to secure vehicle systems

Security-related Management Systems

Cybersecurity Management System (CSMS)

A Cybersecurity Management System (CSMS) is required by the UNECE R155 regulation. The primary goal of a CSMS is to ensure that automotive manufacturers have a systematic approach to managing cybersecurity risks throughout the lifecycle of a vehicle.

This includes the following:

- **risk assessment:** identifying potential cybersecurity threats and vulnerabilities
- **mitigation strategies:** implementing measures to reduce or eliminate identified risks
- **monitoring and detection:** continuously monitoring for new threats and detecting any cybersecurity incidents
- **response and recovery:** having plans in place to respond to and recover from cybersecurity incidents

The CSMS must be integrated into the product development process to ensure that cybersecurity is considered at every stage, from design to decommissioning.

Software Update Management System (SUMS)

A Software Update Management System (SUMS) is required by the UNECE R156 regulation. The main objective of a SUMS is to manage the software updates for vehicles in a secure and efficient manner. Of course, also safety aspects play a crucial role.

This includes the following:

- **update planning:** ensuring that software updates are planned and managed throughout the vehicle's lifecycle
- **security measures:** implementing security measures to protect the integrity and authenticity of software updates
- **deployment:** ensuring that updates are deployed in a controlled and secure manner,
- **monitoring and verification:** continuously monitoring the update process and verifying the success of updates

Like CSMS, SUMS must also be integrated into the product development process to ensure that software updates are managed effectively and securely.

Both CSMS and SUMS are crucial for maintaining the cybersecurity and software integrity of modern vehicles, ensuring they remain safe and secure throughout their operational life.

As automated and connected cars become the norm, functional safety and cybersecurity increasingly overlap. For instance, a cyber breach in the braking system affects both security and safety, requiring a coordinated approach to handle both risks effectively.

It is recommended to implement a common product development process that guides automotive innovators in achieving compliance with both ISO 26262, ISO 21448 (Safety of the Intended Functionality), and ISO 21434 (cybersecurity), ensuring systems are both safe and secure:

- **comprehensive risk management:** assess both safety and cybersecurity risks, ensuring compliance across the vehicle platform
- **tailored compliance and conformity solutions:** seamless integration of safety and security solutions for full protection
- **framework development:** align ASIL (Automotive Safety Integrity Level) and TARA (Threat Analysis and Risk Assessment) requirements for consistent protection, keeping everyone compliant and safeguarded

To assure this, a common maturity model is essential, along with the synchronization of engineering steps to ensure consistency, completeness, and correctness across all safety and security disciplines. A common safety and security management system covering all the risk would be highly recommended.

4.4 Challenges in Today's Development Process

In the past, vehicles were classified based on their purpose, so that even in M1 class vehicles, you could find convertibles, station wagons, limousines, SUVs or other. Due to the reduction in complexity and costs, the vehicle manufacturer starts to develop platforms or module kits. The legal regulations define requirements just on a high-level view and they are often assigned to certain functions, features or any specific system. The legal requirements and the various purposes of the vehicle alone already lead to a huge matrix of allocations that need to consolidate within one platform. Even if the OEMs try to produce configurable platform concepts, the different purposes of the vehicle and the different regulations significantly drive the architecture and design of the different vehicles (Figure 4.2).

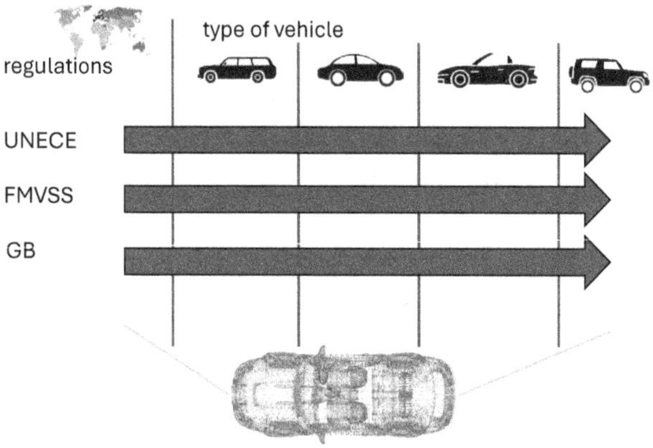

Figure 4.2
Vehicle platform

4.5 Holistic Risk Analysis

A holistic risk analysis must comprehensively consider all potential risks. This raises important questions: Who is the affected party, and who decides which risks are acceptable and which are not? Different stakeholders perceive and prioritize risks differently. For instance, an insurance company evaluates risks differently than a pedestrian in a busy city.

Particularly for vulnerable road users such as children, the elderly, or people with disabilities, the risks differ significantly compared to cyclists, scooter riders, or other motorists. These groups face unique challenges and hazards that must be carefully assessed and mitigated.

For insurance companies, it is crucial to determine which risks they should cover. Assessing the point at which environmental pollution, such as car emissions, poses a direct health threat to people in public spaces is particularly challenging. An insurance company is unlikely to offer coverage to a car manufacturer for health damage caused by car emissions due to the complexity and scale of such risks.

Under product liability laws, the manufacturer or distributor is responsible for any damage caused by their product. This means that if a product causes harm, the entity that placed the product on the market is liable for the damage. This legal framework ensures that manufacturers are accountable for the safety of their products and encourages them to minimize risks through rigorous testing and compliance with safety standards.

A holistic risk analysis must consider the diverse perspectives and priorities of all stakeholders involved. It should aim to protect the most vulnerable while ensuring that manufacturers and insurers can effectively manage and mitigate risks. This comprehensive approach helps create a safer environment for everyone.

For product safety and to assure for the manufacturer a common degree of product integrity, the European Commission introduced a "New Approach" in 2013. The motivation behind the "New Approach" includes the following:

- **innovation and technological flexibility:** allowing manufacturers to innovate and use the latest technologies while ensuring product safety
- **simplification of legislation:** reducing the complexity of regulations and making it easier for manufacturers to comply
- **enhanced market surveillance:** improving the efficiency of market surveillance to protect consumers and ensure fair competition
- **harmonization across the EU:** ensuring consistent safety standards across all EU member states, facilitating the free movement of goods

This "New Approach" was introduced to improve the safety and compliance of products within the EU. It is based on several key principles:

- **Essential requirements:** Legislation focuses on essential safety requirements that products must meet to be placed on the market. These requirements ensure the free movement of products within the EU.
- **Harmonized standards:** Technical specifications are developed by standardization organizations (e. g., CEN) and are not mandatory but provide a presumption of conformity with the essential requirements.
- **Flexibility for manufacturers:** Manufacturers can choose to comply with harmonized standards or demonstrate that their products meet the essential requirements through other means.
- **Market surveillance:** Authorities are responsible for ensuring that products on the market comply with the essential requirements and can take actions if products are found to be non-compliant.

The EU Commission has published a document entitled "EU General Risk Assessment Methodology (Action 5 of the Multiannual Action Plan for the Surveillance of Products in the EU (COM(2013)76)" (European Commission 2015). The document describes an EU-wide methodology for the risk assessment of products covered by harmonization legislation. It extends existing guidelines by taking into account risks beyond health and safety, such as economic damage or environmental damage, and reflects specific legal requirements for harmonized products. The methodology includes a step-by-step risk analysis process, from product definition to risk level determination, and provides recommendations for market surveillance authorities based on the identified risk. It serves as a guideline for deciding on necessary measures, such as recalls or sales bans, and increases the transparency of official decisions. The methodology is explicitly not intended for manufacturers, but it is an interesting approach for manufacturers to assess their risks in the same way as they are assessed by the authorities.

This procedure involves 8 steps (Figure 4.3):

- **Step 1:** In this step, the product is identified and the relevant legal requirements and other provisions, such as harmonized standards, are determined.
- **Step 2:** Identify hazards. This is where the dangers that can emanate from the product are determined. These can result from the physical, chemical, biological or ergonomic properties of the product.
- **Step 3:** Identify at-risk individuals. In this step, the people or other subjects (e. g., animals, environment) who could be harmed by the hazards of the product are identified.
- **Step 4:** Description of how the hazard may harm the subject, i. e., the relevant damage scenario. Here, a "damage scenario" is developed that describes how the danger could affect the person at risk.
- **Step 5:** Determine the potential harm. This step identifies what and who could be harmed by the event, or in some cases, what "loss" could occur as a result of the event.
- **Step 6:** Determine the severity of the damage. Here, the degree of harm that could be inflicted on the subject is evaluated.
- **Step 7:** Determine the probability of damage. In this step, the probability of the damage occurring is estimated.
- **Step 8:** Determine the level of risk by combining the severity of the damage and the likelihood of that damage occurring in the scenario described. In the final step, the severity of the damage and the likelihood of it occurring are combined to determine the overall risk level.

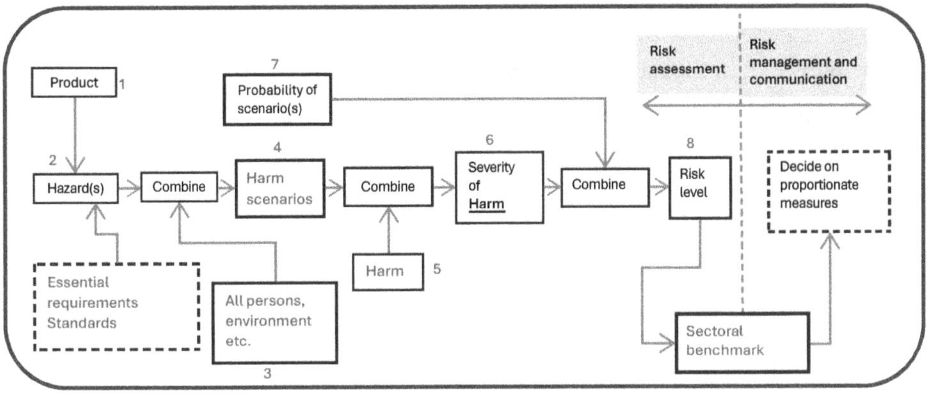

Figure 4.3 Risk methodology according to EU Commission (source: European Commission 2015)

This methodology helps market surveillance authorities assess the conformity of products and ensure that they comply with EU rules. What is interesting about this approach is that it is not based on static structures and risk classes, but on scenarios that meet the requirements that the identified causes do not lie in a specific component but in weaknesses in the dynamic architecture of the new and innovative products.

4.5.1 Risk and Safety

Risks can have different effects on different groups of people and be perceived differently. Therefore, one must also distinguish between danger and hazards. Danger refers to a situation or condition that poses an immediate and obvious risk of harm, injury, or even death. It is a clear and present threat that requires immediate action to avoid negative consequences. Hazard refers to a potential source of danger or harm. It is a condition or element that has the potential to cause harm but may not necessarily pose an immediate risk. Hazards often have latent characteristics, meaning their harmful effects are not immediately apparent but can manifest under certain conditions or over time. Risk is the likelihood or probability that a hazard will actually cause harm. It involves assessing both the presence of a hazard and the potential for it to become a danger. For example, a chemical substance might be a latent hazard because it is not dangerous unless it is inhaled or comes into contact with skin. The risk of harm from this chemical depends on factors such as exposure level and safety measures in place. If the chemical is properly stored and handled, the risk remains low. However, if safety protocols are ignored, the latent hazard can become an immediate danger, leading to harm.

The following definitions could be considered:

- **Danger** is an immediate and obvious threat.
- **Hazards** are a potential source of harm with latent risks.
- **Risk** is the likelihood of the hazard causing harm.
- **Harm** is the actual injury or damage that can result.
- **Latency** refers to the hidden or delayed nature of some hazards.

For example, the risk of people walking past the window with a flowerpot depends on several factors (Figure 4.4). For the owner of the flowerpot, there are also risks, including a minor risk of a broken flowerpot. However, a much greater risk is the owner's liability for any damage caused by the falling flowerpot. The extent of the damage depends on the height of the window, the weight of the flowerpot, and where exactly the flowerpot hits the person. If the flowerpot falls beside the person, only the flowerpot is damaged. One could then assess the probability of the flowerpot causing a fatal injury, a severe injury, or a minor injury to the person. This example clearly illustrates that both the likelihood of the negative event occurring and the actual extent of the damage are very difficult to assess.

Figure 4.4
Risk in the context of danger and hazards

The owner of the flowerpot, acting with foresight, will take measures to secure the flowerpot. For example, they might install a board in front of the windowsill to prevent the flowerpot from falling off (Figure 4.5). While it could be suggested that people wear helmets when walking past houses with flowerpots, such a measure is likely unreasonable.

This brings us to a critical point:

- What measures are reasonable and acceptable for the individuals who could be harmed, and what measures are reasonable and acceptable for the owner to secure their product?
- Where are the boundaries of the duty to secure a product?

Figure 4.5
Risk reduction measures

Reasonable measures for potential victims:

- **Practicality:** Measures should be practical and not overly burdensome for the individuals. For example, expecting pedestrians to wear helmets is impractical and unreasonable.
- **Awareness:** Providing clear warnings or signage can help inform people of potential risks without imposing unreasonable demands on them.

Reasonable measures for the owner:

- **Preventive actions:** The owner should take preventive actions that are feasible and effective, such as securing the flowerpot with a board or other safety devices.
- **Cost and effort:** The measures should be proportionate to the risk. Simple and cost-effective solutions that significantly reduce risk are generally considered reasonable.

Boundaries of product safety obligations:

- **Proportionality:** The safety measures should be proportional to the potential harm. Extreme measures that are costly or impractical may not be required if the risk is minimal.
- **Feasibility:** The measures should be technically and economically feasible for the owner to implement.
- **Legal standards:** Compliance with existing safety regulations and standards is essential. These regulations often define the minimum safety requirements that must be met.

The duty to secure a product involves balancing the practicality and reasonableness of safety measures for both the potential victims and the owner. The goal is to implement effective safety measures without imposing undue burdens on either party.

4.5.2 Holistic Safety Analysis

We could consider the following key characteristics of holistic safety analysis from a high-level perspective:

- **System behavior not as expected:** Holistic safety analysis considers scenarios where a system behaves unexpectedly or deviates from its intended behavior. This includes identifying abnormal responses, failures, or unpredictable interactions.
- **Inadequate system purpose:** A system may not align with its intended purpose. Holistic safety analysis examines whether the system's design, functionality, and features adequately address the requirements and goals set for it.
- **Performance mismatch:** Systems must meet performance demands. If a system's performance falls short of application requirements (e. g., response time, throughput, reliability), it poses safety risks. Holistic analysis assesses performance gaps.
- **Lack of transparency in use:** Users and those in the system's environment should understand how to interact with it safely. If the system's behavior is opaque or its operation is unclear, holistic safety analysis highlights this as a concern.
- **Potential for misuse:** Systems can be misused intentionally or unintentionally. Holistic safety analysis identifies vulnerabilities that could lead to misuse, such as improper handling, unintended actions, or unauthorized access.
- **Motivation for distraction:** Distractions can compromise safety. Holistic analysis considers scenarios where users are highly motivated to divert their attention from critical tasks, potentially endangering themselves or others.

- **Data security and manipulation:** Unauthorized access to data or manipulation of system information can have safety implications. Holistic safety analysis examines data protection mechanisms and potential vulnerabilities.
- **Context-dependent design:** The system's context matters. If the application context (environment, user demographics, usage scenarios) is inadequately considered during design, safety risks may arise.
- **Lifecycle alignment:** The system's lifecycle—from development to integration and operation—must align with safety requirements. Holistic analysis ensures that the system's life cycle adequately supports safety goals.
- **Acceptance and integration:** Acceptance into the operational environment is critical. Holistic safety analysis evaluates whether the system is ready for integration, considering safety, reliability, and compatibility.

Holistic safety analysis involves a comprehensive view, addressing not only technical aspects but also human factors, organizational processes, and environmental considerations. It aims to create safer systems by identifying and mitigating risks across the entire lifecycle.

4.5.3 Safety versus Risk

Safety generally refers to the prevention of accidents and harm to people. In the automotive industry, this includes measures to protect drivers, passengers, and pedestrians from injuries or fatalities, for example the following:

- crash tests to ensure vehicles can withstand collisions
- airbags and seatbelts to protect occupants during accidents
- Advanced Driver Assistance Systems (ADAS) like automatic emergency braking to prevent accidents

Security, on the other hand, focuses on protecting against unauthorized access and malicious activities. In the automotive context, this includes the following:

- cybersecurity measures to protect the vehicle‘s software from hacking
- anti-theft systems to prevent vehicle theft
- data protection to secure personal information stored in the vehicle‘s systems

Risk in the Automotive Industry

Technical risks are related to the vehicle's design, manufacturing, and operation.

Examples include the following:

- mechanical failures such as brake or engine malfunctions
- software bugs in the vehicle's control systems
- battery issues in electric vehicles, which could lead to fires or reduced performance

In contrast to technical risks, **commercial risks** involve financial and market-related uncertainties which could lead to monetary damages.

Examples include the following:

- market acceptance of new technologies, such as electric or autonomous vehicles
- supply chain disruptions affecting the availability of critical components
- regulatory changes that could impact production costs or marketability

Another aspect is **material losses** that refer to physical damage to the vehicle or its components, which can result from the following:

- accidents causing damage to the vehicle
- natural disasters like floods or earthquakes affect vehicle storage or transportation
- vandalism or theft leading to loss or damage of the vehicle

Here are some typical examples, how the terms are used in different contexts:

- **Safety example:** A car manufacturer conducts rigorous crash tests to ensure that their vehicles provide maximum protection to occupants in the event of a collision. This is a safety measure aimed at reducing the risk of injury or death.
- **Security example:** The same manufacturer implements robust cybersecurity protocols to protect the vehicle's software from being hacked, ensuring that the car's systems cannot be tampered with remotely.
- **Technical risk example:** During the development of a new electric vehicle, engineers identify a potential issue with battery overheating. They address this risk by redesigning the cooling system to prevent fires.
- **Commercial risk example:** The company faces uncertainty about how well the new electric vehicle will be received in the market, which could affect sales and profitability.
- **Material loss example:** A dealership's inventory is damaged by a severe hailstorm, leading to significant financial losses due to the need for repairs or replacements.

By understanding these distinctions and examples, automotive companies can better manage and mitigate various types of risks to ensure both the safety and security of their vehicles and operations. Especially the material losses are a factor of particular attention for the insurance companies.

4.5.3.1 Hazard or Danger in the Relation to Risk

Danger and hazard are two closely related but distinct concepts in the field of safety and risk assessment.

- **Danger:** A danger is a potential source of harm or undesirable events. It is often latent, meaning it exists regardless of whether it actually manifests. For example, a chemical can be a danger because it is toxic.
- **Hazard:** A hazard occurs when a danger is present in a specific situation or context and there is a real possibility of harm. The hazard depends on the duration and frequency of exposure to the danger. For example, a chemical becomes a hazard when someone works with it without taking the appropriate safety precautions.
- **Risk:** Risk is the relationship between danger and hazard. It describes the probability and extent of potential harm. The risk is influenced by the frequency of exposure and the duration of exposure to the danger in a particular situation or context. A high risk exists when the probability of harm is high and the potential consequences are severe.

In the automotive industry the terms could be defined as follows:

- **danger:** potential source of harm (latent)
- **hazard:** concrete possibility of harm in a specific situation (dependent on duration and frequency)
- **risk:** probability and extent of harm (dependent on context, situation, duration, and frequency)

Here are some areas from other industries or domains that show how important the definition of terms is. These terms are often used as follows:

- **Occupational safety:** It means identifying and mitigating hazards in the workplace. In the automotive industry it refers more to vehicle occupancies.
- **Environmental health:** This means assessing the risks associated with exposure to pollutants. In the automotive industry it is considered in a similar way and focuses on using regenerative energy resources.
- **Product safety:** It means evaluating the potential dangers of consumer products. In the automotive industry the end user product is the entire vehicle. Except for the aftermarket, all supplier products refer to B2B business between OEMs and tiers.

4.5.3.2 Safety Integrity versus Risk

The IEC 61508 is an international standard that sets out the requirements for ensuring the functional safety of electrical, electronic, and programmable electronic systems. Its primary purpose is to provide a framework for managing and reducing risks associated with these systems to an acceptable level, ensuring they perform their intended safety functions reliably. IEC 61508 was considered the basis for the development of ISO 26262.

Part 5 of IEC 61508 provides informative methods and guidelines for implementing the standard. It offers detailed explanations and practical approaches to achieving compliance with the safety requirements outlined in the standard. This part is particularly useful for understanding how to apply the principles of functional safety in real-world scenarios.

Annex A of Part 5, published in 1998, specifically addresses the relationship between safety integrity and risk. It is divided into seven subchapters, each detailing different aspects of this relationship: This annex provides information on the underlying concepts of risk and the relationship between risk and safety integrity in the context of safety-related electrical/electronic/programmable electronic systems (E/E/PES). The first section introduces the purpose of Annex A, which is to provide information on the basic concepts of risk and how it relates to safety integrity. It sets the stage for the following sections, which will elaborate on these concepts.

Necessary risk reduction is defined as the reduction in risk that must be achieved to meet the tolerable risk for a specific situation (Figure 4.6). The tolerable risk can be stated qualitatively or quantitatively. Determining tolerable risk involves defining what is considered reasonable concerning the frequency (or probability) of a hazardous event and its consequences.

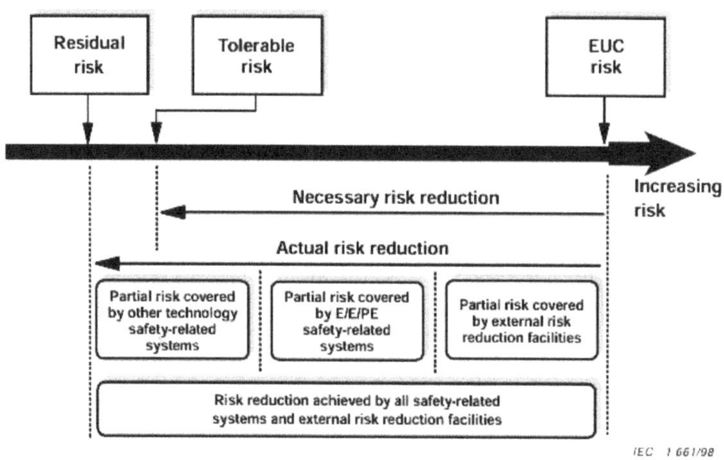

Figure 4.6 Risk reduction according to IEC 61508 (source: IEC 61508:2001, Figure A.1, "Risk Reduction: general concepts")

Several factors influence the tolerable risk level:

- severity of potential injury
- number of people exposed to the risk
- frequency and duration of exposure to the hazard
- perception and opinion of those exposed to the risk

To define the tolerable risk for a specific application, the following inputs are considered:

- guidelines issued by safety regulatory authorities
- discussions and agreements with the various parties involved in the application
- industry standards and guidelines
- international discussions and agreements (National and international standards play an increasingly important role in establishing appropriate safety criteria for specific applications.)
- relevant independent advice from advisory bodies representing industry, experts, and scientists
- legal requirements, both general and those specifically relevant to the application

The next section describes the role of E/E/PES safety-related systems. E/E/PES safety-related systems contribute to achieving the necessary risk reduction to ensure the tolerable risk level.

A safety-related system

- implements the required safety functions to achieve or maintain a safe state of the controlled equipment, and
- achieves, either on its own or in conjunction with other safety-related systems (E/E/PES or based on other technologies) or external risk reduction facilities, the required Safety Integrity Level (SIL) for implementing the safety functions.

It's important to note that a person could be an integral part of an E/E/PES safety-related system. For example, an operator might receive information about the controlled equipment's status from a display and perform a safety action based on that information.

The next section defines the idea of "safety integrity" (Figure 4.7). Safety integrity is defined as the probability that a safety-related system will satisfactorily perform the required safety functions under all the stated conditions within a stated period of time. It focuses on the performance of safety-related systems in carrying out their intended safety functions as outlined in the safety functions requirements specification.

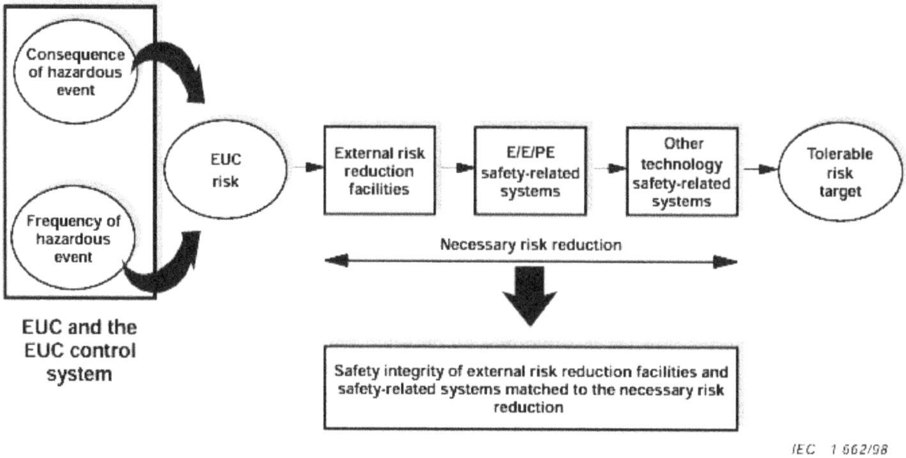

Figure 4.7 Risk and safety integrity according to IEC 61508 (source: IEC 61508:2001, Figure A.2, "Risk and safety integrity concepts")

Safety integrity comprises two key elements:

- **Hardware safety integrity:** This element relates to random hardware failures that lead to a dangerous failure mode. Achieving the specified hardware safety integrity level can be estimated with reasonable accuracy, allowing the requirements to be distributed among subsystems using probability combination rules. Redundant architectures might be necessary to achieve adequate hardware safety integrity.
- **Systematic safety integrity:** This element relates to systematic failures resulting in a dangerous failure mode. While estimating the average failure rate due to systematic failures is possible, the unpredictable distribution of failures from design faults and common cause failures increases uncertainty in failure probability calculations for specific situations. Therefore, selecting techniques that minimize this uncertainty becomes crucial. It's worth noting that measures taken to reduce random hardware failure probability don't necessarily have a corresponding effect on systematic failure probability. For example, techniques like redundant channels of identical hardware, effective for controlling random hardware failures, have limited impact on reducing systematic failures.

The next section explains "risk and safety integrity." Distinguishing between risk and safety integrity is essential. Risk is a measure of the probability and consequence of a specific hazardous event occurring. It can be assessed for different situations like Equipment Under Control (EUC) risk, risk required to meet the tolerable risk, and actual risk. Tolerable risk is determined at a societal level and considers political and social factors. Safety integrity applies exclusively to safety-related systems and measures the likelihood of these systems satisfactorily achieving the necessary risk reduction for the specified safety functions.

Once the tolerable risk level is defined and the necessary risk reduction is estimated, safety integrity requirements for the safety-related systems can be allocated. This allocation process is iterative to optimize the design to satisfy the various requirements.

Safety Integrity Levels and Software Safety Integrity Levels

To address the wide range of risk reduction requirements that safety-related systems must achieve, a system of Safety Integrity Levels (SILs) is established to meet the allocated safety functions' safety integrity requirements. IEC 61508 defines four SILs, with SIL 4 being the highest and SIL 1 the lowest. Target failure measures for the four SILs are specified in tables 2 and 3 of IEC 61508-1. These targets consider the mode of operation of the safety-related system (low demand mode or continuous/high demand mode).

Allocation of Safety Requirements

Figure A.3 (identical to Figure 6 in IEC 61508-1) illustrates the allocation of safety requirements, including safety functions and safety integrity requirements, to the E/E/PES safety-related systems, other technology safety-related systems, and external risk reduction facilities. The specific methods used to allocate the safety integrity requirements depend on whether the necessary risk reduction is specified numerically or qualitatively, leading to quantitative and qualitative approaches.

4.5.4 Safety Goal Approach

A typical safety goal approach as applied in ISO 26262 aims to ensure that the focus is on safety objectives. However, this approach also has significant drawbacks. The classification into ASIL (Automotive Safety Integrity Level) or SIL (Safety Integrity Level) provides a certain weighting, but it is unclear whether this weighting pertains to the process framework or to quality attributes such as safety integrity or safety capability. In other words, it is not transparent which safety level is achieved by a function that performs according to these attributes. The main disadvantage arises when safety goals apply only to a certain scope, as is the case with ISO 26262. According to the standard, only the malfunctioning behavior of E/E (electrical/electronic) systems or the resulting errors are considered.

Classical ISO 26262 approach

There is a safety goal approach according to ISO 26262, and there is a general risk analysis approach from other industries that tries to address all risks arising from the existence of a dynamic machine in public traffic. According to ISO 26262, there are two types of definitions on how a safety goal can be described:

- Most often, according to the standard, one assumes an EE system error, and the safety goal is to avoid the system error or reduce the risk that arises from the effect of the system error.
- The second approach focuses on the misbehavior of the vehicle, for example, a car, and formulates a safety goal as follows: Avoid a misbehavior (dysfunctional behavior) so that it does not lead to a dangerous traffic scenario.

This explanation provides a high-level understanding of the safety goal approach and risk analysis in the context of dynamic machines in public traffic, as per ISO 26262. It's important to note that these approaches are crucial in ensuring the safe operation of such machines, especially in complex and unpredictable environments like public traffic. The ultimate goal is to minimize the risk of system errors and misbehavior that could lead to dangerous traffic scenarios.

The safety goal approach of ISO 26262 leads to a singular consideration of risks because the various scenarios and thus the combinatorics from the many emerging safety goals lead to a very complex structure that cannot be considered in its entirety by the technical disciplines such as drive, chassis, and occupant protection. If you map the various safety goals to specific traffic scenarios, you can see that you can violate many different types of safety goals in one scenario.

This highlights the complexity and challenges in managing safety goals in the context of dynamic machines in public traffic. The multitude of scenarios and the combinatorial nature of safety goals can create a complex structure that may be difficult for individual technical disciplines to fully comprehend. Furthermore, when mapping safety goals to specific traffic scenarios, it becomes evident that a single scenario can potentially violate multiple, diverse safety goals. This underscores the need for a comprehensive and holistic approach to safety management in such contexts.

4.5.5 Vehicle as Equipment under Control

As a product, the car introduces certain risks to public road traffic due to its existence and the way it is used. These risks must be assessed to determine whether they are acceptable to society, for example as defined by relevant laws and further risk analysis methods. If the risks are not acceptable, appropriate risk-reducing measures must be implemented.

These measures can be electrical and electronic, or they can involve other technologies such as mechanics, hydraulics, or pneumatics. The reduction of risks through electrical and electronic systems falls within the scope of ISO 26262.

External risk measures must also consider the type of road construction and traffic signs (which warn of risks and help prevent accidents through traffic regulations). For example, if certain rules are not followed, even the internal vehicle measures may no longer be effective.

A classic example is that commercial vehicles have a lower maximum speed limit compared to passenger cars. If a commercial vehicle exceeds the allowed maximum speed, its braking system may not be designed for this higher speed and thus may not function effectively as a safety system.

Similarly, speed limits are set because the road is not structurally suitable for higher speeds. Therefore, the chassis and braking systems, which are expected to be used in vehicles homologated for traffic, may not function safely at excessive speeds. For instance, stable braking of the vehicle on poor roads or in rainy conditions may not be guaranteed with the braking system as expected by the valid type approval.

Primarily, in road traffic, a significant risk arises from moving vehicles. A vehicle that is properly parked and secured against rolling away, unauthorized access, and use is considered to be in a safe state or a Minimal Risk Condition (MRC).

While driving (i. e., in motion), there are traditional design measures that mitigate risks associated with vehicle dynamics, for example the following:

- **Lateral movement:** To ensure the driver can control the vehicle based on their training, the vehicle is typically equipped with steering according to the Ackermann principle (steering knuckle on the front axle).
- **Longitudinal movement:** The braking system is designed so that the vehicle can be brought to a stable stop within an acceptable braking distance, considering its engine power, maximum speed, and vehicle weight.

As in the machinery industry or in other safety standards derived from IEC 61508, the road vehicle could be considered as the Equipment Under Control (EUC, see Figure 4.8).

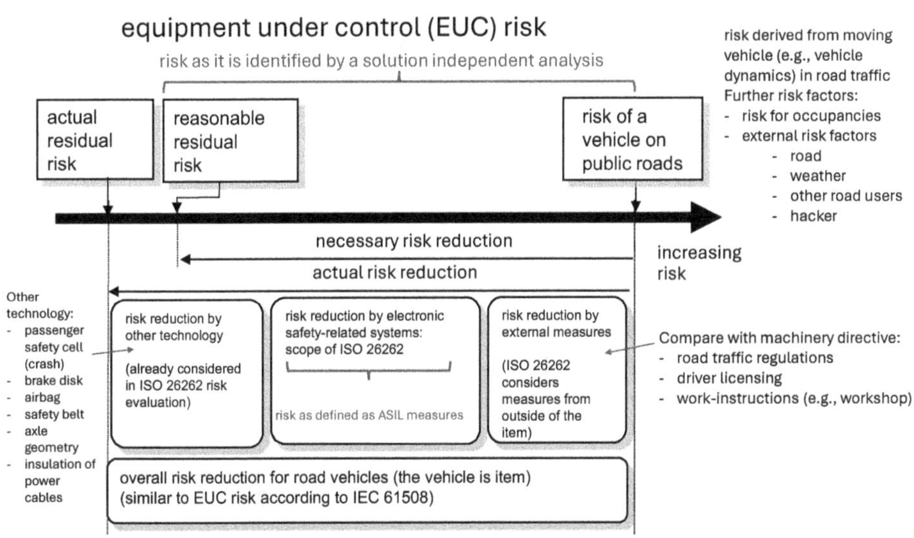

Figure 4.8 Consider the vehicle as Equipment Under Control according to IEC 61508

Besides the moving vehicle, further risk factors should be considered:

- **Risk for occupants:** This includes the risk of injury or death to the driver and passengers in the event of a collision or sudden stop.
- **External risk factors:** These are risks posed by elements outside the vehicle, such as other vehicles, pedestrians, and animals.
- **Road:** The condition and design of the road can significantly impact safety. Poorly maintained roads, sharp curves, and inadequate signage can increase the risk of accidents.
- **Weather:** Adverse weather conditions like rain, snow, fog, and ice can reduce visibility and road traction, leading to higher accident rates.
- **Other road users:** The behavior of other drivers, cyclists, and pedestrians can introduce risks, for example, distracted or impaired drivers, jaywalking pedestrians, and cyclists not following traffic rules.
- **Hacker:** With the rise of connected and autonomous vehicles, cybersecurity threats have become a significant risk. Hackers can potentially take control of vehicle systems, leading to dangerous situations.

The following risk-reducing measures or systems are considered typically in vehicle technology:

- **Passenger safety cell (crash):** This structure is designed to protect occupants by maintaining the integrity of the passenger compartment during a collision.
- **Brake disk or drum brake:** High-quality brake disks are essential for effective braking performance, ensuring the vehicle can stop safely and quickly.
- **Airbag:** Airbags provide crucial protection by inflating rapidly during a collision to cushion and protect occupants from impact.
- **Safety belt:** Safety belts are fundamental for occupant safety, restraining passengers and reducing the risk of injury during sudden stops or collisions.
- **Axle geometry and chassis design:** Proper axle geometry ensures stable and predictable vehicle handling, which is vital for maintaining control, especially during sharp turns or evasive maneuvers.
- **Touch protection and high voltage:** Fuse design and insulating power cables is important to prevent electrical faults, reduce the risk of heat and fire, and ensure the safe operation of the vehicle's electrical systems.

4.5.5.1 Automotive Safety Integrity Level in Road Context

The goal of an automation system is to control and specifically manage the automated process. This applies whether one aims to control the overall behavior of a vehicle in public road traffic or to assist certain functions that assist humans, particularly the driver of a car. The core task is to keep the vehicle stable on the road in accordance with the applicable traffic regulations.

Stable driving can be viewed from two perspectives (Figure 4.9):

1. from the viewpoint of a neutral observer
2. from the viewpoint of the driver

Stable behavior from a neutral observer's perspective is essential so that other road users can adequately assess the behavior of the respective vehicle, allowing them to adjust their own behavior accordingly and assess emerging risks.

Stable behavior for the driver is important so that they can trust the vehicle's behavior, make predictions about how to handle upcoming traffic situations, and mitigate risks and dangers arising from their own driving behavior and that of other road users. The dominant inherent expectation is that all road users adhere to the respective traffic regulations.

However, weather and road conditions are also potentially influencing factors that must affect the driver's appropriate behavior. In bad weather or poor road conditions, the driver must adjust their driving behavior, either by driving more slowly or by avoiding objects on the road.

If we imagine the road as a smooth surface and the risks arising from vehicle or driver misbehavior as holes of varying width or depth, we can set performance requirements for stable behavior on the road. When above the road surface, there are obstacles and risks caused by driving too fast or steering too aggressively, which can make the vehicle unstable or even cause it to veer into oncoming traffic. If a driver steers too intensely to avoid crashing against an obstacle, there is a danger of entering oncoming traffic. Driving too fast can lead to missing obstacles and crash or the vehicle being unable to handle the driving task at high speeds. The driver could also be overwhelmed and unable to adapt their driving behavior to the situation and risks, resulting in loss of control. These are risks associated with excessive performance.

On the other hand, there are risks associated with insufficient performance. This means the driver brakes too late, with too little force, or the vehicle does not respond adequately to the driver's braking request when encountering obstacles, ruts, puddles, or potholes.

There is also a third category of errors where, despite timely and appropriately intense braking or steering, a stabilizing intervention by the Electronic Stability Control (ESC) does not stabilize the vehicle but instead erroneously steers it into oncoming traffic.

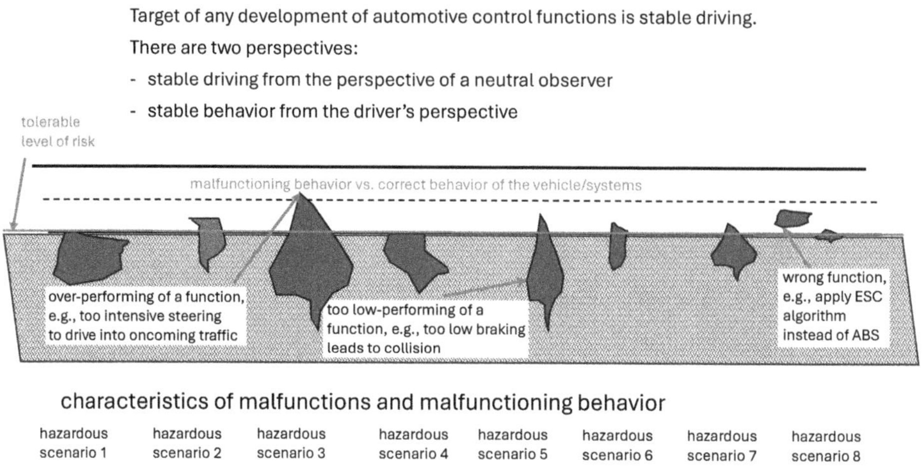

Figure 4.9 Hazardous scenarios versus stable driving

In Figure 4.10, excessive performance is depicted as obstacles above the road surface, while risks and dangerous scenarios due to insufficient performance are depicted as holes below the road surface. Incorrect functions or inappropriate behavior for the situation are shown above the road. This allows us to distinguish whether we want to consider appropriate correct behavior, misbehavior or malfunctioning behavior in the different performance categories.

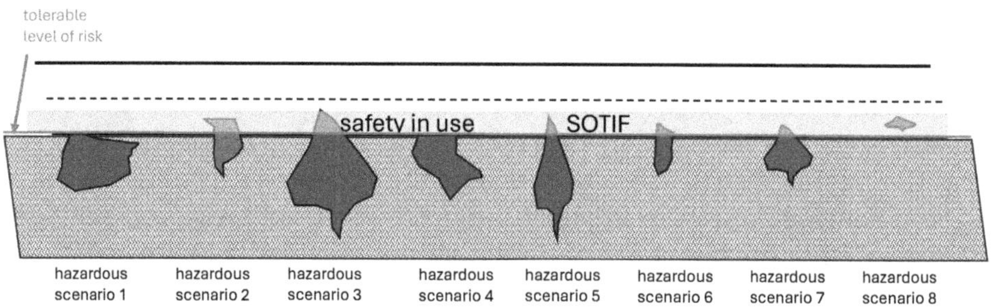

Figure 4.10 Safety categories

Assumption above the road surface: All over-performance characteristics need to be controlled by driver or system functions.

In case of nominal performance issues, the intended function needs to be analyzed by **SOTIF** or **Safety In Use** (SIU) specific analysis.

The 3-circle behavior model of ISO/PAS 21448 describes three risk categories (Figure 4.11). In the operational domain, there is a desired or legally required behavior. There is a risk of misbehavior by other road users, the driver, or technical misbehavior, meaning inadequate or insufficient performance of the respective vehicle system.

At the system level, the expected technical behavior of the vehicle systems should be correctly specified. Fully specifying appropriate behavior for every situation, for all vehicle functions in all driving situations, is a challenging task. Therefore, the risk and likelihood of systematic errors in the specification are very high.

The third risk category is the implemented behavior. Even if the desired or required behavior is correctly specified, there is a significant risk that the implemented behavior does not correspond to the desired or required behavior or does not match the specified behavior.

For safety relevant functions the three circles require a full overlapping. Any incomplete overlap of the three circles represents a residual risk that needs to be identified. If these risks are not identified and mitigated with appropriate measures, the combination of many vehicles and various situations can lead to significant probabilities of accidents.

As a consequence, safety regulations and standards impose high requirements on verification and, in the case of potential behavioral anomalies, require intensive validation.

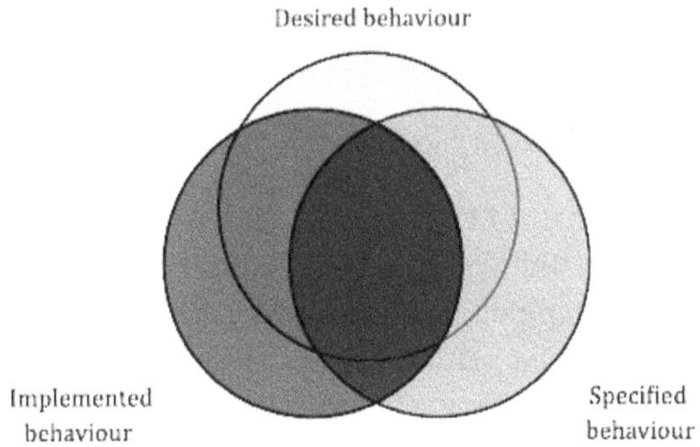

NOTE 1 The significant lack of overlap among the three circles is done for illustrational purposes only and does not imply the real situation.

Figure 4.11 Three circle behavior model (source: ISO/PAS 21448:2020, Figure A.17, "Three circles behavioural model")

Assumption below the road surface: The intended function of the system under consideration and the influence of the actuator is the main input for the HARA.

Known failure conditions from previous applications and a FMEA or HazOp provide failure conditions from the intended architecture and design.

The target is to identify the necessary risk reduction, to define safety goals, and to carry out ASIL assessment, which is the domain of ISO 26262.

The ASIL (Automotive Safety Integrity Level) indicates the degree of necessary risk reduction (Figure 4.12). For a high ASIL (e. g., ASIL D), a higher level of risk reduction is required. The measure of risk reduction is depicted here as the depth of the hole relative to the road surface. This means the deeper the hole below the road, the deeper the risk-reducing measure must be. Consequently, a shallower hole corresponds to a lower ASIL and thus requires a lesser risk-reducing measure.

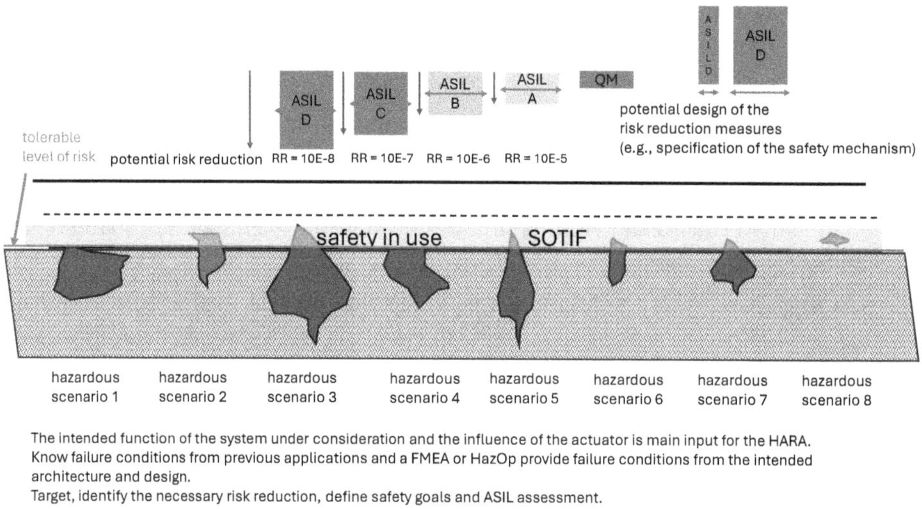

Figure 4.12 ASIL classification

The width of the hole in Figure 4.13 does not address the frequency or probability of the risk occurring. Instead, the width represents the dimension or characteristics of the risk, essentially describing or specifying the risk object. Some risks require only minimal measures and can be sufficiently reduced with simple means. Other risks need extensive measures, requiring a comprehensive specification of the risk and a correspondingly broad implementation of risk-reducing measures.

It is often said that engineering costs increase with the ASIL. However, it is possible to build a very streamlined safety case for a very lean ASIL D function. While measures to manage multiple faults require more analysis, verification, and testing effort than an ASIL A or QM risk, the effort increases with the width of the risk depicted in Figure 4.13. It is important to ensure that the risk is specified very efficiently and clearly so that the risk reduction process can be implemented in a streamlined manner. If the risk description is already very complex, the risk-reducing measures will be disproportionately higher than the risk description itself. Therefore, it is highly recommended that risks be specified very clearly and unambiguously.

Targeted risk analyses and lean safety concepts not only ensure more streamlined engineering processes but also lower costs. Additionally, these targeted risk analyses and lean safety concepts significantly contribute to reducing specification errors and thus systematic errors in the intended function and the risk-reducing measures.

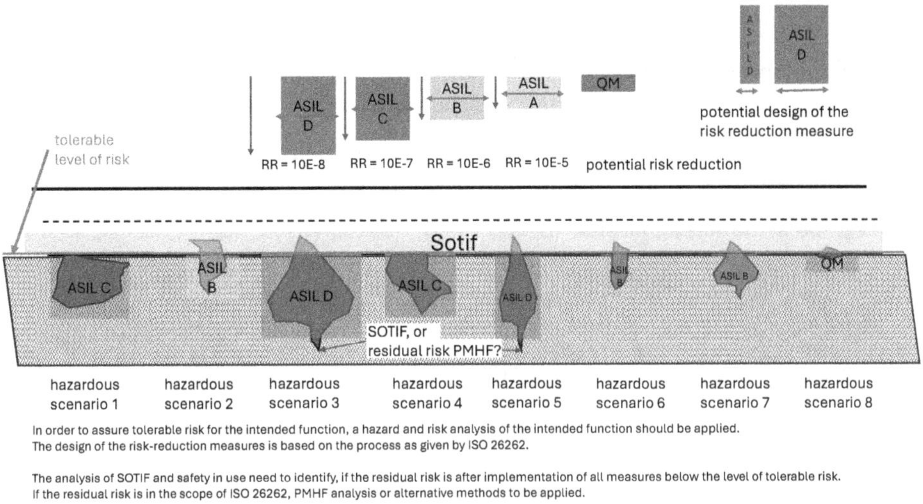

Figure 4.13 ASIL assignment

4.5.5.2 Risk-Based Approach

Unless a safety case is under consideration where a single basic functionality is implemented in a single control unit, the basic functionality and the respective vehicle reaction to an actuator is assigned, one can choose the classic ISO 26262 approach. It is considered that the basic functionality is allocated to the electronic control unit which could be mapped to the item according to ISO 26262. However, if you use by-wire functions, e. g., brake or steering, the loss or significant latencies on the by-wire route are already dangerous risks. Similarly, a hacker can cause such a malfunction so that dangerous behavior of the vehicle can occur in various driving situations.

This highlights the potential risks associated with by-wire functions in vehicles, such as braking or steering. The loss or significant latency in these functions can pose serious risks. Additionally, the potential for hacking introduces another layer of risk, as it could lead to malfunctions and dangerous behavior in various driving situations. Therefore, while the classic ISO 26262 approach can be used for safety-proofing basic functionalities, additional considerations and measures may be necessary when dealing with by-wire functions and potential cybersecurity threats.

Various safety risks addressed from standards, such as ISO 26262, ISO 21448 (SOTIF), ISO 21434 (cybersecurity) and Safety In Use, usability safety, as well as compliance with regulations from UNECE or FMVSS, GB (China) etc., lead to further safety requirements and risks to violate correct behavior demands and potential dysfunctional behavior.

Safety goals for by-wire functions can be formulated as follows:

- **Brake by-wire-system:** Avoid a loss of the vehicle's braking ability. Positively formulated, ensure the braking ability of the vehicle.
- **Steer-by-wire-steering system**: Avoid a loss of the vehicle's steerability. Positively formulated, ensure the steerability of the vehicle.

This can be formulated in a shorter way:

- Assure correct longitudinal control of the vehicle.
- Assure correct lateral control of the vehicle.

This leads to various challenges already within the interfaces:

- Different situations mean different means of correctness.
- Different system states of the vehicle mean different means of correctness.
- Different driving environment means different means of correctness etc.

Further resulting requirements derive from the need of independence of lateral and longitudinal vehicle control.

4.5.5.3 Responsible Vehicle Operator

The fundamental requirement for a responsible vehicle operator comes from the Vienna Convention on Road Traffic. Until 2016, this convention defined the driver or a guide for carriages as the responsible vehicle operator. In 2016, this definition was expanded to include a system. It is essential to always have a responsible vehicle operator whose duty it is to drive the vehicle in accordance with the applicable traffic regulations, traffic and weather conditions, and road conditions. Based on the principles of the Vienna Convention, only a natural person can hold this responsibility. Therefore, in the German regulation (AFGBV) on automated driving (as a supplement to the German Road Traffic Act, StVG), it is stipulated that a natural person must supervise automated driving. The primary safety objective in road traffic must be that the vehicle is always under the control of the responsible vehicle operator. If this operator is a system, the system must always be capable of driving the vehicle in accordance with the given road traffic regulations.

The primary safety objective is as follows: Ensure that the system maintains control over vehicle operation in accordance with the applicable traffic regulations. Negatively formulated as typically for ISO 26262, the primary safety goal is: Avoid loss of vehicle control.

Additionally, since the correct parking of the vehicle and the proper condition of the vehicle are legally regulated, the entire journey of vehicle operation must be subordinated to the primary safety objective. This means that the journey to the destination and the proper completion of the driving mission with the correctly parked vehicle must be subordinated to the safety objective. This also includes the handover to the driver after an automated drive or the transfer from a human driver to a system operation.

Since the Vienna Convention on Road Traffic forms the basis for all legal regulations concerning road traffic, it is essential that technical systems are clearly assigned to a responsible vehicle operator. This can be the traditional view of the driver in the car, the responsible guide of an animal drawn carriage, a remote operator in a control station, or since 2016 a technical system.

From SAE Level 2 onwards, the industry faces a challenge because, above Level 2, assistance systems formally end, and higher automation systems begin. A particular application here is Level 2 Handsfree, where the driver can no longer command vehicle reactions directly through their hands on the steering wheel but must account for latencies similar to remote driving.

The issue of driver takeover in the event of a steering system failure is particularly critical because the driver may not necessarily expect a failure in the steering system. It is fundamentally difficult to argue that if the driver does not have their hands on the wheel, they must be able to control system errors in a timely manner due to their training. Current steering systems are often designed so that the driver in the car serves as a fallback for certain steering system failures. However, it is essential that there is always a clear vehicle operator, whether it be a human or a system.

4.5.6 Risk Factors in Automotive Applications

In the context of automotive mobility solutions in public road traffic, several risk factors play a crucial role in safety and effectiveness. Let's explore these factors and their potential impact.

Speed

Risk factor: inadequate speed

Impact: Inadequate speed reduces reaction time, escalates collision forces, and leads to more severe accidents.

Occupant Safety

Risk factors:

- inadequate design of vehicle safety equipment
- non-use or improper use of vehicle safety equipment such as restraint systems, seat belts etc.

Impacts:

- In case of crashes the occupancies need to be protected, so that possible damages to vehicle occupants is as low as technical possible. It is necessary to consider various crash and accident scenarios.
- Lack of seat belt usage increases the risk of severe injuries or fatalities during accidents.

Inadequate Driver Condition

Risk factors:

- The driver is not in an adequate condition to drive the vehicle. This can be caused by alcohol, medication, or drugs, but fatigue, stress, irritation or emotions can also lead to driver misconduct.
- The driver must be trained to drive vehicles on public roads. A lack of training or insufficient driving experience can lead to misbehavior in traffic.
- The driver should adapt his driving behavior to the traffic environment, operating and weather conditions. Even if the driver is not sufficiently familiar with the traffic environment, he should drive defensively, and anticipatory driving is recommended.

Impact: Impaired judgment, coordination, and reaction time contribute to accidents with severe consequences.

Infrastructure Challenges

Risk factors:

- poor road conditions
- inadequate signage
- lack of proper markings
- weather conditions
- defective traffic lights etc.

Impacts:

- Infrastructure issues contribute to accidents and congestion.
- Driver disturbances, stress, aggression and irritation result in driving misconduct.
- misbehavior of other road traffic participants

Legal and Regulatory Compliance

Risk factors:

- non-compliance with traffic laws and regulations
- intransparent regulatory situational context (which rules are valid, what rules to follow, how to correctly follow rules, etc.)
- mixed traffic, coexistence of different road traffic participants

Impacts:

- Violations can lead to accidents and legal consequences.
- Inadequate behavior leads to accidents and misbehavior of other road traffic participants.

Adequate Safety Equipment Like clothes (e. g., Quads, Motorcyclists, Cyclists)

Risk factors:

- lack of helmet use
- inadequate clothes ór shoes

Impacts:

- loss of control
- increased physical damages
- head injuries can be fatal or cause long-term disabilities etc.
- bad view due to weather conditions

Inadequate Design of Means of Transportation

Risk factors:

- inadequate vehicle control or protection mechanisms (e. g., design of vehicle control systems)
- bad view or ergonomics for drivers or operators
- bad illumination (headlights, rear lights etc.)
- inadequate vehicle conditions (e. g., bad tires)
- bad signaling systems (horn, turn-indicator, hazard lights etc.)

Impacts

- accidents
- danger to other road users
- obstruction and irritation of other road users, which can lead to misconduct

Inadequate Use, Unauthorized Access, Improper Use, Misuse, or Distractive Use

Risk factors:

- unauthorized access to data (hacking)
- inadequate, adverse or distractive use of control functions
- vehicle control or protection mechanisms (e. g., design of vehicle control systems)
- bad view or ergonomics for drivers or operators
- bad illumination (headlights, rear lights etc.)
- inadequate vehicle conditions (e. g., bad tires)
- bad signaling systems (horn, turn-indicator, hazard lights etc.)

Impacts:

- accidents
- danger to other road users
- obstruction and irritation of other road users, which can lead to misconduct

Further risk does not lead to accidents, but, for example, to damages of the environment or late consequences for people's health.

Environmental Risks

Energy efficiency and emissions

Risk factors:

- environmental impact of vehicle emissions
- people health

Impact:

- Balancing mobility needs with environmental concerns is critical.

Lifecycle considerations

Risk factors:

- ignoring sustainability and recycling aspects
- use of toxic materials
- inadequate disposal of waste material (e. g., oil, grease)

The list of risk factors and possible consequences form very complex, cascading and even circular causal chains. Identifying all these risks and implementing appropriate and effective risk reduction measures requires very detailed and transparent analysis. The measures must be coordinated between the development partner and the traffic planner etc. so that a comprehensive safety argument can be made.

4.5.7 Risk Factors Related to Automated Driving

The system risk for automated driving solutions can be described on a lower level of abstraction. Here's a list of risk factors associated with automated driving solutions in public road traffic, along with explanations of their potential resulting damages:

- dysfunctional behavior of the Automated Driving System (ADS) or Advanced Driver Assistance Systems (ADAS)

  Explanation: Malfunctions or failures in system elements (e. g., sensors, actuators) or the control system assigned to ADAS can lead to accidents, collisions, or loss of control since the driver, or the fallback system cannot react adequately and timely to the event of certain system failure.

- system and Software inconsistencies or inadequacies

  Explanation: Inconsistencies between software versions or incompatibility of the software to the system, updates, or configurations may cause unexpected behavior, affecting safety and reliability.

Planning Issues

Explanation:

- Inadequate planning is the root cause of systematic failure that leads to severe events during the operation of assistant or automated systems.
- From application point of view, incorrect route planning, poor decision-making algorithms, or inadequate trajectory planning can result in unsafe maneuvers or collisions.

System Perception Issues

Explanation: Errors in object detection, recognition, or tracking by sensors (e. g., lidar, cameras) or inadequate perception algorithms can lead to misjudgments, collisions, or pedestrian accidents.

Environmental Conditions

Explanation: Adverse weather (rain, snow, fog) or challenging lighting conditions can affect sensor performance, reducing safety.

Unexpected Behaviour of Other Road Users

Explanation: Automated systems must anticipate and respond to unpredictable actions by pedestrians, cyclists, or other drivers. Failure to do so can lead to accidents.

Control Variance

Explanation:

- Instable system behavior or non determined, instable real-time behavior lead to unpredictable effects to the driver and to other road traffic participants. Even a lack of nominal performance in a short time span in a lateral control maneuver could lead to accidents.
- Sudden handover of control from automated to manual mode (or vice versa) can confuse drivers, leading to delayed reactions or unsafe situations.

Human-Machine Interaction (HMI)

Explanation: Poor communication, bad ergonomic coupling or low transparency between the system and the driver (e.g., unclear alerts, confusing displays) can impact safety and user trust.

Data Security and Privacy

Explanation: Vulnerabilities in data transmission, storage, or access can lead to unauthorized control, data manipulation, or privacy breaches.

Legal and Ethical Challenges

Explanation: Compliance with traffic laws, liability issues, and ethical dilemmas (e.g., trolley problem) pose risks that require careful consideration.

Infrastructure Compatibility

Explanation: Inadequate infrastructure (e.g., poorly marked lanes, missing signs) can confuse automated systems, affecting safety.

Human Overreliance on Automation

Explanation: If drivers become overly reliant on automation, they may disengage from monitoring, leading to delayed reactions during critical events.

Cybersecurity Threats

Explanation: Hacking, malware, or cyberattacks can compromise system integrity, leading to accidents or unauthorized control.

Sensor Degradation and Maintenance

Explanation: Sensor calibration, cleanliness, and maintenance are crucial. Sensor degradation can impact accuracy and safety.

Transition Zones (e. g., Construction Zones)

Explanation: Complex or dynamic environments (e. g., work zones) challenge automated systems, risking collisions or unsafe maneuvers.

Emergency Situations

Explanation: Handling emergencies (e. g., sudden obstacles, medical events) requires robust algorithms and quick responses to prevent severe accidents.

Human Factors (Driver Fatigue, Distraction)

Explanation: If the driver is not alert or engaged, automated systems must handle safety-critical situations effectively.

Social Acceptance and Trust

Explanation: Lack of public trust or acceptance may hinder adoption and cooperation with automated vehicles.

Addressing these risk factors through rigorous testing, safety protocols, and continuous improvement is essential but unfortunately not sufficient. A transparent safety analysis as requirement by the homologation, e. g., WP2.9 VMAD (Validation Methods for Automated Driving), are necessary so that all possible risk scenarios could be identified and adequate measures for all the risk factors are addressed and sufficiently reduced on an acceptable level of tolerance.

Chapter 3 shows how to integrate an overall risk management process.

4.5.8 Approach of Tolerable Risk

The concept of tolerable risk can be effectively explained using the Farmer Curve. This curve, developed by F. R. Farmer in 1967, provides a visual representation of risk by plotting the probability of an event against its potential consequences. By doing so, it helps to illustrate how different levels of risk can be assessed and managed. The Farmer Curve is particularly useful in the nuclear industry, where it is used to evaluate the risk of reactor accidents. It allows for a more nuanced understanding of risk by considering both the likelihood of an event and the severity of its impact. This approach helps to determine what level of risk is acceptable or tolerable, balancing safety and practicality. Figure 4.14 shows an F-N diagram, as it had been used by Farmer as a graphical representation of the level of societal risk generated by technology, an activity or a project. The vertical scale (F) shows the frequency of the events represented, and the horizontal scale represents the consequences (typically the number of deaths, N).

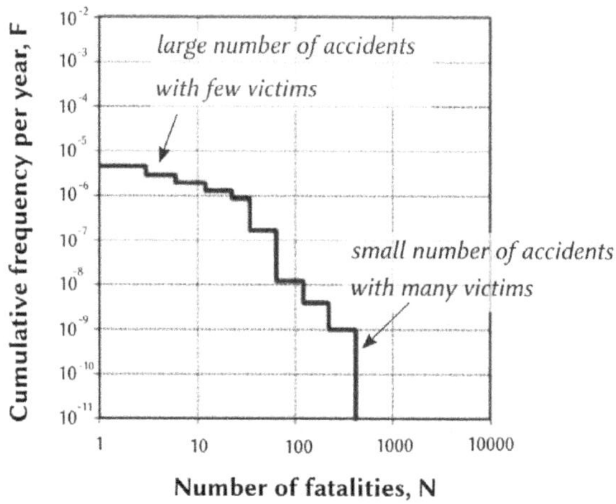

Figure 4.14
Example of a F-N or Farmer Curve (source: Marsden 2022)

Applications

There are three main applications for a Farmer diagram:

- representing the historical record of accidents associated with an activity
- representing the results of a Quantitative Risk Analysis (QRA)
- specifying criteria for assessing the tolerability of risk

The approach of tolerable risk combines the ideas of the three applications and defines a corridor to identify measures to reduce the risk on a tolerable level of risk.

F-N diagrams were initially used to estimate the social tolerability of new hazardous activities. Society must determine what level of risk from a system or technology is acceptable, considering its benefits. One method is to compare the risk of the new system with similar existing systems. If the new system's risk is similar to or lower than existing systems, it is likely acceptable.

The WASH-1400 "Rasmussen Report" (1975) was the first full-scope probabilistic risk assessment (PRA) for nuclear reactors. It compared nuclear risks with other industrial risks using F-N diagrams. The report was controversial, with critics arguing that its executive summary was more political than technical and that it underestimated uncertainties in accident probabilities. The F-N diagrams in the report compared the risks of 100 nuclear reactors with other man-made and natural events, showing that non-nuclear events were much more likely to cause large fatalities and financial losses. This was the first use of F-N diagrams to discuss social tolerability of different risks. When using F-N curves, it's important to consider the scale of activity being compared.

An interesting approach is extending the Farmer Curve by the principles of "ALARP" . ALARP stands for "As Low As Reasonably Practicable." It is a principle used in risk management to reduce risks to a level that is as low as reasonably possible, considering the costs and benefits of further risk reduction. The ALARP region is typically represented in risk matrices or F-N diagrams, where risks are categorized into three regions:

- **unacceptable region:** risks that are intolerable and must be reduced regardless of cost
- **ALARP region:** risks that are tolerable only if further risk reduction is impractical or the cost is grossly disproportionate to the benefit gained
- **broadly acceptable region:** risks that are so low that no further action is necessary

Channel Tunnel Safety Case

The example of the Channel Tunnel Safety Case applies the ALARP principle to ensure the safety of the Channel Tunnel, which connects the UK and France. This involves assessing various risks associated with the tunnel's operation and implementing measures to reduce these risks to ALARP levels. The safety case includes detailed risk assessments, emergency response plans, and continuous monitoring to maintain safety standards.

Relation to Farmer Curve

The Farmer Curve is a type of F-N diagram used to illustrate the relationship between the frequency (F) of an event and the number of fatalities (N) it causes (Figure 4.15). It helps in visualizing societal risk and determining acceptable risk levels. The ALARP principle is often applied using the Farmer Curve to identify which risks fall into the ALARP region and require further mitigation. The ALARP principle and the Farmer Curve are tools used in risk management to ensure that risks are reduced to acceptable levels, balancing safety and practicality.

The Farmer Curve, introduced by F. R. Farmer in 1967, is a graphical representation used in the nuclear industry to assess the risk of nuclear reactor accidents. Farmer proposed this curve to address the limitations of deterministic safety assessments, which were based on design basis accidents. His approach aimed to quantify risk by considering a spectrum of events with associated probabilities and consequences.

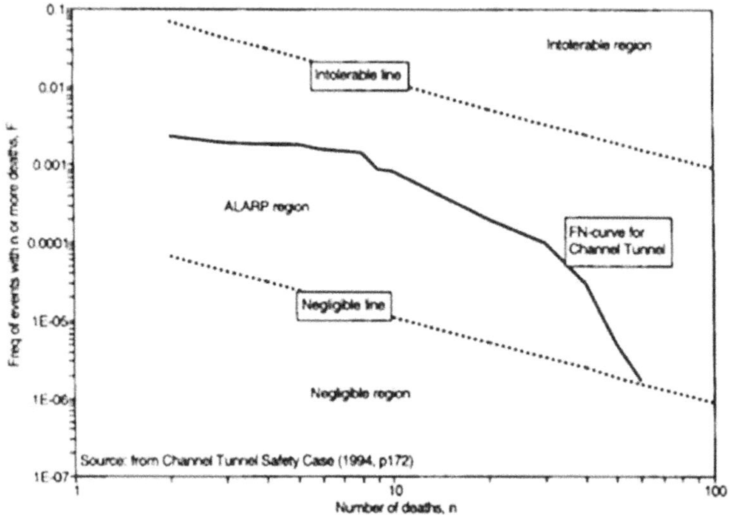

Figure 4.15 ALARP region or Channel Tunnel Safety Case (source: Rasmussen Report 1975)

Understanding Risk within the Context of Risk Reduction

Risk can be understood through several key factors:

- **severity of damage:** the potential impact or harm that an event can cause (Figure 4.16)
- **frequency:** how often an event is expected to occur
- **probability of occurrence:** the likelihood that a specific event will happen
- **duration of exposure:** the length of time individuals or systems are exposed to the risk scenario

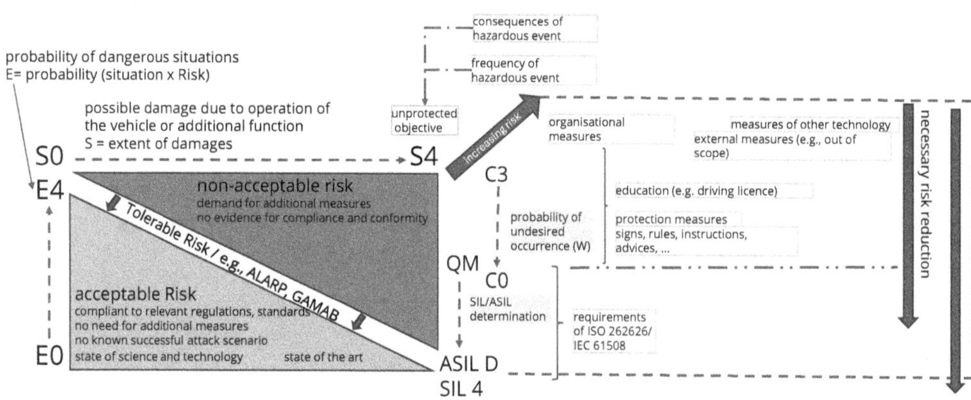

Figure 4.16 Approach of risk reduction

Hazard vs. Risk

Hazard: a potential source of harm or adverse effect, for example a chemical substance that is toxic

Risk: the likelihood that the hazard will actually cause harm, considering both the severity of the potential harm and the probability of occurrence

Safety and Stability

Safety is a state of being protected from harm or danger. It implies a certain level of stability, which can vary significantly among different individuals and groups. What one person considers safe might not be perceived the same way by another.

Risk Perception

Risk perception is subjective and varies among different groups of people. It is influenced by factors such as personal experiences, cultural background, and individual values. Therefore, the concept of tolerable risk is not universal and must be tailored to the specific context and stakeholders involved.

Examples of the Farmer Curve

The Farmer Curve is often explained using examples from the nuclear industry, such as the following:

- **Core Damage Frequency (CDF):** the probability of an event leading to significant damage to the reactor core
- **Large Early Release Frequency (LERF):** the probability of a large release of radioactive material early in an accident

These examples help illustrate how the Farmer Curve can be used to balance the potential consequences of an event against its likelihood, thereby helping to determine what level of risk is tolerable.

4.6 Development Process for Holistic Safety

The holistic safety approach for the road vehicle solutions shall integrate Functional Safety (ISO 26262), Safety of the Intended Functionality (SOTIF, ISO 21448), cybersecurity (ISO/SAE 21434) and the related legal requirements. Safety is paramount in the automotive industry, especially as we transition toward autonomous vehicles. A comprehensive safety approach is essential to address functional safety, Safety of the Intended Functionality (SOTIF), cybersecurity, and legal compliance.

In an automotive holistic safety approach, the following key elements shall be combined.

ISO 26262: Functional Safety

Purpose: ISO 26262 provides guidelines for ensuring functional safety throughout the product lifecycle.

Scope: It covers automotive electrical and electronic systems, including driver assistance and propulsion.

Objectives:

- address and mitigate hazards caused by malfunctioning systems
- promote a risk-based approach (ASILs) for evaluating risk levels
- lay down requirements for validation and verification processes

Safety of the Intended Functionality (SOTIF)—ISO 21448

Purpose: SOTIF bridges the gap left by ISO 26262, specifically addressing challenges posed by autonomous vehicles.

Focus:

- It ensures safety even when systems perform as designed (without system failure). It covers
  - the insufficiencies of the specification of the intended functionality at the vehicle level or
  - the insufficiencies of the specification or performance insufficiencies in the implementation of electric and/or electronic (E/E) elements in the system.
- reasonably foreseeable misuse

Measures: provides guidance on design, verification, and validation to achieve SOTIF

Security Lifecycle—ISO/SAE 21434

Purpose: ISO/SAE 21434 focuses on cybersecurity throughout the vehicle's lifecycle.

Objectives:

- identify security risks related to vehicle functions
- develop security requirements
- implement security measures
- verify and validate security aspects

Legal Safety Requirements

Sources: Legal requirements vary by region but may include homologation standards, road traffic acts, and other regulations.

Considerations:

- Compliance with legal safety requirements is mandatory.
- You need to address safety aspects specific to each jurisdiction.

A holistic safety approach integrates ISO 26262, SOTIF, ISO/SAE 21434, and legal requirements. By combining these elements, such a framework could ensure safety, security, and compliance throughout the entire automotive product lifecycle.

4.6.1 Process Maturity and Avoidance of Systematic Failure

One key principle of all safety and risk management standards is the "avoidance of systematicfailure." Quality standards and derived safety standards considered a life-cycle approach to assure a systematic product development process. As a result of ISO/IEC/IEEE 12207:2017, "Systems and software engineering – Software life cycle processes", not only ISO/IEC 15288, "Systems and software engineering – System life cycle processes", has established, but also various assessment stands came into consideration, such as "Capability Maturity Model Integration" (CMMI) or ISO/IEC 15504-5:2012 "Information technology – Process assessment". Automotive Spice has developed out of the roots of ISO 15504 too.

Basic Principles and Similarities in Automotive Spice and ISO 26262

While ISO 26262 focuses on functional safety and ASPICE emphasizes process maturity, they share underlying principles crucial for successful automotive systems engineering. These commonalities, often rooted in ASPICE's base practices, form a solid foundation for achieving both safety and quality.

Requirements Origin and Elicitation

Both standards stress the importance of understanding the genesis of requirements. It is important to bring both standards in the right context of the dedicated application. Especially the legal regulations are basic stakeholder requirements, which define and restrict the solution space for automotive mobility applications.

ISO 26262, particularly in its concept phase, mandates defining safety goals, which are essentially high-level requirements. For the intended functionality (see also Safety of the Intended Functionality, SOTIF) there are many constraints and additional requirements that derive from the operation environment. From the basic principles for road traffic, the Conventions for road traffic (e.g., Genova and Vienna) demand that the driver, the guide or in our days a system has always control over the correct vehicle behavior. Due to the requirements derived from the vehicle control requirements, Safety In Use considerations define several high-level safety requirements to assure that the driver could control the vehicle based on the education during driving school. This aligns with ASPICE's emphasis on eliciting stakeholder needs (REQ.1) and ensuring that all requirements, including safety-related ones, are rooted in a clear understanding of user needs and system context. Knowing the why behind each requirement is fundamental.

Scope Definition and Interface Coordination

Defining clear boundaries and managing interfaces is critical for both standards. ISO 26262 defines as part of the "Item Definition" that the item under consideration establishes the system's scope for safety analysis. In addition to that, ISO 26262 requires that the failure conditions and the effecting anomalies or misbehavior of the vehicle or malfunctions of the system functions shall be analyzed. This mirrors ASPICE's focus on understanding the project scope and coordinating internal and external interfaces. Effective communication and collaboration with project partners, whether for software development or safety analysis, are essential for both.

Requirements Development and Allocation

Both standards emphasize the structured development and allocation of requirements which provides the interface between requirements and architecture and supports the control of the interfaces. From ISO 26262's point of view, it is a central measure to avoid systematic failure during development. From an organizational point of view the "Development Interface Agreement" (DIA) addresses the roles and the responsibilities of development partners during the project and between development partners. From a technical point of view, the boundary analysis identifies the safety aspects of the system under consideration. ISO 26262 requires deriving Functional Safety Requirements (FSRs) and Technical Safety Requirements (TSRs) and allocating them to system elements. This mirrors ASPICE's practice of developing system and software requirements (REQ.2) and allocating them to appropriate elements. This ensures that every part of the system is designed to meet specific needs, including safety needs.

Requirements Decomposition and Traceability

A core principle is the structured decomposition of requirements into manageable levels. ISO 26262 stresses traceability from safety goals down to software units and hardware parts, or even deeper considering silicon devices, like embedded microcontroller or ASICS (Application-Specific Integrated Circuit) or FPGA (Field-Programmable Gate Array) etc. This aligns perfectly with ASPICE's emphasis on bi-directional traceability (REQ.4), ensuring that high-level requirements can be traced down to software units and vice versa. This detailed traceability is vital for both verification and validation activities.

Design Decision Documentation

Documenting design decisions is crucial for both standards. ISO 26262 requires transparent architectural and design decisions and the analysis of their impact to new possible hazards and consequences for derived requirements. Verification and validation and their aspects for integration and their technical feasibility shall be also considered during safety analysis. Especially for the intended functionality (SOTIF) the design decisions are crucial for the validation of the system and the vehicle design (e. g., vehicle dynamics). ASPICE also emphasizes documenting design decisions at each level of abstraction (e. g., software architecture design in SW.1), creating a clear audit trail and rationale for design choices. This ensures maintainability and understandability and facilitates impact analysis.

Traceability for Integration and Testing

Traceability also provides the bridge between requirement development and the integration activities. ISO 26262 uses traceability to ensure that safety requirements are verified and validated. ASPICE also emphasizes creating traces for integration to develop suitable test cases (e. g., SW.4). This traceability enables the creation of targeted tests that cover all requirements, including safety-related ones.

Maturity Level Reviews and Consistency

Both standards emphasize the importance of reviewing the maturity of development artifacts. ISO 26262 requires reviews of safety-related work products. ASPICE also highlights the need to ensure that development partners (internal and external) maintain consistency and quality in their deliverables. This shared principle ensures a consistent level of quality and safety across the entire development lifecycle.

Baseline Establishment

Establishing baselines is crucial for managing change. Both standards recognize the need for baselines to provide stable points of reference. ISO 26262, for example, uses baselines for safety analyses. ASPICE also emphasizes establishing baselines for configurations (MAN.3), providing a stable foundation for further development and change management.

Change tracking and impact analysis: Managing change effectively is a shared principle. ISO 26262 requires analyzing the impact of changes on safety. Similarly, ASPICE emphasizes tracking every change and analyzing its impact before implementing it. This ensures that changes, whether related to functionality or safety, are carefully evaluated before being implemented.

By understanding and applying these shared principles, automotive engineers can effectively leverage both ISO 26262 and ASPICE to develop safe, high-quality automotive systems. The underlying commonalities provide a strong foundation for a holistic approach to systems engineering.

4.6.2 Hierarchical Level of Abstraction

There are many possible solutions how to structure such a lifecycle approach. In order to follow the typical lifecycle approaches from ISO 26262 for functional safety and ISO/SAE 21434, it is recommended following a top-down approach, which assures traceable requirements for the entire realization process for road worthy motor vehicles. In Chapter 7 on systems engineering, an alternative approach, or better said an alternative view, will be provided based on the process pyramid from the industrial process industry, which could provide some advantages but lead to further organizational measures. Such a pyramid approach is not compliant by nature to the system engineering stands like ISO 15228 or to Automotive Spice etc. This approach follows the box-in-box approach as proposed by SAE J3049 but also extends to an operational level, in order to consider the breakdown of requirements derived from the driving environment.

This standard considers the motor vehicle as a framework for the ground vehicle system (Figure 4.17). In order to assure the systematic development of requirements that are assigned to the vehicle under consideration, a higher level of abstraction needs to be defined that regards the vehicle in the operational context.

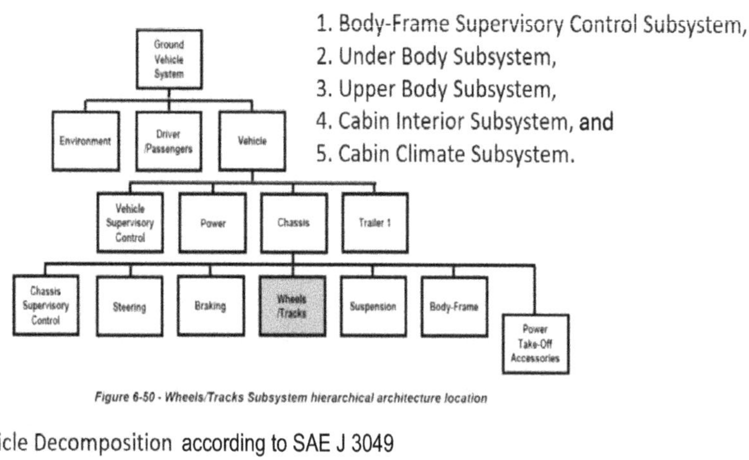

Figure 4.17 Vehicle system decomposition according to SAE J 3049 (source: SAE J 3049:2015, Figure 6-50, "Wheels/Tracks Subsystem hierarchical architecture location")

Considering the vehicle as a system for a physical view, you see different elements of a vehicle as a box-in-box approach. Figure 4.18 is not complete. Control units, harness, illumination, displays seats are not shown, but all these details could have an impact on safety or security.

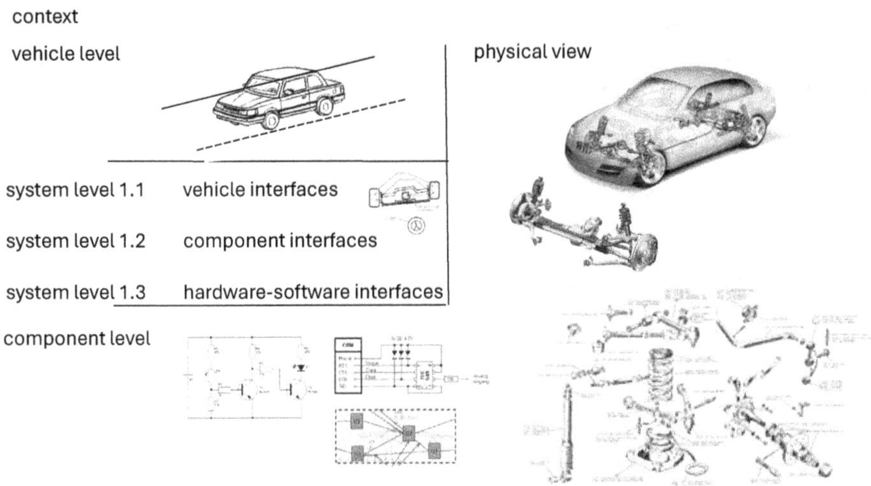

Figure 4.18 Physical view of the level of abstraction

4.6.2.1 Operational Level

All development processes start with a product idea and an elicitation of relevant stakeholder requirements. In order to develop the relevant use cases and an identification of the various roles which are involved in the product definition and the realization, these stakeholder requirements are elaborated in an operational level.

The so-called ego vehicle or the vehicle under development is considered as a black box within the different operational domains (Figure 4.19). The operational domain is the operation environment for which the vehicle is considered to be operated. For the holistic safety approach, we consider a standard passenger car as per M1 class vehicle according to REGULATION (EU) 2018/858.

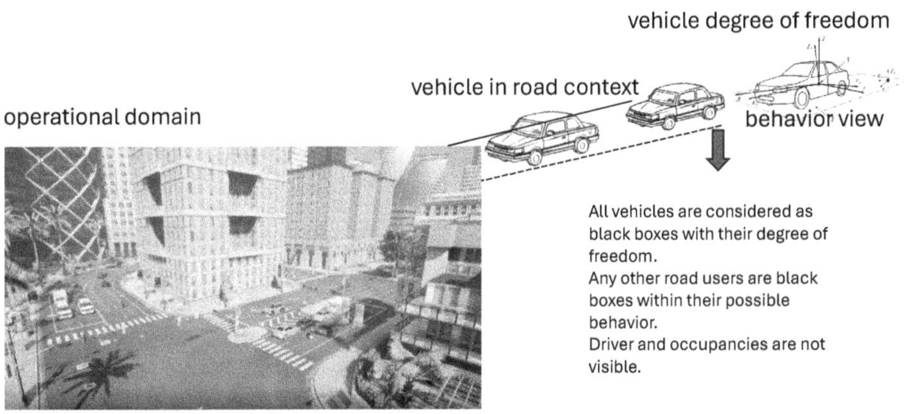

Figure 4.19 Vehicle as a black box within the operational context

In the operational level, you can see all potential other road traffic participants including other road vehicles. This level is necessary in order to identify and develop all the stakeholder requirements. Beside the requirements view, you already need

- the geometrical view,
- the behavioral, functional, logical view, and
- the technical physical view.

These views are necessary in order to work out

- use cases,
- operational scenarios, as well as
- architecture and design constraints.

A major input for all these parameters is a use case description. A typical example for a use case description could be as follows:

- The vehicle under consideration is an M1 class car, designed with a focus on family use and comfort. It is equipped with five seats, including the driver's seat. The three backseats are fully equipped to accommodate adults and children, complete with the legally required child seat safety features.
- The car is designed to be versatile and adaptable to various needs. It includes a roof rack that can be fitted with a roof box, providing additional storage space for long trips or large families. Additionally, the vehicle is equipped with a tow bar that can be used to mount a bicycle rack capable of carrying up to three bicycles. This feature enhances the vehicle's utility for outdoor activities and family outings.
- Accessibility and convenience are key features of this vehicle. The trunk is easily accessible via a tailgate and is designed to have no loading edge, making it easier to load and unload items. The car's design is ergonomic, ensuring a comfortable and enjoyable ride for all passengers.
- The vehicle's driving behavior is designed to be comfortable even at high speeds on German motorways. This is achieved through a combination of advanced engineering and design principles. The vehicle is equipped with a Highway Pilot up to SAE Level 3, providing semi-autonomous driving capabilities and enhancing safety and comfort on long journeys.

Base on such a use case description the following analysis and evaluations need to be done:

- state of the art or state from science and technology
- typical industry requirements or domain requirements
- regulation, legislation, also judicial practices and penalty regulations
- human capability, education, competences and training demands

- well-proven design solutions
- typical market solution for comparable solutions
- typical use case and operational scenarios
- typical risk scenarios and accident statistics
- ergonomic studies, usability and handling principles

From all these, source safety and security requirements could be evaluated which lead to mayor constraints for the architecture, design, and feasible safety and security solutions.

4.6.2.2 Risk on Operational Level

The Geneva Convention on Road Traffic (1949) and the Vienna Convention on Road Traffic (1968) established clear rules for the interaction of road users in public traffic. These conventions outline the fundamental principles for road design, the purpose of roads, as well as signage and markings. The **Geneva Convention on Road Traffic** (1949) aimed to promote the development and safety of international road traffic by establishing uniform rules among the contracting parties. It addressed the minimum mechanical and safety equipment required for vehicles and defined identification marks to indicate the origin of the vehicle. The **Vienna Convention on Road Traffic** (1968) further facilitated international road traffic and increased road safety by standardizing traffic rules among the contracting parties. It included provisions for road design, the purpose of roads, and detailed descriptions of road signs and markings. These conventions have been crucial in ensuring a consistent and safe approach to road traffic across different countries.

The Vienna and Geneva Conventions provide the foundation for all global traffic laws. These conventions have been incorporated into national or regional traffic laws in various countries, including Germany's StVO (Straßenverkehrsordnung), as well as in non-European countries like China and the USA. They also set the standards for vehicle characteristics necessary for safe operation in public traffic and form the basis for homologation laws. Additionally, product liability and safety regulations in the automotive sector determine how vehicles should be designed and what performance features they must meet. These legal frameworks constitute the four pillars of order in road traffic systems (Figure 4.20). Unfortunately, not all road users or those responsible for road infrastructure adhere to these laws. This can be due to various reasons, such as lack of awareness, inattentiveness, or deliberate disregard of the rules.

To minimize accidents to an acceptable level for everyone, the following rule was included in the Geneva Convention (Geneva Conventions 1949):

Every driver, pedestrian, or other road user shall conduct themselves in such a way as not to endanger or obstruct traffic; they shall avoid all behavior that might cause damage to persons, or public or private property.

Traffic areas are generally assigned to specific road users. If everyone behaves within their designated traffic area, it is possible to move safely in traffic. The expectation is that everyone will adhere to these rules.

Vienna Convention:
requires continuous control of the moving vehicle by
- driver (inside vehicle)
- guide (animal drawn)
- system (since 2016)

Convention on road traffic (Geneva, 19 September 1949)
Article 7
"Every driver, pedestrian or other road user shall conduct himself in such a way as not to endanger or obstruct traffic; he shall avoid all behaviour that might cause damage to persons, or public or private property."

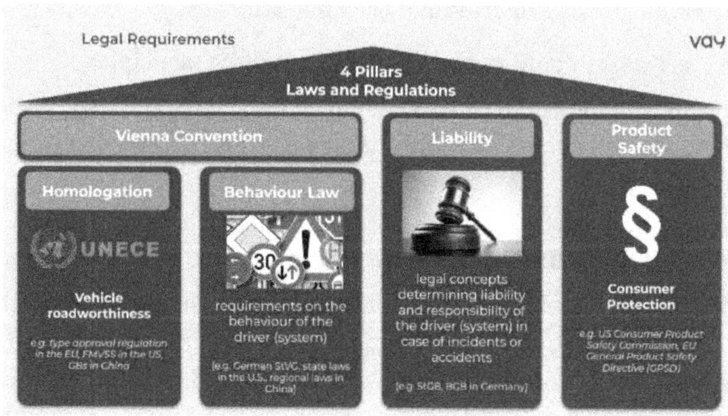

Figure 4.20 Law and regulations in the context of road traffic (source: Hans/Ross 2023)

When considering drivers and people as pedestrians, cyclists, or other vulnerable road users, the following two danger scenarios can be observed:

- **Conflict between vehicles and vulnerable road users:** This scenario occurs when vehicles and vulnerable road users, such as pedestrians and cyclists, share the same space. The risk is heightened when vehicles do not yield to pedestrians at crosswalks or when cyclists are forced to navigate through heavy traffic without adequate bike lanes.
- **Inappropriate behavior within designated areas:** This scenario arises when either drivers or vulnerable road users do not adhere to their designated areas. For example, a pedestrian stepping into the road unexpectedly or a cyclist riding on the sidewalk can create dangerous situations. Similarly, drivers who speed or fail to observe traffic signals can endanger those around them.

These scenarios highlight the importance of all road users following traffic rules and staying within their designated areas to ensure safety for everyone.

Risk derived from people (Figure 4.21):

- moving into the driving corridor
- moving in the wrong corridor at the wrong time, speed, or situation

Potential risk derived from moving vehicles (Figure 4.21):

- leaving the default corridor
- moving in the wrong corridor at the wrong time, speed, or situation

These points highlight how deviations from expected behavior or designated areas can create risks in traffic. Whether it's a pedestrian stepping into a driving lane or a vehicle moving at an inappropriate speed, such actions can lead to dangerous situations. The key to minimizing these risks is for all road users to stay within their designated areas and follow the rules appropriate to their situation.

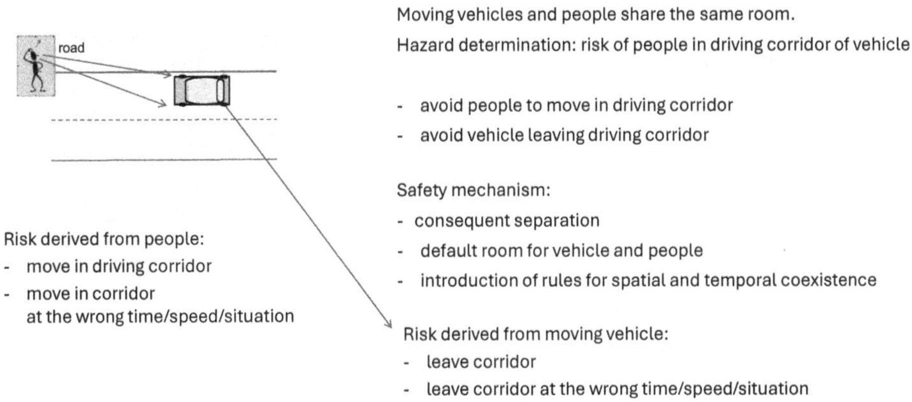

Figure 4.21 Risk derived from people and moving vehicles

In urban environments, moving vehicles, cyclists, and pedestrians often share the same space, leading to potential hazards. The primary risk arises when people enter the driving corridor of vehicles, creating dangerous situations.

To mitigate these risks (Figure 4.21), it is crucial

- To prevent people from moving into the driving corridor of vehicles, and
- to ensure vehicles remain within their designated driving corridors.

Possible safety mechanisms (Figure 4.21):

- **Consequent separation:** Establish clear and consistent separation between vehicles, pedestrians, and cyclists to minimize interactions and potential conflicts.
- **Default spaces:** Designate specific areas for vehicles, pedestrians, and other road users to ensure everyone has a safe and defined space.
- **Rules for coexistence:** Introduce and enforce rules that govern the spatial and temporal coexistence of different road users. This includes traffic signals, pedestrian crossings, and dedicated bike lanes to ensure safe and orderly movement.

By implementing these safety mechanisms, we can create a safer and more efficient traffic environment for all road users.

Such safety mechanisms need to be implemented using various technologies and efficiency goals. In areas with large numbers of pedestrians or where many children move around randomly, structural measures are necessary for protection. Guardrails help prevent vehicles from leaving the roadway. Road markings and traffic light systems are more effective than simple rules learned in driving school.

Here, the principle of product safety applies:
Construction takes precedence over instruction.

However, it is socially unacceptable to separate pedestrians from cars using tunnels or fences. Therefore, it is necessary to find a societal balance, which can vary from city to city and country to country.

The following rules should be considered for ensuring safety for people (Figure 4.22):

- **Make moving corridors transparent for people and vehicles.**

  Ensure that the paths and areas designated for movement are clearly visible and understandable for both pedestrians and drivers. This can be achieved through proper signage, road markings, and lighting.

- **Make the presence of vehicles transparent.**

  Vehicles should be easily noticeable to pedestrians and other road users. This can be facilitated by using lights, reflective materials, and audible signals, especially in areas with high pedestrian traffic.

- **Make people detectable for vehicles.**

  Implement technologies that allow vehicles to detect the presence of pedestrians and cyclists. This includes sensors, cameras, and Advanced Driver-Assistance Systems (ADAS) that can alert drivers to the presence of people in their vicinity.

- **Instruct and educate people.**

  Provide education and training for all road users on safe practices and the importance of adhering to traffic rules. This can be done through public awareness campaigns, school programs, and driver education courses.

Safety obligations for correct vehicle behavior (Figure 4.22):

- **Assure a correct driving corridor.**

  Ensure that vehicles stay within their designated driving lanes to prevent accidents and maintain orderly traffic flow.

- **Make the presence of people transparent.**

  Implement measures to make pedestrians and other vulnerable road users visible to drivers. This can include reflective clothing, pedestrian crossings, and adequate street lighting.

- **Make vehicles detectable for people.**

  Equip vehicles with features that make them easily noticeable to pedestrians, such as headlights, taillights, and audible signals.

- **Instruct and educate vehicle operators.**

  Provide comprehensive training and education for drivers on safe driving practices and the importance of being aware of their surroundings.

Coexistence of moving vehicles and people (Figure 4.22):

- **Analyze given directives, rules, etc.**

  Review existing traffic regulations and guidelines to understand the current safety standards and requirements.

- **Identify analogies to other industries.**

  Look for safety practices and principles from other industries that can be applied to road traffic safety (for example, safety protocols from the aviation or railway sectors).

- **Define and adapt rules.**

  Develop and implement new rules or adapt existing ones to enhance the safety of both vehicles and pedestrians. This can include updating traffic laws, improving infrastructure, and introducing new technologies. This is important to find a proper adaptation for automated driving solutions especially in urban environment.

By following these safety obligations and principles, we can create a safer environment for both vehicles and pedestrians, ensuring their coexistence on the roads.

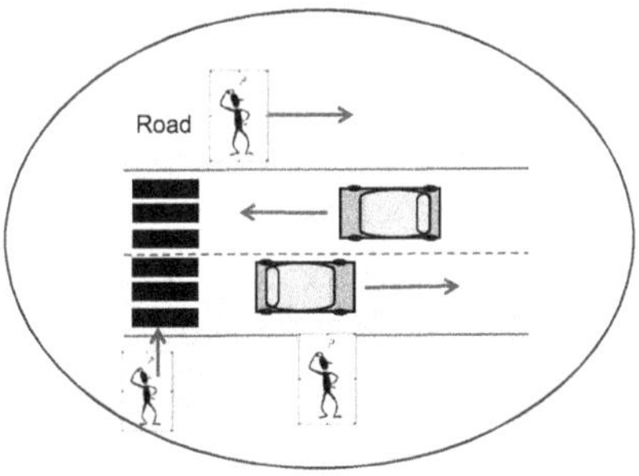

Figure 4.22 Aspects of operational safety

4.6.2.3 Vehicle Level Safety Requirements Derived from Legal Sources

In the vehicle level, you investigate the vehicle, and all the inner parts could become visible. Vehicles consist of technical parts, which usually only become transparent when a vehicle is systematically analyzed. Typically, these include the chassis, including braking and steering systems, drive systems, and the vehicle interior. Of course, many details are included, such as wheels, axles, vehicle underbody, seats, lighting, displays, windshield, doors, etc. All these elements, which now become transparent on the vehicle side, can be subject to safety requirements. In particular, the registration regulations provide various requirements for this:

- **UNECE regulations:** The UNECE World Forum for Harmonization of Vehicle Regulations (WP.29) provides a legal framework allowing member countries to establish regulatory instruments concerning wheeled vehicles and their equipment. These regulations include performance-oriented test requirements, as well as administrative procedures for type approval, conformity of production, and the mutual recognition of the type approvals granted by contracting parties.
- **EU's General Safety Regulation (GSR):** This legislative framework makes advanced safety features mandatory for new vehicles sold into the European market. New commercial vehicles need to possess certain Advanced Driver Assistance Systems (ADAS) features starting from mid-2022 (for homologations) and mid-2024 (for registrations) respectively. These features include Moving Off Information System (MOIS), Blind Spot Information System (BSIS), Reversing Information System (REIS), Intelligent Speed Assist (ISA), Driver Drowsiness & Alertness Warning (DDAW), and Tire Pressure Monitoring System (TPMS).
- **National Highway Traffic Safety Administration (NHTSA):** The regulation focuses on the US market and associated markets. The NHTSA has requirements for manufacturers of motor vehicles and motor vehicle equipment.

The Automotive Regulatory Guide from ACEA provides tables with links to the various regulations relevant for motor vehicles (ACEA 2022).

Visual Capitalist published an impressive graph from the development of automobile safety (Figure 4.23, Lu 2021). It shows the decrease of fatalities and the introduction of legal requirements in the homologation standards from UNECE and NHTSA.

These homologation standards grow different for the classes of vehicle. The vehicle classes are today referencing to the UNECE publication "Consolidated Resolution on the Construction of Vehicles (R. E.3)" (UNECE 2023):

- **L:** two- or three-wheel vehicles and quadricycles
- **M:** motor vehicles with at least four wheels
- **N:** used for the carriage of goods

- **O:** trailers[2] (including semi-trailers)
- **G:** off-road vehicles
- **S:** special purpose vehicles

The classes are subdivided based on gross weight and speed limits.

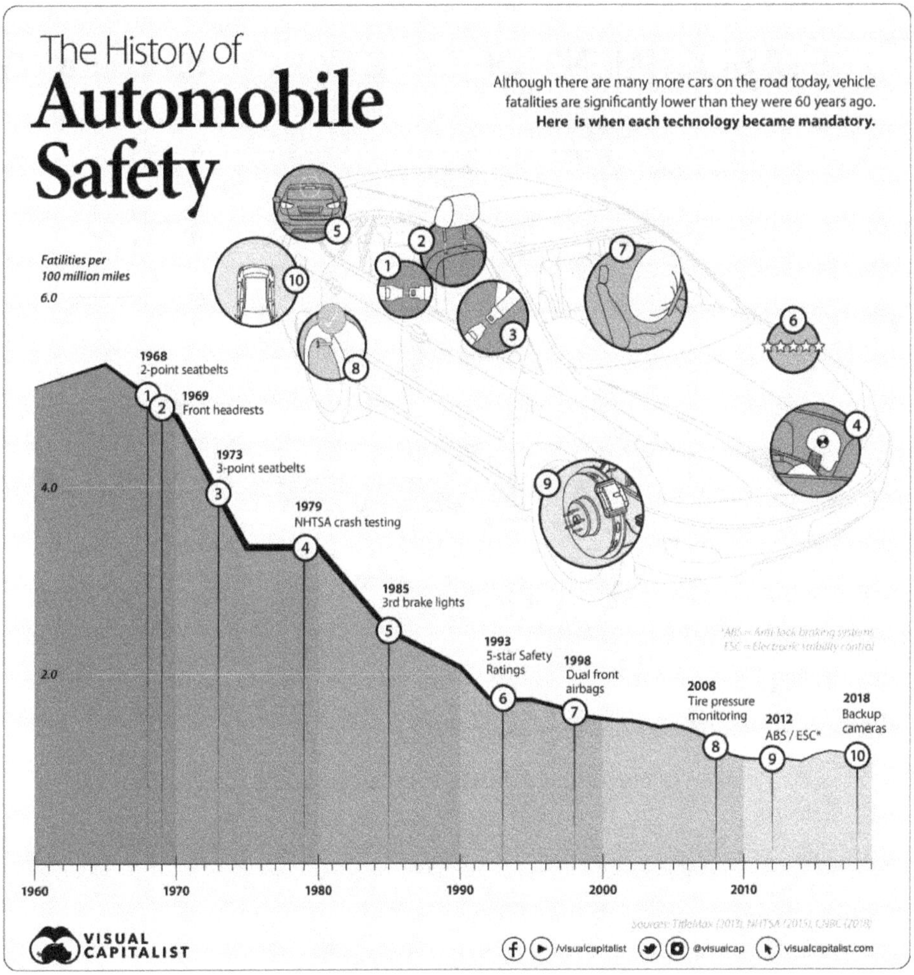

Figure 4.23 Development of legal safety requirements (source: Lu 2021)

2 *https://en.wikipedia.org/wiki/Trailer_(vehicle)*

The table in Figure 4.24 provides some ideas what kind of safety requirements derive from the legal standards. The number of regulations continuously grow. At the end of 2023 UNECE 163, "Vehicle Alarm Systems," was the highest number, but the regulation also changes continuously due to demand. Key drivers are the accident statistics, insights from consumer protection (e.g., NCAP) or new development like automated driving and e-mobility in the past years.

ESTABLISHMENT OF SAFETY REQUIREMENTS

The number of objects of requirements applied to the different vehicle categories

Safety Kinds	Vehicle Category		
	Cars M1	Buses M2,M3	Trucks N1-N3
Active Safety	13	10	13
Passive Safety	16	10	16
Post-crash Safety	2	3	2
Ecological Safety	11	7	11
Total	42	30	42

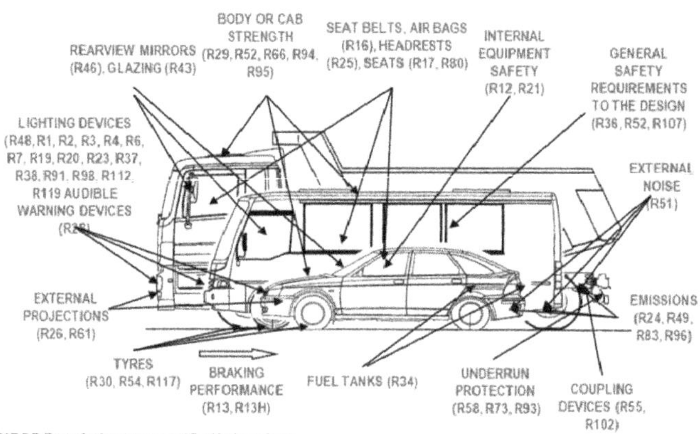

The numbers of the UNECE Regulation are specified In brackets

Figure 4.24 Kinds of safety for the different vehicle categories (source: Kisulenko 2012, Slide 7)

4.6.2.4 Safety and Security Analysis on Vehicle Level

The most prominent vehicle level analysis are the following.

Hazard Analysis (HARA) According to ISO 26262

Methodology: Hazard Analysis, also known as HARA, is a systematic approach used to identify and assess potential hazards in automotive systems. It is a fundamental step in ensuring safety during the development of vehicles. Here's how it works:

- **Identification of hazards:** Engineers analyze the system architecture, components, and interactions to identify potential hazards. These hazards can be related to mechanical, electrical, or software aspects.
- **Risk assessment:** Each identified hazard is evaluated in terms of its severity, exposure, and controllability. The goal is to understand the risk associated with each hazard.
- **Risk mitigation:** Based on the risk assessment, safety measures are proposed to mitigate or eliminate hazards. These measures may involve design changes, redundancy, or safety mechanisms.

Objectives:

- **Safety assurance:** HARA ensures that safety requirements are met throughout the vehicle's lifecycle.
- **Compliance with standards:** It aligns with standards such as ISO 26262, which provides guidelines for functional safety in road vehicles.
- **Risk reduction:** By identifying and addressing hazards early, HARA reduces the likelihood of accidents caused by system failures.

Threat Analysis According to ISO/SAE 21434

Methodology: Threat analysis focuses on security risks, especially in the context of connected vehicles and automotive networks. Here's how it operates:

- **Identifying threats:** Engineers analyze potential threats to the system, including cyberattacks, unauthorized access, and data breaches.
- **Attack feasibility assessment:** Each threat is evaluated for its feasibility. This involves assessing the likelihood of successful attacks and their potential impact.
- **Security measures:** Based on the assessment, security measures are proposed. These may include encryption, intrusion detection systems, and secure communication protocols.

Objectives:

- **Cybersecurity:** Threat analysis aims to protect vehicles from cyberthreats, safeguarding critical functions and sensitive data.
- **Compliance with standards:** It aligns with standards like ISO/SAE 21434, which addresses cybersecurity in road vehicles.
- **Risk mitigation:** By understanding threats, automotive systems can be designed to withstand attacks and minimize vulnerabilities.

Unfortunately, the scope of the various standards and methods are not harmonized and agreed yet. Therefore, many other risks need to be considered.

There are many additional sources of risk for further cause of technical risk, that could lead to accidents with motor vehicles (Figure 4.25).

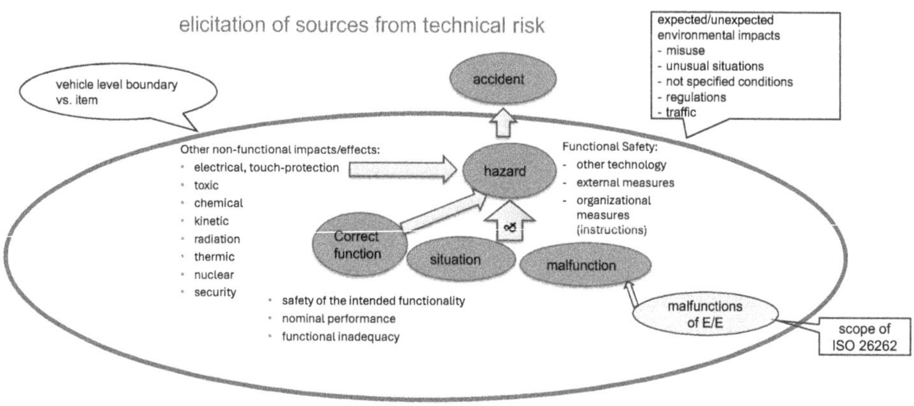

Figure 4.25 Sources of technical risks

4.6.3 Conceptual Solutions on Vehicle Level

There are some developments like Tesla, where people developed a vehicle as a whole, but the traditional motor vehicle brands focus their development on sustainable brand management. A Porsche 911 has another image than a Mercedes S-Class, and such upper-class vehicles differ significantly in their ergonomic and handlings principles from a vehicle in the VW Golf class or even a smaller class like a Smart.

Usually, any brand develops their own concepts for their type of vehicles. All classes of vehicles and types of vehicles from the various brands have different concepts, such as the following:

- usability concept
- HMI concept

- propulsion concept
- charging or fueling concept
- power supply concept
- network concept
- V2x concept
- climate and air-conditioning concept
- temperature and heat management
- entry concept
- start-stop concept, etc.

Since all those concepts are different, they can also affect people in very different ways.

4.6.4 Safety Objectives and Derived Safety Goals

At the highest level of the operational layer, errors are not considered. This level describes the correct behavior of the vehicle within its respective traffic context. It is crucial to provide a complete description of the use cases at this level. Any missing or incorrect description of the use cases leads to change requests throughout the entire system engineering process, extending to both hardware and software implementation.

At the operational level, only the behavior of the vehicle should be considered as a black box (Figure 4.26). Here, the vehicle is viewed solely in terms of its axes of movement. ISO 8855:2011 provides a coordinate system to describe the vehicle's axes of movement. Heading, pitch, and roll angle are considered in the coordinate system. ISO 8855:2011 provides a good basis for the complete vehicle movements and also provides a mathematical basis for the dynamic functions and their configuration parameters.

When deriving requirements from the operational level to the vehicle level, the entire operational context is missing for the derivation of requirements. Without the operational context, the correct derivation of requirements and the necessary design parameters cannot be verified. The design parameters are included in many requirements. They can be thresholds for design constraints, controller parameters, as well as specifications for nominal performance.

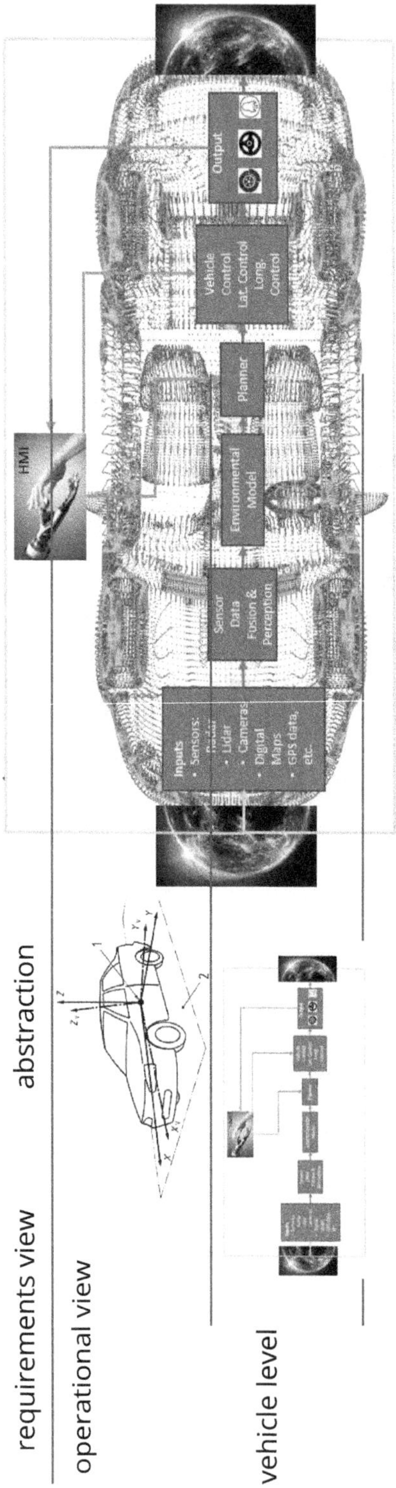

Figure 4.26
Operational and vehicle level

To assess the correct and complete derivation of requirements, a verification method must be defined that considers both the positive and negative perspectives (Figure 4.27). The basis is the desired or intended behavior at the operational level. The negative perspective provides a dysfunctional and malfunctioning view of the positive intended functional perspective. This serves not only as a basis for error consideration and anomaly identification but also as a basis for assessing foreseeable misuse, abuse, sabotage, and improper handling and use of the respective product, feature, or functionality.

In the positive perspective at the vehicle level, the following requirements are needed:

- functional concept
- legal requirements
- architectural assumptions including
  - states and conditions
  - design constraints/decisions
  - effect chains
  - boundary conditions and their specifications

These elements should be part of the architecture and design specification within the requirements specification.

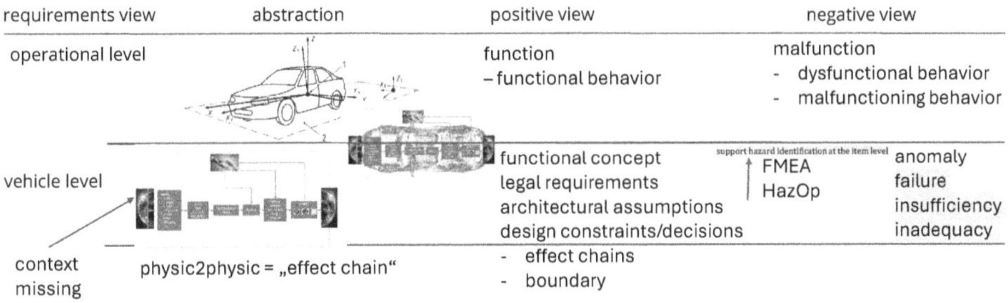

Figure 4.27 Positive and negative view in abstractions

Possible misuse scenarios, sabotage, and improper use should be analyzed using safety methods. Potential unauthorized data access and other cybersecurity threats should also be analyzed using appropriate methods. The corresponding malfunctions and errors for functional safety should be identified based on inductive methods such as FMEA (Failure Modes and Effects Analysis) or HazOp (Hazard and Operability Study), as suggested by ISO 26262.

In this process, not only errors from EE (Electrical and Electronic) systems should be considered, but also general misbehavior from physical stimuli to the physical reaction of the vehicle across various effect chains. The effect chain should consist of quantifiable functional or system blocks, which should include not only the physical effects of the various effect chains but also their logical behavior and the control loops to be considered.

Errors are anomalies, and these anomalies should also consider inadequacies and insufficiencies, so that the nominal performance can be examined and assessed for correctness, over-performance, and under-performance.

Safety objectives should be formulated at various levels. At the higher levels, safety objectives should be formulated positively. The reason for this is that at higher levels of abstraction, the correct and complete description of the intended function should be specified. Therefore, the term "safety objective" should be used for positive safety goals. Another reason is that negative requirements cannot be verified. Through positive safety objectives, verifiable derivations of requirements can be ensured.

The highest operational level is dominated by various legal regulations. In these legal frameworks, the following safety objectives are considered as follows (Figure 4.28):

- **Assure vehicle control**, which is the focus of homologation.
- **Behave according to the road traffic act**, which is the focus of road traffic law.
- **Avoid harm**, which is the focus of product safety and product liability.

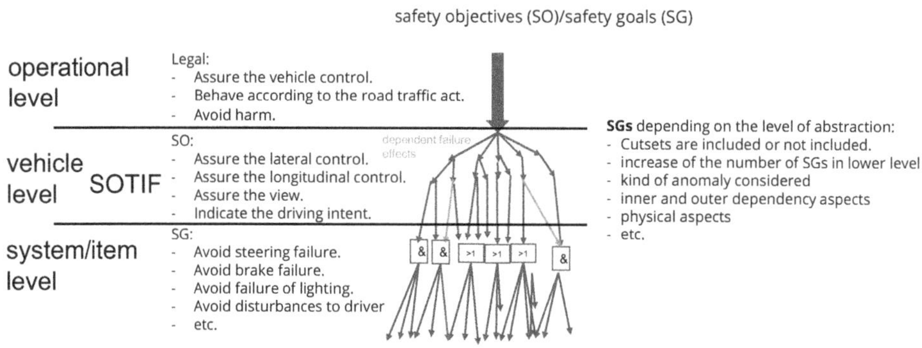

Figure 4.28 Safety objectives and goals on various levels of abstraction

At the vehicle level, the primary focus is on the physical effect chains. This is often considered in the context of SOTIF (Safety of the Intended Functionality). Methodologically, to assess aspects such as correct behavior, correct specification, and correct implementation according to ISO 21448 with respect to uncertainties, such positive safety objectives are useful. These also introduce the level of systems engineering and should be derivable from the goals of legal requirements. These typically include the following (Figure 4.28):

- Assure lateral control.
- Assure longitudinal control.

- Protect against unauthorized use.
- Protect against unauthorized data access.
- Protect against unauthorized access of the vehicle (e. g., theft protection).
- Assure view.
- Indicate driving intent.

In detail, most of the safety-relevant functions and features could be clusters under these safety objectives.

At the system or item level, as formulated in ISO 26262, the primary focus is on negative safety objectives. Therefore, they should also be referred to as safety goals, as in ISO 26262. If these are formulated positively, the usual ASIL (Automotive Safety Integrity Level) assignment cannot be implemented. Typical safety goals include the following (Figure 4.28):

- Avoid steering failure.
- Avoid brake failure.
- Avoid failure of lighting.
- Avoid disturbances to the driver.
- etc.

At this point we are in the scope of ISO 26262:2018.

To fully consider anomalies and also address inadequacies and insufficiencies from SOTIF (Safety of the Intended Functionality) or nominal performance, the safety goals should take the following aspects into account (Figure 4.28):

- **Safety goals (SGs) depending on the level of abstraction:** The level of detail and specificity of safety goals may vary based on the abstraction level.
- **Cutsets are included or not included:** It should be clarified whether or not cutsets (combinations of component failures leading to system failure) are considered.
- **Increase in the number of SGs at lower levels:** The number of safety goals typically increases at lower levels of abstraction.
- **Kind of anomaly considered:** The types of anomalies should be addressed.
- **Inner and outer dependency aspects:** Dependencies that exist within the system and from external systems should be addressed.
- **Physical aspects:** Physical characteristics and constraints of the system should be addressed.

From axiomatic design, we know that functions (features or function groups) must be independent. Therefore, it is important to ensure a clear separation of elements derived from different levels. All elements must have clearly identified boundaries so that functional and logical dependencies can be identified. If these functional and logical dependencies are not traceable, then verification of the effect chains and physical effects cannot be assessed. In such uncertainties, negative consideration will no longer be feasible. Specifically, a Dependent Failure Analysis (DFA) according to ISO 26262 leads to infinite dependencies and makes it impossible to consider the negative or dependent failures.

4.7 Organizational Aspects of Safety

The relationship between safety and organizational structure is multifaceted and deeply interconnected. Safety within an organization is not just about the absence of accidents or incidents; it's about creating an environment where safety is ingrained in every aspect of the organization's culture and operations. Here's how organizational structures play a crucial role in safety from the perspective of the entire organization:

- **Leadership commitment:** Leaders set the tone for safety priorities. When leaders demonstrate a commitment to safety, it cascades down through the organizational hierarchy, influencing the attitudes and behaviors of all employees.
- **Clear communication:** Organizational structures must facilitate clear and open communication channels. This ensures that safety policies are well-understood and that employees feel comfortable reporting safety concerns without fear of retribution.
- **Defined roles and responsibilities:** A well-structured organization clearly defines roles and responsibilities related to safety. This clarity helps in ensuring accountability and that everyone knows their part in maintaining a safe work environment.
- **Training and development:** Organizational structures should support continuous learning and development programs that empower employees with the knowledge and skills necessary to perform their jobs safely.
- **Resource allocation:** Effective organizational structures ensure that adequate resources are allocated for safety initiatives, such as personal protective equipment, safety training, and technology that enhances safety.
- **Continuous improvement:** Organizations with strong structures have mechanisms in place for continuous improvement. They regularly review and update safety procedures based on feedback and incident investigations.
- **Employee empowerment:** When employees are empowered to take ownership of safety issues and participate in safety solutions, it fosters a proactive safety cul-

ture. This empowerment is facilitated by the organization's structure, which should encourage and reward such behavior.

- **Safety performance metrics:** Organizations need to have structures that allow for the tracking and analysis of safety performance metrics. This data-driven approach helps in identifying trends and areas for improvement.

Organizational structures are vital for safety because they provide the framework within which safety culture thrives. They enable the implementation of safety strategies, ensure compliance with regulations, and foster an environment where safety is everyone's responsibility. By integrating safety into the organizational DNA, companies can achieve not only a high level of safety for all employees but also improve their overall performance and resilience.

4.7.1 Conway's Law

The challenge between organizational interfaces and the product designed by this organization had been highlighted by Melvin Edward Conway in 1962. The essence of his observation has become widely recognized and is now referred to as Conway's Law.

Melvin Edward Conway, an American computer scientist, made a profound observation about the interplay between organizational communication and the resulting system design. His core statement, which later became known as Conway's Law, can be summarized as follows: *"Any organization that designs a system (defined broadly) will produce a design whose structure is a copy of the organization's communication structure."*

We can consider the following aspects from Conway's law. Let's break down the key points:

- Organizational interfaces as communication channels:
  - Every interface within an organization—whether it's between teams, departments, or individual members—serves as a communication channel.
  - These interfaces are where team members exchange information, coordinate efforts, and align their work.
- Coordinated interfaces lead to coherent systems:
  - Effective communication and coordination at these interfaces are crucial.
  - When teams work together seamlessly, the resulting system tends to exhibit a coherent structure.
  - Conversely, poor communication can lead to disjointed, fragmented system designs.

- Integration and stakeholder requirements:
  - Interfaces are not just about communication; they also represent integration points.
  - Ensuring that subsystems or components fit together seamlessly is essential.
  - The end product must align with stakeholder requirements, which are often reflected in these interfaces.
- Deriving consistent context:
  - The context for the overall product or system must be systematically derived across all subgroups.
  - Consistency ensures that the system behaves predictably and meets user needs.
- Emergence and context sensitivity:
  - Conway's Law acknowledges that emergence—the behavior of a system as a whole—depends on context.
  - Different conditions, situations, or circumstances may lead to varying emergent properties.
  - All system engineers must consider these contextual factors during design and development.

A typical example from system engineering could be considered as follows:

- In system engineering, interfaces between hardware and software components, subsystems, or modules play a critical role.
- Coherent interfaces lead to well-integrated systems, while disjointed interfaces can result in operational challenges.

In Chapter 7 on systems engineering the challenges will be highlighted in detail.

Conway's Law serves as a reminder that effective communication and organizational structure are not just about efficiency—they profoundly shape the products and systems we create. By understanding these dynamics, we can design better systems that align with stakeholder needs and exhibit emergent behavior consistent with their intended context.

Conway's Law Reversed: Structured Design and Integration

Conway's Law as an observable issue can be reversed for managing complexity. In addition to its original implications, Conway's Law can indeed be applied in a different direction.

- Defining a deliberate structure:
  - Instead of passively reflecting an organization's communication structure, we can proactively define a deliberate structure for the desired product or system.
  - This structured design involves planning interfaces and their communication channels intentionally.

- Managing interfaces and communication:
  - Interfaces are not just technical connectors; they represent communication pathways.
  - By managing these interfaces effectively, we can ensure clear communication between different parts of the system.
- Structured integration from bottom to top:
  - Rather than letting the system emerge haphazardly, we can structure the integration process.
  - Start from lower-level components or subsystems and gradually build up.
  - This approach ensures that the system's architecture aligns with the intended structure.
- Contextual validation:
  - The integrated system should be validated within the context of its application and use cases.
  - Stakeholder requirements and real-world scenarios guide this validation process.
- The V-Model connection:
  - The V-Model, widely used in system engineering, aligns with this reversed application of Conway's Law.
  - It emphasizes structured development, testing, and validation, mirroring the inverted V shape.

By actively shaping the system's structure, managing interfaces, and validating within context, we can harness Conway's Law to create well-organized, coherent systems. Whether we follow the original direction or its reverse, understanding these dynamics empowers us to design better products that meet stakeholder needs and exhibit consistent behavior.

In recent years, there has been a growing awareness of the interplay between safety and security. Combining these analyses allows for a holistic approach, ensuring that vehicles are not only safe but also resilient against security threats.

4.7.2 Challenges in a Learning Organization

For organizations that supply subsystems for vehicles, ensuring a comprehensive overview of all necessary safety aspects is challenging. The highly heterogeneous legal safety requirements across different countries already pose a significant challenge. Vehicle manufacturers, or those responsible for placing the vehicle on the market in terms of product liability, are currently responsible for a homologation cable vehicle that meets the necessary safety requirements as part of type approval.

The driver is responsible for their adequate driving training. While there are recognized standards in the automotive industry for functional safety and Safety of the Intended Functionality (SOTIF), there are no universally accepted standards for Safety In Use.

The Pegasus project has developed a comprehensive understanding of the operational environment and its influence on vehicle design and the layout of functions and features. Various standards normatively describe the ideas from Pegasus. Additionally, there are legislative frameworks like VMAD (Validation Methods for Automated Driving by UNECE) and standards that attempt to bring about standardization in testing and validation within the Operational Design Domain (ODD).

Challenges and Aspects

Heterogeneous legal requirements: Different countries have varying safety regulations, making it difficult for organizations to ensure compliance across all markets. For example, a braking system that meets European standards might need modifications to comply with US regulations.

Integration of safety standards: Incorporating a vast array of safety standards into a single organization is complex. Each design decision can be interpreted differently based on the specific requirements of various standards. For instance, the ISO 26262 standard for functional safety might have different implications compared to the SOTIF guidelines.

Comprehensive safety management system: Implementing a thorough safety management system that scales across an entire traffic system is a significant challenge. This involves not only ensuring that all subsystems meet safety requirements but also that the overall system functions cohesively. For example, integrating Advanced Driver Assistance Systems (ADAS) with existing traffic infrastructure requires meticulous planning and coordination.

Operational environment understanding: Understanding the operational environment and its impact on vehicle design is crucial. The Pegasus project has highlighted the importance of considering environmental factors in vehicle design. For instance, a vehicle designed for urban environments might need different safety features compared to one intended for rural areas.

Standardization efforts: Efforts to standardize testing and validation processes, such as those by VMAD, are essential but challenging. These efforts aim to create uniform testing protocols that can be applied globally, for example, standardizing the testing of Autonomous Vehicle (AV) systems to ensure they perform safely in various conditions.

Balancing societal acceptance: Finding a societal balance that is acceptable in different cities and countries is necessary. While structural measures like guardrails and traffic lights are effective, it is not socially acceptable to separate pedestrians from vehicles using tunnels or fences. For example, urban planners must design pedestrian-friendly spaces that also accommodate vehicle traffic safely.

By addressing these challenges and incorporating the relevant aspects, organizations can work towards creating safer and more efficient traffic systems.

A Safety Management System (SMS) is a solution for an organization to ensure that all safety aspects are systematically managed. However, ensuring that everyone in a traffic environment adheres to the same rules and that safety measures are coordinated requires territorial agreement and the enforcement of these requirements.

This means that for safety measures to be effective across a traffic system, there must be a consensus among the relevant authorities within a specific territory. This consensus involves agreeing on the rules and regulations that govern road use and ensuring that these rules are consistently enforced. Only through such territorial agreements and enforcement can we achieve a harmonized and safe traffic environment where all road users follow the same safety protocols.

In Germany, efforts are being made with the Autonomous Vehicles Approval and Operation Regulation (AFGBV, Verordnung zur Genehmigung und zum Betrieb von Kraftfahrzeugen mit autonomer Fahrfunktion)) and the Remote Road Traffic Regulation (StVFernLV, Straßenverkehr-Fernlenk-Verordnung) to ensure that all manufacturers and other road users adhere to the new rules. However, aligning everyone with these new regulations requires significant effort. The AFGBV provides the legal framework for the approval and operation of vehicles with autonomous driving functions in designated areas. It outlines the requirements for obtaining operating permits, market surveillance, and the responsibilities of manufacturers and vehicle owners. Similarly, the StVFernLV aims to regulate the use of remote-controlled vehicles, providing a legal basis for their testing and operation. This regulation is intended to support the transition to new mobility concepts and ensure that remote-controlled vehicles can be integrated safely into public traffic.

Despite these regulations, achieving full compliance and coordination among all stakeholders remains a challenge. It requires continuous efforts to educate and instruct all road users and manufacturers on the new rules and ensure that safety measures are harmonized across the traffic system.

References

ACEA – European Automobile Manufacturers' Association: The Automotive Regulatory Guide. 2022. *https://www.acea.auto/files/ACEA-Regulatory-Guide-2022.pdf*

ARPANSA – Australian Radiation Protection and Nuclear Safety Agency: What is holistic safety? 2024. *https://www.arpansa.gov.au/regulation-and-licensing/safety-security-transport/holistic-safety/learn-about-holistic-safety*

European Commission: EU General Risk Assessment Methodology (Action 5 of the Multiannual Action Plan for the Surveillance of Products in the EU (COM(2013)76). 2015. *https://ec.europa.eu/docsroom/documents/17107/attachments/1/translations/en/renditions/pdf*

Geneva Conventions: Convention on Road Traffic. Article 7. Geneva, September 19, 1949

Hans, O./Ross, H.-L.: Law and Regulations. Presentation. Safetronic 2023

IEC 61508: Functional Safety of Electrical/Electronic/Programmable Electronic Safety-related Systems. Geneva

ISO/PAS 21448:2020: Road vehicles – Safety of the intended functionality

Kisulenko, B. V.: The UNECE World Forum For Harmonization Of Vehicle Regulations (Wp.29) – Recognized Leader In Development Of Safety Requirements For The Road Vehicles. Presentation. APEC Automotive Dialogue, Kazan 2012. *https://www.unece.org/fileadmin/DAM/trans/doc/2012/wp29/WP29-157-03e.ppt*

Lu, M.: How Has Car Safety Improved Over 60 Years? visualcapitalist.com, December 4, 2021. *https://www.visualcapitalist.com/how-has-car-safety-improved-over-60-years*

Marsden, E.: Farmer's diagram, or F-N curve. Representing society's degree of catastrophe aversion. risk-engineering.org, 24, 2022. *https://risk-engineering.org/concept/Farmer-diagram*

Rasmussen Report/WASH-1400. The Reactor Safety Study. 1975

SAE J 3049:2015: Model Architecture and Interfaces Recommended Practice for Ground Vehicle System and Subsystem Dynamical Simulation

UNECE – United Nations Economic Commission for Europe: Consolidated Resolution on the Construction of Vehicles (R.E.3). 2023. *https://unece.org/transport/vehicle-regulations/wp29/resolutions*

5 From Human Driving to Robot Driving

Over the past 150 years, cars have been designed for human drivers who sit inside the vehicle to control it. Traffic regulations have been established based on this premise, and roads have been optimized accordingly. The primary goals of these optimizations have been twofold:

- Firstly, to enhance traffic flow, ensuring that the transportation of goods and people is as time-efficient as possible from the starting point to the destination, and
- secondly, to achieve a high level of safety, with safety often being measured by the frequency of accidents.

> In contemporary times, the ecological aspect has gained significant importance. Consequently, a fundamental objective is to fulfil transportation tasks in a manner that minimizes harm to nature and the environment.

If we now aim to automate vehicle behavior in traffic, or even introduce autonomous driving functions or systems, we must acknowledge that other road users also have rights. Automated driving solutions must adhere to established and familiar rules and rights. Therefore, we need to understand how humans drive vehicles and how they interact with their environment as drivers. This understanding is crucial so that we can replicate observed behaviors using modern digital technologies. This applies on a higher level to the behavior of the vehicle and on a technical level to comparing how a responsible driver operates a vehicle and how a system can optimally perform this task.

5.1 Human in the Loop

The human is more than the sum of his parts. Human beings are complex systems consisting of different levels that interact and influence each other. These levels range from the molecular and cellular levels to the organic and physiological levels to the psychological and social levels. Each level has its own structures, functions, and processes that can adapt to the requirements of the environment and respond to changes. To understand humans as a system, one must look at both the individual levels and their interactions. Different approaches can be chosen depending on which aspect is to be emphasized. For example, humans can be viewed as a biological system consisting of various organ systems that all work together functionally. These organ systems contain several organs, which in turn are made up of tissues, cells, and molecules. Or you can view humans as a psychological system that consists of different parts of the personality, all of which process meaning. These personality parts can be, for example, the inner child, the superego, or the ego, which can harmonize with one another or come into conflict. Or one can view humans as a social system consisting of different groups of people who exert an influence on the behavior of other people. These groups can be, for example, the family, the circle of friends, or society, each of which has its own rules, norms, and values. Another possibility is to view humans as a system consisting of various operations that are separated from their environment. These operations can be, for example, communication, perception, or action, which are self-referential and autopoietic. This approach is based on the systems theory perspective of Niklas Luhmann, who sees the world as fundamentally consisting of autopoietic systems. As already explained, there are many ways of looking at people as a system. Each approach has its advantages and disadvantages, its strengths and weaknesses, its opportunities and limitations. There is no one right or wrong way, just different perspectives that are more or less suitable depending on the context and objectives. Humans as a system is a fascinating topic that encourages further research and reflection.

5.1.1 The Human Driver

Human drivers are not only passive users of vehicles, but also active participants in the driving system. They interact with the vehicle, the road, and the traffic environment through various sensory, cognitive, and motor processes. These processes enable the human driver to perceive, decide, and act on the driving situation, and to adapt to changing conditions and challenges. However, human drivers also have limitations and vulnerabilities, such as fatigue, distraction, or error-proneness that can compromise their performance and safety. Therefore, it is important to understand how human drivers function as technical control systems, and how they can be supported by Advanced Driver Assistance Systems (ADAS) that augment or automate some aspects of the driving task. The human driver can be seen as a technical control system which is a system that operates based on communication, perception, and action, and that has a specific goal and context.

Driving is a complex and dynamic activity that involves the interaction of the human driver, the vehicle, and the environment. The human driver is not a passive observer, but an active agent who controls the vehicle and adapts to the changing road and traffic conditions. The human driver can be seen as a technical control system, that is, a system that operates on the basis of communication, perception, and action, and that has a specific goal and context. According to Luhmann, the human system and the context of the systems and the environment can be separated, as they have different modes of operation and observation (Luhmann 2011). The human system operates on the basis of meaning, while the environment operates on the basis of causality.

The human driver processes information through sensory and motor channels, which are independent but interrelated. The human brain is lateralized and symmetrical, meaning that the right and left hemispheres have different functions and abilities, but also cooperate and communicate with each other. The right hemisphere is more involved in spatial, holistic, and intuitive processing, while the left hemisphere is more involved in verbal, analytical, and logical processing. The human driver uses both hemispheres to perform the driving task, depending on the strengths, weaknesses, chances, risks, and limitations of each hemisphere, as well as the contexts and objectives of the driving situation. The human driver also communicates and cooperates with other drivers, passengers, pedestrians, and road authorities, using verbal and non-verbal signals, such as gestures, eye contact, and horn.

The ultimate goal of the human driver is to survive and to avoid or minimize risk. The human driver uses the historical patterns stored in the brain to predict and anticipate the future events and outcomes, and to prioritize and execute the actions that are most likely to ensure safety and efficiency. However, the human driver is not perfect, and can make errors, mistakes, or violations, due to factors such as fatigue, distraction, stress, or overconfidence. Therefore, it is important to understand how the human driver functions as a system technical control system, and how the human driver can be supported by Advanced Driver Assistance Systems (ADAS) that augment or automate some aspects of the driving task. This chapter reviews the current state of the art in modelling human driver behavior and performance. It discusses the implications for designing and evaluating ADAS that aim to enhance human-vehicle interaction and cooperation.

5.1.2 Human Control System

The human control system is a complex and adaptive system that enables humans to interact with their environment and perform various tasks (Figure 5.1). The human control system is not limited to the brain, but involves the whole body, including the spinal cord, the nervous system, the muscles, the tendons, the blood vessels, and the nerves. These components form a distributed regulatory system that allows communication between the brain and the rest of the body, and controls reflexes and autonomic functions, such as breathing, heartbeat, and digestion. The human control system also uses the muscles, tendons, blood vessels, and nerves to transmit energy and information, and to generate forces and movements.

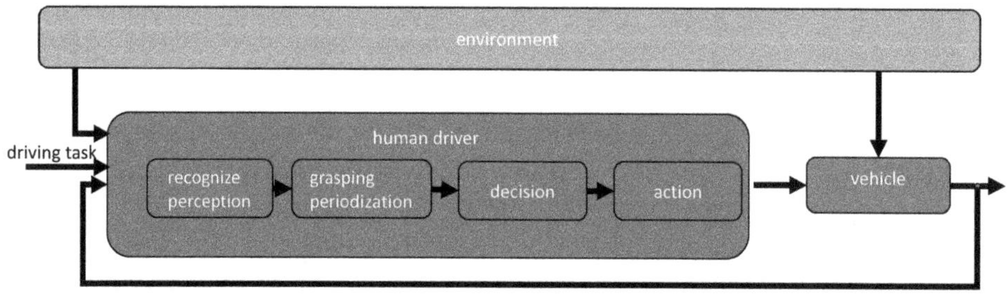

Figure 5.1 Human control loop

The human control system can be seen as a distributed control system, in which different parts of the body have different roles and responsibilities but also cooperate and coordinate with each other. For example, reflexes are fast and involuntary responses to stimuli that are mediated by the spinal cord and the peripheral nervous system, without involving the brain. Reflexes help to protect the body from harm and to maintain homeostasis. However, reflexes are not fixed but can be modulated by the brain and by learning, depending on the context and the goal of the task. Similarly, other control functions, such as posture, balance, locomotion, and manipulation, involve the interaction of multiple levels of the human control system, from the spinal cord to the brain, and from the sensory receptors to the motor effectors.

The human control system is also influenced by various factors, such as training, inherent processes, genealogy, mode, experiences, situational context, situational awareness, and the ability to perceive the environmental impact. These factors affect the performance and the adaptability of the human control system and can lead to different outcomes and behaviors. For example, training can improve the skills and the efficiency of the human control system, while inherent processes and genealogy can determine the innate abilities and the limitations of the human control system. Mode, experiences, situational context, situational awareness, and the ability to perceive the environmental impact can affect the decision making and the action selection of the human control system and can also trigger emotional and motivational states that influence the human control system.

The human control system would be affected by various factors. Human drivers are not only passive users of vehicles, but also active participants in the driving system who controls the vehicle and adapts to the changing road and traffic conditions. They interact with the vehicle, the road, and the traffic environment through various sensory, cognitive, and motor processes. These processes enable the human driver to perceive, decide, and act on the driving situation, and to adapt to changing conditions and challenges. However, human drivers also have limitations and vulnerabilities, such as fatigue, distraction, or error-proneness, that can compromise their performance and safety. Therefore, it is important to understand how human drivers function as technical control systems, and how they can be supported by Advanced Driver Assistance Systems (ADAS) that augment or automate some aspects of the driving task. The human driver can be seen as a technical control system which is a system that operates based on communication, perception, and action, and that has a specific goal and context.

According to Luhmann, the human system and the context of the system and the environment can be separated, as they have different modes of operation and observation. The human system operates based on meaning, while the environment operates on the basis of causality (Luhmann 2009).

The human driver continuously and simultaneously monitors the driving environment and the vehicle behavior within it and adapts the execution of the driving task accordingly.

5.1.3 Information Processing of the Human Driver

The human driver processes information through sensory and motor channels, which are independent but interrelated. The human brain is lateralized and symmetrical, meaning that the right and left hemispheres have different functions and abilities, but also cooperate and communicate with each other. The right hemisphere is more involved in spatial, holistic, and intuitive processing, while the left hemisphere is more involved in verbal, analytical, and logical processing. The human driver uses both hemispheres to perform the driving task, depending on the strengths, weaknesses, chances, risks, and limitations of each hemisphere, as well as the contexts and objectives of the driving situation. The human driver also communicates and cooperates with other drivers, passengers, pedestrians, and road authorities, using verbal and non-verbal signals, such as gestures, eye contact, and horn. In general, information processing can be explained within multi-level models, whereby these models can be classified into three categories:

- differentiation according to types of behavior
- differentiation according to types of driving task
- differentiation according to control processes

Jens Rasmussen's skill, rule, knowledge-based (SRK) taxonomy is often used to differentiate between types of behavior. The levels within the Rasmussen model are classified according to the degree of automation of the behavior, which ranges from highly integrated, often sensorimotor skills to regular linking of stimulus and response to complex problem-solving behavior. Within the model, signals trigger automatized skills, symbols initiate processes of action selection, and signs are linked to actions.

Human Performance and Errors at the Human-Machine Interface

According to Rasmussen, human performance and related errors can be categorized into the three levels of cognitive control (Figure 5.2):

- **Skill-based:** Errors at this level are typically slips or lapses.
- **Rule-based:** Errors usually result from the incorrect application of rules.
- **Knowledge-based:** Errors are caused by incomplete mental or conceptual models.

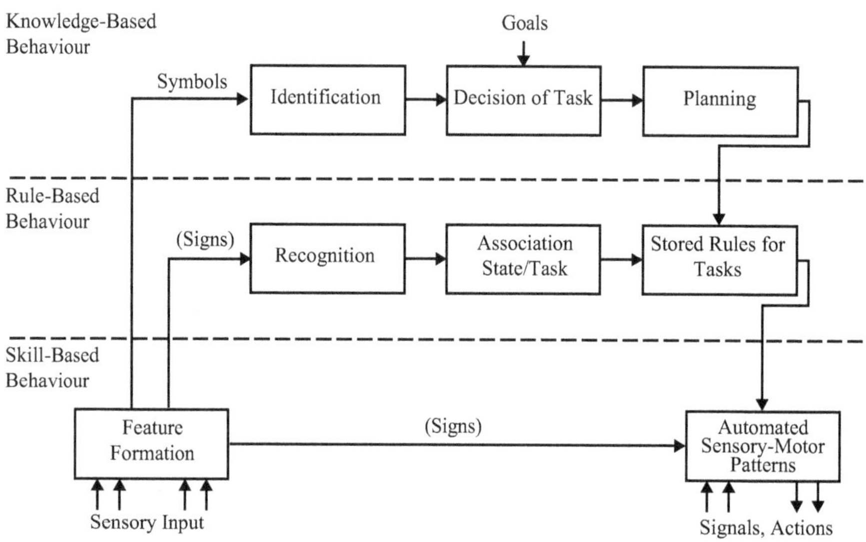

Figure 5.2 Differentiation of behavior according to Rasmussen (source: Wentink et al. 2003)

Operators of systems exhibit the following characteristics related to these three classes:

- **Error rates and knowledgeable operators:** Especially in new or complex scenarios, operators tend to function at the knowledge-based level. Although their expertise can be a great tool for problem-solving, they may still be prone to errors due to cognitive load or incomplete information.
- **Inexperienced operators:** They rely more on skill-based or rule-based behaviors. If processes are well-designed and followed, they may be less prone to errors but are less adaptable to new situations. This approach might be more suitable for individuals with prior knowledge and experience.

Analyzing errors of beginners and experts could lead to the following observations:

- **High error rates among beginners:** Inexperienced operators lack sufficient skills and knowledge, leading to a higher reliance on rule-based behavior.
- **Decreasing errors with experience:** As operators gain experience, they develop more skill-based and rule-based behaviors, which can lead to a reduction in errors.
- **Slight increase in errors among experts**: Experienced operators often work at the knowledge-based level in unfamiliar scenarios. This can increase cognitive demands and lead to errors due to incomplete mental models. Additionally, overconfidence, negligence, reduced perception of complexity, recklessness, and ignoring risk factors are possible causes of higher error rates among senior experts.

A human driver must manage various driving tasks while operating a vehicle. These tasks can be categorized into three classes based on the driver's attention and awareness of the respective activities:

- **Operational tasks:** These are the basic, routine tasks that require minimal conscious attention. Examples include steering, accelerating, and braking. These tasks are often performed automatically through muscle memory and do not require continuous conscious thought.
- **Tactical tasks:** These tasks involve short-term decision-making and require a moderate level of attention. Examples include changing lanes, navigating through traffic, and adjusting speed based on road conditions. Tactical tasks demand more awareness and cognitive processing than operational tasks.
- **Strategic tasks:** These are higher-level tasks that involve long-term planning and decision-making. Examples include route planning, responding to unexpected events, and managing overall trip goals. Strategic tasks require the highest level of attention and awareness, as they involve complex thinking and problem-solving.

Each level of driving task could correspond to different types of behavior as defined by Rasmussen.

Operational Driving Tasks

Operational driving tasks involve the basic control of the vehicle, such as steering, braking, and accelerating. These tasks are typically skill-based, meaning they are performed automatically and with little conscious thought due to extensive practice and familiarity.

Skill-based behavior: An example of skill-based behavior in operational tasks is a driver smoothly changing gears or maintaining a steady speed on a highway. These actions are performed almost instinctively, without the need for conscious decision-making.

Possible misconduct: Errors at this level often involve slips and lapses, such as accidentally driving in the wrong gear level. Pressing the accelerator instead of the brake pedal could only be observed by novice drivers, especially elder driving novices.

Tactical Driving Tasks

Tactical driving tasks involve more complex decision-making processes, such as navigating through traffic, overtaking other vehicles, and responding to road signs. These tasks are primarily rule-based, relying on learned procedures and rules.

Rule-based behavior: An example of rule-based behavior in tactical tasks is a driver deciding to change lanes after checking mirrors and signaling, following the learned rules of safe driving. Another example is stopping at a red light or yielding to pedestrians at a crosswalk.

Possible misconduct: Misinterpretation or misapplication of rules can lead to errors, such as failing to yield the right of way.

Strategic Driving Tasks

Strategic driving tasks involve long-term planning and decision-making, such as planning a route for a long trip or deciding when to leave to avoid traffic. These tasks are knowledge-based, requiring a deeper understanding of the driving environment and the ability to adapt to new situations.

Knowledge-based behavior: An example of knowledge-based behavior in strategic tasks is a driver planning a route using a GPS, considering factors like traffic conditions, weather, and road closures. This requires the driver to integrate various pieces of information and make informed decisions.

Possible misconduct: Errors at this level can occur due to a lack of knowledge or incorrect assumptions, such as underestimating travel time or failing to account for road construction.

Understanding driving tasks through the lens of Rasmussen's behavior model helps in identifying the types of behaviors involved and the potential errors that can occur at each level. Skill-based behaviors dominate operational tasks, rule-based behaviors are crucial for tactical tasks, and knowledge-based behaviors are essential for strategic tasks. Recognizing these distinctions can aid in improving driver training and reducing errors on the road. The mapping of behavior models to driving tasks is a broad consideration and depends on many additional factors. To take a closer look, we will consider the individual driving tasks as vehicle functions.

This means that while we can categorize driving tasks into operational, tactical, and strategic levels and map them to skill-based, rule-based, and knowledge-based behaviors respectively, this is a simplified view. In reality, driving behavior is influenced by a variety of factors such as the driver's experience, the driving environment, and the specific context of each driving situation.

By examining driving tasks as vehicle functions, we can better understand the complexities involved. For example, the function of maintaining lane position involves not just skill-based behavior (keeping the vehicle centered) but also rule-based behavior (following lane markings) and knowledge-based behavior (understanding the importance of lane discipline in different traffic conditions). This approach allows us to see how different types of behaviors interact and overlap in real-world driving scenarios, providing a more nuanced understanding of driver behavior and potential areas for improvement in driver training and vehicle design.

The typical operational driving tasks could be classified based on their attention by the driver. Differentiations according to the type of driving task are usually divided hierarchically into primary, secondary and tertiary driving tasks. These are the primary driving tasks and the actual driving process and thus the execution of the journey (e. g., steering, braking). econdary driving tasks are actions that are directly or indirectly related to the primary driving task (e. g., indicating or honking). Actions relating to driving comfort are defined as tertiary driving tasks. The driving tasks require different levels of attention, which can change due to the situational impact of the driving environment.

The following basics could be considered regarding the driver's attention:

- **primary driving task:** requires continuous attention
- **secondary driving task:** requires attention in specific conditions
- **tertiary driving tasks:** comfort functions

Primary driving tasks are based on continuous and simultaneous monitoring of the position and dynamics of the vehicle in the environment.

The navigation system is considered as a basic information system. The role and importance of the navigation system in automated driving system would significantly change.

In the merged model from Jens Rasmussen and Edmund Donges we could recognize the cascading control circuits. The entire impact from the scenery as a dynamic actor of the environment continuously provides input to the driver.

In human driving systems the driver acts as the navigator for the driving mission, so that the navigation system assists the driver only with the driving task (Figure 5.3).

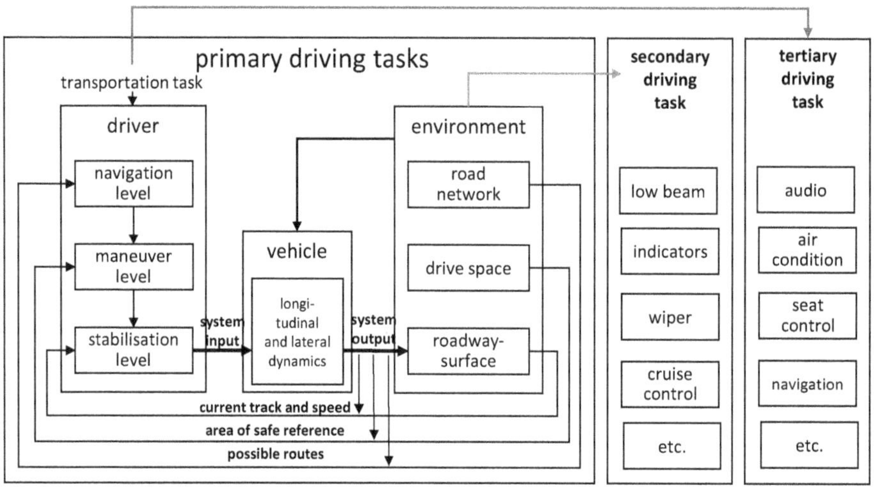

Figure 5.3 Classified driving tasks

The driver's tasks could be considered as a continuous loop with the following (Figure 5.4):

- **navigation:** continuously updating the driving mission with the scenery of the driving environment
- **guidance:** commanding the vehicle according to the driving mission
- **stabilization:** assuring that the vehicle is driving continuously towards the target

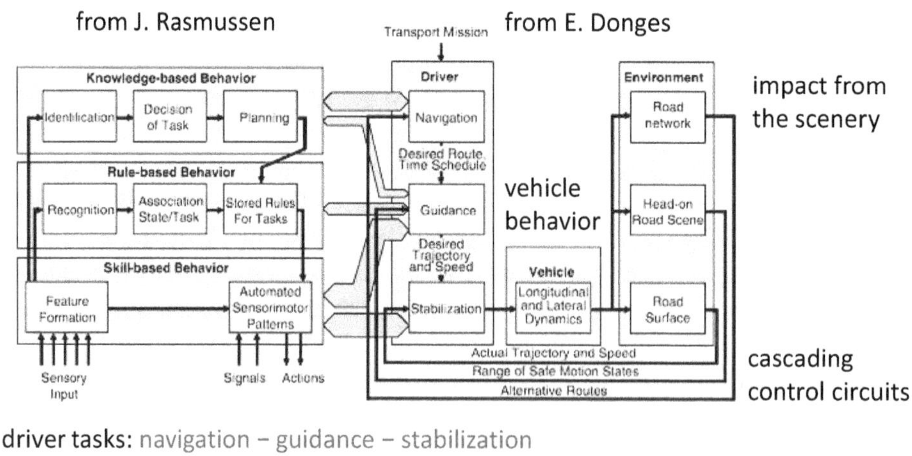

Figure 5.4 Merger of Rasmussen and Donges driver models (source: Donges 1978)

The primary driving task can be hierarchically divided into different control processes, as described by researchers such as Edmund Donges and John A. Michon. These control processes differ in terms of decision-making and the level of consciousness required during task execution. At the navigation level or strategy level, high-level driving decisions are made. This includes choosing the driving route or planning the timing of the journey, while also evaluating costs and risks. Actions at this level require conscious thought and are knowledge-based, as defined by the Skill-, Rule-, and Knowledge-based (SKR) taxonomy. The maneuvering level or guidance level involves tasks such as adhering to speed limits, following traffic regulations, interacting with other road users, and avoiding obstacles. Actions at this level are consciously accessible, meaning they can be consciously controlled but often become routine. Behavior here is knowledge-, rule-, and predominantly skill-based. At the control level, also called the stabilization or regulation level, highly automated actions such as steering and braking take place. These actions occur unconsciously and are skill-based, meaning they are performed through learned automatisms without active conscious control.

5.1.4 Roles of the Human Control System during Driving

The human driver is in various roles during active driving such as perception of vehicle status and condition orientation, perception of the environment, commanding according to mission and situational needs and providing energy and forces to control and switch specific actions. While actively engaged in driving, the driver assumes multiple roles and undertakes a diverse array of tasks.

- **Monitoring vehicle status and orientation:** The drivers consistently monitor various aspects of the vehicle's condition, including speed, fuel level, temperature, and warning indicators. Additionally, they maintain awareness of the vehicle's position on the road, its direction, and its placement within the lane.
- **Observing surroundings:** Constantly scanning the road, the drivers remain vigilant for obstacles, traffic signs, other vehicles, and pedestrians. They adjust their driving approach according to prevailing traffic conditions, such as inclement weather or heavy congestion.
- **Executing commands as per mission and situational demands:** Adhering to traffic regulations, the drivers navigate the vehicle towards the intended destination. They promptly respond to dynamic traffic scenarios, such as making turns, overtaking, or coming to a halt when necessary.
- **Applying energy and forces to execute specific actions:** Operating the accelerator, brake, and steering wheel, the drivers maintainscontrol over the vehicle's movements. They also manage gear shifts, utilize turn signals, and manipulate other control mechanisms to ensure smooth operation.

In addition to the mentioned activities, the drivers perform these tasks within different contexts which result in mainly two different actions:

- **Subconscious local control actions:** Amidst the flow of driving, the drivers rely on reflexes and ingrained rhythms to navigate unexpected situations. Whether it's a sudden need to brake or swerve, the driver reacts instinctively, drawing upon years of honed skills and ingrained responses.
- **Overarching motivation guiding conscious movements:** Beyond the immediate demands of driving, the drivers are propelled by broader objectives. These could include reaching a destination punctually, ensuring safety for all occupants, or enhancing overall comfort during the journey. In pursuit of these goals, the drivers consciously make decisions, such as selecting the optimal route or adopting considerate behavior towards fellow road users.

These activities could be considered as a cascading control system, where the drivers act simultaneously in various roles.

5.1.5 The Human Control System as a Cascading Control System

Understanding the fundamental principles of control systems is essential for optimizing performance and ensuring safety. In this section, we explore the concepts of basic closed-loop control and human multi-cascading control systems, drawing parallels with the behavior of a human driver:

- Closed-loop control systems maintain the desired output by continuously adjusting the system based on feedback. The basic components include a sensor (to measure the output), a controller (to compute the error), and an actuator (to adjust the input). The feedback loop ensures stability and accuracy.
- Human drivers exemplify multi-cascading control systems. They process sensory information (e.g., road conditions, traffic signals) and adjust their actions (steering, braking, accelerating) based on feedback (Figure 5.5). The human brain acts as a complex controller, integrating multiple feedback loops.

Understanding the interplay between basic closed-loop control and human multi-cascading control systems is crucial for designing efficient and safe control systems. By studying the human driver as an example, we gain insights into the complexity of human-machine interactions and the need for adaptive control strategies.

The control cascades can be simplified into two basic control cascades: the higher-level control loop and the underlying control circuits. The higher-level control loop is the conscious control loop where the driver consciously perceives external events, such as cars at a red light. Normally, the driver can remember such events even after the trip.

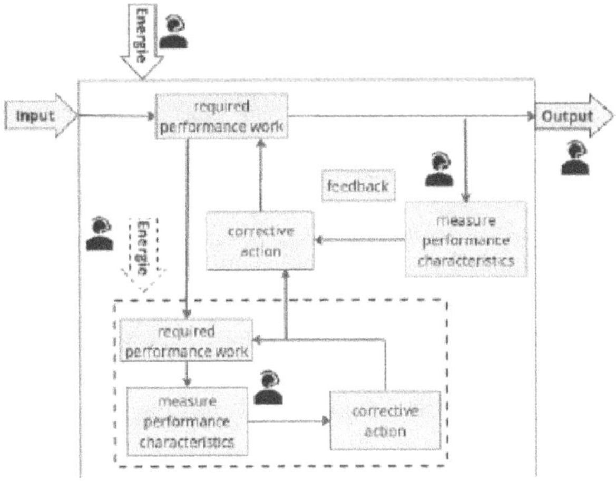

Figure 5.5
Driver in various roles in a cascading control system (source: Ross 2021)

The underlying control circuits are the unconscious functions that are necessary for driving a car, which the driver of a vehicle must learn, for example, in driving school so that the driver can drive a car appropriately in traffic. Typical examples include driving around a curve or uneven road, compensating for crosswinds, or adjusting the speed when driving up or downhill. These functions of the driver are unconscious, and the driver will most likely not remember these details after the journey.

In both control circuits, the driver perceives certain effects of the environment or the behavior of the vehicle and compensates for the behavior of the entire vehicle according to the driver's driving intention. Not only does it act as a closed control loop, it also consciously or unconsciously provides the energy for the necessary action or reaction. As an example, the driver will remember an emergency brake and the driver will also know that the brake pedal was pressed with significantly greater force. If the driver goes down a hill without Adaptive Cruise Control (ACC), the driver will also brake to maintain speed but will not remember the braking event in detail later. The driver continuously measures the performance characteristics achieved, corrects them in the event of deviations and gives a certain performance characteristic to the actuator so that the target deviation remains within the tolerable range.

In these control loops, the human operation can be compared with a Proportional-Integral-Derivative (PID) controller. A PID controller is a type of control system that aims to achieve various objectives such as short transient responses and high stability through design and tuning. Through the proportional component, the driver enacts the intended actions. The integral component involves actions based on summarized, fused, or conscious information. The differential component encompasses trained, reflexive, and/or intuitive actions.

In a human system, all interfaces operate with certain latencies and only partially in parallel. These latencies must be considered as delays within the human control system.

Building upon these principles, various scenarios like

- comparing an in-vehicle driver with a remote driver,
- contrasting an in-vehicle driver with an automated control function,
- evaluating a remote driver against an automated control function,

become feasible to benchmark.

While this may seem like a purely theoretical exercise, it holds immense potential for enhancing ergonomic and learning efficiency. By developing effectiveness, success, and benchmark metrics, we can gain valuable insights. Moreover, it offers significant implications for driver take-over in automated driving scenarios, as well as transitions from remote driving to automated and in-vehicle driving environments.

5.1.6 Model Extension Through "Situational Awareness"

The human driver's capability cannot be assessed by observation of the behavior of the human driver within the operational context. The behavior of the human driver depends on the context, so that the behavior of the driver can only be assessed by the combination of the scenery in the context and the adequate behavior of the driver within the given scenery. In order to make all the control cascades of a human driver transparent, we consider Mica Endsley's situational awareness model as a 3rd driver model in addition to driver models from Jens Rasmussen and Edmund Donges. The current scenery of the road, its surface conditions, the weather, and the behavior of other road traffic participants all contribute to the risk assessment for determining the appropriate driving behavior of the driver. Endsley put the situational awareness in the center of her driver model (Figure 5.6).

Endsley focuses on the physiology of the driver and describes how a person perceives a situation under various influencing factors. This model is not necessarily derived from the experience of a human driver in the automotive industry but from the perspective of a fighter pilot in a military aircraft, where there have been many studies on Endsley's model. The model demonstrates very well that human attention is crucial in determining what the driver perceives and how they contextualize it. Based on the amalgamation of three driver models, a discernible array of control cascades emerges, elucidating the intricate orchestration through which a human driver engages in closed or open control loops across various roles.

Numerous factors impinge upon human perceptual acuity, prioritization, and categorization of environmental stimuli, as well as the decision-making process and the intensity of responses in diverse scenarios. Consequently, an ADS cannot be simply delineated into a linear signal chain encompassing environmental perception, environmental modelling, drive strategy planning, vehicle motion planning, and odometry, speed, and trajectory planning, as observed in conventional frameworks. The human control system comprises multiple internal control cascades, each exhibiting distinct decision-making dynamics within the aforementioned context.

In the behavior Model, in particular, the human driver operates many different perception functions to which he reacts consciously or unconsciously. The reactions to the perceptions do not only depend on the external events and the current vehicle behavior. They are also dependent on his emotional states and will selectively lead to a reaction or action with different priority depending on the situational context and thus influence the driver's behavior.

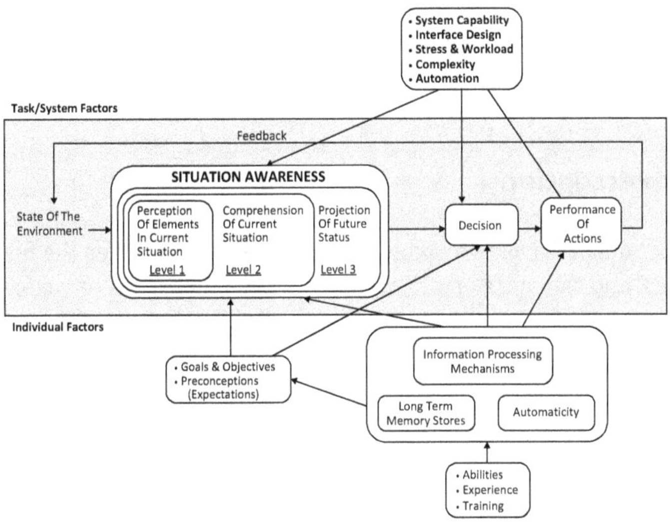

focus on "situation awareness"

scenery becomes dynamic state of the environment

Figure 5.6 The driver model from Endsley focuses situational awareness (source: Endsley/Garland 2000).

Selective perception and vestibular perception are integral components of multisensory perception functions. Selective perception is a cognitive process that allows individuals to attend to specific stimuli while ignoring or suppressing irrelevant sensory inputs. This process is governed by both strategic and automatic attentional processes, with physical salience, statistical learning, and reward learning playing key roles in automatic prioritization in working memory. Vestibular perception, on the other hand, contributes significantly to self-motion perception. The vestibular system, composed of specialized cells called neurons, integrates signals from multiple sensory systems to refine self-motion estimates and facilitate behavioral responses. This system plays a crucial role in maintaining stable vision during head movements and a stable posture.

The control and prioritization of information in perception are closely linked processes. Perceptual processes involve the reception and processing of information via sensory channels, while attentional processes prioritize certain perceptual contents by selectively amplifying some pieces of information and weakening others. This prioritization is essential due to our limited conscious processing capacity. In the nervous system, neurons serve as integration units that transform all received information into new information. The neurons generate unitary electric pulses of invariant form and duration called action potentials or spikes, which serve as the basis for information transfer. The prioritization of information in the nervous system is a complex process that involves both the selective strengthening of relevant signals and the suppression of irrelevant ones.

In conclusion, selective and vestibular perceptions are critical multisensory perception functions that involve the control and prioritization of information. These processes are essential for our interaction with and understanding of the world around us. Further research is needed to fully elucidate the underlying mechanisms and their implications for behavior and cognition.

5.1.7 Human Risk Perception

The ultimate goal of the human driver is to avoid or minimize risk. However, the human driver is not perfect, and can make errors, mistakes, or violations, due to factors such as fatigue, distraction, stress, or overconfidence. In addition, human risk perception in driving is not always accurate or consistent with the objective risk level. Sometimes, drivers may overestimate or underestimate the risks or have biased or distorted perceptions of the risks. This can lead to inappropriate or ineffective responses, such as underestimating the dangers of speeding, overestimating the benefits of seatbelts, or ignoring the warnings of road signs. Therefore, it is important to understand and improve human risk perception in driving, as it can have significant implications for road traffic safety and the prevention of road crashes.

Some possible ways to enhance human risk perception in driving are the following:

- providing accurate and timely information and feedback about the risks and the consequences of driving behaviors, such as through education, training, campaigns, or technology
- increasing the salience and visibility of the risks and the hazards, such as through road design, signage, lighting, or markings
- strengthening the enforcement and deterrence of the traffic rules and laws, such as through fines, penalties, or sanctions
- encouraging positive and responsible attitudes and norms towards road safety, such as through social influence, persuasion, or incentives

In the realm of human-machine interaction, understanding and modelling behavior is crucial for developing intelligent systems. The Skill-, Rule-, and Knowledge-based Behavior (SRK) model, originally proposed by Rasmussen, provides a comprehensive framework for categorizing different levels of human behavior. This model can be adapted to a more technical perspective, where each behavior type is interpreted through the lens of modern neural networks and artificial intelligence. The following sections delves into the specifics of skill-based, rule-based, and knowledge-based behaviors, illustrating how these can be consolidated into three key phases (Figure 5.8): comprehension of the current situation, detection based on neural networks, and projection on future status. These phases collectively enhance the system's ability to perceive and respond to various scenarios effectively.

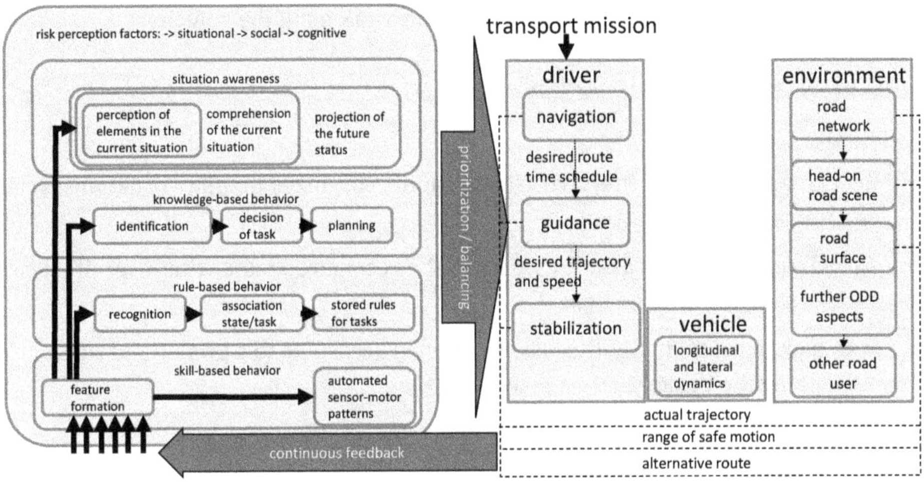

Figure 5.7 Merger of the models and incorporation with risk perception

The combined model can be converted into a more technical view where the Skill-, Rule-, and Knowledge-based Behavior (SRK model) from Rasmussen would be interpreted differently (Figure 5.8).

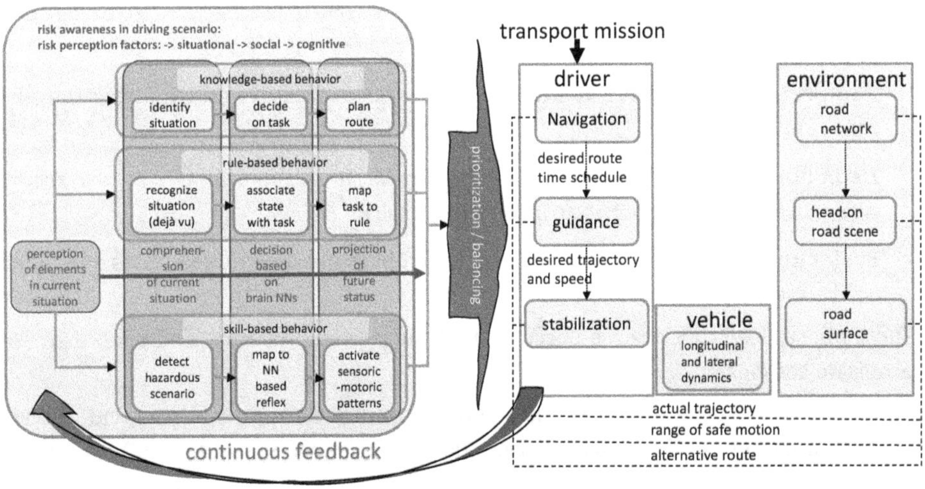

Figure 5.8 Enhanced risk perception model (source: Bauer/Ross 2024a)

Skill-based Behaviour (Figure 5.8)

Comprehension/recognition:

- **Detect hazardous scenarios:** The system automatically detects potentially dangerous situations, such as a suddenly appearing obstacle on the road.
- **Map to neuronal network-based reflexes:** These detected scenarios are immediately matched with predefined reflex patterns in the neural network.
- **Activate sensoric and motoric patterns:** Accordingly, sensory and motor responses are triggered, such as the immediate braking of the vehicle.

Rule-based Behaviour (Figure 5.8)

Comprehension/recognition:

- **Recognize situations (déjà vu):** The system recognizes familiar situations, such as a traffic light intersection.
- **Associate state with task:** It associates the recognized state with a specific task, such as stopping at a red light.
- **Map task to rule:** Finally, the task is executed according to a predefined rule, such as stopping the vehicle.

Knowledge-based Behaviour (Figure 5.8)

Comprehension/recognition:

- **Identify situations:** The system analyzes and identifies complex situations that are not covered by simple rules, such as navigating in an unfamiliar city.
- **Decide on tasks:** Based on the analysis, the system decides which tasks need to be performed, such as finding an alternative route.
- **Plan route:** It plans the best route to reach the destination safely and efficiently.

These behaviors can be consolidated into three phases (Figure 5.8):

- **Comprehension of current situation:** The system captures and analyzes the current conditions.
- **Detection based on brain neuronal networks:** It uses neural networks to recognize patterns and trigger appropriate responses.
- **Projection on future status:** The system projects possible future states and plans accordingly.

These phases lead to a precise perception of the current situation and enable an appropriate response.

The entire driving function, from the perception of the traffic environment, the respective situations, and the possible risk states and risk scenarios, to a stable reaction to these influencing factors in the respective traffic context, must be continuously balanced by the automation system.

Otherwise, other driving functions can be negatively and possibly dangerously affected. Even infotainment systems must be designed in such a way that the driver can always perceive warnings and signals from the traffic environment, such as a siren.

5.2 Deeper Dive into Human Brain Functions

Our central nervous system processes the information that we perceive from the environment or our body. This can be divided into the brain and the spinal cord. The spinal cord lies in the spine and is approximately 40–50 cm long. It transmits commands from the brain to the muscles and consists of an internal gray matter and an external white substance. The gray matter contains the nerve cell bodies and the white matter contains the nerve pathways.

> In comparison to a distributed control system, the gray matter are distributed control elements, and the white matter are communication data lines.

In everyday language, the gray matter is always described as the gray brain cells. We have to imagine the white matter not only as a conductor of information but also as a guard rail for the data paths, so it also protects against unwanted communication paths; it also provides a stable white matter around the individual areas. The elements of the brain also ensure that people can be resilient, more robust and more concentrated. White matter disruption leads to malfunctions in the human system.

Data transfer in the human nervous system should not be thought of as a coded data line; perhaps a mirrored waveguide is a better abstraction. In other words, the information is passed along the white matter, like a tube with a mirror on the inside. If you compare this with a solid glass fiber, the attenuation of the light by the glass material is eliminated and the light is optimally focused in the center of the mirror tube and moves out like a laser beam (Figure 5.9).

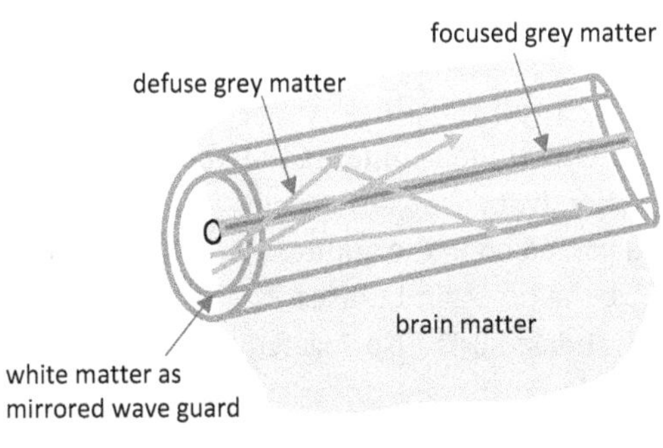

Figure 5.9
Data transfer in human nervous system

The brain can initially be divided into two large areas: the cerebrum and the brainstem (Figure 5.10). At this point, it is important to mention that there are many different classifications of the brain depending on the textbook and working group.

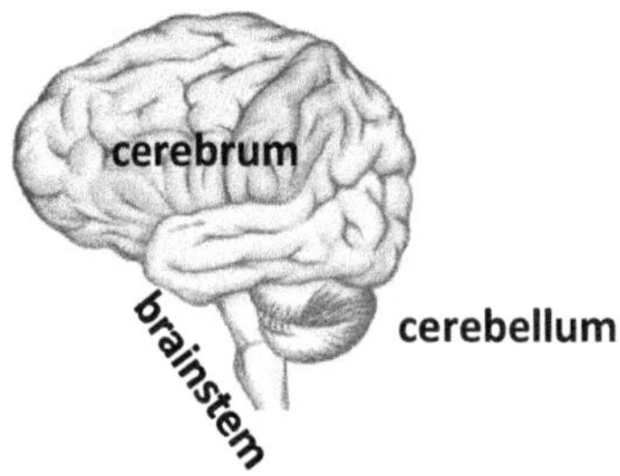

Figure 5.10
Main parts of the brain

Cerebrum

The cerebellum is assigned to the brainstem. The cerebrum is the youngest part of the brain and the most developed. It can be divided into different subunits. These are roughly the four lobes, which can be seen by looking at the brain from the outside.

Occipital Lobe

Roughly described, the occipital lobe is responsible for processing visual (i. e., visual stimuli). This means that when we see something with our eyes, the information is passed on to the occipital lobe and this connects and interprets the incoming information in a meaningful way so that we recognize objects, people, and places, etc.

The occipital lobe could be compared with the visual perception system.

Parietal Lobe

The parietal lobe connects to the upper occipital lobe and is primarily responsible for attention processes and sensory sensations. Sensory sensations refer to information that is received through the senses of sight, hearing, smell, taste, and touch. As a result, damage can, on the one hand, lead to us no longer being able to feel sensory sensations in certain areas of our body and, on the other hand, to us no longer being able to concentrate as well.

The parietal lobe could be compared with the sensor fusion.

Temporal Lobe

The temporal lobe that also adjoins the occipital lobe contains the auditory cortex. This is responsible for processing information from acoustic stimuli. This means that when our ear picks up a sound signal, it is passed on to the auditory cortex. It decodes the information and allows us to recognize the tones and noises. The Wernicke area, which is responsible for language comprehension, is also located in the temporal lobe. If this area is damaged, it is difficult for us to understand what other people are saying.

The temporal lobe could be compared with an acoustical perception system. The Wernicke area could be considered as a synthesizer, that support decoding of voice.

Frontal Lobe

The frontal lobe has many different tasks. On the one hand, it is important for motor skills, that is, for the movement of various muscle groups. Broca's area is also located in the frontal lobe. Broca's area is responsible for speech production. The frontal part of the frontal lobe is often demarcated. It is called the prefrontal cortex and is the youngest part of the cortex, the outer cortex of the brain. The cortex is significantly larger in humans than in other mammals. The prefrontal cortex is responsible for the so-called control of our actions (executive functions). This refers to complex mental functions, such as the planning of movements and actions, or the inhibition of certain actions.

The frontal lobe seems to be the high-level control system and could be compared with a motion controller. The Broca's area is a voice synthesizer. The prefrontal cortex is a motion controller that initiates the overall movement strategy.

However, it is important to be aware that the brain always works in networks, which means that, for example, damage to the parietal lobe does not always lead to an attention deficit disorder, but damage to another area (e. g., the frontal lobe) can also lead to an attention deficit disorder.

The Brainstem

The brainstem lies beneath the cerebrum and, like the cerebrum, can be divided into different areas.

Hindbrain

The hindbrain connects directly to the spinal cord. That's why the first structure is also called the medulla oblongata. The medulla oblongata controls, among other things, blood circulation, breathing and various reflexes such as swallowing, sneezing and coughing reflexes. The pons is located above the medulla oblongata. This is important for our sense of balance.

The medulla oblongata can be compared with the control of the utilities for all the main functions.

The cerebellum is another important part of the hindbrain. The cerebellum is located below the occipital lobe. It is the second largest part of the brain after the cerebrum. Roughly speaking, it is responsible for motor skills, i. e., the control, coordination and fine-tuning of movements. It is also considered to have important roles in the area of learning movement sequences and regulating the sense of balance.

The cerebellum could be compared with the learning memory. It could be considered as that part of the ADS where we implement all the learning algorithms.

Midbrain

The midbrain is responsible for the reflex movements of the eyes and ocular motor skills, and on the other hand, it is an important region for the hearing system. Acoustic stimuli are processed here so that they can later be consciously perceived. The midbrain is also important for pain perception, movement control and voluntary motor skills. You can see that although the midbrain is only a very small structure, there are many nerve fibers running here that are responsible for a wide variety of processes.

The midbrain could be compared with basic control system that controls underlaying processes, but also very fast processes like an interrupt handler.

The diencephalon

The main part of the diencephalon is the thalamus. The thalamus is made up of many individual nuclei and is one of the most complex structures in the central nervous system. The thalamus is also known as the gateway to the cerebral cortex. It filters incoming information according to its importance and decides whether it reaches the cortex and thus becomes conscious or remains below the level of consciousness. Only the information received olfactorily (via smell) is not processed in the thalamus.

The thalamus can be compared with a gateway, a scheduler, or data flow manager.

The Reticular Formation

There is another substance in the brain that we would like to briefly describe: the reticular formation. This is a not well-defined area of the brain. Rather, the reticular formation is a collection of nerve cells and associated fibers that runs through various areas of the brain. The reticular formation is found in the brainstem, cerebellum, medulla oblongata, and spinal cord.

Here the information collected via the nerves is sorted and filtered. It “decides” what is important enough to be passed on to the cerebrum. We become “conscious” of the information passed on to the cerebrum. Information that is not transmitted to the cerebrum can lead to unconscious reactions through impulses from the reticular formation. It is the most important component of the body’s own “alarm system,” the Ascending Reticular Activating System (ARAS). The ARAS sends signals into our consciousness. We immediately become aware of these signals because of their extraordinary importance. The ARAS ensures that we concentrate our attention on this one important action and ignore trivialities. However, it must be said that the ARAS is not an anatomical structure in the brain, but rather a kind of explanatory model for why we behave in dangerous situations in a way that ensures our survival.

The Ascending Reticular Activating System (ARAS) is comparable with an emergency operation system, which controls and prioritizes movement and reaction in critical situations.

We hope that you have now gained a good initial overview of how the brain works its connections among the individual structures and its structure. There are of course many more structures and subdivisions in the brain, but this would go too far here. With the information provided so far, we have hopefully made it a little bit easier for you to understand why damage to a part of the brain led to the problems that your relative or you have developed.

Based on this information, we could see that the human brain and the nervous system is a very complex control system, with many intervenable functions that influences our behavior and reactions on different stimuli.

There are some other functions in our brain and nervous system which are comparable with firm algorithms which play a major role in our behavior and reaction.

Hippocampus

This structure is particularly important for neuropsychology. To put it simply, the hippocampus ensures the transfer of memory content from short-term memory to long-term memory.

Fornix

The fornix is part of the hippocampus, which has already been described here, and is primarily responsible for transferring things that we want to remember from short-term memory (remembering time here is a maximum of 2 minutes) to long-term memory.

Amygdala

The amygdala is central to processing emotions, particularly anxiety and fear. Environmental information is checked by the amygdala to determine whether it is dangerous to us or not.

Insula

The insula has not yet been fully deciphered as to its functions. It is assumed that, among other things, it is involved in processing unconscious body sensations and especially taste stimuli. It also seems to be involved in some processes in the perception of pain. It is the most important part of the so-called viscero-sensitive cortex (sensitivity to (unconscious) sensations of the internal organs).

Basal ganglia

The basal ganglia control our intentional (so-called voluntary) movements, for example when we want to pick up an object. They are therefore responsible for fine-tuning the movement sequences.

Lateralization of the Brain

As described above, many tasks can be assigned to individual brain regions. What is important here is that there is lateralization of the brain. This means that the right side of the body is controlled by the left hemisphere and the left side of the body is controlled by the right hemisphere, as shown in a somewhat simplified manner in Figure 5.12. The connection between the two hemispheres is called the corpus callosum or bar. Here the information travels via the fiber connections from right to left, i. e., from one hemisphere to the other.

The corpus callosum plays a crucial role in coordinating conscious movements. The more developed and efficient the corpus callosum, the better a person can synchronize actions between the right and left-brain hemispheres. This coordination is essential for tasks like walking, writing, and other complex motor activities. The corpus callosum ensures seamless communication between the brain's hemispheres, allowing us to perform coordinated movements and integrate various functions.

When Is An AI-Based System Better Than Humans?

The question of whether Artificial Intelligence (AI) can surpass human capabilities is multifaceted. Let's explore this by considering various aspects:

Computation and speed

- AI excels in processing vast amounts of data quickly. It can perform complex calculations, analyze patterns, and make decisions at speeds far beyond human capacity.
- For tasks involving repetitive computations, AI outperforms humans due to its computational efficiency.

Memory and recall

- AI systems have near-perfect memory recall. They can store and retrieve information without degradation over time.
- Humans, on the other hand, rely on memory, which can be fallible and subject to forgetting or distortion.

Specialization

- AI is superior in specialized domains. For instance, AI models trained in medical diagnostics can analyze medical images more accurately than human radiologists.
- Humans, while versatile, may lack the precision and consistency of AI in specific areas.

Data processing and pattern recognition

- AI algorithms excel at recognizing patterns in large datasets. They can identify subtle correlations that humans might miss.
- Humans rely on intuition and experience, but AI can process more data and discover hidden insights.

Objective decision-making

- AI systems make decisions based on predefined rules and data. They are not influenced by emotions, biases, or external factors.
- Humans, despite their reasoning abilities, can be swayed by emotions, personal beliefs, and cognitive biases.

Efficiency and repetition

- AI tirelessly performs repetitive tasks without fatigue or boredom.
- Humans may become fatigued, leading to errors or reduced efficiency.

Deep learning and neural networks

- AI models, inspired by the brain's neural networks, can learn from vast datasets and generalize patterns.
- Humans learn through experience, but AI can process more examples and adapt faster.

Parallel processing

- AI systems can process multiple tasks simultaneously, leveraging parallel computing.
- Humans focus on one task at a time, limiting their parallel processing capability.

Risk and safety

- AI can handle dangerous or high-risk environments without risking human lives.
- Humans may be vulnerable in hazardous situations.

Creativity and imagination

- Humans excel in creativity, imagination, and abstract thinking.
- AI lacks true creativity; it generates based on learned patterns rather than original ideas.

Social and emotional intelligence

- Humans understand social cues, emotions, and context. They empathize, negotiate, and build relationships.
- AI lacks emotional intelligence and struggles with nuanced human interactions.

Adaptability and learning

- AI can adapt rapidly to new information and learn from feedback.
- Humans adapt through experience, but AI can learn faster.

In summary, AI surpasses humans in specific domains, such as computation, memory, and specialized tasks. However, it lacks human qualities like creativity, intuition, and emotional understanding. The ideal scenario involves collaboration, where AI complements human abilities, enhancing efficiency and addressing limitations.

5.2.1 Human Time Management

Time management in our brain, nervous system, and actions is a complex process involving various regions and mechanisms. Here's an explanation of how it works.

Synchronization of Right and Left Limbs

The synchronization of our right and left limbs is managed by the brain's motor cortex, which sends signals through the spinal cord to the muscles. The corpus callosum, a bundle of nerve fibers, connects the two hemispheres of the brain and allows them to communicate and coordinate movements. This ensures that both sides of the body work together harmoniously.

Internal Clock

Our internal clock, or circadian rhythm, is regulated by the suprachiasmatic nucleus (SCN) in the hypothalamus. This "master clock" synchronizes with external cues like light and darkness to maintain a roughly 24-hour cycle. It influences sleep-wake cycles, hormone release, and other bodily functions.

Senses Involved in Time Management

Several senses contribute to our perception and management of time:

- **Vision:** Light exposure helps regulate our circadian rhythm.
- **Hearing:** Auditory cues can help us estimate the passage of time.
- **Proprioception:** This sense of body position and movement helps us coordinate actions and maintain balance.
- **Interoception:** This involves the perception of internal bodily states, such as hunger and temperature, and is crucial for time perception.

Continuous Monitoring and Protective Mechanisms

The brain continuously monitors actions through feedback loops involving the sensory and motor systems. Protective mechanisms, such as reflexes, are in place to prevent harm. For example, if you touch something hot, sensory neurons send a signal to the spinal cord, which immediately triggers a reflex to withdraw your hand, bypassing the brain to ensure a rapid response.

5.2.2 Human Time Management and Coordination of Actions

The human brain processes events that have already occurred and uses this information to make decisions for future actions. This involves several senses and complex neural mechanisms to ensure precise and timely responses.

Processing Past Events for Future Actions

When an event is perceived, it is already in the past. The brain quickly processes this information to make decisions about future actions.

This involves the following:

- **Visual system:** The eyes capture visual information, which is processed by the visual cortex to understand the scene.
- **Auditory system:** Sounds are processed by the auditory cortex, providing additional context.
- **Proprioception:** This sense helps the brain understand the position and movement of the body.

The sense of balance, or equilibrioception, plays a crucial role in timing and synchronizing our actions. This sense is primarily managed by the vestibular system, located in the inner ear, which detects changes in head position and movement. Here's how it influences our actions and involves various senses.

Influence on Timing and Synchronization

- **Coordination of movements:** The vestibular system helps coordinate movements by providing the brain with information about the body's position and motion. This ensures that actions are timed correctly and executed smoothly.
- **Postural control:** Maintaining balance requires constant adjustments to our posture. The vestibular system works with the proprioceptive system (which senses the position of our limbs) and the visual system to keep us upright and stable.
- **Reflexes:** The vestibule-ocular reflex (VOR) stabilizes our vision by adjusting eye movements to compensate for head movements. This reflex has a very short latency, allowing us to maintain clear vision while moving.

Senses Involved

- **Vestibular system:** It detects changes in head position and movement, providing critical information for balance and spatial orientation.
- **Proprioception:** It senses the position and movement of our limbs, helping to coordinate actions and maintain balance.
- **Vision:** It provides visual cues that help the brain understand the body's position in space and detect potential obstacles or changes in the environment.
- **Hearing:** Auditory cues can also contribute to spatial awareness and timing, especially in activities like dancing or playing sports.

Applying Actions and Movements to the Future

To apply actions and movements to future scenarios, the brain relies on the following:

- **motor cortex:** plans and executes movements
- **cerebellum**: coordinates timing and precision of movements
- **basal ganglia:** involved in movement regulation and decision-making

Example: A Soccer Player and Goalkeeper

- **Predicting the ball's curve:** The soccer player uses visual cues to predict the ball's trajectory. The brain processes the speed, angle, and spin of the ball.
- **Goalkeeper's reaction:** The goalkeeper observes the striker's movements and the ball's flight. The visual and motor cortices work together to plan and execute a response.
- **Executing the kick:** The player calculates the force and direction needed to curve the ball past the goalkeeper. This involves the motor cortex and cerebellum for precise control.
- **Goalkeeper's defense:** The goalkeeper anticipates the ball's path and moves accordingly. If the ball's trajectory changes, the brain quickly updates the plan and adjusts the movement.

Neural sequence and synchronization:

- **Perception:** The visual and auditory systems capture the scene.
- **Processing:** The brain analyzes the information, involving the visual cortex, auditory cortex, and proprioceptive inputs.
- **Decision making:** The prefrontal cortex evaluates options and decides on the best action.
- **Action planning:** The motor cortex and cerebellum plan the movement.

- **Execution:** The motor cortex sends signals to the muscles to perform the action.
- **Feedback:** Sensory feedback from the movement is processed to adjust and refine the action.

Involved organs and senses:

- **eyes:** capture visual information
- **ears:** capture auditory information
- **muscles and joints:** provide proprioceptive feedback
- **skin:** detects temperature and pressure, contributing to the sense of touch

Intensity and precision:

In order to determine how hard to kick the ball, the brain integrates the following:

- **visual information:** distance to the goal and position of the goalkeeper
- **proprioceptive feedback:** position and movement of the player's body
- **motor planning:** calculating the necessary force and angle

This complex interplay of sensory inputs, neural processing, and motor execution allows both the player and the goalkeeper to perform precise and coordinated actions, demonstrating the remarkable capabilities of the human brain and nervous system.

5.2.2.1 Markov Models to Explain Dynamic Scenarios

Let's use the example from soccer to explain the scenes using Markov models. Markov processes are effective tools for modelling and predicting future actions based on current states. They rely on the principle that the future state depends only on the present state, not on the sequence of events that preceded it. This property, known as the Markov property, allows us to create models that can predict outcomes in dynamic environments, such as a soccer game.

States and Transitions in a Soccer Scenario

In a soccer scenario involving a striker and a goalkeeper, we can define various states and transitions (Figure 5.11).

States:

- **S1:** striker dribbling towards the goal
- **S2:** striker preparing to shoot
- **S3:** goalkeeper positioned to block
- **S4:** ball in flight
- **S5:** goalkeeper diving to save
- **S6:** ball in the net (goal)

- **S7:** ball saved by the goalkeeper
- **S8:** defender saves the ball
- **S9:** ball is out beside the goal

Transitions:

- **T1:** probability of the striker moving from dribbling **(S1)** to preparing to shoot **(S2)**
- **T2:** probability of the goalkeeper positioning to block **(S3)** when the striker is preparing to shoot **(S2)**
- **T3:** probability of the ball being in flight **(S4)** after the striker shoots **(S2)**
- **T4:** probability of the goalkeeper diving to save **(S5)** when the ball is in flight **(S4)**
- **T5:** probability of the ball ending in the net **(S6)** or being saved **(S7)** after the goalkeeper dives **(S5)**
- **T8x:** defender attacks striker
- **T9x:** ball passes to out beside the goal

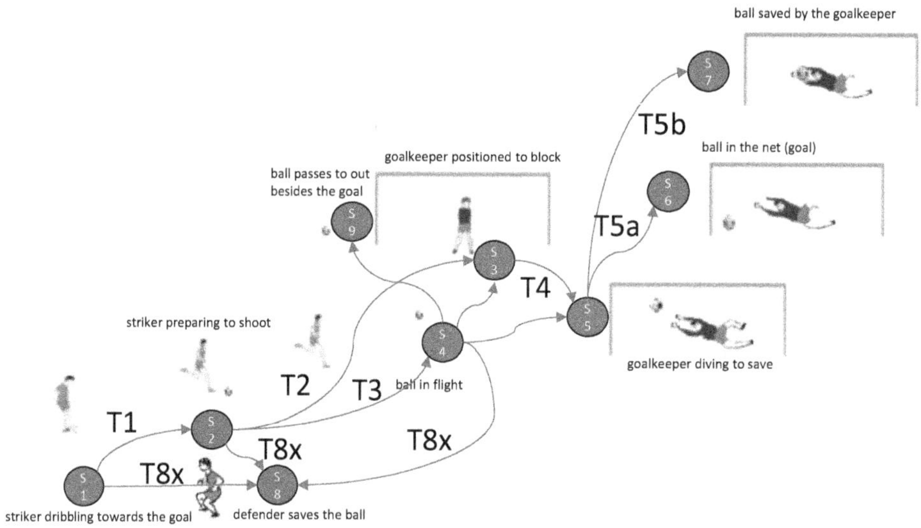

Figure 5.11 States and transitions in a soccer game

Evaluating and Predicting Probabilities

By using Markov diagrams, we can visualize these states and transitions, assigning probabilities to each transition. This allows us to evaluate and predict the likelihood of different outcomes. For example, if the striker is dribbling towards the goal **(S1)**, we can calculate the probability of each subsequent state, leading to either a goal **(S6)** or a save **(S7)**. There are always certain possibilities that a defender saves the ball **(S8)** or that the ball passes beside the goal (**S9**). There are multiple transitions possible that the scene does not end with a successful goal. These transitions also need to go into consideration.

Factors Influencing Transitions

Several factors influence the transitions between states, including the following:

- **synchronization:** the coordination between the striker's movements and the goalkeeper's reactions
- **timing:** the precise moment when the striker decides to shoot and the goalkeeper's response
- **intensity:** the force and speed of the shot, which affect the ball's trajectory
- **skill Assessment:** the striker's ability to accurately judge the optimal point to strike the ball, considering factors like shot power and spin

Example sequence:

- **Initial State (S1):** striker dribbling towards the goal
- **Transition (T1):** striker prepares to shoot (**S2**) with a certain probability
- **Transition (T2):** goalkeeper positions to block (**S3**) based on the striker's action
- **Transition (T3):** ball in flight (**S4**) after the striker shoots
- **Transition (T4):** goalkeeper dives to save (**S5**) when the ball is in flight
- **Final State (S6 or S7):** the ball either ends in the net (goal) or is saved by the goalkeeper

Predicting Success

To predict the success of a goal or a save, we need to consider the following parameters:

- **shot power:** the force with which the ball is kicked
- **ball spin:** the spin applied to the ball, affecting its trajectory
- **optimal strike point:** the precise point on the ball where it is struck
- **goalkeeper's reaction time:** how quickly the goalkeeper can respond to the shot
- **positioning:** both the striker's and goalkeeper's positions at the time of the shot

By incorporating these parameters into the Markov model, we can calculate the probabilities of each transition and predict the most likely outcome. For instance, if the striker strikes the ball with optimal power and spin, the probability of the ball reaching the net (**S6**) increases. Conversely, if the goalkeeper has excellent reaction time and positioning, the probability of a save (**S7**) increases.

Maybe the example with the football game is a bid complex, but it shows how to define certain states and how to identify influence factors that could lead to various possible different transitions. The probability for the various transitions could be dependent on the abilities, of the striker, the defender and the goalkeeper or their experiences, their skills, but also depend quality of the football ground and many more influence factors.

Markov processes and diagrams provide a structured way to model and predict the dynamic interactions between a striker and a goalkeeper. By evaluating the probabilities of different transitions and considering factors like synchronization, timing, intensity, and skill, we can make informed predictions about the likelihood of scoring a goal or making a save. This approach highlights the importance of precise coordination and skill assessment in achieving successful outcomes in soccer.

5.2.2.2 Markov Models in Road Traffic Scenarios

Let's convert the soccer example into a driving maneuver on the road and compare it to making a left turn with oncoming traffic. The goal is to turn left at an intersection without causing an accident, accurately estimating the speed of oncoming traffic, and determining the necessary speed of your own vehicle. This involves the driver's assessments and decisions in various states, such as approaching the intersection, observing oncoming traffic, crossing the left lane, estimating the correct steering angle, and adjusting the vehicle's speed. The driver must also move their body to ensure the car follows the correct trajectory at an appropriate speed. The success is a safe left turn without disturbing or hindering other traffic. Let's create a suitable Markov model, describe the states, transitions, performance characteristics, and the precision and synchronization of the driver's activities in each transition.

Markov Model for Left Turn Manoeuvre

States:

- **S1:** approaching the intersection
- **S2:** observing oncoming traffic
- **S3:** initiating the left turn
- **S4:** crossing the left lane
- **S5:** completing the turn
- **S6:** successfully turned left without incident
- **S7:** potential collision or near-miss

Transitions:

- **T1:** probability of moving from approaching the intersection **(S1)** to observing oncoming traffic **(S2)**
- **T2:** probability of moving from observing oncoming traffic **(S2)** to initiating the left turn **(S3)**
- **T3:** probability of moving from initiating the left turn **(S3)** to crossing the left lane **(S4)**
- **T4:** probability of moving from crossing the left lane **(S4)** to completing the turn **(S5)**
- **T5:** probability of moving from completing the turn **(S5)** to successfully turned left **(S6)** or potential collision **(S7)**

Performance Characteristics and Synchronization

Approaching the intersection **(S1)**:

- **performance characteristics:** speed control, maintaining lane position
- **synchronization:** coordinating visual inputs with speed adjustments

Observing oncoming traffic **(S2)**:

- **performance characteristics:** estimating speed and distance of oncoming vehicles
- **synchronization:** integrating visual information with decision-making processes

Initiating the left turn **(S3)**:

- **performance characteristics:** timing the turn, signaling intention
- **synchronization:** coordinating hand movements on the steering wheel with foot movements on the pedals

Crossing the left lane **(S4)**:

- **performance characteristics:** maintaining appropriate speed and steering accuracy
- **synchronization:** balancing visual inputs with motor actions to ensure smooth lane crossing

Completing the turn **(S5)**:

- **performance characteristics:** adjusting speed and steering to align with the new lane
- **synchronization:** ensuring the vehicle follows the intended trajectory without abrupt movements

Example sequence:

- **Initial state (S1):** The driver approaches the intersection.
- **Transition (T1):** The driver observes oncoming traffic (**S2**) with a certain probability.
- **Transition (T2):** The driver initiates the left turn (**S3**) based on the assessment of oncoming traffic.
- **Transition (T3):** The driver crosses the left lane (**S4**) while maintaining control.
- **Transition (T4):** The driver completes the turn (**S5**) and aligns with the new lane.
- **Final State (S6 or S7):** The driver either successfully turns left (**S6**) or encounters a potential collision (**S7**).

Predicting Success

To predict the success of the left turn, we need to consider the following parameters:

- **speed estimation:** accurately estimating the speed of oncoming traffic
- **timing:** choosing the right moment to initiate the turn
- **steering control:** maintaining precise control over the steering wheel
- **speed adjustment:** modulating the vehicle's speed to match the traffic conditions

By incorporating these parameters into the Markov model, we can calculate the probability of each transition and predict the most likely outcome. For instance, if the driver accurately estimates the speed of oncoming traffic and times the turn correctly, the probability of successfully completing the turn (**S6**) increases. Conversely, if the driver misjudges the speed or timing, the probability of a potential collision (**S7**) increases.

Markov processes and diagrams provide a structured way to model and predict the dynamic interactions involved in making a left turn with oncoming traffic. By evaluating the probabilities of different transitions and considering factors like synchronization, timing, intensity, and skill, we can make informed predictions about the likelihood of a successful maneuver. This approach highlights the importance of precise coordination and skill assessment in achieving safe driving outcomes.

Learning from the Human System

The ability to capture a scene or task from different perspectives, continuously monitor actions, and implement independent protective mechanisms is essential for deriving safe actions. In this area, we can learn a lot from the human system. Without incorporating such architectural elements into our intelligent automation systems, it will be challenging to develop safe automation solutions for our future mobility needs.

5.2.3 Processes in Human Control Systems

In the human body, information is transmitted from the brain to the limbs through the spinal cord in a time-deterministic process involving the activation of nerve cells, or neurons. This process is supported and enhanced by both chemical and physical mechanisms, which work together to ensure efficient and accurate information flow and its conversion into various forms of actuation.

Chemical Processes

Neurotransmitters are chemical messengers that transmit signals across synapses, the gaps between neurons. For example, adrenaline increases heart rate and energy levels, preparing the body for "fight or flight" responses. Dopamine, often referred to as the "pleasure hormone," plays a key role in reward and motivation. Serotonin regulates mood, appetite, and sleep, while endorphins act as natural painkillers and mood elevators. Hormones like cortisol, known as the stress hormone, help the body respond to stress, and oxytocin, often called the "love hormone," promotes bonding and social interactions.

Physical Processes

Temperature changes can affect nerve function, with cold temperatures slowing down nerve conduction and warmth enhancing it. Neurons communicate through electrical impulses generated by the movement of ions, such as sodium and potassium, across the neuron's membrane. This electrical activity is crucial for the rapid transmission of information. Physical pressure or touch can activate sensory neurons, sending signals to the brain about the external environment.

Integration of Processes

These chemical and physical processes support the functions of neurons, including both white and grey matter, to regulate information flow and convert it into appropriate actuation. Synaptic plasticity, the ability of synapses to strengthen or weaken over time in response to activity, is influenced by both chemical signals and electrical activity. Neurogenesis, the process of generating new neurons, can be influenced by hormones and physical activity.

Actuation

Actuation refers to the body's ability to convert neural signals into actions. This includes not only movement but also speech, sweating for cooling, and regulating tension (such as an excess of adrenaline leading to anxiety sweat or hot flashes). These processes result in varying speeds and intensities of actions or reflexes. For instance, sweating helps to cool the body and regulate stress, while hot flashes can lead to rapid, intense actions or reflexes.

The intricate interplay of chemical and physical processes supports the functions of neurons, ensuring that information is transmitted efficiently and accurately throughout the body. This enables the body to perform a wide range of actions, from movement and speech to regulating temperature and stress responses.

5.2.4 Dynamic Information and Energy Management

In the human body, the transmission of information from the brain to the limbs involves complex interactions of chemical and physical processes, which can be examined through the lens of thermodynamics. The laws of thermodynamics provide a framework for understanding how energy and matter interact, and these principles can be applied to the body's information management systems.

First Law of Thermodynamics

The first law, also known as the law of energy conservation, states that energy cannot be created or destroyed, only transformed. In the human body, this principle is evident in the way neurons convert chemical energy from neurotransmitters into electrical impulses. This transformation of energy is crucial for the transmission of information across the nervous system. For example, the energy stored in molecules like ATP (adenosine triphosphate) is used to power the ion pumps that maintain the electrical gradients necessary for nerve impulses.

Second Law of Thermodynamics

The second law states that the entropy, or disorder, of an isolated system will tend to increase over time. In biological systems, maintaining order requires a constant input of energy. The human body achieves this through metabolic processes that convert food into usable energy. This energy is then used to maintain the highly ordered state of the nervous system, allowing for precise and efficient information processing. The release of neurotransmitters and the subsequent reuptake and recycling processes are examples of how the body manages entropy to maintain function.

Third Law of Thermodynamics

The third law states that as the temperature of a system approaches absolute zero, the entropy approaches a constant minimum. While the human body operates far above absolute zero, temperature regulation is still crucial. Neurons function optimally within a narrow temperature range, and the body employs various mechanisms, such as sweating and shivering, to maintain this range. These processes ensure that the biochemical reactions necessary for nerve function proceed at appropriate rates.

Bernoulli's Principle

Bernoulli's principle, which relates the pressure, velocity, and height of fluid flow, can also be applied to the flow of blood and other fluids in the body. Changes in blood pressure and volume can affect the delivery of oxygen and nutrients to neurons, influencing their ability to function. For example, during physical exertion, increased blood flow ensures that neurons receive the energy they need to sustain heightened activity.

Philosophical and Sociological Aspects

From a philosophical perspective, the laws of thermodynamics can be seen as metaphors for social dynamics. The first law's conservation principle can be likened to the conservation of social energy within a group, where the collective effort is redistributed among members. The second law's tendency towards disorder parallels the natural entropy in social systems, where maintaining order requires continuous effort and energy. The third law's approach to a constant minimum at low temperatures can be seen as a metaphor for the stability and predictability in tightly knit communities. In terms of sociology, the body's temperature regulation mechanisms can be compared to the way social groups manage collective stress and tension. Just as the body sweats to cool down and maintain homeostasis, social groups develop mechanisms to release tension and maintain cohesion. The interplay of pressure, volume, and temperature in Bernoulli's principle can also be seen in social interactions, where changes in group dynamics can influence individual behavior and overall group performance.

Muscle Actuation

Muscle actuation involves converting chemical energy into mechanical work. This process requires energy for both muscle maintenance and specific actions. ATP (adenosine triphosphate) is the primary energy source for muscle contraction. During muscle contraction, ATP is hydrolyzed to ADP (adenosine diphosphate), releasing energy that powers the interaction between actin and myosin filaments, resulting in muscle shortening and force generation.

To ensure actions are quick, powerful, precise, and synchronized, the body employs several mechanisms:

- **Motor unit recruitment:** The nervous system activates different numbers and types of motor units (groups of muscle fibers controlled by a single neuron) depending on the required force and precision.
- **Synchronization with internal events:** Hormones like adrenaline are released in response to stress, increasing heart rate and energy availability, which enhances muscle performance.
- **Coordination with external events:** Sensory feedback from the environment helps adjust muscle activity in real-time, ensuring appropriate responses to external stimuli.

For example, during a stressful situation, the release of adrenaline prepares the muscles for rapid and powerful actions. This is complemented by increased blood flow and oxygen delivery, ensuring that muscles have the necessary energy to perform. Additionally, the body's thermoregulatory mechanisms, such as sweating, help maintain optimal muscle function by preventing overheating.

In conclusion, the principles of thermodynamics provide a valuable framework for understanding the complex interactions of chemical and physical processes in the human body. These processes not only support the functions of neurons but also ensure the efficient transmission of information and its conversion into various forms of actuation. By examining these principles through both scientific and philosophical lenses, we gain a deeper appreciation of the intricate balance that sustains life and social harmony.

5.2.5 Adaptation of Human Control Systems to Technical Systems

In animals, these processes often differ significantly from those in humans. Systems must be ergonomically optimized for humans to initiate safe and understandable reactions, while also avoiding fear. Additionally, for people with disabilities, understanding the physical functioning of the human control system is crucial to enable communication with the environment and to perform actions quickly and precisely using their available means.

For instance, dogs have a much stronger sense of smell than humans. Many animals do not rely as heavily on vision as humans do; instead, they use other senses. Smell, as a chemical and physical process, is highly developed in many species. Hearing, which involves acoustic processes, is prioritized and focused on by many animals when performing actions. Bats, for example, have specialized ultrasonic sensors that allow them to navigate through forests quickly, precisely, and safely.

In the field of bionics, we are still far from utilizing the optimal mechanisms that humans or other living beings have at their disposal to perform meaningful and high-quality actions. Understanding these differences and the underlying physical processes is essential for developing systems that can effectively support human needs, especially for those with disabilities. This knowledge helps in creating devices and interfaces that allow for precise and synchronized actions, whether it's movement, speech, or other forms of interaction with the environment.

5.2.5.1 Human-Machine Interface (HMI) and Safety Engineering

In safety engineering, it is crucial to design Human-Machine Interfaces (HMI) that are optimally adapted to human needs and capabilities. This involves creating ergonomic and learnable processes that can be seamlessly integrated into intrinsic movement patterns. By doing so, we ensure that users can interact with machines and systems in a way that is intuitive, reducing the risk of errors and enhancing overall safety.

Ergonomically designed HMIs take into account the physical and cognitive abilities of users, making it easier for them to perform tasks efficiently and safely. For example, controls and displays should be positioned within easy reach and view, minimizing the need for awkward movements. The interface should also provide clear feedback, helping users understand the system's status and their actions' outcomes.

Incorporating learnable processes into HMI design means that users can quickly become proficient in using the system. This is achieved through consistent and intuitive design principles, such as standardized symbols and controls, which reduce the learning curve and improve user confidence. By aligning these processes with natural human movements and behaviors, we can create systems that feel like an extension of the user's own body, enhancing both performance and safety.

5.2.5.2 Limitations of Touch Displays

Touch displays, while responsive to touch, do not provide haptic feedback, which is crucial for human operators who rely on feedback loops to confirm that their desired actions have been registered by the system. Without this feedback, users may be uncertain whether the system has accepted their input, whether it has been processed dynamically, or whether it should initiate a temporary or continuous event with a specific course of action.

Importance of Sensory Feedback

Examples of where sensory feedback is essential include the following:

- **Driving:** Drivers rely on haptic feedback from the steering wheel, pedals, and seat to control the vehicle. The accelerator pedal (floor-mounted) and brake pedal (suspended) are designed differently to optimize human capabilities in critical situations. The arrangement of the steering wheel in relation to the seat position is crucial for visibility and reaction time.
- **Industrial controls:** Operators in industrial settings need tactile feedback from control panels to ensure precise adjustments and safe operation of machinery.
- **Medical devices:** Surgeons use instruments that provide tactile feedback to perform delicate procedures with precision.

Integration of Feedback Mechanisms

Incorporating haptic, acoustic, and temperature feedback into HMIs can significantly enhance the user's ability to control dynamic systems. For instance, haptic feedback can provide a physical sensation that confirms an action has been registered. Acoustic feedback, such as beeps or tones, can indicate different states or errors. Temperature feedback can be used in environments where heat or cold is a factor, providing additional sensory input to the operator.

In conclusion, the principles of thermodynamics provide a valuable framework for understanding the complex interactions of chemical and physical processes in the human body. These processes not only support the functions of neurons but also ensure the efficient transmission of information and its conversion into various forms of actuation. By examining these principles through both scientific and philosophical lenses, we gain a deeper appreciation of the intricate balance that sustains life and social harmony. Additionally, optimizing HMIs for human use through ergonomic design and learnable processes is essential for ensuring safety and enabling effective interaction with technology, particularly for individuals with disabilities. Understanding the limitations of touch display and the importance of sensory feedback can lead to the development of more effective and intuitive interfaces, enhancing both performance and safety in various applications.

5.2.6 Consequences for Automation Technology Based on the Brain Functions

The brain's white and grey matter have fundamental roles. The white matter serves as the information highways in the brain and the human nervous system, while the grey matter acts as the thinking cells where decisions are made. Essentially, the white matter is responsible for information control and also ensures independent processing cells, where information processing can be carried out protected from parasitic effects.

In our current technical architectures, we often overlook the systematic control of information and the necessary requirements for independence. Additionally, we neglect the measures needed to create secure and protected areas for logical data processing. This oversight can lead to vulnerabilities and inefficiencies in our systems, highlighting the importance of designing architectures that prioritize both information control and security.

The human brain is divided into two hemispheres, each responsible for different functions and types of thinking.

The left hemisphere is often associated with the following:

- **Logical thinking:** It processes information in a linear and logical manner, making it crucial for tasks involving logic and reasoning.
- **Language:** This hemisphere is dominant in language processing, including speaking, reading, and writing.
- **Analytical skills:** It excels in tasks that require analysis, such as mathematics and problem-solving.
- **Detail orientation:** The left hemisphere focuses on details and is adept at tasks that require precision and accuracy.

The right hemisphere, on the other hand, is known for the following:

- **Creativity:** It is responsible for creative thinking and artistic abilities, such as drawing and music.
- **Spatial awareness:** This hemisphere helps in understanding spatial relationships and visualizing objects in space.
- **Holistic thinking:** It processes information in a holistic and intuitive manner, making it essential for recognizing patterns and understanding the big picture.
- **Emotional processing:** The right hemisphere plays a significant role in processing emotions and non-verbal cues, such as facial expressions and body language.

Together, these hemispheres work in harmony, allowing us to perform a wide range of cognitive tasks and navigate the complexities of daily life. While each hemisphere has its specialties, they constantly communicate and collaborate through a structure called the corpus callosum, ensuring a balanced and integrated approach to thinking and problem-solving.

There is, of course, no such thing as a purely left-brained or right-brained person, just as there are no purely auditory or visual learners. The interplay between the right and left hemispheres of the brain is of great importance to us. The right hemisphere allows us to fully grasp the meaning of sentences, not just what is said, but what is meant.

The left hemisphere of the brain provides structure and tasks, while the right hemisphere contributes to the elaboration and creativity. This interplay can be seen in three-step sequences across the hemispheres, such as: understand, interpret, and create. This three-step process is similar to the actions derived from Rasmussen's model of Skill-, Rule-, and Knowledge-based Behavior (SRK model), where we observe the steps: comprehend/recognize, decide, and act.

The two hemispheres of the brain are connected by the corpus callosum. This structure significantly develops its data pathways until the human reaches puberty. The corpus callosum acts as a data highway, linking the two distinct thinking areas and ensuring that decision-making information is consolidated and transmitted to the spinal cord. This consolidation of information is crucial for coordinated actions.

Optimal decision-making occurs when the right and left hemispheres of the brain evaluate a situation from two different perspectives, each leveraging their respective strengths (Figure 5.12). They then send the best decision, or the one with the highest priority, to the spinal cord for action. This ensures that the limbs can perform a physical action, or the mouth can be prompted to produce speech.

Within the spinal cord and all other nerve cells, there are both white and grey matter, which can further influence and direct these decisions. The white matter facilitates the transmission of information, while the grey matter processes and integrates this information, ensuring that the final action is well-coordinated and effective.

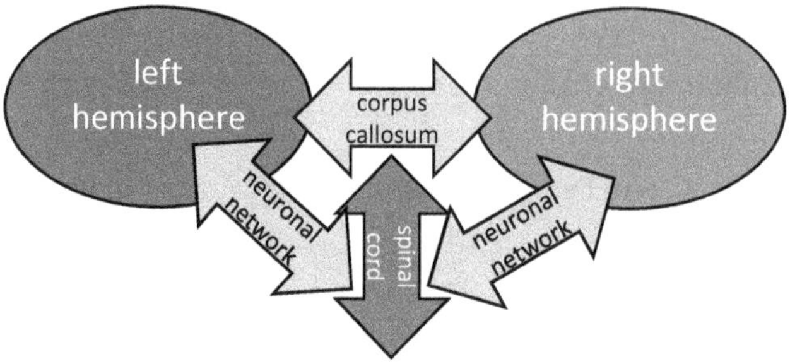

Figure 5.12
Technical brain architecture

Process example

The left hemisphere of the brain states that water boils at 100 degrees Celsius. The right hemisphere observes that when bubbles rise, the boiling point has been reached. The consolidated result is to remove the pot from the heat. This consolidation concludes that the pot must be taken off the fire. The action is then initiated. However, if our hand feels that the temperature is too high, a reflex will pull the hand back to prevent a burn.

The example illustrates that many decisions are often considered from multiple perspectives and then consolidated into a single action. In parallel, protective mechanisms are in place that monitor and accompany the action, leading to a protective response designed to avert potential dangers.

In today's AI solutions, a lot of synchronization, principles for decision making and mapping of scenarios into different context are missing. Even with quantum computing power it would be questionable, if we could find the ideal perception and motion planning algorithms since a lot of the technical sensors are not yet in the quality available as the human perceptions and data management systems.

The ability to capture a scene or task from different perspectives, continuously monitor actions, and implement independent protective mechanisms is essential for deriving safe actions. In this area, we can learn a lot from the human system. Without incorporating such architectural elements into our intelligent automation systems, it will be challenging to develop safe automation solutions for our future mobility needs.

5.3 Fuzzy Logic Could Bridge Human Perception and Automation

Fuzzy logic is a form of reasoning that deals with approximate rather than fixed and exact reasoning. Unlike traditional binary logic, where variables must be either true or false, fuzzy logic allows for degrees of truth. This is particularly useful in situations where information is imprecise or uncertain.

Key concepts of fuzzy logic:

- **Fuzzy sets:** These are sets without a clear boundary. For example, the concept of "tall" can be considered a fuzzy set because there is no precise height at which a person becomes tall.
- **Membership functions:** These functions define how each point in the input space is mapped to a membership value between 0 and 1. For instance, the degree to which a person is considered "tall" can vary gradually rather than abruptly.
- **Fuzzy rules:** These are if-then statements that use fuzzy logic to map inputs to outputs. For example, "If the temperature is high, then the fan speed should be high."

Application in Human Perception

Fuzzy logic mirrors human reasoning by allowing for a range of possibilities. For example, the transition from daylight to darkness is gradual and subjective. Someone accustomed to short twilight periods might perceive darkness differently than someone used to long twilight periods. Similarly, the perception of colors like orange, which is a blend of red and yellow, can vary based on individual perception.

Fuzzy Logic in AI and Automation

In AI and automation, fuzzy logic is used to handle uncertainty and make decisions based on imprecise data. For example, in autonomous driving, decisions such as whether to accelerate or brake, turn left or right, or overtake another vehicle are influenced by numerous fuzzy perceptions like road conditions, traffic density, and the behavior of other drivers. These fuzzy inputs are processed to produce clear, logical actions.

Human Decision-Making

Humans often make decisions based on context, skills, experience, and knowledge. Under stress, the influence of alcohol or medication, or at different times of the day, decisions can vary. For instance, driving decisions on an empty highway differ from those in a busy city. Each decision, whether to drive faster or slower, brake hard or gently, or turn left or right, is based on fuzzy perceptions that lead to clear, logical actions.

Risk Assessment

Fuzzy logic also plays a role in risk assessment. Depending on an individual's risk tolerance, the thresholds for perceiving danger can be higher or lower. This affects decisions in various contexts, from everyday activities to complex strategic planning.

5.3.1 Comparison of Artificial Intelligence and Machine Learning in Automated Driving

Artificial Intelligence (AI) is a broad field that encompasses various technologies and approaches aimed at creating machines capable of performing tasks that typically require human intelligence. These tasks include reasoning, learning, problem-solving, perception, and language understanding. A typical example in automated driving solutions could be as follow: AI systems in autonomous vehicles can interpret traffic signs, recognize pedestrians, and make complex decisions based on the environment. For instance, an AI might decide the best route to take based on real-time traffic data.

Machine Learning (ML) is a subset of AI that focuses on the development of algorithms that allow computers to learn from and make predictions or decisions based on data. ML models improve their performance over time as they are exposed to more data. A typical example in automated driving solutions could be as follow: ML algorithms can be used to predict the behavior of other vehicles on the road. For example, an ML model might learn to recognize patterns in how drivers typically behave at intersections and use this information to anticipate and react to potential hazards.

Fuzzy logic is a form of logic that deals with reasoning that is approximate rather than fixed and exact. It allows for degrees of truth rather than the traditional binary true/false logic. This is particularly useful in situations where information is imprecise or uncertain. A typical example in automated driving solutions could be as follow: Fuzzy Logic can be used to handle uncertain or imprecise data, such as determining the appropriate speed in varying traffic conditions. For instance, instead of having a strict rule like "if the road is wet, reduce speed by 20%," a fuzzy logic system might consider the degree of wetness and adjust the speed more smoothly and appropriately.

Here are some typical examples for each of the technologies in different contexts of application.

Decision making:

- **AI:** An AI system might decide when to change lanes based on a combination of sensor data, traffic conditions, and predictive models.

- **ML:** An ML model could learn from past driving data to predict the safest time to change lanes.
- **Fuzzy logic:** Fuzzy logic might be used to handle the uncertainty in sensor data, such as determining the exact moment to change lanes based on the perceived distance and speed of nearby vehicles.

Obstacle detection:

- **AI:** AI can integrate data from various sensors (cameras, lidar, radar) to detect and classify obstacles.
- **ML:** ML algorithms can improve obstacle detection accuracy by learning from vast amounts of labeled data, recognizing patterns that indicate the presence of obstacles.
- **Fuzzy logic:** Fuzzy logic can manage the uncertainty in sensor readings, such as determining the likelihood of an object being an obstacle based on partial or noisy data.

Adaptive cruise control:

- **AI:** AI systems can adjust the vehicle's speed based on real-time traffic conditions and predictive models.
- **ML:** ML can enhance adaptive cruise control by learning from historical driving data to predict the optimal following distance in different traffic scenarios.
- **Fuzzy logic:** Fuzzy logic can provide smooth adjustments to the vehicle's speed by considering factors like the gradual change in traffic density and the varying speeds of surrounding vehicles.

5.3.2 Fuzzy Control in Automation

Fuzzy control is a method of controlling systems based on fuzzy logic. Unlike classical logic, which only knows the values "true" or "false," fuzzy logic allows for smooth transitions and the processing of imprecise information. This resembles human thinking, which often deals with vague terms and inaccurate data.

Basics of Fuzzy Control

Fuzzification: This process converts the input values of the system (e.g., temperature, speed) into fuzzy sets. A fuzzy set assigns a membership degree between 0 and 1 to each value. For example, the speed "fast" could be described by a fuzzy set that assigns a membership degree of 0 at 100 km/h, 1 at 130 km/h, and intermediate values accordingly.

Rule base: The system is controlled based on if-then rules formulated in fuzzy logic. An example would be: “If the speed is ‘fast’ and the distance to the car in front is ‘short,’ then brake hard.”

Inference: The current input values are linked with the rules in the rule base to determine an output value (e. g., braking force).

Defuzzification: The fuzzy output value is converted into a concrete value that is passed to the system.

Comparison with Human Driving Characteristics

Human perception and reaction while driving can be well compared with the principles of fuzzy control:

- **Sensory organs as sensors:** Our eyes, ears, and sense of balance perceive the environment and provide information about speed, distance, traffic situation, etc. This corresponds to the sensors in a technical system.
- **Fuzzy perception:** Human perception is often fuzzy. We do not always estimate distances or speeds accurately. Fuzzy logic can handle this inaccuracy, similar to our brain.
- **Experience-based rules:** Over time, we learn to react appropriately in certain situations. These experiences manifest as if-then rules. For example: “If it rains and the road is wet, then drive slower.” This corresponds to the rule base in a fuzzy control system.
- **Flexible reaction:** Humans react flexibly to different situations. They can adjust their driving style, e. g., drive more cautiously in heavy traffic. Fuzzy control systems are also able to react flexibly due to their fuzzy logic.

Examples in Automotive Automation

Adaptive Cruise Control (ACC): Modern cars often use fuzzy control for Adaptive Cruise Control. The system measures the distance to the car in front and automatically adjusts the speed. It also considers fuzzy factors such as traffic density or the driving behavior of the car in front.

Anti-Lock Braking System (ABS): ABS prevents the wheels from locking during braking to assure steerability. It uses particularly the wheel speed sensors to monitor wheel speed and adjusts the brake pressure to achieve optimal braking performance. Fuzzy logic measures the road friction, the oil pressure with pressure sensors, lateral and longitudinal acceleration, and yaw rate with the inertial sensors. The wheel speed sensors measure the individual rotational speed of all four wheels. This allows for the estimation of slip and road friction. In combination with longitudinal and lateral acceleration sensors and the yaw rate sensor, it is also possible to infer the centrifugal forces and the vehicle’s slip angle. By combining these estimates, the lateral forces per wheel can be determined, which is crucial information for the vehicle to initiate stabilization interventions via the Electronic Stability Control (ESC) function and serves as a fundamental parameter for torque vectoring. Torque vectoring aims to distribute the engine torque to the wheels individually to achieve maximum cornering speeds.

Advantages and Disadvantages of Fuzzy Control

The following advantages could be considered for fuzzy control:

- **Handling fuzzy information:** Fuzzy control can handle vague and inaccurate data, which is advantageous in many real-world applications.
- **Simple implementation:** Compared to complex mathematical models, the implementation of fuzzy control systems is often simpler.
- **Flexibility and adaptability:** Fuzzy systems can adapt to changing conditions and react flexibly.
- **Unambiguous and traceable behavior:** The fuzzy control differs und solves the uncertainty of measurements, but it could lead to deterministic behavior as a consequence of the measured uncertainty.

The following disadvantages need to be considered:

- **Lack of formal analysis:** Analyzing and formally proving the stability of fuzzy systems is often difficult.
- **Dependence on expert knowledge:** Designing the rule base requires expert knowledge of the system to be controlled.

Real-time and determinism in the measurement is not that precise as in embedded control solutions. Therefore, fuzzy control operates parallel to the embedded control circuits and adapts just the set-points based on the interpretation uncertainties. Real-time and determinism are far better, though, than AI solutions like machine learning etc.

In automation, especially in real-time systems, determinism and the requirements of cascaded control loops are crucial. Temporal and spatial separation between perception and dissimilar algorithms is necessary to ensure reliable and predictable behavior. Fuzzy logic, with its ability to handle imprecise data, can be integrated into these systems to enhance decision-making processes while maintaining the necessary real-time and deterministic properties.

Fuzzy logic provides a framework for making decisions in situations where information is uncertain or imprecise, closely mimicking human reasoning and perception. It plays a crucial role in AI and automation, particularly in applications like autonomous driving.

5.4 Human versus Automated Solutions

An Automated Driving System (ADS) is a system that can perform the entire or parts of the dynamic driving task of a road vehicle on a sustained basis. To achieve this, an ADS should be able to mimic the human driving control system, which is a complex and adaptive system that involves the interaction of the human driver, the vehicle, and the environment. The human driving control system can be modelled using different approaches, such as the models from Endsley, Rasmussen, and Donges:

- Endsley's model is a cognitive model that describes three levels of situation awareness in human performance: perception, comprehension, and projection. These levels correspond to different stages of information processing and decision making of the human driver, ranging from sensing the relevant elements of the environment, to understanding their meaning and significance, to anticipating their future status and consequences.
- Rasmussen's model is a qualitative model that describes three levels of human behavior in goal-directed tasks: skill-based, rule-based, and knowledge-based behavior. These levels correspond to different degrees of cognitive involvement and adaptation of the human to the task situation, ranging from routine to novel and critical scenarios.
- Donges' model is a quantitative model that describes two levels of human steering behavior in motor vehicles: guidance and stabilization. These levels correspond to different aspects of the driving task: Guidance is the anticipatory open-loop control of the vehicle's lateral and longitudinal position, while stabilization is the compensatory closed-loop control of the vehicle's lateral and longitudinal dynamics.
- The "merged model" combines aspects from the three models—skill-based, rule-based, and knowledge-based behaviors—and extends them by incorporating risk perception and a logical sequence for situational awareness. This integrated approach ensures that actions are not only based on predefined skills and rules but also on a comprehensive understanding of the situation, including potential risks. By enhancing situational awareness, the model allows for more informed and adaptive decision-making, leading to safer and more effective outcomes.

Using Markov models, predictions for performance characteristics in states and transitions can be planned and incorporated into the design. These models allow us to evaluate the probabilities of different outcomes and understand how various factors influence transitions between states. By doing so, we can design systems that optimize performance and ensure safety by anticipating and mitigating potential risks. This approach enhances the overall effectiveness and reliability of the system by integrating predictive insights into the design process.

In order to achieve safe automated driving solutions, more and more remote driving solutions come into the picture as bridging technology. Even in validation methods, remote driving technology is known as the driver in the loop validation method.

5.4.1 Functions of Human Control Systems Derived for Automated Systems

Human risk perception is the subjective judgment of the probability and severity of negative outcomes that may occur as a result of one's actions or inactions in a given situation. In the context of driving on public road traffic, human risk perception can influence drivers' decisions, behaviors, and attitudes towards road safety. The following factors can affect human risk perception in driving:

- There are **personal characteristics**, such as age, gender, driving experience, personality traits, driving style, and emotions. For example, young drivers do not have as much experience as older drivers, which results in their reckless driving behavior. Another study found that younger drivers reported significantly longer annual mileage than older drivers.
- There are **situational factors**, such as road conditions, traffic density, speed limits, weather, visibility, and time of day. For example, drivers may perceive higher risks and drive more carefully on wet, dark, or congested roads than on dry, bright, or empty roads.
- There are **social factors**, such as norms, rules, laws, enforcement, peer pressure, and communication. For example, drivers may perceive lower risks and drive more recklessly if they observe other drivers violating traffic rules, if they face weak or inconsistent enforcement, or if they are influenced by their passengers or other road users.
- There are **cognitive factors**, such as knowledge, skills, awareness, attention, and memory. For example, drivers may perceive higher risks and drive more safely if they have adequate knowledge and skills about driving, if they are aware of the potential hazards and consequences of their actions, if they pay attention to the road and the traffic, and if they remember their previous experiences or lessons.

5.4.2 Human in the Loop Concept in Remote Driving Systems

The outstanding advantage of a Remote Driving System (RDS) is the continuous and active involvement of humans in the control system, the so-called "human in the loop" concept (Figure 5.13). Unlike ADSs, where the vehicle makes all decisions independently, in an RDS the Remote Driver (RD) always remains a crucial part of the control process. This ensures both greater flexibility and higher availability, especially in complex or unforeseen situations. The human RD has several central functions. In certain scenarios, ADS can reach its limits, whether due to unforeseen events or a lack of sufficient data to make an informed decision. In such moments, the RD plays a key role.

Humans possess the unique ability to respond creatively and flexibly to new or unexpected situations, while automated systems rely on predefined data and algorithms. An RD can recognize unforeseen obstacles and react accordingly, even if the system fails to interpret them correctly. The RD can take control of the vehicle in real-time when ADS are unable to make appropriate decisions.

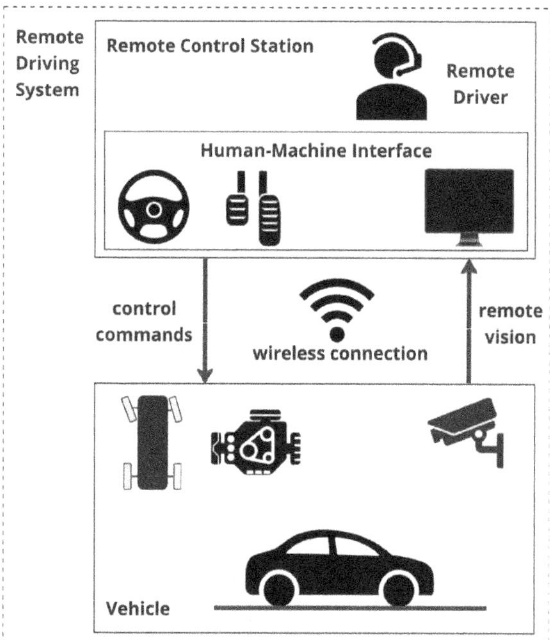

Figure 5.13
Simplified remote driving approach (Hans et al. 2025)

Despite the precise data processing enabled by advanced sensors and algorithms in RDS, human intuition remains an invaluable factor in decision-making. Based on their experience, RDs can quickly and efficiently respond to environmental factors that go beyond mere technical data. Subtle social interactions on the road, such as non-verbal communication between drivers, can only be accurately interpreted by humans.

The collaboration between human control and advanced technology offers the best of both worlds. RDs benefit from the precise data and analyses provided by sensor systems, giving them a comprehensive and detailed view of the situation. This enhances human perception and minimizes errors due to limited visibility or information gaps.

Additionally, RDS can be designed to learn from the interventions and decisions of human Remote Drivers. These learning processes can be used to improve future automated decisions, gradually evolving the system to better align with the Operational Design Domain (ODD) (Hans/Adamy 2025, Hans et al. 2025).

5.4.3 Enhanced Considerations for an Automated Driving System

In the previous sections, an Automated Driving System (ADS) was considered as a system capable of taking over either a portion or the entirety of the driving task from the driver. To achieve this, an ADS should be able to mimic the human driving control system, which is a complex and adaptive system that involves the interaction of the human driver, the vehicle, and the environment. In addition to that, an ADS should be able to replicate the functionality of the three drivers' models, and to integrate them in a coherent and consistent way. This requires the ADS to have several components and subsystems, such as the following:

- **Perception system:** This is the system that collects and processes the sensory data from the environment, such as road geometry, traffic signs, lane markings, obstacles, vehicles, pedestrians, etc. The perception system should be able to provide accurate and reliable information to the other components of the ADS, and to handle uncertainty and noise in the data.

  Similar to the human vestibular perception system that is located close to the ear and the sense of balance, an ADS also needs a sensor set to perceive the position, orientation and the dynamic behavior. In ADS, it is the fusion of the positioning system and the inertial sensor system, which provides information about current accelerations in lateral and longitudinal directions.

- **Environmental system:** This is the system that models and predicts the state and behavior of the environment, such as road conditions, weather, traffic flow, etc. The environmental system should be able to capture the dynamics and variability of the environment, and to anticipate the possible changes and events that may affect the driving task.

- **Vehicle behavior planning:** This is the system that determines the high-level goals and strategies of the ADS, such as the desired route, speed, lane, and maneuvers. The vehicle behavior planning should be able to consider the situation awareness of the ADS, the environmental system, and the human factors, such as the preferences and expectations of the driver and the passengers, and the social norms and rules of the road.

- **Motion planning:** This is the system that generates the low-level commands and trajectories for the vehicle, such as the steering angle, throttle, brake, and gear.

The motion planning should be able to optimize the performance and safety of the vehicle, and to account for the vehicle dynamics, the physical constraints, and the feedback from the other components of the ADS.

- **Odometry:** This is the system that estimates the position and orientation of the vehicle, using the data from the sensors, such as the GPS, the IMU, the wheel encoders, etc. The odometry should be able to provide accurate and consistent information to the other components of the ADS, and to correct the errors and drifts that may occur over time.
- **Speed and trajectory planning:** This is the system that controls the speed and the trajectory of the vehicle, using the data from the motion planning, the odometry, and the perception system. The speed and trajectory planning should be able to follow the commands and trajectories generated by the motion planning, and to adjust them in real time according to the feedback from the other components of the ADS.
- **Actuation:** This is the system that executes the actions and movements of the vehicle, using the data from the speed and trajectory planning, and the actuators, such as the steering wheel, the pedals, the gearbox, etc. The actuation should be able to perform the actions and movements smoothly and precisely, and to handle the possible failures and malfunctions of the actuators.

An ADS should also be able to coordinate and communicate with the human driver and the other road users, using various modalities, such as visual, auditory, haptic, etc. The ADS should be able to provide meaningful human control and responsibility over the driving task, and to ensure the safety and well-being of the driver and the passengers, as well as the other road users.

5.4.4 Conclusion of the Conversion of the Models

Understanding the human control system in driving as a distributed control system provides critical insights into the complex interaction between human drivers, their vehicles, and the surrounding environment. The human driver operates as an adaptive control system, with roles that range from monitoring vehicle status to executing driving tasks based on situational demands. The cascading control system of human drivers—comprising both conscious and unconscious actions—mirrors fundamental principles of control theory, including closed-loop feedback and PID control mechanisms.

These roles and control processes are influenced by factors such as situational awareness, sensory processing, and risk perception. ADSs aim to augment or replace some of these human control functions, but the complexity of human decision-making, especially in real-time and high-stakes scenarios, presents significant challenges. Human drivers integrate multiple feedback loops and dynamically adapt to unpredictable conditions, a capability that current ADSs still struggle to replicate fully. The addition of RDSs, which incorporate human-in-the-loop control, emphasizes the value of human intuition and flexible decision-making in complementing automated systems.

Overall, further exploration into the interplay between human control systems and technological automation will be essential for developing safer and more efficient driving systems. As we continue to refine and integrate models like Endsley's situational awareness and the cascading control approach, both ADSs and RDSs will benefit from a deeper understanding of human perceptual and decision-making processes. This knowledge can inform the design of more adaptive and resilient automated systems that enhance road safety while maintaining driver involvement where necessary.

5.5 Human Driving in Application

The PDCA cycle, also known as the Deming Cycle, is an iterative four-phase model for continuous improvement and problem-solving. The four phases are the following:

- Plan: Identify and analyze a problem or task and develop a plan to improve or solve it.
- Do: Implement the plan and execute the necessary steps.
- Check: Monitor and evaluate the results to determine if the desired improvements have been achieved.
- Act: Based on the results, make adjustments and repeat the cycle to achieve continuous improvements.

This cycle is commonly used in quality control and project management to optimize processes and ensure continuous improvement.

The PDCA cycle can also be applied to human driving. It provides a foundation to define the various phases of driving and then iteratively break them down into each individual action. Here's how the PDCA cycle can be applied to driving:

- **Plan:** The first phase involves planning the trip. For example, if you want to drive from Frankfurt to Darmstadt, you need to consider whether you will need fueling or charging along the way and where you can park at your destination.
- **Do:** Decide on the route you will take, the time you will start, and when you aim to reach your destination.
- **Check:** Before starting, check which car you will use and its availability. Determine if it is an Internal Combustion Engine (ICE) vehicle or an Electric Vehicle (EV). Assess the weather conditions and anticipate any potential traffic jams. Just

before departure, ensure the vehicle is in good condition and has sufficient fuel or charge.

- **Act:** Finally, follow your plan and start driving.

The PDCA cycle can be applied in various forms behind any systematic process that relies on planning. By using this cycle, you can ensure that each task is well-planned, efficiently executed, and continuously improved.

Example

Planning the trip from Frankfurt to Darmstadt is necessary (Figure 5.14). A driver who is familiar with the route might check for traffic jams in advance. A driver who does not know the route will plan to use a road map or set the destination address in the navigation system and then start the journey.

He will **plan** the route, **decide** which route to take, **check** if his car is in good condition or if traffic jams are expected, and whether the batteries are sufficiently charged, or if the fuel tank is full enough. He might also check for traffic jams again, and then he will **act** and start the journey.

Figure 5.14
Mission planning

During the drive, he will perform various monitoring tasks. These tasks can be categorized into primary, secondary, and tertiary driving tasks.

Primary driving tasks involve direct control of the vehicle, such as steering, accelerating, and braking. Secondary driving tasks include activities that support the primary tasks, like checking mirrors, monitoring the speedometer, and adjusting the radio or climate controls. Tertiary driving tasks are those that are not directly related to driving but can still impact on the driver's attention, such as using a navigation system or conversing with passengers (Figure 5.15).

The second planning phase is the trajectory planning. The human driver is no longer aware of trajectory planning. He knows most of the scenes, risks, and possible influences on his driving behavior from many years of driving experience or from driving school training. The appropriate speed is also determined more by emotions, stress in the respective situation, or by one's social awareness.

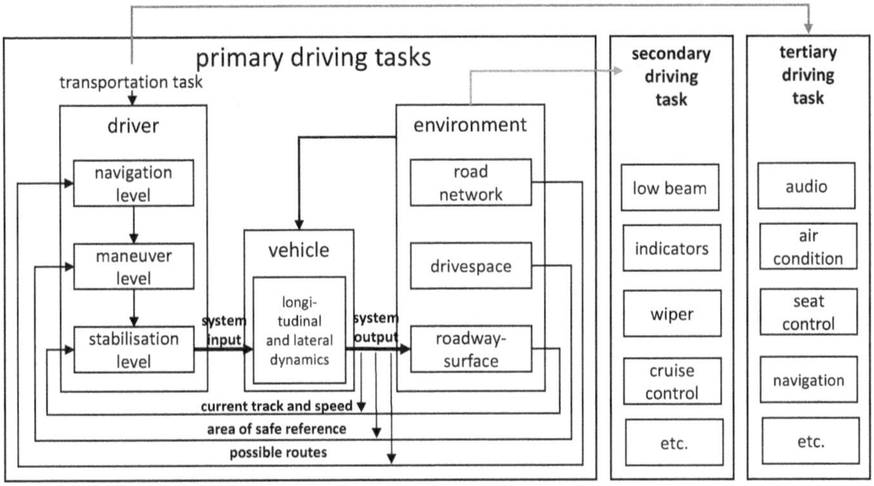

Figure 5.15 Primary, secondary and tertiary driving tasks

To understand these tasks better, we can refer to the driver models by Rasmussen, Donges, and Endsley as we did in detail in previous sections:

- **Rasmussen's model:** This model focuses on different levels of cognitive control in driving, from skill-based (automatic actions) to rule-based (following procedures) and knowledge-based (problem-solving in unfamiliar situations).
- **Donges' model:** Donges emphasizes the importance of anticipatory and compensatory control in driving. Anticipatory control involves planning and predicting future states of the driving environment, while compensatory control deals with immediate corrections based on feedback.
- **Endsley's model:** Endsley is known for her work on situational awareness, which is crucial for safe driving. Situational awareness involves perceiving the environment, comprehending its meaning, and projecting future states to make informed decisions.

By examining these models, we can gain a deeper understanding of the various tasks involved in driving and how drivers manage them. In this section we look deeper into the three categories of driving tasks. Engineering must transfer all three driving tasks from the human driver to an Automated/Autonomous Driving System (ADS), especially when the vehicle excludes the human driver from the driving task.

The primary driving task is three-fold: stabilization of the vehicle, application of the right driving maneuvers, and executing the passenger-intended navigation (Figure 5.16).

- **Stabilization of driving:** This is the basic characteristic that any human driver supporting feature from SAE J3016 Level 1 onward needs to demonstrate. Instability leads to stress for the driver and causes anxiety and discomfort for passengers. Therefore, all road traffic acts worldwide have corresponding requirements for stable driving in various traffic situations.
- **Well-defined maneuvers:** These need to underline features with an increase in taking over responsibility from the human driver. The higher the SAE J3016 level, the more detailed the features underlying maneuvers must be defined. This is again a requirement from all road traffic acts worldwide.
- **Navigation:** The concatenation of defined driving maneuvers is a long-range function. When taken from the human drivers from SAE J3016 Level 3 onward, it also needs a clear definition as combinations of possible maneuvers for complying with legal requirements when entering the autonomous driving world.

The main difference between SAE J3016 Level 3 and Level 2 is not the transfer of driving responsibility from the human driver but rather that the driver cannot over-steer the Autonomous Vehicle (AV) activities.

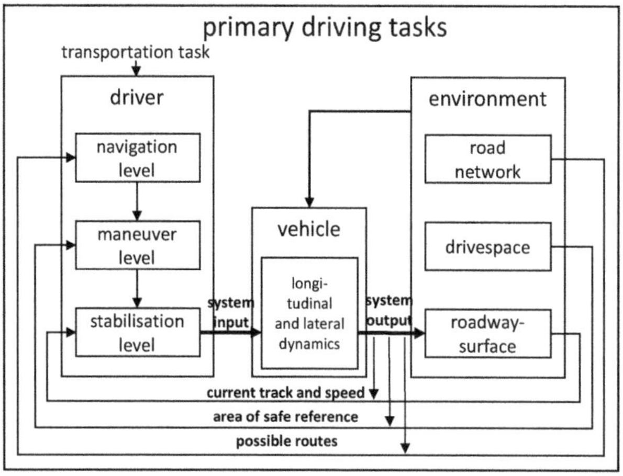

Figure 5.16
Primary driving task

The PDCA (Plan-Do-Check-Act) cycle can be extended to incorporate the functions of Automated Driving (AD) or Advanced Driver Assistance Systems (ADAS) into a corresponding cycle:

Sense—Perceive & Fuse—Interpret—Plan—Stabilize & Activate

The activities could be considered as follows (Figure 5.17):

- **Sense:** This involves collecting data from various sensors on the vehicle, such as cameras, radar, and lidar. These sensors gather information about the vehicle's surroundings, including other vehicles, pedestrians, road signs, and obstacles.
- **Perceive & fuse:** In this step, the data collected from different sensors are combined (fused) to create a comprehensive understanding of the environment. This process helps in reducing uncertainties and improving the accuracy of the information.
- **Interpret:** The fused data is then analyzed to interpret the current situation. This involves recognizing objects, understanding their behavior, and predicting their future movements. For example, the system might identify a pedestrian crossing the road and predict their path.
- **Plan:** Based on the interpreted data, the system plans the vehicle's actions. This includes determining the best route, deciding on speed adjustments, and planning maneuvers such as lane changes or stops. The plan must ensure safety, efficiency, and compliance with traffic rules.
- **Stabilize & activate:** Finally, the planned actions are executed. This involves controlling the vehicle's steering, acceleration, and braking to stabilize the vehicle and carry out the planned maneuvers. The system continuously monitors the execution to ensure it aligns with the plan and makes adjustments as necessary.

By extending the PDCA cycle in this way, AD and ADAS functions can be systematically integrated into the vehicle's operation, ensuring a structured approach to automated driving.

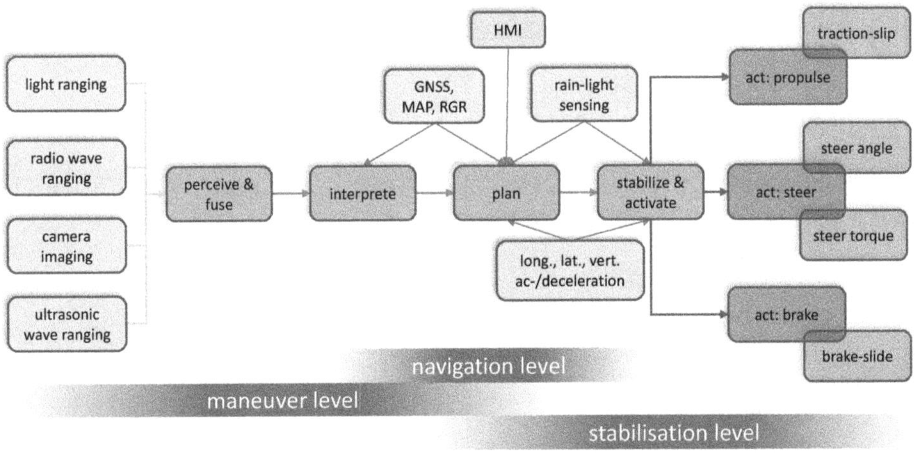

Figure 5.17 Primary functions in AD/ADAS architecture (source: Bauer/Ross 2024b)

The system breaks down the task to the level of trajectory planning for lateral movement, and in terms of longitudinal movement, it adjusts the speed according to the environmental conditions (Figure 5.18).

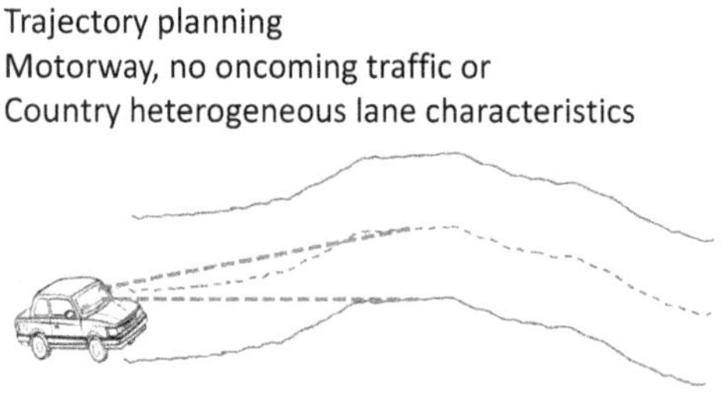

Figure 5.18
Trajectory planning

The secondary driving tasks of the vehicle allow for choosing or maintaining the appropriate interaction with the vehicle environment (Figure 5.19).

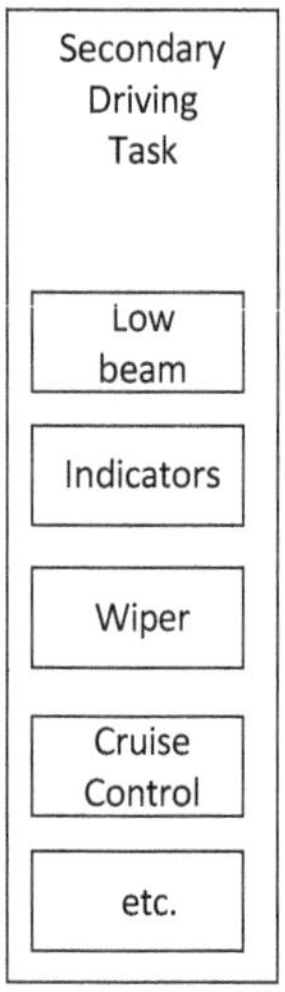

Figure 5.19
Secondary driving tasks

Function category 1: This category enables the interpretation of intended actions by traffic participants. It includes the following:

- direction lighting indicators
- emergency lights
- brake lights
- general lighting
- rear fog lights
- emergency signal and sound detection

Function category 2: This category allows maintaining the interaction of the human driver with the environment. It includes the following:

- windshield wipers
- rear window heater
- window anti-glare

Depending on the ADS feature implemented, these functions may be automated. This includes both the detection of the start and end of their need (e.g., by rain/light sensors) as well as their appropriate performance.

The tertiary driving tasks are usually for comfort and normally not directly safety relevant (Figure 5.20). However, with increasing automation, more and more functions may become part of the overall features and thus become safety relevant.

In the classical split of dashboard and infotainment system, warnings and alarms with safety relevance were indicated in the dashboard. This may change when the dashboard and the infotainment system merge with one meta system. Then, certain functions like indications, warnings, and alarms may become part of these meta systems and need to fulfill safety characteristics.

In highly automated driving features and functions, beginning with SAE J3016 Level 3, a navigation system may become an essential component of the navigation level and thus is safety related.

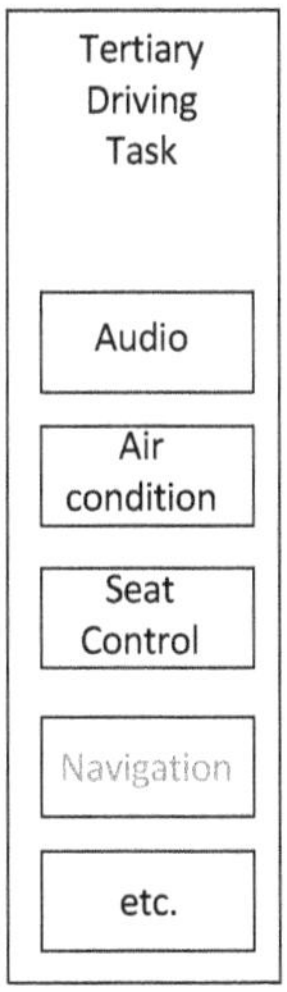

Figure 5.20
Tertiary driving tasks

Not every function relies on neural networks. Specifically, stabilization tasks and motion planning are very deterministic driving tasks. These tasks should be implemented using highly reliable real-time algorithms in the appropriate hardware.

In the guidance phase, which includes light ranging, radio wave ranging, camera imaging, and ultrasonic wave ranging, as well as in their perception and fusion, neural networks can be very effectively utilized. This is especially true when algorithms are designed to capture sensor data from various perspectives with different focal points. Neural networks are also very helpful in the interpretation and planning stages.

You unconsciously check how well you are driving in your lane and whether the traffic situation allows you to drive faster and whether you can cut the curves. The weather conditions and the behavior of other road users are also continuously monitored. Our subconscious also usually checks whether there are external risks, such as accidents on the route or road users behaving unusually. Experienced drivers trust the condition of the car while driving and only react if the vehicle displays any error messages or warnings. Your hearing also comes into play, when changes in the rolling noise indicate flat tires, wet roads or another driver honking to alert us to risks.

In every driving task, the driver goes through the following three phases based on their perception (Figure 5.21):

1. guidance
2. navigation
3. stabilization

This behavior pattern is learned by the driving student during driving school. When the driver consistently goes through these three phases for each driving task, especially for primary tasks like driving, braking, and steering, they can drive stably.

Guidance phase: In this phase, the driver transforms the operational task into tactical or strategic goals. For example, deciding to change lanes or planning a route.

Navigation phase: During navigation, the driver makes decisions about specific actions, such as accelerating, decelerating, and steering. This involves making fundamental decisions on how to interact with the driving environment.

Stabilization phase: In this phase, the driver checks whether the initial decision from the navigation phase leads to the correct behavior. They assess if the intensity of their actions was too strong or too weak and whether their guidance achieved the desired outcome.

These three phases form the basis for the control loops in various driving tasks like driving, braking, and steering. When mapping to automated driving, it is sensible to initiate a feedforward control during the navigation phase. In simple terms, feedforward control means a control based on a prediction or a model of the final actuator behavior. The observed behavior generates appropriate feedback, which then stabilizes the driving behavior. Significant deviations in feedforward control can lead to major accidents that feedback alone cannot compensate for. Therefore, the gradients of feedforward control must be chosen as precisely and conservatively as possible.

Inexperienced drivers often drive slower because these control loops have not yet become part of their implicit and intuitive behavior. The same principle applies to handling uncertain information in automated driving. Experienced drivers drive more proactively and anticipate situations, allowing them to reach the stabilization phase more quickly.

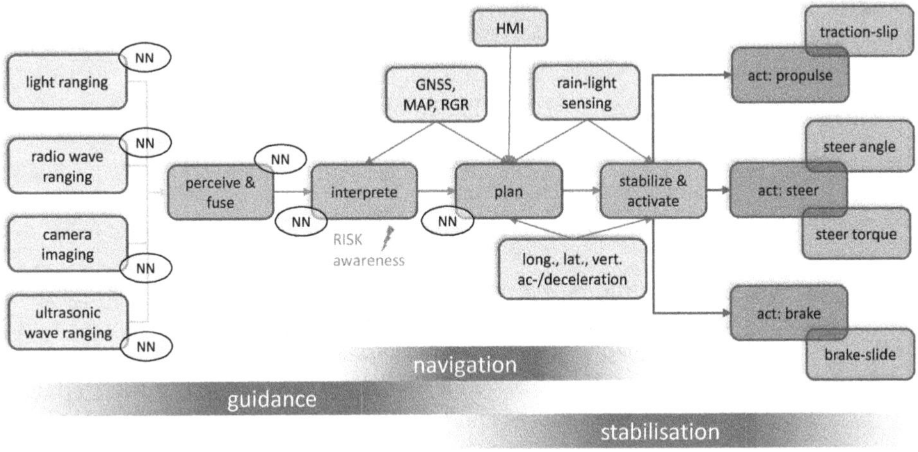

Figure 5.21 Map for future neuronal network solutions

Driving is a matter of continually checking and reacting to events and situational insights that affect the driver while driving. Trajectory planning, which is still partially conscious, transitions into motion planning depending on familiarity with the environment, frequency of unexpected events, and the driver's physical condition (Figure 5.22). Usually, the motion planning mostly takes place in the subconscious. If you ask the driver why he cut the curve at a certain point or whether he noticed a certain oncoming car, he cannot answer. Many of the movements he performs, though, are carried out by underlying functions in his brain, for which his nervous system and musculoskeletal system are trained accordingly.

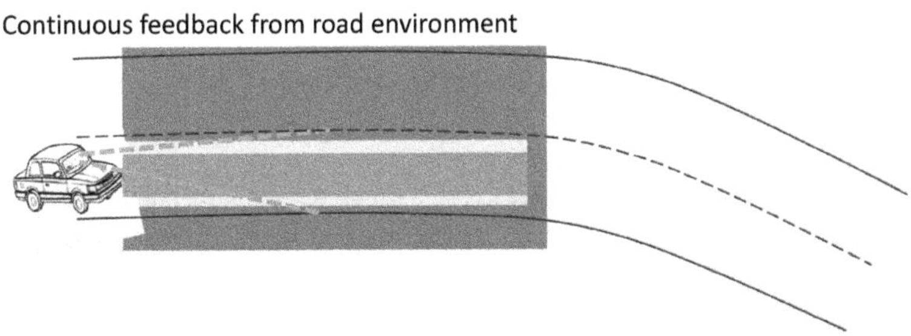

Figure 5.22 Motion planning

The basis for the limits of AI in automotive AD (Automated Driving) solutions stems from the theoretical foundation that AI relies on non-deterministic, probabilistic mathematics (stochastics).

Safety does not inherently require deterministic causality, but it benchmarks against existing systems, the human driver community, and socially accepted ethics. These benchmarks include principles such as GAMAB (Globalement Au Moins Aussi Bon), GMAE (Globalement Moins d'Accidents et d'Erreurs), PRB (Principle of Reasonable Balance), ALARP (As Low As Reasonably Practicable), ENFLI (Ethically Neutral Failure Impact), and MEM (Minimum Endogenous Mortality).

The performance of AI Neural Networks (NNs) highly depends on training, as well as the number of nodes and layers. We can explore whether performance metrics derived from these two parameters can help demonstrate safety evidence.

An evolutionary approach can be employed:

- **Use rule-based causal mechanisms:** These allow for cross-checking feasible AI solutions.
- **Gradual substitution:** With the increase of AI systematicity, gradually substitute causal mechanisms with sophisticated, matured AI neural networks.

This approach ensures that as AI systems become more systematic and reliable, they can progressively take over more complex tasks, while maintaining safety and compliance with established benchmarks.

References

Bauer, B./Ross, H.-L.: Enhanced Risk Perception. Safetronic 2024

Bauer, B./Ross, H.-L.: Primary Functions in AD/ADAS Architecture. Safetronic 2024

Donges, E.: A Two-Level Model of Driver Steering Behavior. In: Hum. Factors J. Hum. Factors Ergon. Soc., Dec. 1978, Vol. 20, No. 6, pp. 691–707 (DOI: 10.1177/001872087802000607)

Endsley, M. R./Garland, D. J. (eds.): Situation awareness analysis and measurement. Lawrence Erlbaum Associates, Mahwah, NJ, USA 2000

Hans, O./Adamy, J.: Learning from disengagements. An analysis of safety driver interventions during remote driving. In: 2025 IEEE Intelligent Vehicles Symposium (IV), IEEE, 2025

Hans, O./Walter, B./Adamy, J.: Evaluation of remote driver performance in urban environment operational design domains. In: IEEE Open Journal of Intelligent Transportation Systems, Vol. 6, 2025, pp. 722–737 (DOI: 10.1109/OJITS.2025.3574692)

Luhmann, N.: Einführung in die Systemtheorie. Herausgegeben von Dirk Baecker. 5. Auflage. Carl Auer 2009

Luhmann, N.: Soziale Systeme. Grundriß einer allgemeinen Theorie. Original publication date: 1984. Frankfurt am Main 2001 (ISBN 3-518-28266-2)

Ross, H.-L.: Safety for Future Transport and Mobility. Springer 2021

Wentink, M/Stassen, L./Alwayn, I./Hosman, R. J. A. W./Stassen, H. G.: Rasmussen's model of human behavior in laparoscopy training. In: Surg Endosc 17, 2003, pp. 1241–1246

6 Artificial Intelligence

Krystian Radlak

6.1 What Is Artificial Intelligence (AI)?

Artificial Intelligence (AI) is a field of computer science that focuses on creating intelligent machines that can perform tasks that typically require human intelligence. AI refers to the simulation of human intelligence processes by machines, especially computer systems. These processes include learning, reasoning, problem-solving, perception, and language understanding. The importance of AI cannot be overstated. It has the potential to revolutionize numerous fields, including healthcare, education, transportation, and more, by automating tasks, providing insightful data analysis, and enabling innovative solutions (Rashid/Kausik 2024).

Unfortunately, there is no single definition of AI. Looking for a definition, we can refer to the Latin noun "intelligentia" and the verb "intelligere," which denotes to understand or perceive and the ability to understand and perceive.

We can also refer to a definition provided by American mathematician and AI researcher John McCarthy, who defines AI as follows: *"Intelligence is the computational part of the ability to achieve goals in the world."* (McCarthy 2007)

The father of computer science, Alan Turing, was one of the first to delve into the concept of AI. In his seminal work, "Computing Machinery and Intelligence," published in 1950, he asked the question, "Can machines think?" and offered the famous "Turing Test" to distinguish between a computer and human text response (Turing 1950).

Another interesting definition was provided by Minsky: AI is *"the science of making machines capable of performing tasks that would require intelligence if done by [humans]."*

Nowadays, AI has been successfully applied in various industries and domains. Every day we can find some information about novel potential applications of AI to make our work more efficient or even fully automated. The AI algorithms are present in science since the beginning of the 20th century when the basics of statistics were introduced, however, recently these algorithms, thanks to Deep Neural Networks (DNNs), achieved the performance that was not achievable for other AI algorithms for decades. The latest AI algorithms have a strong ability to mimic aspects of human intelligence and leverage immense computing power to analyze large datasets quickly, identify patterns, analyze the data from complex sensors, understand text and support decision-making processes. Therefore, AI is now considered as a crucial tool for gaining advantages in industry. Consequently, companies perceive AI as a significant development lever, investing substantial financial and organizational resources into AI research, believing that they will provide better and more advanced products into the market (Harari 2017). AI is expected to make one of its most transformative impacts in the field of transportation, particularly in mobility solutions. The high expectations are particularly focused on Connected Autonomous Vehicles (CAVs) and Advanced Driver Assistance Systems (ADAS). CAVs are considered as a potential game-changer in transportation and mobility solutions. While an individual Autonomous Vehicle (AV) may not yet surpass the abilities of an experienced human driver, a network of self-driving cars is likely to be much safer and more efficient than the unpredictable behavior of human drivers on today's roads. Currently, approximately 1.25 million people die in car accidents each year (World Health Organization 2015), with over 90% attributed to human error (Fagnant/Kockelman 2015). However, Autonomous Driving (AD) systems, which use highly complex AI algorithms as a black box, still raise more questions than answers.

This chapter explores the application of AI in mobility solutions, highlighting critical concerns such as identifying hazards and risks in AI system behavior, managing unpredictable actions in scenarios beyond the AI's training, and addressing the challenge of transparency, which makes it difficult to interpret and validate the decisions made by autonomous vehicles. AI systems may also encounter ethical dilemmas (Gerdes 2020), like choosing between two negative outcomes, e. g., when a crash cannot be avoided and a machine will analyze the likelihood of one individual being harmed against the risk of multiple others being harmed. These concerns exist and they are still not fully resolved. Unfortunately, recent academic and industry experience reveals a lack of understanding regarding the limitations of various AI algorithms, particularly in terms of low-level technical specifics. For example, many AI developers are not fully aware of safety limitations, risks and system contexts of the developed AI algorithms, while functional safety practitioners often lack a solid foundation in technical details of AI algorithms. This chapter aims to piece together the complex AI puzzle related to mobility solutions by clarifying the role of different AI algorithms in AD and ADAS, offering recommendations for both functional safety and Machine Learning (ML) professionals, and outlining future research directions. In this chapter, we avoid complex mathematical definitions, which can be easily found in most of the academic books (e. g., Goodfellow et al. 2016, Russell/Norvig 2022). Instead, we concentrate on clarifying fundamental terms—especially those related to safety—using simple sketches that automotive practitioners from diverse academic backgrounds can easily understand.

6.1.1 Types of AI

Artificial Intelligence (AI) has permeated various industries, revolutionized processes and creating value in unprecedented ways. The term AI is sometimes misleading. Therefore, we start our considerations from attempt to define what is AI and what algorithmic solutions can be hidden behind this term. Starting from the highest level, we can define two main paradigms of AI: symbolic AI and sub-symbolic AI (Ilkou/Koutraki 2020).

Symbolic AI, also known as Good Old Fashioned Artificial Intelligence (GOFAI), was the most important part in early development of Artificial Intelligence. It uses high-level "symbolic" (human-readable) representations of problems. The main assumption is that common-sense problems can be solved by logical reasoning. Symbolic methods include propositional and predicate logic, rule-based knowledge representation, ontologies, and semantic networks. The methods are associated with formal and knowledge-based methods and programing languages such us Prolog or Mercury. These methods typically work well for simple static problems and are favored in areas where clear, human-understandable explanations for decisions are crucial. However, in many complex real-world problems symbolic AI methods may not provide sufficient performance or it is not easy to define logical statements for the analyzed problem.

Sub-symbolic AI in comparison to symbolic AI does not assume any form of knowledge representation. Its goal is to find correlations between input and output variables. Sub-symbolic models are trained to identify such correlation using the provided datasets. Typically, such relations have a high complexity, require large volumes of the data to create the models and the results provided by the models are considerably more difficult to interpret than symbolic methods. Sub-symbolic AI includes statistical Artificial Intelligence, which allow to resolve problems such as regression, classification or clustering. Regression and classification problems can be resolved by *supervised learning methods*. Supervised learning methods require that labeled training datasets are provided. These methods also include very popular Machine Learning (ML) methods, and they subgroup Deep Learning (DL) methods. An exemplary classification task can be an image recognition problem, in which each image (input) is paired with a label (output) indicating the object in the image. Another example of object detection task is a combination of the regression and classification problem, in which the goal is to localize the position of the object within an image and assign this object to a predefined class. In the case of unlabeled data, we can apply *unsupervised learning methods*, which allow to cluster the data and discover hidden structures, patterns and correlations in the available data. Common algorithms in this group are clustering methods such as k-means, DBSCAN or dimensionality reduction techniques such as Principal Component Analysis (PCA) and t-Distributed Stochastic Neighbor Embedding (t-SNE). Exemplary problems resolved by unsupervised learning methods are customer segmentation based on purchasing behavior or anomaly detection, in which unusual data points or outliers are detected. We can also distinguish *Reinforcement Learning* (*RL*), which is a type of learning where an agent learns to make decisions by interacting with an environment and receiving feedback. Its advantage is its ability to learn complex behaviors without explicit supervision, making it suitable for tasks like game playing or robot navigation. Compared to other AI types, RL is unique in its focus on sequential decision-making and exploration-exploitation trade-off. In RL, an agent learns to behave in an environment, by performing actions and discovering errors or rewards. In space technology, RL is used to train satellites and rovers to navigate unfamiliar terrains with limited human intervention.

The relationship between symbolic and sub-symbolic AI and some of their concepts is illustrated in Figure 6.1. Unfortunately, the existing AI methods have highly specialized applications that can only solve concrete, dedicated tasks and typically are not applicable to other domains. These methods are also known as **weak AI** or **narrow AI**. The problem of achieving so-called **strong AI** remains unresolved. Strong AI, which aims to mimics human-like cognitive abilities, is a significant challenge. It goes beyond being merely a tool for studying intelligence, suggesting that a properly programmed computer is, in essence, a mind itself. Despite recently published advancements, researchers have not yet developed systems that fully emulate human reasoning and understanding or being able to teaching itself to solve for new problems. Current AI, including advanced models like ChatGPT, only process information based on patterns, not actual comprehension or reasoning like humans. Strong AI would need to have self-awareness, emotions, and independent thought, which no existing AI possesses. AI remains a topic of controversy because “weak AI” has advanced significantly, becoming increasingly powerful and capable. Meanwhile, “strong AI” has made little progress and remains nearly as limited as it was decades ago (Liu 2021).

Some of the challenges that AI aims to solve include communication planning, knowledge, reasoning and perception. They are classified as weak AI. Perception involves interpreting the environment through artificial senses, processing raw data, and its pre-processing. Communication encompasses interactions using artificial or natural language with humans and other machines. Planning is a cognitive process that identifies goals and develops strategies to achieve them. Knowledge involves conceptualizing, understanding, and representing the world, while reasoning is crucial for problem-solving. AI-related terms are frequently confused and used interchangeably, despite having completely distinct meanings.

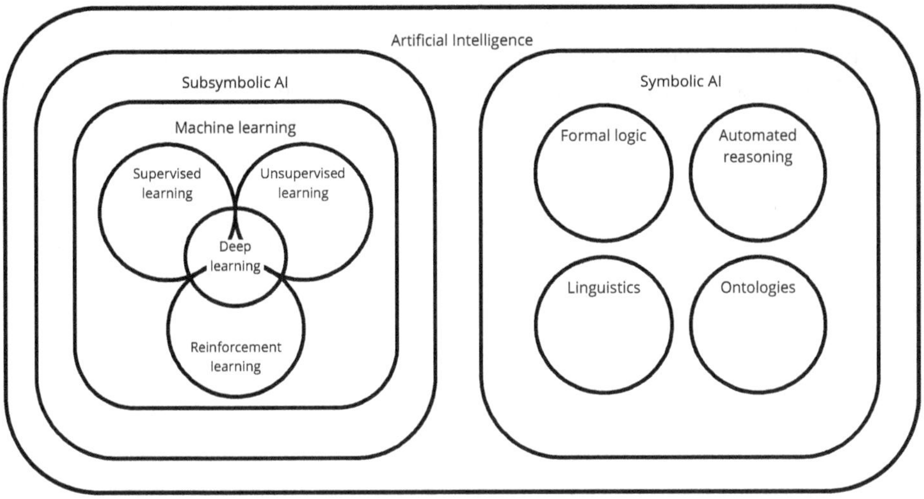

Figure 6.1 Relationship among symbolic and sub-symbolic AI concepts

In this section, we list some fundamental concepts within AI and their applications across different sectors (Figure 6.2), which are discussed in detail in the following sections:

- **Machine Learning (ML)** is a subset of sub-symbolic AI that uses statistical techniques to enable machines to improve with experience. It is extensively used in the automotive industry for autonomous driving. Cars equipped with ML can learn to recognize traffic signs, avoid obstacles, and make safe navigation decisions.
- **Deep Learning (DL)** is a subset of ML, uses Neural Networks with many layers (hence "deep") to model and understand complex patterns. Most of the state-of-the-art algorithms applied in autonomous driving vehicles use Deep Neural Network architectures. DL-based architectures have revolutionized Computer Vision (CV) and Natural Language Processing (NLP), and it would be easier to list the domains in which DL was not yet applied. In medicine, DL algorithms are used for diagnosing diseases by identifying patterns in medical images that may be invisible to the human eye. Nowadays, it is hard to find a discipline in which DL is not applied and it allows to achieve spectacular results.
- **Generative AI (GenAI)** refers to a class of Artificial Intelligence models designed to create new content, such as text, images, music, or even video, based on the data they have been trained on. Unlike traditional AI systems that are programmed to recognize patterns and make predictions, generative AI can produce original outputs that resemble human creativity. It operates by learning from large datasets and using that knowledge to generate new, coherent, and relevant content in response to prompts or questions.

- **Explainable AI (XAI)** refers to the group of models designed to provide clear, understandable explanations for the decisions, predictions, and behaviors of AI models. The goal of XAI is to make the inner workings of complex AI models, particularly those based on Deep Learning (DL) or Machine Learning (ML), more transparent and interpretable to humans. This helps users, developers, and stakeholders understand why and how an AI system arrived at a particular outcome.
- **Natural Language Processing (NLP)** helps machines understand and respond to human language. It's widely used in customer service to power chatbots and voice assistants, providing customers with instant responses and 24/7 support. To the NLP methods belong classical approaches such as bag-of-words model or byte pair encoding, but also ML and DL algorithms, including the recently very popular ChatGPT (Generative Pre-trained Pransformer).
- **Computer Vision (CV)** enables machines to "see" and interpret visual information. In aviation, CV is used for automatic inspection of aircraft components, ensuring safety and reliability. In autonomous driving is used in perception components, which allow to perceive the environment.
- **Robotic Process Automation (RPA)** is the use of software robots or "bots" to automate highly repetitive routine tasks. In finance, RPA is used to automate mundane tasks such as data entry and invoice processing, increasing efficiency and reducing errors.

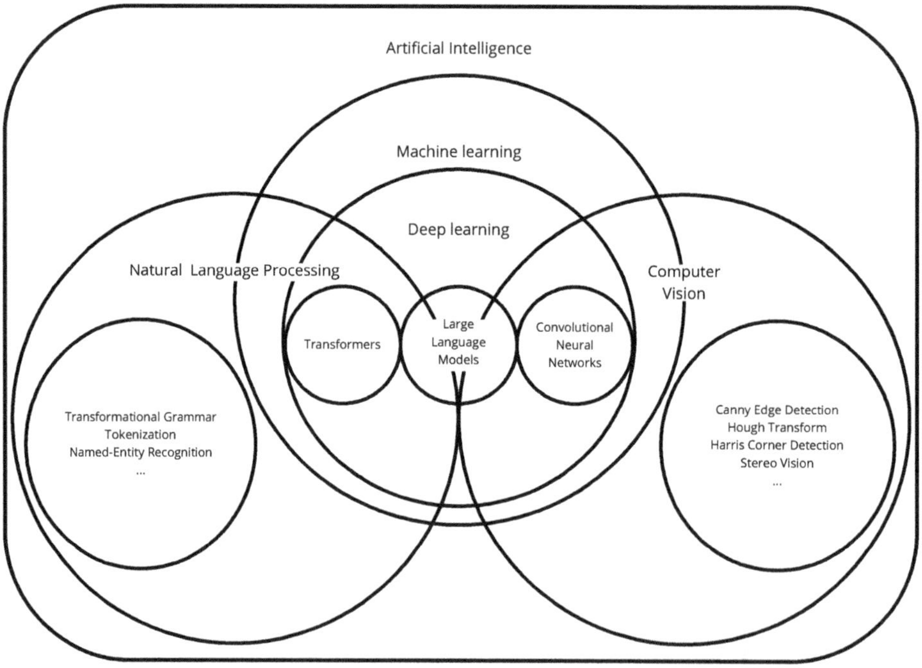

Figure 6.2 The relationship between Natural Language Processing, Computer Vision and Machine Learning concepts

The relationships between AI concepts and paradigms can be highly complex, often with significant overlap. Some algorithms may be applicable to many fields, e.g., very popular transformer models can be used to resolve tasks specific for both Natural Language Processing and Computer Vision. The same problem or task can be also resolved by many different AI algorithms. For example, road lane detection problems can be solved using advanced Deep Learning techniques as well as simple Computer Vision (CV) methods, such as the Canny edge detection algorithm.

Depending on their application context, we can identify the following properties of any of the AI types above:

- ML uses statistical techniques to enable machines to learn from data and improve over time. The key advantage of ML is its ability to process large amounts of data and make predictions or decisions based on patterns it identifies. Compared to other AI types, ML is particularly effective when dealing with vast datasets where manual analysis would be impractical.
- DL uses artificial Neural Networks with multiple layers to model and understand complex patterns. Its strength lies in its ability to learn from unstructured data like images or text. Compared to ML, DL can handle more complex tasks and achieve higher accuracy, but it typically requires more data and computational resources.
- GenAI has applications across various fields, from creative industries (art, design, writing) to more technical domains like software development, drug discovery, and content creation for marketing. It can produce original content such as text, images, music, and designs, enabling faster content creation in various industries. It can also create synthetic data to improve training datasets for Machine Learning models, especially when there is not enough real-world data or collection of such data is very expensive.
- XAI enhances trust, confidence and accountability in AI systems by making their decisions and reasoning processes transparent and understandable to humans. This is crucial in areas where AI models are used to make high-stakes decisions, such as healthcare, finance, or law, and where users need to understand and trust the AI's recommendations and ensure ethical and fair use.
- NLP enables machines to understand and respond to human language. Its advantage is its ability to automate and enhance interactions between humans and machines, making technology more accessible and user-friendly. While ML and DL can be used in NLP tasks, NLP also involves specific techniques for understanding language structure and context, setting it apart from other AI types.
- CV enables computers and systems to derive meaningful information from sensors such as cameras, lidars, and radars, and take actions or make recommendations based on that information. Its advantage is its ability to automate tasks that

require visual understanding, such as image recognition or object detection. While ML and DL techniques are often used in CV, CV also involves specific techniques for image processing and feature extraction. Exemplary classical CV approaches are Local Binary Patters (LBP) or Histograms of Oriented Gradients (HOG) for feature extraction used together with Support Vector Machines (SVM) for classification.

- RPA involves using software robots to automate repetitive tasks. Its advantage is its ability to increase efficiency and reduce errors in business processes. Unlike other AI types that focus on learning and decision-making, RPA focuses on automating rule-based tasks, making it more straightforward to implement and control.

The term "AI" is now commonly used to describe a wide range of algorithms that mimic human intelligence or automate tasks previously performed manually. However, in most cases, these references are specifically directed at Deep Neural Networks. In our view, the term "AI" is often overused in discussions about its potential applications, particularly in the mobility sector, where the focus tends to be on DNN-based solutions. Therefore, in this chapter, we concentrate on Deep Learning (DL) algorithms and provide an explanation of the fundamental concepts related to Neural Network models that can be applied to address various challenges in the automotive industry.

6.2 Deep Learning (DL)

6.2.1 What Is the Difference between AI and DL?

Artificial Intelligence (AI) is a field of computer science that focuses on creating intelligent machines that can perform tasks that typically require human intelligence. Machine learning (ML) is a subset of sub-symbolic AI that involves training algorithms to learn from data, without being explicitly programmed. Exemplary algorithms in this group can be decision trees, Support Vector Machines or linear regression (Hastie et al. 2009). Neural Networks also belong to this group, which creates the subgroup called Deep Learning (DL). Nowadays the most sophisticated and efficient AI algorithms belong to the DL subgroup.

The term "deep" refers to using multiple layers of neurons in a single network. A Neural Network is a mathematical model that mimics how the human brain and the nervous system work and is responsible for processing and transmitting information through electrical and chemical signals. The basic unit of the animal nervous system is a neuron. The nervous system can be understood as a network of the connected neurons, which exchange the energy between them. To put it simply, the neuron accumulates the electric energy, and if it exceeds a certain threshold, then it sends an impulse to other connected neurons, increasing or decreasing their potential. Artificial Neural Networks are working in a similar manner and use neurons as the basic building blocks unit in its architecture. These artificial neurons mimic the behavior of biological neurons by taking inputs, processing them, and generating an output. The simplest artificial Neural Network model is called a perceptron. The neuron converts the neuron inputs to the neuron outputs, multiplying the inputs by neural weights, adding biases, and modifying by an activation function. The most popular activation functions are linear, sigmoid, logistic and ReLU (Rectified Linear Unit) (Nair/Hinton 2010). The neurons can be connected into layers and a network can contain many layers. The layer with neurons which are inputs to other layers are called hidden layers and the last layer is called output layer. One of the most useful networks is the two-layer perceptron due to its simplicity and very good approximation properties. In general, the Neural Network can resolve regression or classification problems. The core part of the network are its weights and the whole knowledge of the model is stored in these weights. A schematic view of the artificial neuron is presented in Figure 6.3.

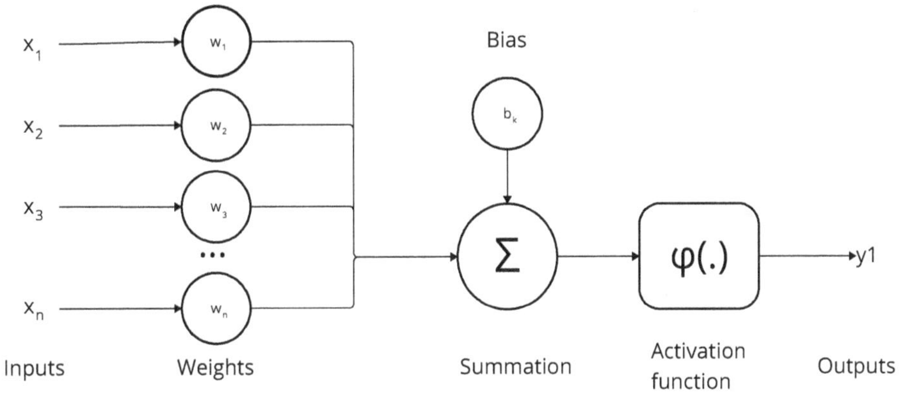

Figure 6.3 Schematic structure of artificial neuron

Historically, the concept of the artificial Neural Network was introduced by Warren McCulloch and Walter Pitts in 1943 (McCulloch/Pitts 1943). Until 2012, Neural Networks have been considered as models which have the tendency for overfitting and were less popular than other Machine Learning algorithms and other algorithms such as Support vVctor Machines were chosen as a better option. The turning point was, when in 2012, Alex Krizhevsky, Ilya Sutskever, and Geoff Hinton introduced AlexNet and won the Large Scale Visual Recognition Challenge (LSVRC), achieving a 10% reduction in errors compared to the previous best in image recognition tasks (Krizhevsky et al. 2012). AlexNet has rediscovered Convolutional Neural Networks (CNNs), which were introduced by Yann LeCun in 1989, but did not become popular at that time (LeCun et al. 1989).

Before considering the potential risks for deep Neural Networks, let's start with introducing the basic terms related to classical Neural Networks. In comparison to a classical approach, in which an engineer defines an algorithm, the Neural Network similarly to other ML algorithms creates its own decision logic based on the examples provided to the model during the so-called training procedure. In classical programing, humans provide both the rules (a program) and the data to be processed, resulting in specific answers. In contrast, Machine Learning reverses this approach: Humans supply the data along with the expected answers, and the system derives the rules (Figure 6.4). These learned rules can then be applied to new data to generate original outcomes (Chollet 2018).

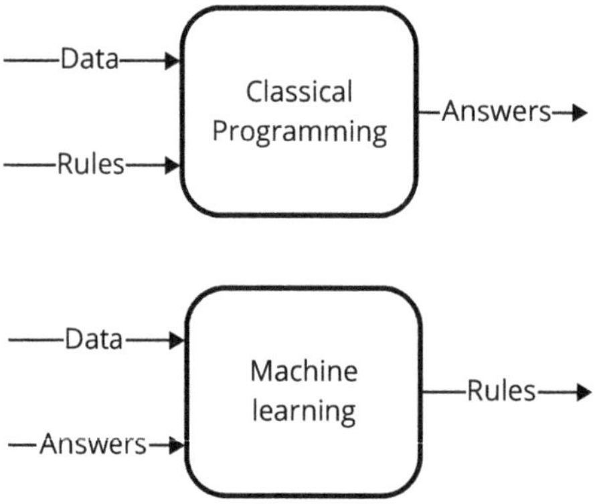

Figure 6.4
Comparison of classical programing and the Machine Learning approach

Such data could be, for example, images with a class assigned to every instance for image recognition problems, or some numerical vectors which represent specific properties of the collected object. For example, this could be a patient with its blood parameters (white blood cells, red blood cells, hemoglobin, hematocrit) and a class assigned to this pattern if they are sick or healthy. During training, the network adjusts its weights as it learns to recognize patterns, allowing the model to form its own internal representation of the data. Once the training is complete, the model's learned weights can be used to make predictions on new, unseen data, such as in a production environment. This process is known as inference. In Figure 6.5, we present an example where a Neural Network is trained to classify an input image into one of three categories: bicyclist, pedestrian, or vehicle. During training, the entire dataset is divided into smaller subsets known as mini-batches. In each forward pass, the model processes these mini-batches and predicts the output classes. These predictions are then compared to the actual, known classes (ground truth) from the training data, and the error is calculated. This error is used to update the model's weights in the backward pass. One complete cycle through the entire training dataset is called an epoch. Typically, the network is trained for multiple epochs, such as 50, to improve accuracy. The obtained model weights are then used in the target environment to analyze new unseen samples in the target environment. In the typical industry setup, the DL models are trained on the dedicated powerful servers, because the model training is computationally expensive, and then the final models are deployed to the target environment, e. g., via Open Neural Network Exchange (ONNX) format.[1]

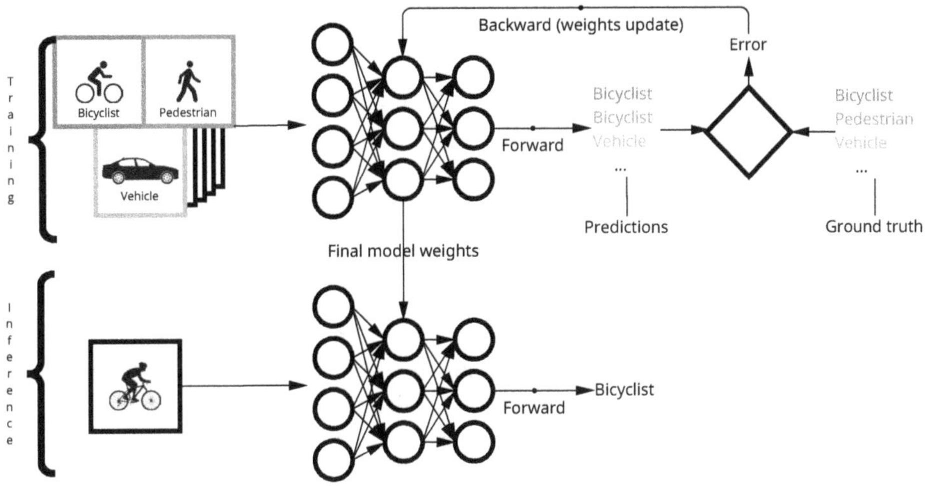

Figure 6.5 The difference between Neural Network training and inference

The primary goal of training is to minimize the error between the predicted outputs and the ground truth labels. The obtained model should generalize well to new, unseen data. The Neural Networks can be considered as a multidimensional approximation function and can share the same problems as those in approximation theory. The approximation theory tries to resolve a problem to represent a complex function using a simpler one. A common method is polynomial approximation, where the degree of the polynomial is chosen. If the selected degree is too low for a highly complex function, underfitting occurs, meaning the approximation fails to capture the function's intricacies. Conversely, choosing a degree that is too high can result in overfitting, where the model fits the observed data too closely but performs poorly on new data. Similarly, training a Neural Network involves approximating an unknown complex function based on observed data. A typical Neural Network can have millions of parameters, and during training, it may also experience underfitting, proper fitting, or overfitting, depending on how well it learns the patterns in the data. Figure 6.6 compares the overfitting and underfitting effect.

[1] ONNX – Open Neural Network Exchange. 2017. *https://github.com/onnx/onnx*

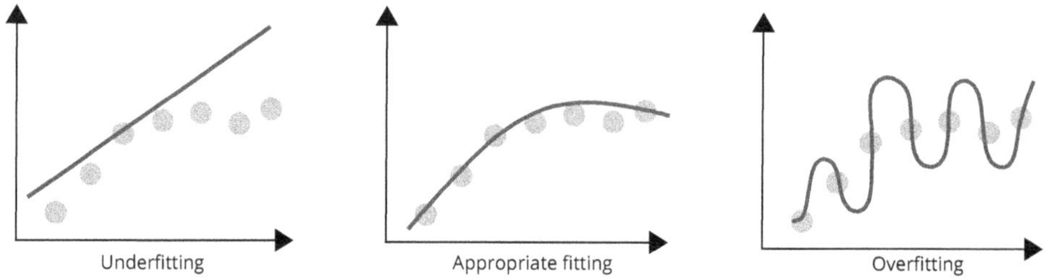

Figure 6.6 Comparison of overfitting and underfitting

The simplest method to detect overfitting is to split our training dataset into two parts: training and validation parts. Then the training is performed on training split, and the model performance is measured on the unseen during the training validation part. During the training, the optimization algorithm allows to decrease the error rate on training dataset, but after some epochs the error rate on the validation split starts increasing. The point in training when the error rate is increasing on the validation split is a moment when to stop the model training in order to avoid an overfitting. The visualization of an early stopping point in the model training is presented in Figure 6.7. Other techniques to prevent overfitting can be dropout, weights regularization, model simplification or data augmentation. Dropout is a regularization technique used in Neural Networks to prevent overfitting. During training, it randomly "drops out" (i. e., sets to zero) a fraction of the neurons in each layer, effectively forcing the network to not rely too heavily on any single neuron. This encourages the model to learn more robust and generalized patterns, improving performance on unseen data. Weight regularization is a technique used to prevent overfitting in Neural Networks by discouraging excessively large weights in the model. It involves adding a penalty term to the loss function, typically using either L1 regularization (which promotes sparsity) or L2 regularization (which penalizes the sum of squared weights). This encourages the model to learn simpler, more generalizable patterns rather than memorizing the training data. Model simplification refers to techniques used to reduce the complexity of a Machine Learning model while maintaining its performance. By reducing the complexity of the model, it becomes less likely to memorize the training data and instead focuses on learning the underlying patterns. This can include strategies like pruning unnecessary neurons in Neural Networks, reducing the number of features (feature selection), or using simpler models (e. g., linear models instead of deep Neural Networks). Simplification helps improve model interpretability, reduces overfitting, speeds up training, and lowers computational costs. Another technique to prevent overfitting is data augmentation, which is used to provide artificially created variations of the same input data to the training dataset. Typical transformations include rotations, flips, scaling, translations, and color adjustments for images, or noise addition and pitch changes for audio. This helps models generalize better by exposing them to varied inputs during training.

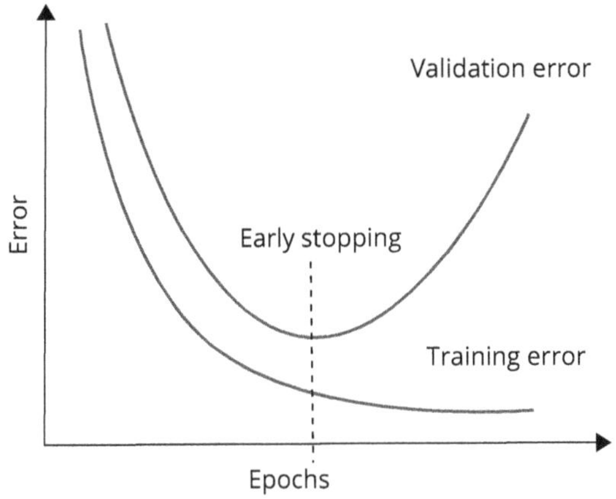

Figure 6.7
Overfitting in Deep Neural Networks: Early stopping allows to identify the best point to stop the training of the model.

In order to achieve a useful DL model in non-safety critical applications, we need to perform at least the following tasks:

1. data collection and normalization
2. definition of the Neural Network architecture and their hyperparameters
3. model training, evaluation and deployment including definition of a criterion for weights adjustment to the data, weights initialization and weights update using the selected optimization algorithm.

In a typical academic course, the system context of the DL model is ignored. Students typically are working on the existing open source datasets and their main task is to optimize the hyperparameters or pretrained models to resolve more complex tasks, e. g., applying pretrained object detection model to automatically measure soccer ball possession. Very often the system context is not analyzed, and DL engineers are focused on a single model and single dataset trying to achieve the best performance for the proposed task running the model on their local computers or rented servers in the cloud. However, the system design and understanding the requirements is one of the key points when we would like to apply DL models in mobility applications. In one of the first works dedicated to ML applications in road vehicles, we argued that every AI/ML model including DL can be developed according to the well-known V-model proposed in the ISO 26262 safety standard (Radlak et al. 2020). Development of DL-based components should not start on a software level but should be designed similarly to other features in the vehicle on the concept level, starting with the item definition and following the V models for system, hardware and software development. In Figure 6.8, we present the V model proposed in the ISO 26262 safety standard adopted for DL-based components development.

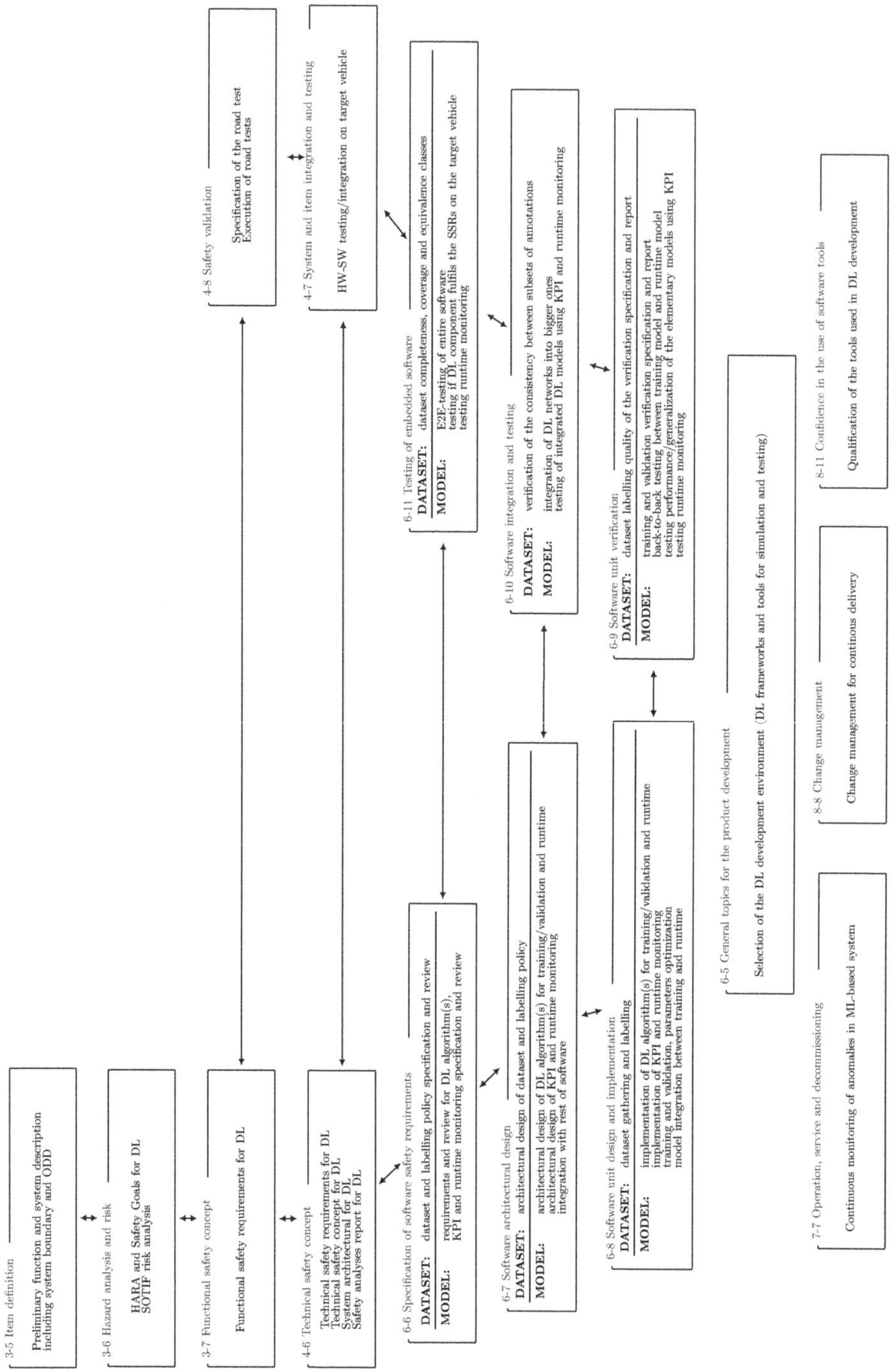

Figure 6.8 The most relevant work products which should be carefully analyzed when the DL-based component is developed in the context of mobility solutions

The proposed model recommends to start the development of DL components on concept level with the identification of related hazardous scenarios. Next, based on the identified safety goals, the system architecture and requirements should be derived including functional, technical safety requirements and safety analyses. Every DL-based component should be considered as a classical ISO 26262 component, but some additional requirements specifics for DL should be considered, because the current version of the ISO 26262 does not cover specific properties of the AI/ML/DL algorithms. Unfortunately, the biggest problem that still seems to be not resolved is the following: Which requirements should be added to ensure that DL-based components are safe enough to be deployed or used in road vehicles or how do you qualify the DL-based tool to not introduce additional risks to classical systems? It should be a clear distinction between algorithms used in perception stack, and algorithms used for trajectory planning and decision making. Recently, the first attempt was done with the release of the new ISO/PAS 8800 safety standard, which tries to address the risk of undesired safety-related behavior at the vehicle level due to output insufficiencies, systematic errors and random hardware errors of AI elements within the vehicle. At the other extreme, we can find tools which support the development and testing activities such as simulation frameworks, static and dynamic code analyzers, IDE, etc. Therefore, similarly like in ISO 26262, the selection of the development environment and tool qualification steps are applicable. Finally, from system level we should perform the hardware and software level related processes and work products.

We can highlight key arguments explaining why the standardization of the AI, ML and DL components for road vehicles should follow ISO 26262 and why this is feasible. Firstly, the ISO 26262 is a widely recognized safety standard with clear terminology, extensively utilized by the automotive industry and functional safety professionals since 2011, which defines well established processes and work products. Secondly, each AI-based component is connected with other non-AI components via well-defined interfaces and moreover each AI-based component is also a specific software which is run on the dedicated hardware. The core difference between AI-based components and classical components is that even correctly implemented AI-based software could fail when outlining input data are provided to the component. From a software perspective, Neural Networks are nothing other than a multiplication of the input signals with weights obtained during the training process of Neural Networks and those weights are the primary source of learned knowledge. Even if software that contains DL components is free of systematic faults and there are no random hardware faults, the DL model still can fail detecting a pedestrian due to specification or performance insufficiencies such as insufficient sensor resolution or a dirty camera lens. These sensors and algorithms insufficiencies are addressed in the ISO 21448 safety standard and are referred to as the Safety Of The Intended Functionality (SOTIF).

6.2.2 Data Collection and Normalization

The data collection process is typically one of the most tedious and expensive steps in ML/DL model creation. This step is not defined in the current version of the ISO 26262 standard. However, the data collection is one of the core steps in DL-based component development. Various Machine Learning models including Deep Neural Networks require thousands or even hundreds of thousands of the examples to create a meaningful model. Most of the students, engineers, and researchers have never created such datasets, and typically the existing and publicly available datasets focus on building sophisticated models that are able to achieve the highest possible performance. However, most of the ML and DL models are only as good as the data.

Many teams in Machine Learning often overlook critical data issues, focusing instead on fine-tuning hyperparameters and optimizing model architectures. This approach can lead to wasted time and suboptimal outcomes, as the quality of data is a key determinant of model performance. Evaluating data quality involves essential activities such as checking the accuracy of annotations, analyzing data distribution, and identifying missing values or noise. Addressing these issues early can unlock greater value from the data and improve overall results. Prioritizing data quality ensures a more efficient and effective model development process. It is also common that such datasets contain various systematic errors made during data collection projects. The latest research presented that even up to 5% in the state-of-the-art datasets for image classification and object detection may contain various noisy labels, which significantly decrease the performance of the models (Chachula et al. 2023). Researchers and students who are interested in development algorithms for vehicle perception stacks typically work on publicly available datasets. Several datasets have been developed specifically for autonomous driving applications. The most popular datasets are summarized in Table 6.1. The last column specifies the number and types of sensors used in the dataset: Camera (C), LiDAR (L), Radar (R). It also indicates whether an IMU was included—'Y' for Yes, and 'N' for No.

Table 6.1 Summary of the existing datasets for autonomous driving applications

Dataset	Location	Scenes	Images	Point clouds	Annotated frames	Number of 3D annotations	Classes	Weather	C/L/R/IMU
Kitti	Karlsruhe, Germany	22	15k	15k	15k	200k	8	Sun, daylight	4/1/0/Y
BDD100K	4 cities, USA	100k	100M	0	100k	0	10	Different weather conditions, day and night	1/0/0/Y

Dataset	Location	Scenes	Images	Point clouds	Annotated frames	Number of 3D annotations	Classes	Weather	C/L/R/IMU
Apollo-Scape	10 cities, China	-	144k	~ 62k	144 images, ~ 12k PC	70k	28	various weather, daytime	2/2/0/Y
Argov-erse	Pittsburgh Miami	113	490k	44k	22k	993k	15	various weather, daytime night, seasons	7/2/0/Y
nuScenes	Boston, Singapore	1k	1.5M	400k	40k	1.4M	23	various weather, daytime night	6/1/5/Y
Waymo	San Francisco, Phoenix, Mountain View	1150	1.15M	230k	230k	12M	4	various weather, daytime	5/5/0/N
Ford	Michigan	30	-	-	0	0	-	various weather	7/4/0/Y
A2D2	3 cities, Germany	-	-	-	-	12k	-	various weather	6/5/0/Y
PandaSet	San Francisco El Camino San Mateo	103	49k	16k	8k	-	28	daylight and night	7/2/0/Y
Soda10M	32 cities, China	-	10M	0	20k	0	6	different weather conditions, daylight and night	1/0/0/Y
Kitti-360	Karlsruhe, Germany	11	300k	80k	80k	68k	19	-	4/2/0/Y
CODA	Singapore, Germany, China	1.5k	1.5k	0	1.5	0	30	-	edge cases selected from existing datasets

Table 6.1 Summary of the existing datasets for autonomous driving applications *(continued)*

Dataset	Location	Scenes	Images	Point clouds	Annotated frames	Number of 3D annotations	Classes	Weather	C/L/R/IMU
aiMotive	California, USA, Austria, Hungary	176	106k	26.5k	26.5k	427k	14	different weather conditions, daylight and night	4/1/2/Y
Zenseact	14 European countries	1473	100k +1473	100k +1473	100k +1473	100k +1473	18	different weather conditions and seasons, day and night	1/3/0/Y

Most of the analyzed databases provide data from at least three sensors: cameras, lidar, and GPS. Only two databases provide radar data. None of the existing databases provide data for ultrasonic sensors. Additionally, nuScenes and Zenseact provide vehicle data collected from a vehicle CAN bus. These data were collected under different weather conditions and different light conditions including day and night. This is important because without this variety, the algorithms learned from them could not be prepared for real driving conditions. The authors of the research papers not always provide all metadata, therefore it is hard without direct usage of the data to determine which weather conditions are collected. Sometimes authors also provide only aggregated characteristics such as "We collected more than one billion of lidar points." Very often the authors do not provide the details about data calibration and synchronization process, and labeling policy according to which the data were annotated, and how they evaluated the quality of the provided annotations.

Most databases provide both images and point clouds. The observed trend is to increase the number of available frames that are annotated. At the same time, the number of annotation methods is increasing. Rectangles and perpendiculars are the basic way to mark objects in the image and point cloud, respectively. We can also observe an increasing number of annotations that assign a class to each pixel or point for scene segmentation tasks. The process of annotating such a large number of frames is very time-consuming and expensive, which is why it is increasingly supported by various algorithms and other solutions. Some authors used an automatic method to generate preliminary annotations and later these annotations are approved or corrected manually by human annotators. Another problem is the consistency of the data annotation, e. g., when only a human hand is visible then the bounding box should contain only this hand or the bounding box should be drawn around the occluded person. We can find in almost every dataset different number of classes. Probably, in the future the data labeling policy should be somehow standardized across countries in order to address the following questions: How are the data annotated? How do you achieve consistent annotations and how do you avoid the presence of the noisy labels in the training datasets? It should be highlighted that when training data contains systematic faults then the models will be affected by these incorrect annotations. In this respect, better defined processes for data collection and annotation could help. Therefore, the data collection and annotation process should be treated as another component that is developed for a specific mobility solution. Organizations which use this data should firstly define the requirements and answer several questions such as the following: What type of the data should be collected? How much data is required? How do you split data into training, validation and test split? What should be the distribution of the data considering the objects distribution? etc.

The data collection process can be supported via the PEGASUS framework (Bock et al. 2018). The main goal of the **PEGASUS six-layers model** is to define safety-relevant scenarios, which use systematics to generate a comprehensive scenario catalog described by six layers which include the following: street layer, traffic infrastructure layer, temporal modification layer, movable object layer, environment condition layer, and data and communication layer.

Within these layers, we can define various types of objects, roads, obstacles, traffic infrastructure, lighting, and weather conditions that need to be accounted for in the collected data. It is important to understand that ML/DL models typically do not perform well on data that differs significantly from what was observed in the training dataset (Wang et al. 2020). One potential issue is that the data collected for training may differ significantly from the data encountered during the inference phase, a challenge known as data distribution shift. This discrepancy can lead to reduced model performance when deployed in real-world scenarios. Another challenge arises when new classes of data appear during inference—data that the model has not encountered before during training but must recognize and incorporate into its knowledge base. This issue can be particularly problematic in mobility applications.

In designing system architectures, it is also crucial to consider that the model may be unable to analyze a given input sample and should have mechanisms to switch the system into a safe state. In DNN architectures, additional functionality can be implemented to detect whether new, previously unseen classes or objects are present. These newly identified classes can then be added to the training dataset, enabling continuous improvement in the model's adaptability and accuracy.

Unfortunately, the current DNN models work in such a way that the model's predictions are unpredictable for data from different distributions. The existing DNN models often struggle with predictions when faced with data that comes from a different distribution than what they were trained on. As a result, there is a significant trend in the research community to quantify **model uncertainty** and assess whether an input sample is in-distribution or out-of-distribution. When an out-of-distribution sample is detected, the system can transition to a safe state and send an unknown sample to a data center for thorough analysis and annotation. Once these samples are carefully labeled, they can be incorporated into the training dataset, allowing the model to be retrained. This iterative process is part of an approach known as **active learning**, which is illustrated in Figure 6.9.

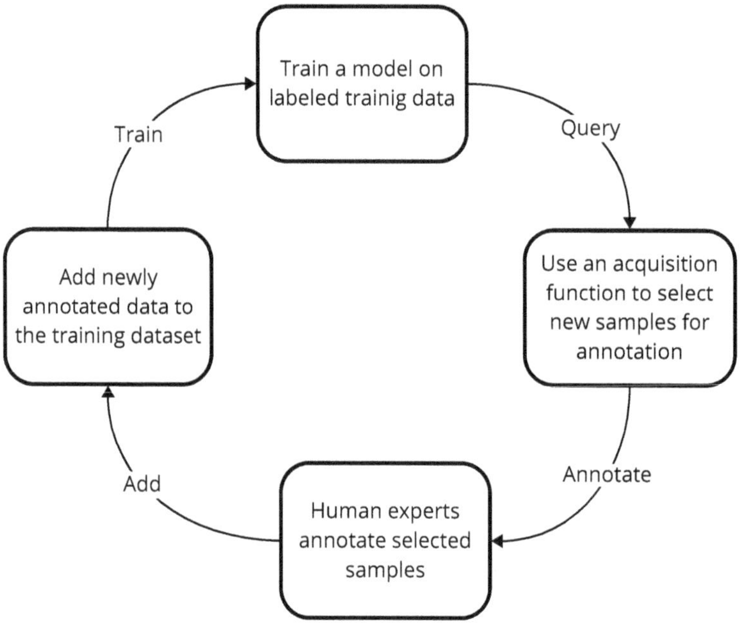

Figure 6.9 Active learning steps

In the active learning approach, the model is initially trained and deployed for using the first version of the training dataset. In the next step, the model works together with a supervisory mechanism that continuously monitors and automatically flags instances where the model's predictions exhibit low confidence. These uncertain samples are then sent to the data center for thorough analysis and annotation. Once new annotated data is available, the model is retrained to incorporate these insights. This process is repeated iteratively, refining the model's performance until the frequency of low-certainty samples is significantly reduced. In that way, the model could also be continuously updated in case of test driver intervention or when edge case scenarios are observed, e. g., incidents or accidents.

In this context, one of the key challenges that can arise during the model retraining process is **catastrophic forgetting**. This phenomenon occurs when a Deep Neural

Network, while being retrained on new data, inadvertently loses the knowledge it previously acquired. Essentially, as the model adapts to new samples—especially those collected from out-of-distribution scenarios—it may degrade its performance on the original training data. This issue is particularly problematic in active learning workflows, where models are iteratively updated with new data. Each time the model is retrained, there's a risk that its ability diminishes to handle previously seen distributions. For instance, while incorporating newly annotated samples to improve its performance on uncertain cases, the model might lose its proficiency on cases it handled well before. Addressing catastrophic forgetting requires careful strategies, such as using techniques like experience replay, regularization methods, or incremental learning, where the model is trained to retain its previous knowledge while still adapting to new information. By effectively managing catastrophic forgetting, the system ensures continuous improvement without sacrificing the robustness it has already achieved on earlier distributions.

6.2.3 Definition of the Neural Network Architecture and its Hyperparameters

Neural network architecture and hyperparameters play a crucial role in determining the performance and efficiency of DL models. The architecture defines the structure of the network, including the number of layers, types of layers, and connectivity patterns between neurons. Meanwhile, hyperparameters are the tunable settings that control how the network learns, influencing aspects such as training stability, convergence speed, and model generalization.

Selecting an appropriate Neural Network architecture requires a deep understanding of the problem domain, data characteristics, and computational constraints. From simple feed-forward networks to complex convolutional and recurrent architectures, each design choice impacts the model's ability to extract meaningful patterns from data. The selection of an appropriate architecture strongly depends on these problems:

- Fully Connected (Dense) Networks (e. g., multi layers perceptron) can be a good choice for tabular data.
- Convolutional Neural Networks (CNNs) are suitable for image data.
- Recurrent Neural Networks (RNNs) are appropriate selections for sequential data.

Similarly, hyperparameters such as learning rate, batch size, number of epochs, and regularization techniques to prevent overfitting must be carefully optimized to ensure effective training. A well-chosen combination of architecture and hyperparameters can lead to high accuracy, efficient learning, and robust performance, while poor choices can result in underfitting, overfitting, or computational inefficiencies.

The selection of hyperparameters is influenced by the chosen Neural Network architecture. However, many hyperparameters remain common across different architectures. The most widely used hyperparameters are summarized in Table 6.2.

Table 6.2 Exemplary hyperparameters that can be defined for DL model

Hyperparameter	Description
number of layers	refers to the total count of computational layers that process data
number of neurons or number of filters	defines the computational power and feature extraction capability of a neural network
activation function	determines whether a neuron should be activated or not, helping the network learn complex patterns
L1/L2 regularization	used to prevent overfitting by adding a penalty to the loss function
dropout rate	prevents overfitting by randomly deactivating (or "dropping") a fraction of neurons during training
optimizer	updates the weights of a neural network in order to minimize the loss function during training
learning rate	determines the step size of updates in the direction of the gradient to minimize the loss function
learning decay	is used to gradually reduce the learning rate during training
number of epochs	defines the number of complete passes through the entire training dataset during training
batch size	determines how many training samples are processed in one forward and backward pass before updating the model's weights
weights initialization strategy	defines the strategy how the weights are initialized

Neural Network architecture consists of several key components that influence a model's learning ability, computational efficiency, and generalization performance. Understanding these fundamental elements is essential for designing effective Deep Learning models.

The number of layers in a Neural Network refers to the total count of computational layers that process data, including the input layer, one or more hidden layers, and the output layer. This parameter significantly impacts the model's learning capacity, computational complexity, and ability to generalize. Similarly, the number of neurons refers to the total count of individual processing units (nodes) within a layer, primarily in fully connected (dense) layers.

An activation function is a mathematical function applied to the output of a neuron to introduce non-linearity, allowing the network to learn complex patterns. It determines whether a neuron should be activated by transforming input signals before passing them to the next layer. ReLU, along with its variants like Leaky ReLU, is commonly used in the hidden layers of deep networks due to its ability to efficiently handle vanishing gradient issues. Sigmoid activation is typically applied in the output layer for binary classification tasks, as it maps outputs to a probability range between 0 and 1. Tanh is often used in recurrent neural networks, as it provides zero-centered outputs, which can help with gradient flow. For multi-class classification, the Softmax activation function is preferred in the output layer, as it converts logits into probability distributions across multiple categories.

To prevent overfitting, regularization techniques impose constraints on the learning process, improving generalization. L1 and L2 regularization add a penalty to the loss function to discourage large weights, making the model more stable. Dropout is another regularization method that randomly deactivates a fraction of neurons during training, forcing the network to learn more robust and generalized features.

An optimizer is an algorithm that updates the weights of a Neural Network to minimize the loss function during training. Gradient-based optimization methods—such as Stochastic Gradient Descent (SGD), RMSProp, AdaGrad, AdaDelta, and ADAM—iteratively adjust parameters using computed gradients to improve convergence.

The Learning Rate (LR) is a crucial hyperparameter that controls how much the model adjusts weights during training. A well-chosen learning rate ensures fast convergence and stability, while an improper choice can lead to slow training or divergence. Learning rate decay (or annealing) gradually reduces the Learning Rate over time, helping the model converge more effectively by taking large steps initially and smaller steps as it nears an optimal solution.

Other key hyperparameters include the number of epochs—which defines how many complete passes through the training dataset occur—and batch size, which determines how many training samples are processed in one forward and backward pass before updating model weights. These parameters influence memory usage, training speed, and overall convergence behavior.

Finally, weight initialization refers to assigning initial values to the network's weights before training begins. Proper weight initialization is critical for convergence speed, stability, and model performance. Poor initialization can lead to vanishing or exploding gradients, making training ineffective.

By understanding these core concepts, practitioners can build more efficient and reliable Deep Learning models while optimizing performance and generalization. From a safety perspective, incorrect selection of hyperparameters and architecture can degrade network performance and lead to issues with reproducibility. Therefore, it is crucial to use appropriate tools that enable traceability between selected hyperparameters and the resulting models. Additionally, carefully selecting the right parameters and understanding their impact help define optimal hyperparameter ranges and initial values, ensuring more stable and effective training.

6.2.4 Training, Evaluation and Deployment of DNN Models

The DNN models are models that are typically trained on a selected dataset to solve specific tasks. Evaluation of the quality of AI models is necessary in order to verify their suitability for the applications that they are intended to serve. Training and validating a DNN model is a multi-step process designed to ensure the model learns effectively from data and generalizes well to unseen examples. To achieve this, a structured approach must be established to support the training and validation process while aligning with the defined requirements. The ML training and validation approach should include the following:

- entry and exit criteria for both training and validation
- strategies for hyperparameter tuning and optimization
- methods for dataset creation and modification
- specifications for the training and validation environment

To implement this approach, data must be carefully selected from the ML data collection and assigned to appropriate training and validation sets. The ML model is then developed in accordance with the defined ML architecture and trained using the designated datasets, ensuring compliance with the ML training and validation strategy, performance requirements, and exit criteria. This structured methodology enhances the efficiency, reliability, and reproducibility of the training of DNN models. After training, the next step is to create a test dataset from the data collection, which will be used to evaluate both the trained ML model and the deployed model. The deployed model is derived from the trained ML model in accordance with the ML architecture. Once deployed, the model undergoes testing to ensure it meets the required performance criteria before being integrated into the software system. It is important that the test data set shall not be used for training purposes because it can build a false confidence about the model performance.

The model's performance can vary across various tasks. The performance of the state-of-the-art Neural Network architecture varies because general object detection cannot be optimal when applied to a face detection problem. Even more, the performance may significantly differ when the model is trained on one dataset (e. g., autonomous driving dataset collected in Germany) and tested on a different dataset (e. g., autonomous driving dataset collected in USA). Therefore, one of the most important steps in preparing ML models is their continuous evaluation on new tasks or when a data distribution shift is observed after the model deployment. The data distribution shift refers to a change in the statistical properties of the data between training and the deployment of (real-world) environments. This shift can lead to a significant decline in a DL model's performance, reliability, and generalization ability, as the model may no longer correctly interpret new data.

Evaluating Neural Network performance requires a set of well-defined data metrics that measure accuracy, generalization, and robustness. These metrics help assess model effectiveness by analyzing predictions, detecting biases, and guiding improvements in training and deployment. The primary way to assess classification quality is by measuring the classification error on a set of examples. This error reflects how frequently the model's predictions for the examples deviate from their true classes. In other words, it shows the rate at which the model makes incorrect predictions. A lower classification error indicates better model performance.

Typically, the error distribution is presented using a confusion matrix, which is a table used to evaluate the performance of a classification model, but it can be also applied for an object detection task. It shows the number of correct and incorrect predictions made by the model, broken down by each class. The matrix typically has four components:

- **True Positive (TP):** The model correctly predicts the positive class. In the case of the object detection problem, it denotes how many objects were correctly detected by a model.
- **True Negative (TN):** The model correctly predicts the negative class. In the case of the object detection problem, it denotes that the model correctly identifies the area without objects and can be treated as background or drivable area.
- **False Positive (FP):** The model incorrectly predicts the positive class (also known as a Type I error). In the object detection problem, it indicates that the model incorrectly predicts the presence of an object in a location where no object actually exists.
- **False Negative (FN):** The model incorrectly predicts the negative class (also known as a Type II error). In the object detection problem, it denotes that an object was present in the scene, but the model failed to detect it.

For multiclass classification, the confusion matrix expands to represent the predictions for all classes, making it a useful tool for understanding the model's performance across different categories. In literature we can find various metrics that are used to measure model performance. The most popular metrics for a classification problem are accuracy, precision, recall, F1-score, and AUC ROC (Area under the Receiver Operating Characteristic) curve.

For object detection problems, the key metric is mean Average Precision (mAP). This metric calculates the Average Precision (AP) for each class and then takes the mean across all classes. Precision measures how many of the predicted objects are correct, while AP evaluates precision at different recall levels. mAP provides a comprehensive assessment of a model's ability to detect objects accurately, considering both precision and recall. A higher mAP score indicates better model performance in detecting and classifying objects correctly.

Which metric should be used depends on the problem and the data that are analyzed. For instance, in medical diagnosis, it is often more appropriate to use recall as the primary metric. Recall emphasizes the model's ability to identify as many sick patients as possible, minimizing false negatives. False negatives denote cases where a sick patient is mistakenly classified as healthy. A high recall score indicates that the model can correctly identify most sick patients, even if it occasionally misclassifies healthy individuals. For example, if 100 patients are sick and the model identifies 95 of them correctly, the recall is 95%, demonstrating its effectiveness in catching most cases. Missing a diagnosis can lead to delayed treatment and potentially severe consequences. Therefore, it is generally preferable to classify a healthy person as sick, because additional tests can subsequently confirm the correct diagnosis. However, if a sick person is classified as healthy, there may be no opportunity to conduct further testing, leading to missed diagnoses and untreated conditions.

In mobility solutions, false negative and false positive detections are safety critical and can have severe consequences. A false negative, where the system fails to detect an object, can result in the vehicle not stopping when necessary, potentially causing a collision with a vehicle ahead or a pedestrian. On the other hand, a false positive, where the system detects an object that isn't actually present, can also have negative consequences. This could trigger unnecessary braking, increasing the risk of a rear-end collision. Both types of errors compromise safety and highlight the importance of achieving a balanced and accurate detection system in mobility applications as increasing the number of true positive detections may also increase the number of false positive detections. Exemplary results of object detection algorithm predictions are presented in Figure 6.10.

Figure 6.10 The exemplary results of the object detection algorithm: The dashed green boxes denote manually annotated vehicles and the blue boxes are predicted as object by a model. Incorrect model prediction denotes false positive (FP) prediction.

6.2.5 Impact of the Dataset Size on ML/DL Performance

The size of a dataset plays a pivotal role in shaping the performance of ML/DL models, as larger datasets typically provide more diverse and representative examples for training. Insufficient data can lead to underfitting and poor generalization, while a well-sized dataset enhances the model's ability to learn complex patterns and perform accurately in real-world scenarios. In Figure 6.11, we present the impact of the size of the dataset on the image classification using the very popular image classification benchmark CIFAR-10[2] that contains 60,000 32 × 32 color images in 10 classes: airplane, automobile, bird, cat, deer, dog, frog, horse, ship, and truck. The experiment was performed using a simple Convolutional Neural Network which contains three convolutional layers and two fully connected layers. As the dataset size expands from 100 to 50,000 samples, a clear trend of improving model performance emerges. With only 100 samples, the model struggles to detect meaningful patterns, resulting in low training and validation accuracies. Overfitting is also more pronounced with this smaller dataset. However, as the dataset size increases, the model's performance gradually improves. When the dataset reaches 50,000 samples, the model achieves its best performance. Both training and validation accuracies show significant improvement at this stage, reflecting superior generalization. This progression highlights the critical role of dataset size in Deep Learning. Larger datasets enable the model to learn more robust patterns and ultimately perform more effectively.

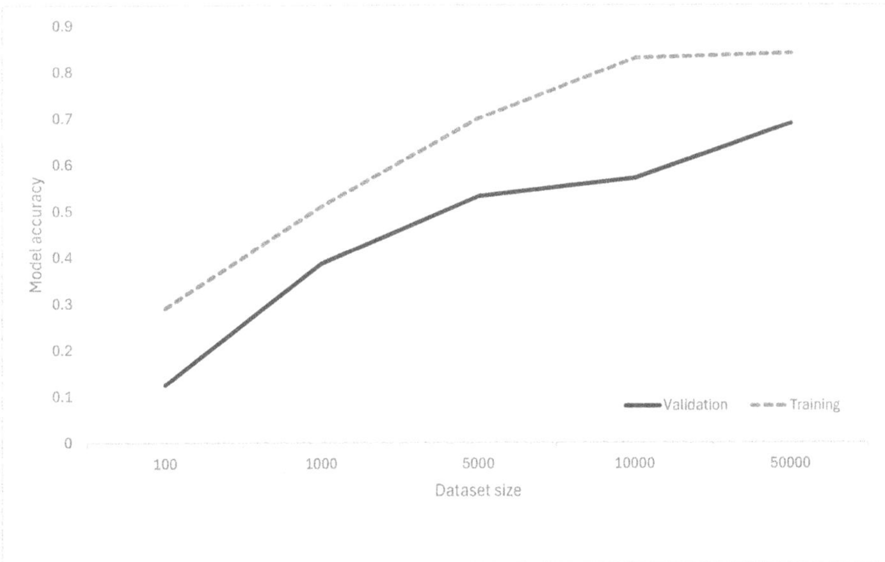

Figure 6.11 Impact of the number of samples in the CIFAR-10 dataset on the final model performance

[2] CIFAR 10 dataset: *https://www.cs.toronto.edu/~kriz/cifar.html*

However, the size of the dataset is not the only critical factor. Different models have different abilities to achieve the optimal performance. Large Deep Neural Networks models are typically capable to achieve better performance on the same amount of data than smaller Neural Networks. Small neural networks are typically better than traditional Machine Learning models, however it depends on the tasks. In some problems, non-neural network-based models are able to achieve better performance (Figure 6.12).

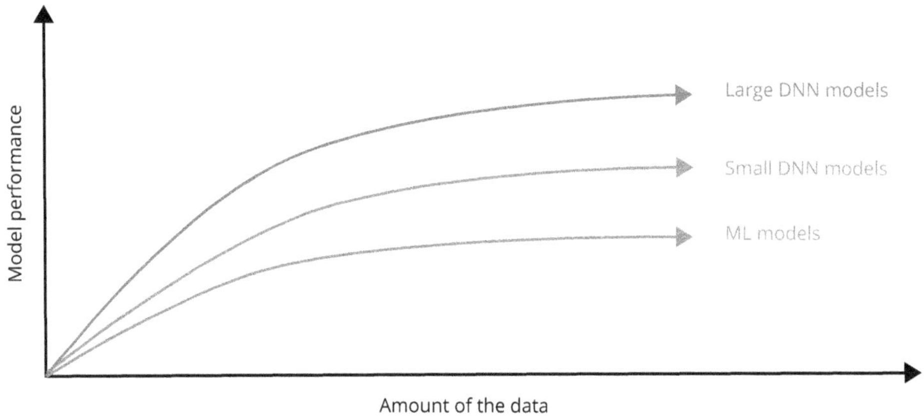

Figure 6.12 Impact of the dataset size on the Machine Learning model performance

6.3 Deep Neural Networks in Mobility Applications

The term AI is widely used in the context of mobility applications. However, most of the algorithms used in AD systems use neural network-based algorithms. Some of the ADAS systems are using classical Computer Vision algorithms, but the main challenge for the automotive industry are Deep Neural Networks. So, most of the standards and regulations try to propose solutions for AI-based systems while in practice the main scope is dedicated for Deep Neural Networks. In this section, we present the most popular tasks and the most popular models that are used to resolve various problems related to mobility applications.

The first approach that applied Neural Networks for Autonomous Vehicle navigation was ALVINN (Autonomous Land Vehicle in a Neural Network) method proposed in 1989. ALVINN used camera images and laser range finders as inputs and was able to drive 0.5 meters per second (Pomerleau 1989). The network was trained using only 1200 simulated road images at resolution 30×32 without sophisticated GPU hardware.

The autonomous vehicles can be built following the robotic paradigm: Sense—Plan—Act. According to this paradigm, autonomous robotic systems can be divided into three subsystems: perception subsystem, planning subsystem, and control subsystem. Nowadays, DL algorithms can be applied in all subsystems, but the most common approach in the past was to apply DL models only for perception tasks. Planning and control subsystems were designed and implemented without DL using classical approaches, in which the existing safety standards such as ISO 262626 or IEC 61508 can be directly applied. The simplified overview of how Sense—Plan—Act architecture that can be applied in road vehicles is presented in Figure 6.13.

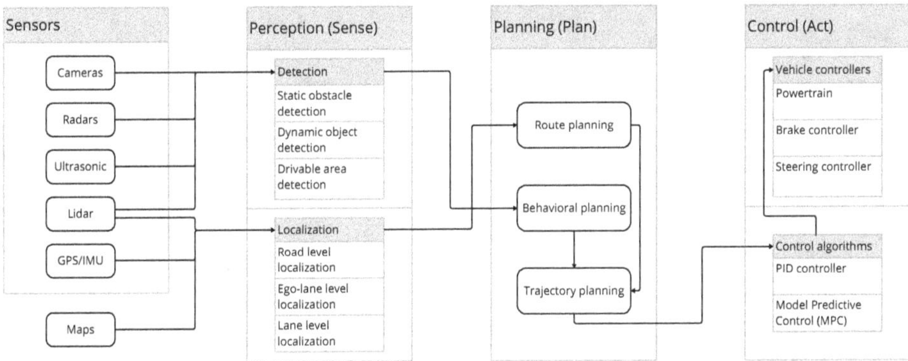

Figure 6.13 Simplified overview of the Sense—Plan—Act architecture in autonomous driving vehicles

Perception involves transforming raw sensor data into meaningful information. It enables the detection of obstacles, vehicles, and vulnerable road users such as pedestrians, cyclists, and motorcyclists around the ego vehicle, determining their distance, size and speed. Another role of perception is the identification of road lane lines. To apply any algorithm, firstly, it is essential to use sensors to capture environmental information. The primary types of sensors used for perception tasks are cameras, lidars, radars, and ultrasonic sensors. To identify potential algorithmic faults, we must first understand how these sensors operate and recognize their limitations. A short summary of the most important properties of the various sensors used in mobility solution is presented in Table 6.3.

Table 6.3 Summary of the sensor properties used in mobility solutions

Attribute	Camera	Lidar	Radar	Ultrasonic
spatial resolution	high	medium	low	low
type of data	2D or 3D array	point cloud	array/list of detections	list of detec-tions
range	1–200 m	2–400 m	0.2–300 m	0.05–2 m
price	low	high	medium	low
weather condition	high impact	high Impact	low impact	low impact
night driving	high impact	low impact	low Impact	low impact
object classification	good	good	poor	poor

Not all sensors can accurately perceive the environment under varying lighting and weather conditions. As a result, understanding the Operational Design Domain (ODD) is essential when designing and developing new systems. The ODD defines the specific conditions—such as weather, lighting, road types, and traffic scenarios—under which an Autonomous Vehicle (AV), a semi-autonomous vehicle or ADAS system can safely operate. Different sensors have varying strengths; some perform better in adverse weather, while others may struggle with tasks like recognizing traffic lights. Therefore, a combination of sensors is necessary to achieve a comprehensive understanding of the environment. It's also important to emphasize that if a sensor fails to accurately capture an object's characteristics, any AI algorithms relying on that data will not be able to interpret it correctly. Thus, robust sensor performance is critical to the overall effectiveness of autonomous systems.

Another challenge is sensor reliability. To prevent collisions and ensure road safety, automotive sensors must be highly reliable, requiring rigorous and repeated testing of their performance. This involves thorough diagnostics to confirm that sensors are functioning correctly. One method for assessing sensor performance is calibration at specialized service stations. However, unlike offline calibration, a more proactive approach involves continuously monitoring the health of automotive radar sensors in real time. By tracking changes in the performance of sensor components, fault prediction systems can detect any deviations. This real-time monitoring allows for dynamic assessment and management of sensor performance, promptly identifying faults and addressing their progression. Such systems should provide ample warning and data for timely fault prevention and resolution. Additionally, they should be capable of forecasting the sensor's future condition, estimating its remaining lifespan, scheduling maintenance, and ensuring the availability of replacements.

Another key challenge for automotive sensors is conducting a thorough hazard and risk analysis. A deep understanding of how each sensor functions is essential for recognizing its limitations and identifying potential faults. For this reason, we provide a detailed discussion of each sensor type to highlight these aspects.

One of the cheapest sensors that are widely used in mobility applications are **cameras**. In the market, we can find various types of the camera setups, which allows to achieve different properties. Cameras can be used for object detection tasks in AV but also to capture environment in remotely controlled vehicles. Monocular cameras that capture 2D images over single lens are used for object detection, lane detection, and recognition tasks. Stereo vision cameras are paired cameras that mimic human binocular vision to capture 3D depth information by comparing the images from two slightly different perspectives. The last group are omnidirectional cameras that provide a 360-degree view around the vehicle, and that useful for comprehensive situational awareness. Recently, also event cameras were introduced which are imaging devices that react to local variations in brightness. Unlike traditional cameras that capture images with a shutter, event cameras function differently. Each pixel within an event camera works independently and asynchronously, signaling changes in brightness as they happen and remaining inactive otherwise. These cameras are not yet popular in mobility applications.

State-of-the-art automotive image sensors typically have resolutions ranging from 1 megapixel (MP) to 8 MP. Higher resolution allows to capture more detailed images of the environment and enables the detection of objects from greater distances. The image data is typically returned by the camera as 2D (grayscale image) or 3D array (RGB color image) where each element represents a pixel. Unfortunately, the research presents that if the image data for Deep Learning are not provided in the same manner during training and inference, this can negatively affect the model performance. Therefore, it is necessary to understand more closely how a camera typically creates a digital image. The camera image acquisition pipeline is presented in Figure 6.14 (Li 2010).

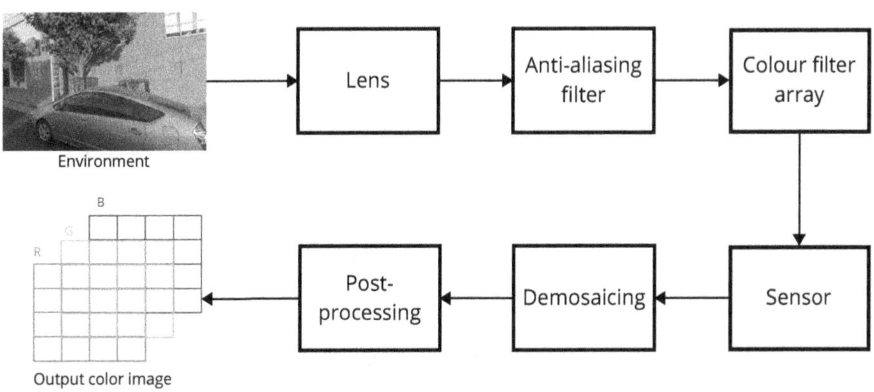

Figure 6.14 Image acquisition pipeline for color images

The light reflected from the objects in the environment passes over lens and anti-aliasing filters and reaches the Color Filter Array (CFA). CFA is a mosaic or pattern of color filters placed over the individual photodiodes of an image sensor to capture color information, because individual photodiodes in image sensors can only detect the intensity of light, not its color. Every photodiode captures only a single color component (R or G or B component per pixel), and the two remaining components are reconstructed via a demosaicing algorithm. The most common CFA pattern is the Bayer filter, which uses a 2×2-grid with two green, one red, and one blue filter (because the human eye is more sensitive to green light, so green is given more representation). Common factors that can influence the camera performance are the following: lens occlusion or soiling, weather conditions (e. g., snow, fog, rain), light conditions (e. g., sunrise, sunset, night), optical faults, sensor faults, and visibility distance. Possible defects such as blooming, smear, picture element defects or thermal noise can significantly affect the quality of the collected and processed visual data. It is also important to understand that a model trained on high-quality data will not work if camera data are degraded. Therefore, during the design of the system which uses camera data, their quality should be considered as an important factor that can limit the performance of the whole system.

LIDAR (Light Detection and Ranging) combined with cameras is currently the fundamental sensor used in autonomous driving systems. Lidar uses laser beams to measure distances to objects and creates detailed 3D maps of the environment. This sensor is highly accurate in providing depth information and identifying the shape and size of objects. It is very effective in a wide range of lighting conditions, including complete darkness. The lidar outputs contains 3D point clouds represented as X, Y, Z coordinates that correspond to the scanned environments and the reflectivity that corresponds to the strength of the reflected laser signal. Sometimes additional information like intensity, timestamp, and ring number may be also included depending on the manufacturer. A simplified representation of the lidar operation principle is shown in Figure 6.15. The lidar measures distance in a specific direction using direct or coherent methods via a scanning system, involving a transmitter (Tx) and receiver (Rx) (Li/Ibanez-Guzman 2020).

A lidar system can be divided into two main components: the laser rangefinder system and the scanning system. The laser rangefinder includes the following elements: **(1)** a laser transmitter that illuminates the target using a modulated wave; **(2)** a photodetector that converts the reflected photons into an electronic signal through optical processing and photoelectric conversion; **(3)** the optical components that collimate the outgoing laser beam and focus the reflected signal onto the photodetector; and **(4)** signal processing electronics that calculate the distance between the laser source and the reflecting surface based on the received signal. The scanning system is responsible for directing the laser beams across various azimuths and vertical angles, defining the beam's pointing direction. The laser rangefinder can use two types of signals:

pulsed lasers, where their Time of Flight (ToF) can be measured, and direct-detection laser rangefinders. The laser signal can also be a Frequency-Modulated Continuous Wave (FMCW), which indirectly measures the distance and velocity from the Doppler effect, and which are known as coherent-detection laser rangefinders. The most popular solution used by most lidar manufacturers is the ToF lidar.

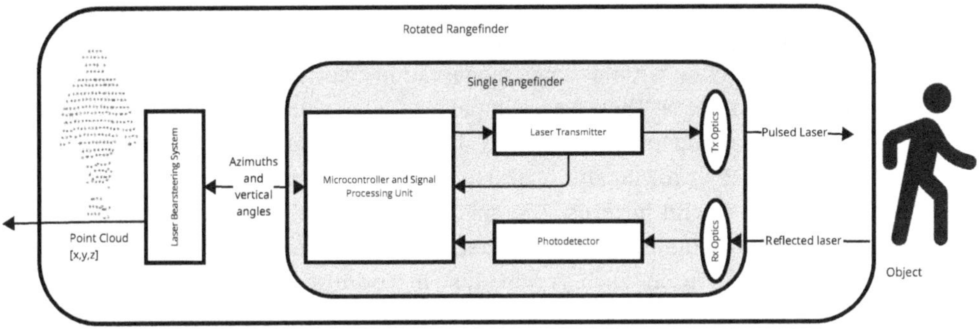

Figure 6.15 Schematic for lidar rangefinder

On the automotive market, we can find various types of lidar sensors that use different scanning systems. Mechanical spinning lidars were the first type of lidars introduced into mass-produced cars, which steer the laser beams through a rotating assembly (e. g., mirror, prism, and so on) controlled by a motor to create a large FoV. Recently, to reduce cost and improve robustness, many companies focus on solid-state scanning systems, which have a scanning system without the rotating parts. These lidars behave as a camera, in which a single laser is spread by an optical diffuser to illuminate the whole scene at once.

Although advanced lidar systems are built for outdoor use and offer high precision, their mechanical components and optics remain sensitive to shocks. It is essential to position these devices securely and strategically on the vehicle. Ideally, lidar units should be installed in areas with high ground clearance and minimal surrounding of vehicle components to prevent obstructions in their field of view. Proper maintenance is also vital, as damage can occur due to impacts from foreign objects, shocks, or vibrations caused by collisions or driving on rough terrain. The lidar's optical filter is particularly critical because any damage to it can significantly reduce the accuracy of the measurements, compromising the system's reliability. Similarly, as with the cameras, the object detection algorithms cannot deal with the low-quality of the lidar data or data from different sensor manufacturers or collected in different geographical locations. A recent study explored the use of lidar technology for object detection problems, focusing on testing systems developed in Germany under different conditions in the USA (Wang et al. 2020). The research aimed to assess the effectiveness of state-of-the-art 3D object detectors trained across diverse environments represented by various AV databases, particularly addressing challenges like varying weather conditions, terrain, and infrastructure differences. The research showed that the performance of the model trained on the data from one database (e.g., Kitti database recorded in Germany) cannot be directly used on the data from a second country (e.g., Waymo database recorded in USA). The performance of the model (Average Precision metric) trained and tested on the Kitti database was 82.5%, while the model trained on the Waymo database and tested on Kitti achieved only 13.1%. The performed experiment clearly showed that the model trained on one dataset cannot be directly used in a new domain without domain adaptation algorithms or without the preparation of a new database. Therefore, companies that develop perception algorithms in one geographical region cannot directly deploy their models in new regions. The usage of lidars in AV or ADAS require also to detect and mitigate potential failure modes that can cause several hazards. The engineers who use the lidars in their systems should also clearly understand the limitations of the lidar which are for example adverse weather conditions such as rain, dust, fog, or dust. Additionally, the farther the object is away from the lidar, the smaller is the number of points that represent the object in the point cloud. At closer ranges, more lidar points typically hit the pedestrian, making it easier to recognize them. At greater distances, fewer points may be captured. The object detection algorithms to detect an object need a certain number of points. Typically, field tests and empirical tuning are essential to determine the precise point density needed for a specific system or application. Some use cases may require to use more expensive lidars with a higher point cloud density or combine lidar data with other sensor modalities like cameras or radars.

The next sensor widely used in mobility applications is **RADAR** (Radio Detection and Ranging). Radars are much cheaper than lidars, but they offer much lower resolution. Radar is a type of sensor that uses radio waves to detect the range, speed, and angle of objects. It emits radio waves to be reflected by an object and the signal runtime is measured to approximate radial velocity using the Doppler effect phenomenon (Gamba 2020). The radar is very robust in various weather conditions like fog, rain, and dust, making it reliable in adverse conditions. The radar is very useful for detecting the speed of moving objects and for long-range detection. However, the radar's low resolution makes it challenging to accurately classify different types of objects. Radars are typically used in adaptive cruise control systems, and collision avoidance systems to determine the information about the distance to the objects and their speed. A simplified schematic of the radar system is illustrated in Figure 6.16.

The waveform generator creates a signal, typically a Frequency-Modulated Continuous Wave (FMCW) or pulsed waveform. The generated waveform determines the radar's operational characteristics such as resolution, range, and Doppler sensitivity. The signal from the waveform generator is typically at a lower frequency and is upconverted to a higher frequency (e.g., 77 GHz) to be transmitted over long distances. The up-converter uses a local oscillator and a mixer to shift the signal to the desired frequency. The power amplifier boosts the up-converted signal's power to ensure it can travel long distances and reflect off objects. This is crucial for detecting objects at long range, such as vehicles or obstacles hundreds of meters ahead. The duplexer allows the same radar antenna to be used for both transmitting and receiving signals. When transmitting, it directs the amplified radar signal towards the antenna. When receiving, it protects the sensitive receiver components from being overwhelmed by the transmitted signal. The transmitted radar waves travel through the environment and reflect off objects such as vehicles, pedestrians, or obstacles. The reflected signals (or echoes) return to the radar system, carrying information about the object's distance, velocity, and angle. The returned weak signals are first amplified by a low-noise amplifier and ensure that the signal remains strong without introducing excessive noise. The amplified signal is then down-converted from the high-frequency radar signal (e. g., 77 GHz) to a lower frequency for easier processing. The down-converted signal is sent to the signal processor, which extracts valuable information such as the following:

- **range (distance to the object):** calculated based on time delay between transmission and reception,
- **velocity (speed of the object):** determined using the Doppler effect,
- **angle (direction of the object):** measured using beamforming or multiple antennas

Advanced Machine Learning and sensor fusion techniques may be used to improve detection accuracy and integrate data from other sensors (e. g., LiDAR, cameras). The HMI displays the processed radar data in a meaningful way, either to the driver (ADAS systems) or to an autonomous driving system.

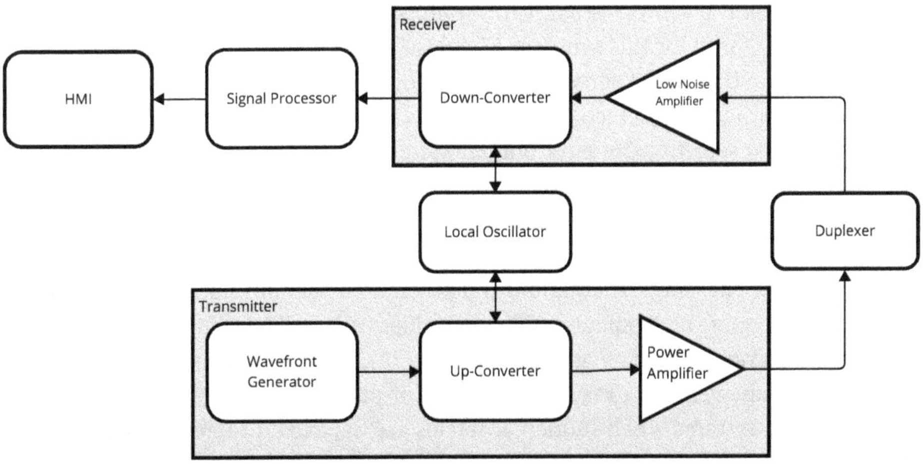

Figure 6.16 Simplified radar system

Various types of radar sensors are available on the market, each designed with specific capabilities. A radar's primary characteristic is its range, or how far it can detect objects. This range varies depending on the type of radar:

- **Short-range radars** typically operate at distances of up to 30 meters and are known for their high resolution. These radars usually operate in the 77 GHz to 81 GHz frequency band, allowing for precise detection of objects at close range.
- **Medium-range radars** cover distances between 30 meters to around 150 meters and balance the resolution and range. They often operate in the 24 GHz or 77 GHz frequency bands.
- **Long-range radars** are designed to detect objects at distances up to 300 meters or more, making them essential for applications requiring early detection and long-range awareness. These radars also typically operate in the 77 GHz frequency band, but with a wider beam and greater range.

In order to correctly detect the objects and their speed using radars, it is also required to detect and control the failure modes related to radar. Radar calibration and run-time health monitoring of the radar are important. Exemplary failure modes that could happen in radars are the following: misalignment during installation or due to physical impact (e. g., collision, vibrations) can cause inaccurate readings or blind spots. Over time, radar components may degrade, leading to reduced performance or failure or electromagnetic interference from other radar-equipped vehicles or nearby devices can result in degraded performance or erratic data. Identifying and addressing these faults is crucial for maintaining the reliability of radar sensors in automotive applications.

The main drawback of the sensors presented above is that they are typically not able to detect the objects that are in close proximity of the ego vehicle, when these sensors are mounted on the vehicle's roof. **Ultrasonic sensors** are one on the cheapest sensors that can be used for this problem. They use sound waves to measure distances to nearby objects. Ultrasonic sensors are used for short-range tasks, such as parking assistance and obstacle detection at low speeds. Unfortunately, they are limited by short-range and sensitivity to environmental noise. They are particularly effective in low-speed scenarios, such as navigating tight spaces or crowded environments, where their ability to detect nearby obstacles enhances safety. One of their primary applications is in parking assistance systems, where they detect curbs, nearby vehicles, and other obstacles, enabling both manual and automated parking maneuvers. Ultrasonic sensors also support blind spot monitoring by identifying objects in areas not covered by cameras or radar, offering additional safety during lane changes or turns. In low-speed environments, such as residential areas or parking lots, they can trigger emergency braking systems when sudden obstacles are detected, providing an extra layer of protection.

Additionally, ultrasonic sensors contribute to pedestrian detection in close-proximity scenarios, further ensuring safety in crowded areas. When integrated with other sensor modalities, such as cameras, lidar, and radar, they enhance the vehicle's environmental awareness through sensor fusion, creating a comprehensive and reliable perception system. Despite their limited range and lower resolution compared to other sensors, ultrasonic sensors excel in low-speed and short-range tasks, making them indispensable for parking assistance, obstacle detection, and low-speed navigation in autonomous driving systems.

Perception sensors such as cameras, lidars, and radars are essential for detecting objects in the environment, providing critical data for autonomous driving systems. However, detecting objects alone is not sufficient for enabling safe and efficient navigation. To generate actionable trajectories, the coordinates of the vehicle and detected objects must be accurately transformed into a real-world coordinate system. This conversion allows the system to understand the relative positions and movements of objects within the broader context of the environment. By aligning sensor data with real-world coordinates, the vehicle can make precise decisions and plan its path effectively. The conversion of vehicle and object coordinates into real-world coordinate systems relies on Global Positioning System (GPS) and Inertial Measurement Unit (IMU) sensors. These sensors play a crucial role in autonomous driving systems by providing accurate localization, navigation, and motion tracking. GPS determines the vehicle's geographic position through satellite signals, enabling global localization, route planning, and geofencing. It also serves as a key input for data fusion with other sensors like lidar and cameras, enhancing localization accuracy and acting as a fallback when these sensors face limitations, such as occlusion. However, GPS can experience reduced accuracy in challenging environments, such as urban areas or tunnels, where signal degradation occurs.

The IMU complements GPS by measuring the vehicle's acceleration, angular velocity, and orientation using accelerometers and gyroscopes. It supports dead reckoning to maintain localization during GPS signal loss, monitors vehicle dynamics, and provides short-term, high-frequency motion data for precise tracking during rapid maneuvers. IMU data is crucial for path prediction and bridging localization gaps when GPS is unavailable.

Together, GPS and IMU data are integrated through sensor fusion algorithms, such as Kalman filters, to deliver highly accurate localization by mitigating the weaknesses of each individual sensor. This combination is essential for lane-level positioning on high-definition maps and ensures real-time navigation even in environments with poor GPS signals. By contributing to smooth trajectory control and precise motion planning, GPS and IMU sensors form the foundation of reliable and safe autonomous driving systems. These sensors can be used to recognize the objects around the ego vehicle in autonomous vehicles or in the ADAS systems. An exemplary perception pipeline for an autonomous driving system is presented in Figure 6.17.

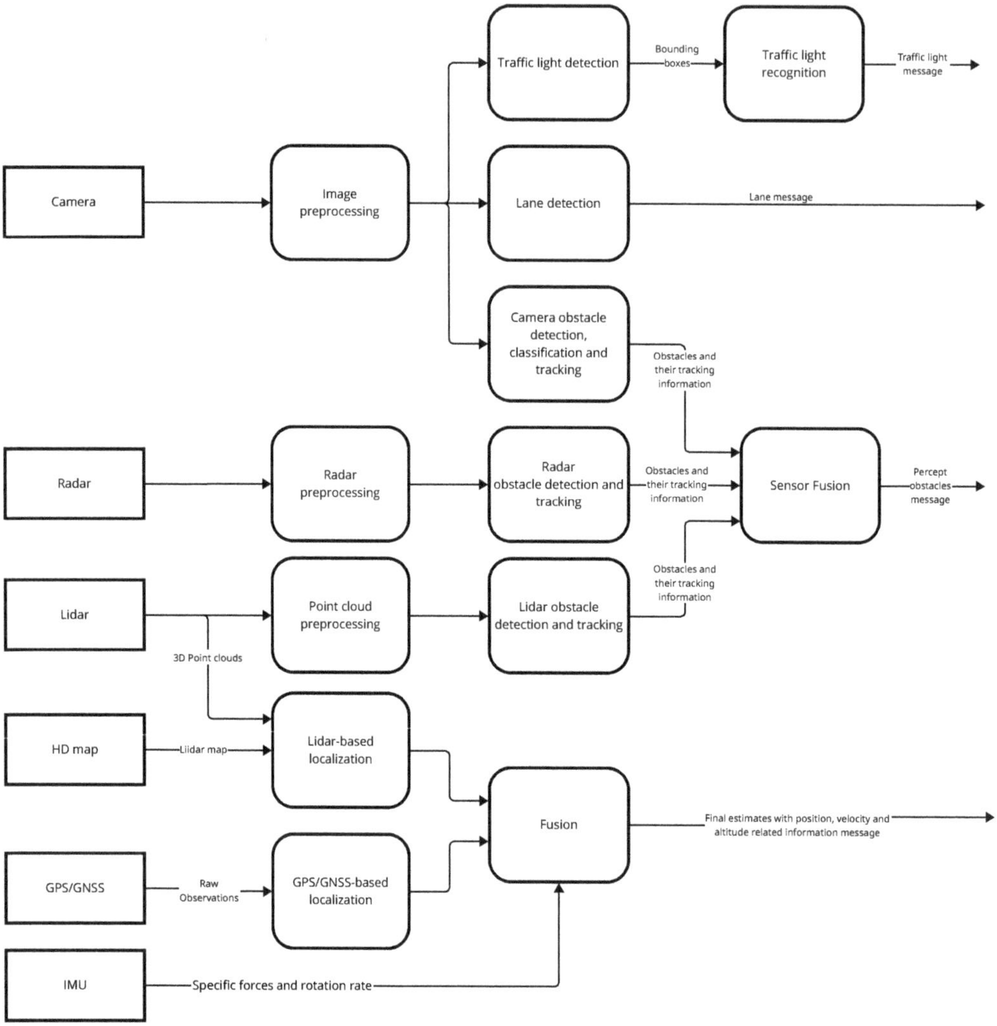

Figure 6.17 An exemplary perception system which could be used in autonomous vehicles

When sensor software and hardware failure modes are not properly detected and controlled, it becomes impossible to ensure that the Machine Learning or Deep Neural Network algorithms used in mobility applications can accurately process sensor data. This failure undermines the ability of these systems to convert sensor information into safe control commands necessary for controlling vehicles reliably. Ensuring the integrity of sensor data is a critical prerequisite for the safe operation of autonomous and semi-autonomous systems. This integrity includes the sensors mounting in the vehicles, their calibration, and recalibration and diagnostics also after crashes. This integrity encompasses several critical aspects, including the proper mounting of sensors on vehicles, ensuring that they are securely and precisely positioned to provide accurate data. Equally important is the regular calibration and recalibration of these sensors to maintain accuracy over time, as misaligned or poorly calibrated sensors can significantly degrade system performance. Diagnostics are also essential, especially after events like collisions or crashes, where sensors may become damaged or dislodged, leading to incorrect readings. Additionally, ongoing health monitoring systems should be in place to detect any gradual performance degradation, such as drift in IMU measurements or wear and tear in radar or lidar components. Ensuring sensor integrity also involves robust environmental protection, as exposure to dust, moisture, extreme temperatures, or vibrations can compromise sensor reliability. Finally, software-level checks, such as self-diagnostics and redundancy mechanisms, are critical to identifying and compensating for potential hardware failures, ensuring continuous and accurate data for vehicle control systems.

In the following sections, we discuss fundamental deep neural network-based algorithms designed to process data from the key perception sensors, specifically cameras and lidars. These sensors typically generate large amounts of high-dimensional data that are well-suited to processing by Deep Neural Networks. For example, Convolutional Neural Networks (CNNs) (LeCun et al. 1989) are commonly used to extract features from camera images, while point cloud data from lidar sensors are often processed using algorithms which requires point cloud voxelization, or other architectures specialized for 3D data representation.

In contrast, other types of signals, such as those from radars, ultrasonic sensors, IMUs, and GPS systems, often rely on classical signal processing techniques rather than DNNs. Radar signals, for instance, are commonly processed using the Fast Fourier Transform (FFT) for range and velocity estimation, and techniques like the Constant False Alarm Rate (CFAR) algorithm (Richards 2005) are used for object detection. IMU data, which include measurements of acceleration and angular velocity, are typically filtered and fused with other sensor data using algorithms like Kalman filters or extended Kalman filters to estimate vehicle pose and motion trajectories. GPS data are often enhanced through Real-Time Kinematic (RTK) (Mannings 2008) processing to improve positional accuracy, while advanced smoothing algorithms, such as the Rauch-Tung-Striebel (RTS) filter (Rauch et al. 1965) can further refine localization estimates by combining historical and real-time data.

These classical algorithms are essential for low-level signal processing and complement the capabilities of DNNs, which excel at higher-level tasks such as object detection, classification, and decision-making. Together, these approaches form a robust pipeline that integrates data from diverse sensors, enabling the reliable perception and control required for safe and effective Autonomous Vehicle operation.

6.3.1 2D Object Detection

One of the most important tasks in mobility applications is object detection. The goal of object detection is to identify location of all objects in the image and their positions. The basic techniques analyzed the images in the iterative procedure called **sliding window** algorithm (Figure 6.18). The sliding window technique in object detection involves dividing an image into multiple segments or grids and performing image classification on each segment to determine if an object is present. A fixed-size rectangular window is defined, which slides across the image from left to right and top to bottom. As the window moves, it crops different regions of the image, and each cropped segment is passed through a pre-trained classifier to identify potential objects. This process is repeated across different scales of the image to detect objects of varying sizes. While effective, the sliding window approach is computationally expensive due to the large number of regions that need to be processed. Additionally, the fixed window size may not perfectly match the aspect ratio of the objects, leading to potential inaccuracies. Despite its limitations, the sliding window technique laid the foundation for various object detection methods like Viola Jones algorithm (Viola/Jones 2001) or Histogram of Oriented Gradients (HOG) plus Support Vector Machines (SVM) (Dalal/Triggs 2005). More advanced methods such as Faster R-CNN (Ren et al. 2015) and YOLO (You Only Look Once) (Redmon et al. 2016) use more efficient techniques such as region proposals or anchors methods.

Figure 6.18 In the sliding window technique a box is moved across the whole image and in each position it is checked whether it contains an object or not.

When the sliding window technique is applied it is very common that the result we obtain contains many overlapping detections candidates around the same object. The next step is to remove these redundant object candidates. This algorithm is called **Non-Maximum Suppression (NMS**, see Figure 6.19). NMS is an essential algorithm used in object detection, particularly in Computer Vision tasks like detecting objects in images. It is designed to reduce redundancy by eliminating multiple overlapping bounding boxes that detect the same object, keeping only the best one. The algorithm starts by taking a set of bounding boxes, each with a confidence score indicating the likelihood that the box contains an object. It also requires a predefined **Intersection over Union (IoU)** threshold, which determines the amount of overlap between boxes that is acceptable.

The algorithm first sorts all the bounding boxes based on their confidence scores in descending order, with the highest score representing the most likely object detection. Then, the box with the highest confidence is selected as the current box and will be included in the final set. Next, the algorithm calculates the IoU for every other remaining box with the current one. IoU is the ratio of the area of overlap between two boxes to the area of their union. If the IoU between the current box and any other box exceeds the predefined threshold, that box is suppressed or discarded, since it likely represents the same object. This process is repeated for the next box with the highest confidence, and so on, until all boxes have either been selected or suppressed.

As a result, the algorithm outputs a reduced set of bounding boxes where overlapping detections are eliminated. NMS is widely used in object detection models such as YOLO, and Faster R-CNN to ensure that overlapping predictions are filtered out. Variants of NMS exist, such as Soft NMS, which decays the confidence scores of overlapping boxes rather than discarding them, and class-specific NMS, which applies the process separately for each object class. Overall, NMS algorithm plays a critical role in enhancing the performance of object detection systems by ensuring that only the most accurate detections are retained.

(a) before NMS (b) after NMS

Figure 6.19 Example of NMS algorithm

One of the most known perception models for object detection problem which uses Deep Neural Networks is the Faster RCNN model. The simplified architecture of Faster RCNN is depicted in Figure 6.20. This model uses two stages to detect objects in the image. In the first state, the **Region Proposal Networks (RPN)** generates the region proposals based on the equally distributed anchors in the image. In the second stage, every proposal is classified into one of the classes used during the training, and the final coordinates of the object are adopted by the regression part of the model. It is worth to mention here, that the model shares the same features extracted using convolutional layers. Finally, the overlying detections are removed by an NMS algorithm. This algorithm is the basis for many more sophisticated 2D and 3D object detection algorithms. Understanding the architecture principles helps to perform a more efficient safety analysis for the algorithm and identifies potential failure modes. A potential failure mode for Faster RCNN model could be that the anchors used to generate a region proposal are too small, and therefore very large objects cannot be efficiently detected.

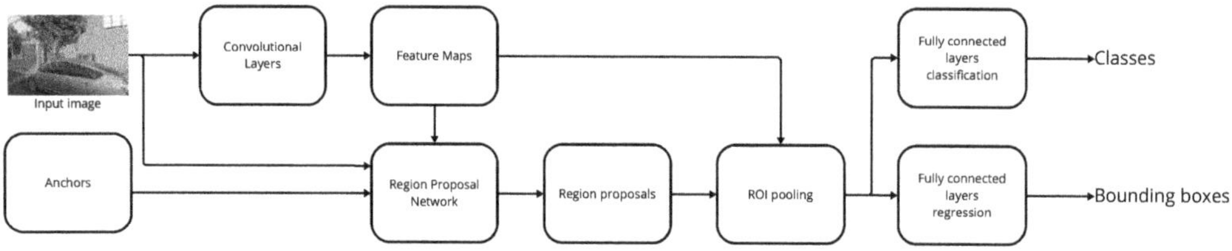

Figure 6.20 Faster RCNN object detection architecture

The second widely used model for object detection is the YOLO (You Only Look Once) algorithm, which is a very effective framework for real-time object detection. The YOLO architecture is presented in Figure 6.21.

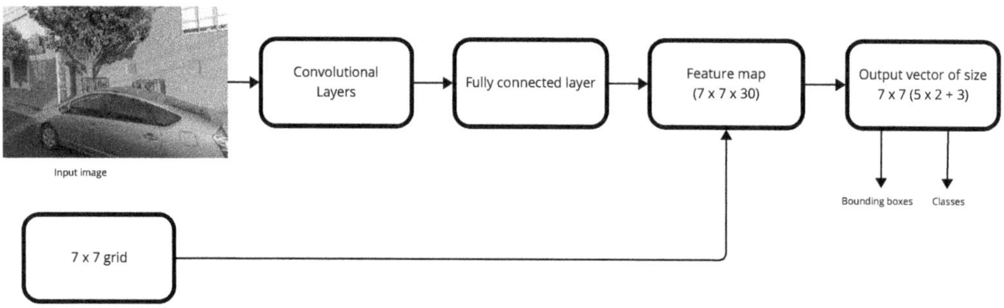

Figure 6.21 YOLO v1 network architecture

The YOLO model sets itself apart from traditional object detection methods by approaching the problem as a single regression task rather than a multi-step classification process. YOLO employs a single Convolutional Neural Network that predicts multiple bounding boxes and their corresponding class probabilities directly from the input image in one pass.

The algorithm divides the input image into a 7×7-grid, with each grid cell responsible for detecting objects whose center falls within that cell. For every grid cell, YOLO predicts a fixed number of bounding boxes along with confidence scores indicating the likelihood of an object being present and the accuracy of the bounding box. Additionally, it predicts class probabilities for each grid cell, showing the chances of various object classes being detected.

After generating these bounding boxes, YOLO uses the NMS algorithm to eliminate redundant boxes, retaining only the most confident predictions. This approach ensures that the final output includes unique detections for each object. A significant advantage of YOLO is its speed; by processing the image in a single pass, it can achieve real-time object detection, making it suitable for applications like autonomous driving, video surveillance, and robotics. Over the years, various versions of YOLO, such as YOLOv2, YOLOv3, and YOLOv5, have been released, each enhancing the original model's accuracy, speed, and functionality.

In case of the YOLO model, a common problem is that YOLO can detect maximally two objects per single cell as presented in Figure 6.22.

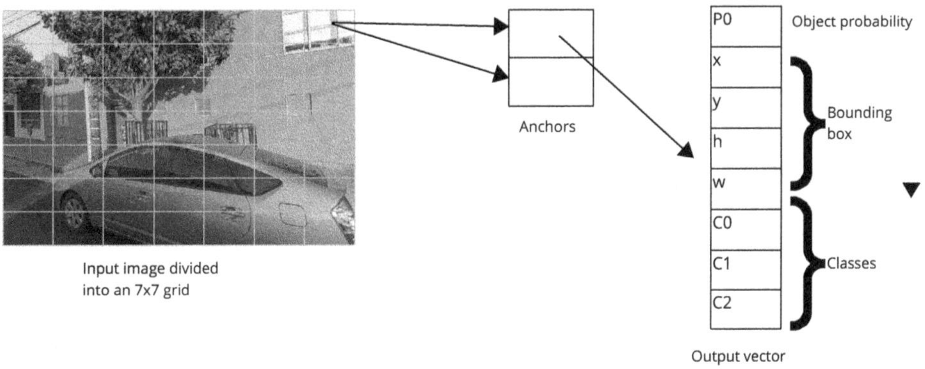

Figure 6.22 In the YOLO v1 architecture, the model generates two objects per grid cell. Every anchor is encoded as vectors that contain information about the probability (P0) that this anchor is a box, about its coordinates (x, y, h, w) and about its class probabilities (C0, C1, C2), where the number of classes in this example is C = 3.

When an image contains more than two images in the single cell then the model will not detect them. Therefore, YOLO v1 is not well designed to work with data where we expect many small objects in single scenarios. The exemplary scene, which could be problematic for the YOLO algorithm defined with grid size 3×3 is presented in Figure 6.23. In this example, the model is able to detect only two pedestrians annotated in dashed green, and the rest of the pedestrians, annotated in red, are ignored by the model. Therefore, correct understanding of the target environment is critical during design decision related to object detection model selection.

Figure 6.23 Exemplary scenario in which YOLO is not able to identify all objects within the cell: The model is able to detect maximally two objects per grid cell in the default configuration.

6.3.2 3D Object Detection

3D object detection is another critical algorithm in autonomous driving that uses lidar sensors to detect and classify objects in the environment by generating highly accurate 3D point clouds. Lidar works by emitting laser pulses and measuring the time it takes for the light to bounce back from surrounding objects. This data is then used to create a detailed 3D map of the vehicle's surroundings, including objects, road features, and obstacles. Lidar's ability to capture depth information allows for precise object detection, even in poor lighting or weather conditions, which are challenging for cameras. By analyzing the 3D point cloud, the system can recognize and track objects such as vehicles, pedestrians, cyclists, and road barriers, providing essential input for path planning and navigation. The spatial resolution and range accuracy of lidar make it indispensable for tasks like obstacle avoidance detection, and scene understanding in AD systems.

There are more than one hundred algorithms for 3D object detection described in the existing literature. One of the most popular algorithms for 3D object detection is PointPillars presented in Figure 6.24 (Lang et al. 2019). PointPillars is a voxel-based architecture designed for the task of 3D object detection. The input point cloud is split into pillars, i. e., columns. The first stage of processing is the Pillar Feature Net, which extracts features from non-empty pillars. The 2-dimensional nature of the pillar location along with the third dimension of features allow for easy further processing using image-based methods. The second stage is the Backbone network, which is a 2D CNN responsible for transforming the pillar features. The third and last stage is the Detection Head in the form of the Single Shot Detector. It is used to predict the location, size, rotation, and category of the objects. PointPillars achieves decent results while also being capable of working in real-time scenarios. In the case of the PointPillars algorithm, a potential limitation could be the voxelization step. Let's imagine a model which is trained in USA, and then the model is deployed in Europe, where a typical vehicle is much smaller than the most popular pickups. The voxelization process used in the PointPillars model is not able to generate the pillars that can be processed independently of the vehicle size. Therefore, the model trained on vehicles with specific distribution will work only on the vehicles with the same size.

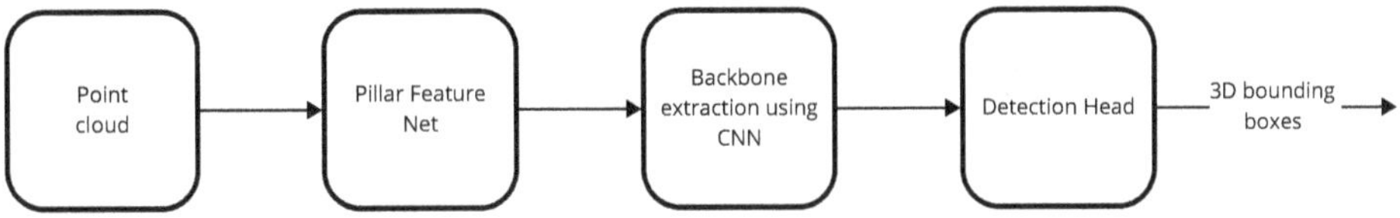

Figure 6.24 PointPillars architecture for object detection on lidar data

3D object detection task can be also conducted without lidar sensors using stereo vision cameras. In stereovision, two or more cameras are used to capture images from slightly different viewpoints, mimicking human binocular vision. By analyzing the disparity between the images, stereo vision systems can calculate depth information and create a three-dimensional representation of the scene. This technique enables the detection and localization of objects such as vehicles, pedestrians, and obstacles in real time. The depth maps generated by stereo vision cameras provide crucial spatial context, allowing for more accurate classification and tracking of objects as they move through the environment. Additionally, stereo vision is advantageous in various lighting conditions, offering robustness against shadows and reflections that can confuse other sensing technologies. This capability makes stereo vision a valuable tool in autonomous driving, enhancing the vehicle's ability to navigate complex and dynamic environments. However, the stereo vision systems share the same limitations as cameras and they can be sensitive to bad weather conditions and low light condition or overexposure.

6.3.3 Sensor Fusion for Multimodal Object Detection

Due to the fact that sensors show different robustness to adverse weather and light conditions, some researchers propose to use algorithms which use sensor fusion with multiple sensors for object detection. Sensor fusion in autonomous driving involves combining data from multiple sensors, such as cameras, lidar, radar, GPS, and ultrasonic sensors, to create a more accurate and comprehensive understanding of the vehicle's environment. Each sensor provides different types of information—cameras capture visual details, lidar offers precise 3D mapping, radar excels in poor visibility, and GPS provides location data. By merging these data streams, sensor fusion compensates for the limitations of individual sensors and enhances overall perception. This approach also introduces redundancy, ensuring that the system remains robust even if one sensor underperforms or fails. By integrating sensor data in real time, sensor fusion may improve the accuracy of object detection and also other tasks in autonomous vehicles. In literature, depending on the process of how the sensor channels are combined, we can distinguish two approaches called late and early fusion.

Late fusion involves processing data from each sensor separately and then merging the individual detection results (Figure 6.25). Typical steps for object detection using one lidar and multiple cameras are the following:

- Perform 3D object detection on the lidar point cloud.
- Execute 2D object detection on the camera images.
- Project the 2D detections into 3D space and integrate them with the 3D detections.

In this approach, the projected 2D detections typically help to classify the 3D detections, which can be challenging to discern from point cloud data alone, such as differentiating between a car and a van. The data of each sensor is processed independently to make a local prediction. These individual results are then combined at a higher level to make the final fused prediction.

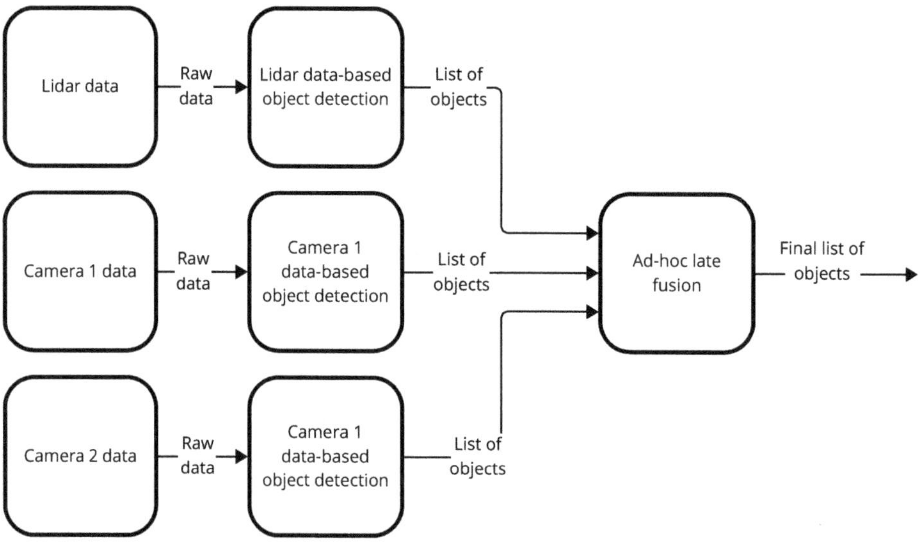

Figure 6.25 Illustration of the late fusion approach

In autonomous driving, **early fusion** is a sensor fusion technique where raw data from different sensors (such as cameras, lidar, and radar) is combined at the initial stages of data processing, before running perception algorithms as presented in Figure 6.26. This approach integrates information at the lowest feature level, allowing the Neural Network to learn relationships and correlations between the data from various sensors directly. Early fusion merges raw data from multiple sensors prior to applying a perception algorithm. There is increasing interest in end-to-end early fusion methods that use a single Deep Neural Network to map raw sensor inputs directly to object detections, eliminating the need for manually designed fusion algorithms. By combining data early, the network can exploit fine-grained correlations across the different sensors, such as matching depth information from lidar with texture and color information from cameras. Early fusion significantly increases the dimensionality of the data, as all raw inputs from various sensors are processed together. This can make the training process more challenging, especially when limited labeled data is available. Since early fusion tightly integrates sensors at the raw data level, any noise or malfunction in one sensor can impact the entire perception pipeline. This can make the system less robust, as sensor-specific issues are more difficult to isolate and correct. Early fusion is popular in end-to-end learning models that aim to map sensor inputs directly to outputs, such as object detections or driving decisions, without requiring separate modules or hand-crafted fusion algorithms.

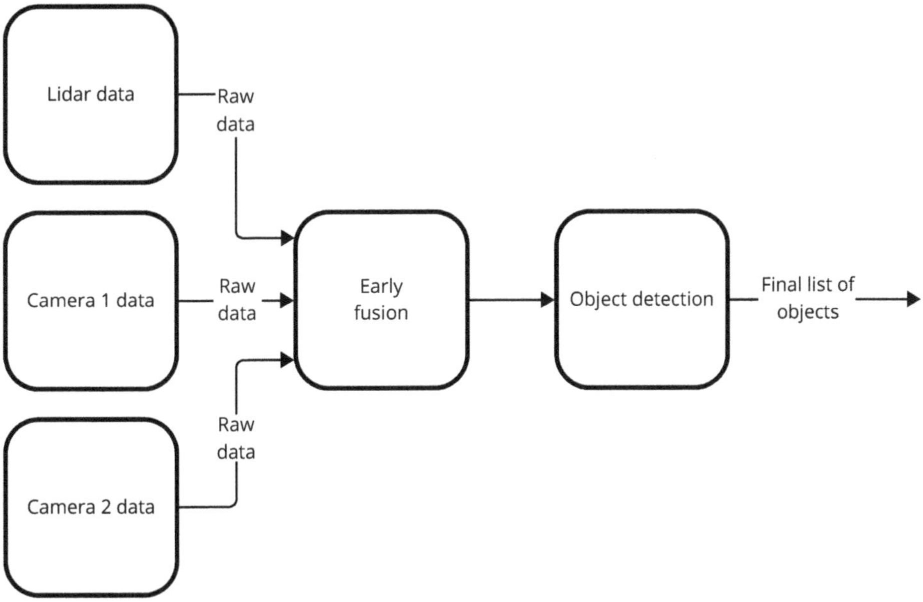

Figure 6.26 Illustration of the early fusion approach

One of the first two algorithms designed for sensor fusion in autonomous driving that use DL are AVOD (Aggregate View Object Detection)) (Ku et al. 2018) and MV3D (Multi-View 3D) (Chen et al. 2017). They use data from lidar and cameras to improve the accuracy of detecting and localizing objects in 3D space.

MV3D, which is presented in Figure 6.27, is a pioneering sensor fusion method that uses data from multiple sensor views, including lidar, camera, and sometimes radar, to detect objects in 3D. The model is designed to benefit from multiple perspectives of the same scene, combining these views into a unified representation. MV3D uses early-stage feature fusion to combine data from lidar and cameras, focusing on fusing features early in the pipeline for better spatial understanding of the scene.

MV3D begins by processing LiDAR data into two representations:

- Bird's-Eye View (BEV) that provides a top-down perspective that captures object shapes and spatial layout
- Front View (FV) that provides a perspective similar to a camera view, focusing on the vertical structure of objects

Additionally, a RGB camera image is used to capture color and texture information, which can be missing from lidar data. MV3D uses an RPN to generate 3D bounding box proposals. The RPN operates on the lidar BEV representation and produces a set of possible object locations, creating bounding boxes around likely objects. Once the bounding box proposals are created, MV3D extracts feature maps from each of the three views (BEV, FV, and RGB). These features are then fused across views, often by concatenating or stacking feature maps. This fused feature map represents the objects in 3D space with improved contextual information from all views. The final fused feature map is used by a bounding box regression network to refine the proposals, adjusting each bounding box to better fit the objects in 3D space. This final prediction step outputs the refined 3D bounding boxes with class labels, indicating object type and confidence level. By integrating information from multiple perspectives, MV3D can detect complex objects and capture more context, such as object height, which is challenging for single-view models.

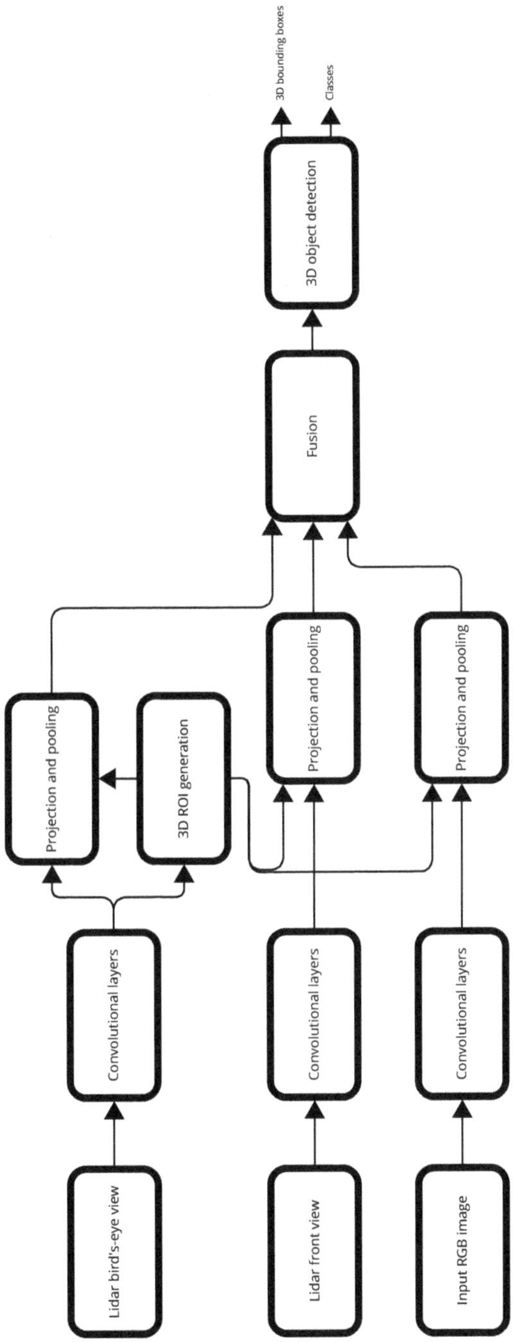

Figure 6.27
Simplified architecture of the MV3D sensor fusion architecture

AVOD, which is presented in Figure 6.28, is a sensor fusion framework designed to detect 3D objects (like cars, pedestrians, cyclists) by combining data from lidar point clouds and camera images. Its primary goal is to leverage the strengths of both modalities—precise depth from lidar and rich texture from images. AVOD uses a two-stage fusion approach, combining features from both lidar and camera data at an intermediate stage to enhance 3D object detection accuracy. AVOD starts with a Region Proposal Network that generates potential 3D bounding box candidates around objects in the environment based on both lidar and camera data. Using lidar's BEV representation, the RPN generates a large set of 3D bounding box proposals around probable object locations. The RPN produces proposals in both the lidar BEV and camera image views. Then, using the 3D bounding box proposals, feature maps are extracted from both views. These features are aggregated, meaning they are combined from both views to create a comprehensive representation for each proposed object. Aggregation combines the accurate spatial features of lidar and the contextual visual information of images. In the final stage, AVOD uses a 3D bounding box regression network to refine the bounding box proposals based on the aggregated features. This network adjusts the proposals to more accurately fit the detected objects, providing a final 3D detection output.

During our experiments with these architectures, we observed that both models completely ignore the camera channels even when the authors showed small improvements thanks to the fusion of the sensors. In our experiments, we took pretrained models and we ran them with images with zeros, and the performance of the models was not decreased in comparison to the scenario in which lidar and correct camera data were used. This experiment shows that even when we used two sensors to train the model, it learned only lidar modality. Therefore, again proper understanding how the model is designed and deriving proper test may help do detect potential failure modes and discover model limitations.

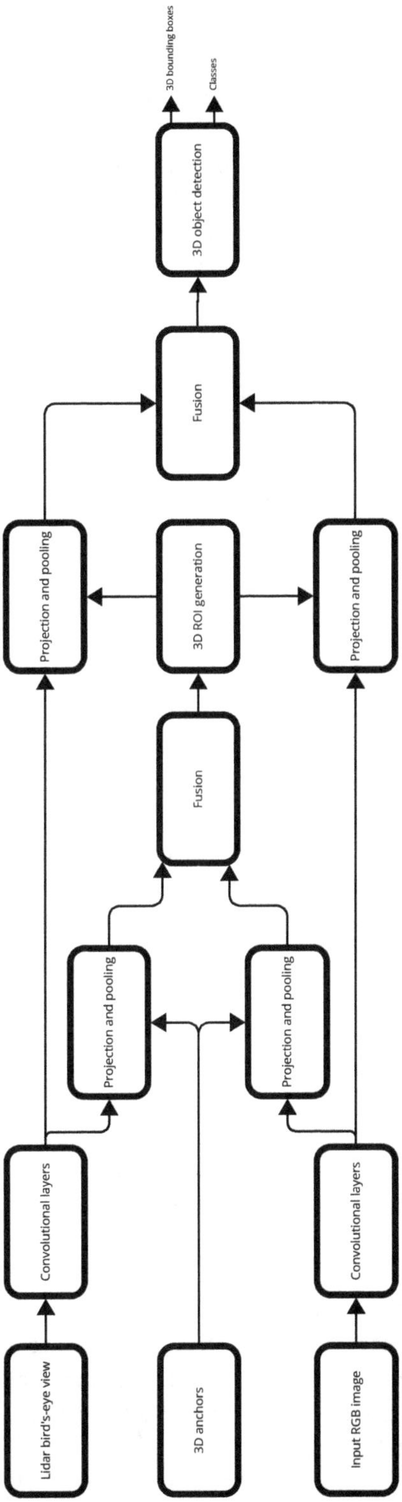

Figure 6.28
Simplified architecture of the AVOD sensor fusion architecture

6.3.4 Scene Segmentation

Scene segmentation in the context of autonomous driving refers to the process of dividing a visual scene captured by sensors, such as cameras or lidar, into distinct regions or objects. This process helps the Autonomous Vehicle understand and interpret its environment by classifying various parts of the scene into meaningful categories like roads, sidewalks, vehicles, pedestrians, traffic signs, and other relevant objects. Comparison of the results from object detection and segmentation tasks are presented in Figure 6.29.

(a) Original image (b) Object detection (c) Semantic segmentation

Figure 6.29 Comparison of the results from object detection and segmentation tasks

There are two primary types of scene segmentation: semantic segmentation and instance segmentation. In **semantic segmentation** each pixel is labeled in an image according to its category (e. g., road, car, pedestrian, etc.), allowing the vehicle to have a detailed, pixel-level understanding of its surroundings. It helps the vehicle recognize which parts of the scene belong to drivable areas, obstacles, or dynamic objects. In addition to labeling each pixel, **instance segmentation** distinguishes between individual objects within the same category. For example, it can identify separate vehicles or pedestrians in a crowded environment, providing more granular information for decision-making.

One of the first algorithm for semantic segmentation that uses DLwas SegNet, presented in Figure 6.30 (Badrinarayanan et al. 2017). SegNet is a DNN architecture, which predicts pixel-level class labels in an image. It follows an encoder-decoder structure where the encoder consists of convolutional layers and pooling operations to extract features and compress spatial dimensions. Unlike other architectures, SegNet saves the max-pooling indices from the encoder during the downsampling process. In the decoder, these saved indices are used to perform non-linear upsampling, restoring the spatial resolution efficiently while preserving spatial details. The final layer predicts the class probabilities for each pixel. SegNet is particularly effective for tasks requiring boundary precision, such as road scene understanding. In scene segmentation problem, we have to deal with potential problems when too many small segments are detected or when the number of detected pixel cluster is too small to reasonable identify a particular object or road in the scene.

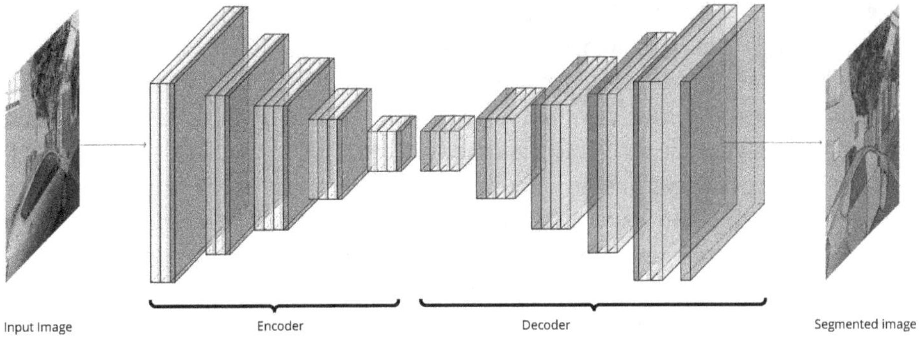

Figure 6.30 Simplified architecture of the SegNet model for semantic segmentation

Scene segmentation is crucial for enabling the vehicle to safely navigate complex environments, and make informed decisions in real time. Latest advancements in segmentation algorithms have enabled full scene segmentation to be applied in autonomous driving as another option to process the inputs from the sensors. In comparison to traditional methods that classify pre-selected region candidates, the deep network-based models can classify the entire image, providing a foundational output for higher-level decision-making algorithms essential for vehicle navigation, such as collision avoidance or lane keeping assistance. It is worth to notice here, that scene segmentation can also be performed for lidar data, not only for the image data.

6.3.5 Object Tracking

In AD and ADAS solutions, simple object detection is not sufficient because these systems operate in dynamic environments where understanding object motion and behavior over time is critical. Object tracking provides temporal continuity by linking detections across consecutive frames, ensuring that objects are not lost between frames. This capability is crucial for predicting the trajectories of moving objects like vehicles, pedestrians, and cyclists. Tracking also helps in filtering out false positives from detection systems, reducing noise and improving reliability. Moreover, it enables advanced functionalities like collision prediction, path planning, and decision-making by providing accurate and consistent information about the speed, direction, and behavior of surrounding objects. Without tracking, these systems would lack the temporal context required to ensure safety and efficiency in real-world scenarios. Therefore, object tracking is another component that is typically used after object detection. The most popular algorithm for object tracking is the Kalman filter, but in literature, we can also find a plethora of deep neural network-based models. To better understand the impact of the object tracking, we have to distinguish single and multiple object tracking. Both tasks can be performed for 2D camera images as well as for lidar and radar sensors.

Single Object Tracking (SOT) is a task that focuses on locating and following a specific object in a video sequence over time. The process begins with initialization, where the tracker is provided with a bounding box or region around the object in the first frame. The tracker then continuously updates the object's position in subsequent frames as it moves. The goal is to maintain the tracking accuracy despite challenges such as motion, changes in object appearance, occlusion, or background clutter. SOT algorithms typically use features like texture, color, or shape to identify and follow the object. They adapt dynamically to changes in the object's state to ensure consistent tracking throughout the video. This capability is essential in applications like surveillance, autonomous driving, and video analysis, where maintaining focus on a single object is critical. A schematic representation of single object tracking is presented in Figure 6.31. In rich literature, we can find many SOT models which use DL such as SiamMask or ADNet (Action Driven Network), but also classical algorithms like Kalman Filter are still used in practice (Soleimanitaleb/Keyvanrad 2022). The main challenges for SOT algorithms can be object occlusions that can lead to drift or loss of object entirely. Another problem can be appearance variations and changes in scale, pose, illumination or viewpoint. The SOT algorithm can also latch onto a background or incorrect object and have problems with initialization and termination of tracks.

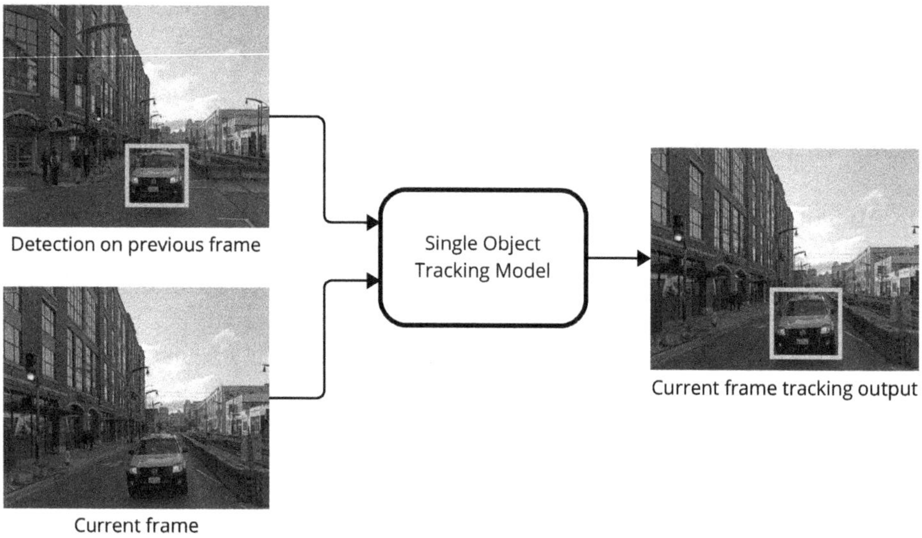

Figure 6.31 Schematic presentation how single object tracking works

In comparison to SOT, **Multiple Object Tracking (MOT)** involves detecting and tracking multiple objects simultaneously in a video sequence. The goal is to maintain unique identities for each object over time, even as they move, interact, or change appearance. MOT starts with object detection to identify the objects in each frame and then associates these detections across frames based on features like position, motion, and appearance. Challenges in MOT include handling occlusions, similar-looking objects, entry and exit of objects from the scene, and changes in object appearance. Effective MOT algorithms use techniques such as data association, motion prediction models, and feature matching to address these challenges. DL algorithms are less popular for MOT problems, but recent approaches that use attention mechanisms or Long Short-Term Memory (LSTM) can improve tracking performance significantly (Luo et al. 2021). Applications of MOT in autonomous driving are essential to understanding the dynamics of multiple entities and ensuring safe navigation and effective decision-making. By maintaining consistent object identities and trajectories, MOT provides critical information for tasks like behavior analysis, collision avoidance, situational awareness and path planning. A schematic representation of single object tracking is presented in Figure 6.32. The most significant challenges associated with MOT algorithms are similar to SOT algorithms, but MOT additionally requires two tasks to be solved: determining the number of objects, which typically varies over time, and maintaining their identities. MOT algorithms can present lower performance in crowded environment, which increases the risk of tracking failures due to overlapping and occlusion.

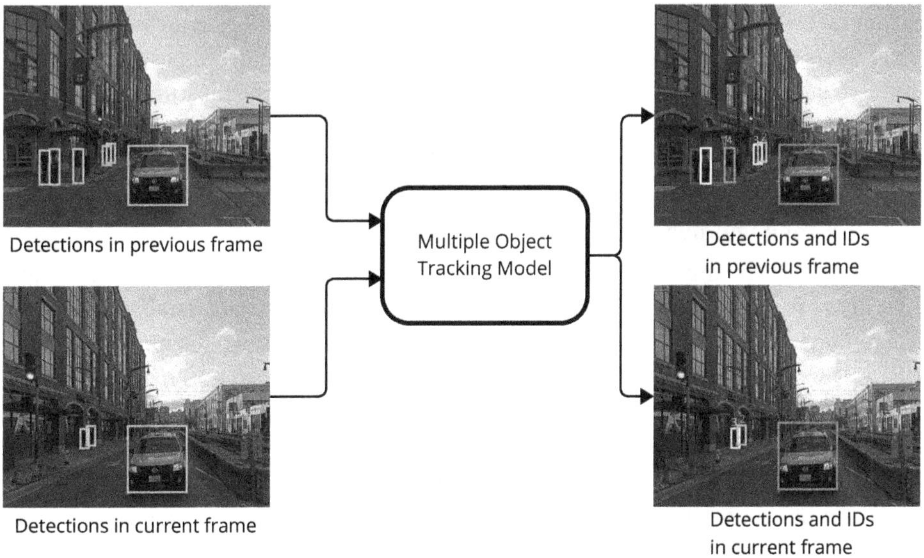

Figure 6.32 Schematic representation of multiple object tracking

6.3.6 Ego Vehicle Localization

The localization subsystem can be divided into (Laconte et al. 2022) the following parts:

- **Road Level Localization** (RLL) helps to localize the road on which the ego vehicle travels.
- **Ego-Lane Level Localization** (ELL) determines the position of the vehicle in the lane in terms of lateral and longitudinal position.
- **Lane-Level Localization** (LLL) localizes the position of the host lane within the road (i. e., the lane on which the vehicle travels).

RLL provides a high-level understanding of the vehicle's route within the road network, identifying the specific road on which it's traveling. It helps the vehicle stay on the correct road segment and make necessary transitions at intersections. RLL primarily uses global positioning to establish which road or highway segment the vehicle occupies. GPS data, road maps, and high-definition (HD) maps are commonly used for this localization. Deep Learning models are often used for interpreting GPS data, map-matching, and fusing information from different sources to determine the road segment. Convolutional Neural Networks or Recurrent Neural Networks can process map data, satellite imagery, and visual data to match the vehicle's position with the road structure. GPS can be noisy or inaccurate in urban canyons, tunnels, or areas with poor satellite visibility. Deep Learning models, especially those using Bayesian or Kalman filters combined with Deep Neural Networks, can help to integrate GPS data with IMU readings, road map data, and even environmental data to correct positioning errors and improve accuracy.

While Deep-Learning-based Computer Vision algorithms can identify drivable areas, classify road types, and interpret road features from camera data, enabling the vehicle to understand road boundaries and on-road/off-road areas, ELL enables the vehicle to assess its lateral (side-to-side) and longitudinal (forward and backward) positions within a lane. This fine-grained localization is essential for lane-keeping, obstacle avoidance, and smooth lane changes. ELL typically utilizes data from sensors such as lidar, cameras, and IMUs, often combined with HD maps, to provide real-time adjustments to the vehicle's exact positioning within the lane. Recurrent Neural Networks and Long Short-Term Memory (LSTM) networks are often employed to handle the sequential data and track vehicle movement over time, helping to predict the vehicle's path. Deep Reinforcement Learning (DRL) can adaptively correct minor position drifts by learning from past errors, which improves positioning accuracy and lane-keeping behavior, especially on winding roads or in heavy traffic.

LLL localizes the vehicle within the road's lane structure, helping the vehicle determine which lane it occupies (e. g., left lane, center lane, or right lane on a multi-lane road). This localization is important for planning maneuvers like lane changes, turns, and managing complex intersections. LLL combines RLL's map-based data with ELL's sensor-based insights to identify the precise lane. HD maps, lane markings, and sometimes vehicle-to-infrastructure data are used to enhance accuracy. Models like Long Short-Term Memory (LSTM) and transformer-based networks are used to maintain temporal consistency across frames, which helps accurately track the vehicle's position across lane boundaries, complex intersections, and highway interchanges. This task is especially dependent on segmentation models, which helps to identify lane types, lane directions, and special road features (e. g., bus lanes, turning lanes, or merge lanes). This information aids in decision-making, such as determining if a lane change is needed or which lane to select for upcoming maneuvers.

In combination, these three localization methods allow autonomous vehicles to understand their position at multiple levels: on the road, within specific lanes, and in relation to other objects, ensuring safe and efficient navigation. Fusion of lidar and radar, GPS, and camera data is essential to reduce uncertainty in localization. The latest models can involve deep neural network-based fusion methods that combine data streams into a cohesive understanding of position. The localization task can be also performed using classical algorithms without Deep Neural Networks.

Analyzing the potential failure modes for localization components shows that they strongly depend on the accuracy of the perception algorithms used in the AD pipeline. Incorrect detection of the objects, incorrect tracking or road segmentation may result in incorrect vehicle localization Therefore, proper modeling of the architecture will be crucial to perform efficient safety analysis of such complex algorithms.

6.3.7 Planning

One of the final steps in the autonomous driving pipeline is calculating the vehicle trajectory based on the determined vehicle localization in order to achieve that the processed information from the perception subsystem is used as input for the planning subsystem to calculate the vehicle's path. This subsystem is typically divided into three modules: route planning, behavior planning, and trajectory generation. In the **route planning** step, waypoints for the desired path are determined, representing the high-level path between two points on a map. During the behavior planning step, using data from the localization module and the IMU sensor, the software decides the necessary maneuvers for the vehicle to stay on course and calculates the final velocity. Finally, the trajectory generation module produces the detailed path the vehicle will follow. In autonomous driving, planning algorithms are responsible for determining the vehicle's path and actions, including trajectory generation, lane changes, overtaking, and decision-making in complex environments. Deep Neural Network models may enhance planning by processing high-dimensional sensory data, predicting potential obstacles or risks, and generating smooth, safe, and efficient driving plans. The vehicle can learn based on the historical trajectories and input data from the sensors how to determine more accurate and safer trajectories.

One of the main tasks for DL models is **behavior prediction and intention estimation** of other vehicles, bicyclists and pedestrians. Recurrent Neural Networks, Long Short-Term Memory networks, and transformer-based models can be used to analyze time-series data from surrounding vehicles, pedestrians, and cyclists to forecast future positions and actions. This prediction helps the vehicle anticipate behaviors, such as when another car might change lanes or a pedestrian might cross.

Another task is to create a smooth, feasible path that adheres to safety, comfort, and efficiency constraints, which is called **trajectory generation**. Generative models like Generative Adversarial Networks (GANs) and Variational Autoencoders (VAEs) are used to generate feasible trajectories. GANs can create diverse trajectory possibilities, which help the vehicle adapt to complex, dynamic scenarios (e. g., traffic jams or dense urban settings). Reinforcement learning models are also used to optimize the trajectory by learning from experience to maximize safety and efficiency while minimizing abrupt changes in speed or direction. The next task can be to determine high-level driving actions such as lane changes, turns, merges, and stopping at intersections. Deep Reinforcement Learning is widely used in decision-making, where the agent learns policies (i. e., strategies) for taking specific actions based on observed states of the environment. Models like Deep Q-Networks (DQNs, see Mnih et al. 2015) and Proximal Policy Optimization (PPO, see Schulman et al. 2015) are trained to navigate through various situations, such as dense traffic or intersections, by learning to balance risk, legality, and comfort. This enables the vehicle to make intelligent, context-aware decisions, allowing it to respond appropriately to real-world driving scenarios like yielding, overtaking, or maintaining a safe following distance. Another important task is to navigate through environments with multiple moving entities, such as urban intersections or dense traffic. Models like Graph Neural Networks (GNNs, see Scarselli et al. 2009) are used to represent and model interactions among multiple agents (e. g., other vehicles, pedestrians).

Deep Learning models contribute to autonomous driving planning algorithms by predicting the actions of surrounding agents, generating optimal trajectories, making high-level decisions, assessing risk, understanding the environment, and managing interactions with other agents. Through these applications, Deep Learning enables the vehicle to make contextually aware planning decisions, however this decision are the black boxes and there is also rather a trend to use classical algorithms for vehicle trajectory planning task.

6.3.8 Control

The vehicle ECUs handle the vehicle control and convert the generated trajectories into low-level commands for throttle, brake, and steering. To ensure the vehicle performs as intended, classical PID or MPC controllers are used, employing a control loop feedback mechanism to manage process variables. However, the latest approaches show that instead of building the whole autonomous driving pipeline connecting independent perception and planning components, we can directly allow that the algorithm makes the decision about steering, braking, and throttle commands based on the raw input sensor data. This approach uses Reinforcement Learning (RL). The objective of RL is to find an optimal control command (e. g., change steering, brake, or acceleration), by exploring the environment in an iterative manner. The environment is rewarding the AVs based on their current behavior to correct their errors in the future. A simplified overview of how RL can be applied for autonomous vehicles is presented in Figure 6.33. The E2E approach for AD is now used by a startup from UK called Wayve.[3)]

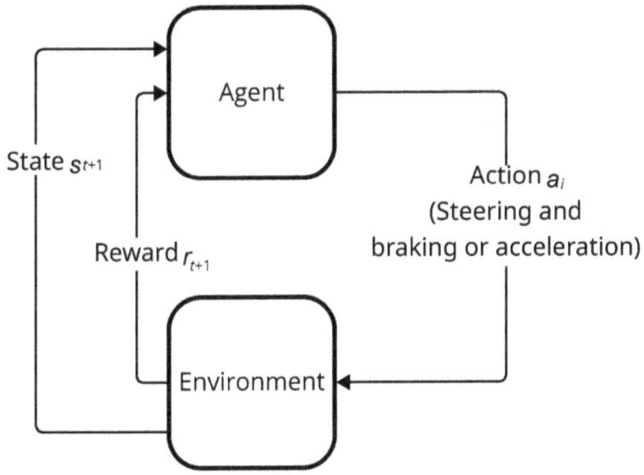

Figure 6.33
Vehicle control system using RL

Despite the significant results of DL and RL-based techniques in autonomous driving solutions, there are many remaining challenges to overcome before full autonomous cars can be deployed to the roads. The major challenge for AVs is to have a good understanding of the complex real world. So far, DL methods have shown significant performance to characterize and recognize objects in traffic environments, but this performance may still be insufficient. Unfortunately, many of the presented algorithms are not presenting the same performance in all conditions such as various traffic types, weather conditions, illumination conditions or different geographical regions. Thus, further research should be conducted to evaluate, e. g., the effects of night-time driving in the snow, and other weather conditions. Furthermore, it is still

[3] Wayve AI Driver: *https://wayve.ai/product/wayve-ai-driver*

not guaranteed if E2E algorithms trained on known data or other collected datasets would be enough for AVs to understand all possible scenes and how these model behave in edge case scenarios.

6.4 Other AI Concepts in Mobility Applications

6.4.1 Generative AI and Large Language Models

Generative AI (GenAI) is a branch of Artificial Intelligence that focuses on creating new content, such as text, images, music, and code, by learning from patterns in existing data. Unlike traditional AI, which is primarily used for tasks like classification, prediction, or recommendation, generative AI has the unique capability to produce novel outputs that resemble the data it was trained on. By leveraging advanced techniques, particularly Deep Learning and neural networks, generative AI models can identify and replicate complex patterns and relationships within large datasets.

One prominent example of generative AI is text generation, where models like GPT (Generative Pre-trained Transformer) (Radford et al. 2018) are able to generate coherent, human-like text based on a given prompt. This enables the creation of essays, responses to questions, and even full conversations. Similarly, image generation is made possible through models like DALL-E[4] and Generative Adversarial Networks (GANs) (Goodfellow et al. 2014), which can produce entirely new images based on input descriptions or learned data, resulting in artwork, realistic photographs, or digital designs.

Generative AI also extends to music and audio creation, where it can compose original music or synthesize voices that imitate specific styles or genres. In addition, models like Codex enable code generation, allowing developers to describe a task in natural language and have the AI generate the appropriate programming solutions automatically.

Generative AI primarily relies on Machine Learning techniques like unsupervised learning, where models learn from unlabeled data, and transformers, which are especially effective for handling sequential data such as text or code. This technology has wide-ranging applications, from creative industries like art and design to practical fields like gaming and software development.

While generative AI opens up new possibilities for creativity and automation, it also brings ethical concerns and challenges related to intellectual property, as well as potential misuse, such as the creation of deepfakes or disinformation. As generative AI continues to evolve, it will be crucial to address these concerns responsibly to ensure

4 DALL-E 3 model: *https://openai.com/index/dall-e-3*

the technology is used for positive and productive purposes. In the following, we present some basic concepts and algorithms related to GenAI.

A **Variational Autoencoder (VAE)** (Kingma/Welling 2014) is a type of generative model in Machine Learning that excels at learning compact representations of data and generating new, similar samples by capturing the underlying data distribution. VAEs combine the strengths of autoencoders with probabilistic modeling, making them valuable for tasks like image generation, data compression, and anomaly detection. The main components of the VAE model are encoder and decoder. The encoder network in a VAE compresses input data, such as images or text, into a lower-dimensional representation known as the latent space. Unlike standard autoencoders, the VAE encoder produces a distribution (typically characterized by a mean and variance) rather than a fixed point, introducing a probabilistic element to the model. In comparison to the encoder, the decoder network takes a sample from the latent space distribution and reconstructs the original input. The goal of the VAE is for the decoder to produce an output that closely resembles the input data, enabling the model to not only compress information but also generate new data from the same distribution. The VAE model also uses so-called latent space. This latent space is a continuous, multi-dimensional space where each point corresponds to a potential data sample. By encoding data as a distribution within this space, VAEs allow for smooth transitions between different data points, which is essential for generating new content. VAEs incorporate a loss function that includes two key components: the reconstruction loss, which ensures that the output is similar to the original input, and the Kullback-Leibler (KL) divergence, which encourages the latent space distribution to approximate a standard normal distribution. This regularization helps maintain a well-organized latent space for generating new samples.

During training, input data is processed through the encoder, which maps it to a distribution in the latent space. A sample is drawn from this distribution and passed to the decoder, which attempts to reconstruct the original data. The VAE is trained to minimize the reconstruction loss while also keeping the latent space well-structured through the KL divergence. This results in a model capable of both accurate data reconstruction and effective sample generation. VAEs are applied in a variety of generative tasks, such as image generation, data compression or data interpolation. VAEs are powerful algorithms for modeling complex data distributions and generating new, meaningful samples. Their combination of autoencoder efficiency and probabilistic modeling creates a flexible and robust framework for a wide range of applications in generative modeling and data representation.

The second type of architecture popular in Generative AI is the **Generative Adversarial Network (GAN)** (Goodfellow et al. 2014). GAN is a type of Machine Learning model designed for generating new data that closely resembles a given training dataset. Introduced by Ian Goodfellow and his colleagues in 2014, GANs consist of two Neural Networks, the generator and the discriminator, which compete against each other in a process known as adversarial training. This competition allows the model to generate highly realistic synthetic data, such as images, audio, or even video. The key components of the GAN model are generator and discriminator. The generator's role is to create new data samples from random noise. It tries to mimic the distribution of the real data and generates outputs that are as close as possible to the actual data in the training set. However, initially, the generator produces very poor-quality data since it learns progressively. The discriminator acts as a judge, attempting to distinguish between real data from the training set and fake data produced by the generator. It outputs a probability indicating whether the input data is real or generated. Finally, the two networks are trained simultaneously in a game-like scenario. The generator tries to fool the discriminator by generating increasingly realistic data, while the discriminator tries to improve its ability to tell real data apart from the fake. Over time, the generator learns to produce data that is indistinguishable from the real data, and the discriminator improves its classification accuracy. The training process of a GAN involves the generator and discriminator working in opposition. The generator takes random noise as input and generates synthetic data. This data is then fed to the discriminator, which evaluates whether the data is real or fake. The discriminator's feedback helps the generator improve its outputs, while the discriminator also improves its ability to identify fake data. This iterative process continues until the generator becomes skilled enough to produce realistic data that the discriminator can no longer reliably distinguish from real data. A simplified architecture of the GAN architecture is presented in Figure 6.34.

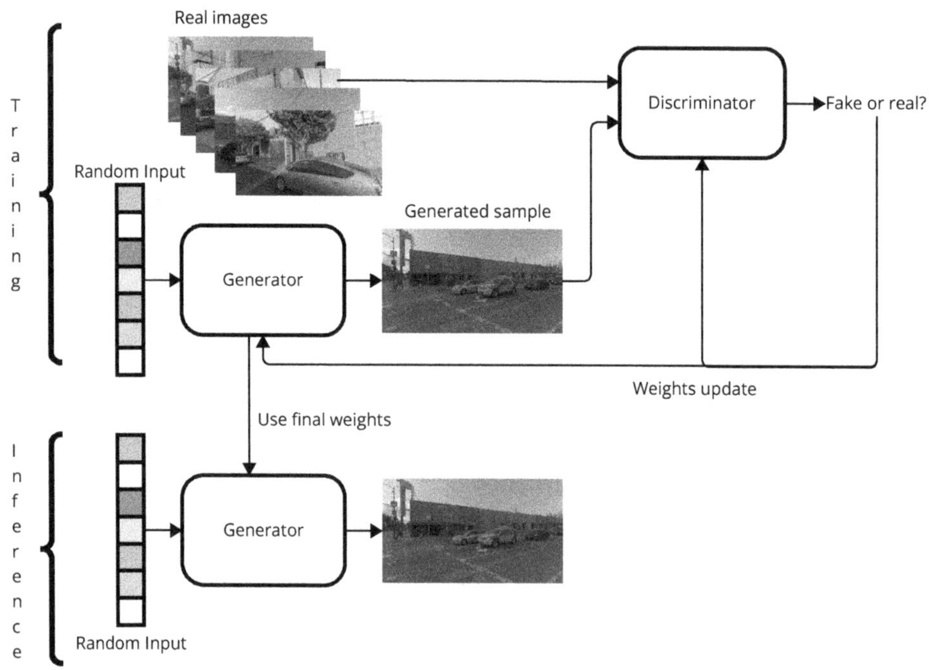

Figure 6.34 Simplified architecture of the GAN algorithms

GANs are widely used to generate realistic images, such as faces, landscapes, or objects. Applications like deepfake technology and tools for artistic image creation rely heavily on GANs. GANs can be used to transfer styles between images, such as turning a photo into a painting or applying a specific artistic style to a photograph. GANs can generate synthetic data to augment datasets, which is particularly useful in fields like medical imaging, where labeled data can be scarce. GANs are used to generate high-resolution images from low-resolution inputs, enhancing the quality of images in applications such as satellite imagery or video enhancement. Advanced GAN models can create images based on textual descriptions, generating realistic visuals that match the input text (Reed et al. 2016).

There are also some challenges when working with GAN models. Training GANs can be complex due to the adversarial nature of the model. If the generator or discriminator becomes too strong or too weak, the training process may collapse, resulting in poor-quality data. Additionally, GANs can sometimes produce realistic-looking but meaningless outputs due to a phenomenon called mode collapse, where the generator learns to produce only a limited variety of outputs.

GANs are a powerful tool for generating realistic synthetic data, revolutionizing fields like Computer Vision, art, and data augmentation. By harnessing the adversarial relationship between a generator and a discriminator, GANs can produce high-quality, believable data, opening up new possibilities for creativity and innovation in AI applications.

Understanding GAN and VAE models is critical because these models are used for image and video generation, and therefore they can directly be used to generate synthetic data for model training or test data for verification and validation of Deep Learning models. For example, the synthetic images can be generated to test performance of the models in edge scenarios.

Another interesting models in the family of GenAI are **Large Language Models (LLMs)**, which are designed to understand, generate, and manipulate human language. These models are built on Deep Learning architectures, particularly transformer networks, which enable them to process vast amounts of text data. By training on diverse datasets that include books, articles, websites, and other written materials, LLMs learn the complexities of language, including grammar, context, and various linguistic nuances.

One of the key features of LLMs is their ability to generate coherent and contextually relevant text based on a given prompt. This capability is achieved through a process, where the model predicts the next word in a sentence based on the preceding words. LLMs have various applications, including but not limited to Natural Language Processing tasks such as text completion, translation, summarization, sentiment analysis, and conversational agents. They can also assist in creative writing, content generation, and coding, demonstrating versatility across multiple domains. However, the deployment of LLMs raises ethical concerns, including issues related to bias, misinformation, and the potential for misuse. Ensuring responsible usage and addressing these concerns is crucial as these models continue to evolve and become more integrated into society. These models are considered a potential game changer in mobility applications because such models can automate the tedious engineering tasks like definition of the requirements, architecture design, unit design implementation and testing, but also be used in traffic forecasting.

As the field of Natural Language Processing advances, researchers are focused on improving the efficiency, interpretability, and robustness of LLMs. Ongoing developments aim to create models that are not only more powerful but also capable of understanding and generating language in a way that aligns with human values and ethics. The impact of Large Language Models on technology, communication, and society is profound, making them a pivotal area of study in Artificial Intelligence.

In the context of Large Language Models (LLMs), the **transformer** architecture has emerged as a ground-breaking development, significantly transforming Natural Language Processing (NLP) and other tasks involving sequential data. Introduced in the influential paper "Attention is All You Need" by Vaswani et al. in 2017, the transformer model distinguishes itself by eliminating the need for recurrent or convolutional layers (Vaswani et al. 2017). Instead, it utilizes a mechanism known as self-attention to effectively process input data. A typical task for transformer could be text translation task, e. g., from English to German as presented in Figure 6.35.

Figure 6.35 Illustration of the high-level overview of the transformer model used for a text translation task

One of the main innovation of transformers is the self-attention mechanism, which enables the model to evaluate the importance of different words in a sentence in relation to one another. By calculating attention scores, the transformer model can selectively focus on specific elements of the input sequence during output generation. This capability allows the model to capture long-range dependencies and contextual relationships among words with great efficacy. For example, the word "apple" can have different meaning depending on the context in the sentence. Let's consider two sentences:

1. I picked an apple from the tree.
2. I bought a new Apple smartphone in the electronic store.

In the first sentence, an apple is used in the context as a fruit. In the second sentence, it clearly denotes a company brand. The attention mechanism allows to generate different feature vectors for the same word focusing on the context.

Transformers also employ multi-head attention, which involves using multiple self-attention heads to discern various types of relationships within the data. Each attention head processes the input independently, and their results are subsequently combined, enabling the model to learn multiple aspects of language simultaneously.

To account for the sequential nature of language, Transformers integrate positional encoding. Unlike recurrent neural networks, which naturally understand word order, transformers do not possess an inherent sense of sequence. Positional encoding is employed to provide information about the position of each word in the input, helping the model grasp the order of words effectively.

Additionally, each output from the attention mechanism is funneled through feed-forward neural networks. These networks, composed of fully connected layers, further refine the representations learned from attention, enhancing the model's understanding of the data.

The architecture also incorporates layer normalization and residual connections to stabilize training and improve gradient flow. This design mitigates issues related to vanishing gradients, which allows for the creation of deeper architectures and more effective learning processes.

Typically, a Transformer model features an encoder-decoder structure, where multiple layers of encoders and decoders are stacked. The encoder processes the input text, generating rich contextual embeddings, while the decoder utilizes these embeddings to produce the output text. This architecture is particularly effective for tasks such as translation and text generation.

The Transformer architecture works in the iterative manner as presented in Figure 6.36. The input to the transformer is input text, which is provided into the encoder part. The decoder part in the first iteration takes the token called Beginning of Sequence (BOS) and the model returns for this input the first word in the translated sequence. Then in second iteration, the token related to this output word is concatenated with the BOS token and the model predicts the second word in the translated sequence. The procedure is repeated until the End of Sequence (EOS) token is returned. Thanks to the iterative process, the transformer model is able to generate a different number of words as the output text. It does not have to be a fixed length as it is assumed in many sequential models.

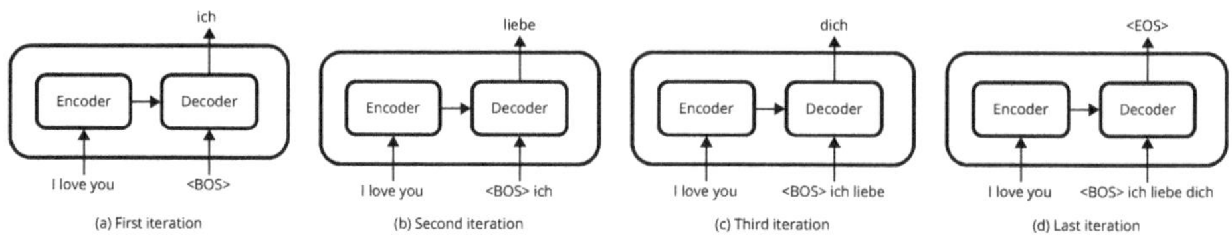

Figure 6.36 Example of the iterative process of the output generation by the Transformer model

The original transformer model consists of an encoder stack and a decoder stack. Each encoder and decoder has multiple layers (often 6 or 12), each with self-attention and feed-forward sublayers. The encoder processes the input sequence, while the decoder generates the output sequence in autoregressive models (e. g., for translation). In some applications, like BERT (Bidirectional Encoder Representations from Transformers) (Devlin et al. 2019), only the encoder stack is used, while GPT models use only the decoder stack. Transformers don't process sequences one step at a time. Instead, they handle entire sequences at once, allowing for much faster processing, especially on large datasets. Since Transformers don't process data in a step-by-step sequence, they need a way to understand the order of tokens. Positional encoding adds information about the position of each token in the sequence. Positional encoding is a set of vectors added to the input embeddings. These vectors use sine and cosine functions with different frequencies, creating unique positional information for each token based on its position in the sequence. This allows the model to distinguish between "The cat sat on the mat" and "On the mat sat the cat." The encoder processes the input sequence by applying self-attention and feed-forward layers across multiple stacked layers. The decoder receives the output from the encoder and generates the target sequence (or prediction). During training, transformers receive both the input and output sequences. The model is trained to maximize the likelihood of the correct output sequence by minimizing a loss function (e. g., cross-entropy loss for text generation). In the inference phase in tasks like text generation, the model generates outputs token by token. For each generated token, it updates the sequence and predicts the next token until a stopping criterion is met (e. g., a special end token). While originally developed for NLP, transformers have expanded to other domains, such as Computer Vision (Vision Transformers) (Dosovitskiy et al. 2021), audio processing, and even autonomous driving, where they're used for interpreting complex sequences in sensory data (Singh 2023). A simplified architecture of the Transformer model is presented in Figure 6.37.

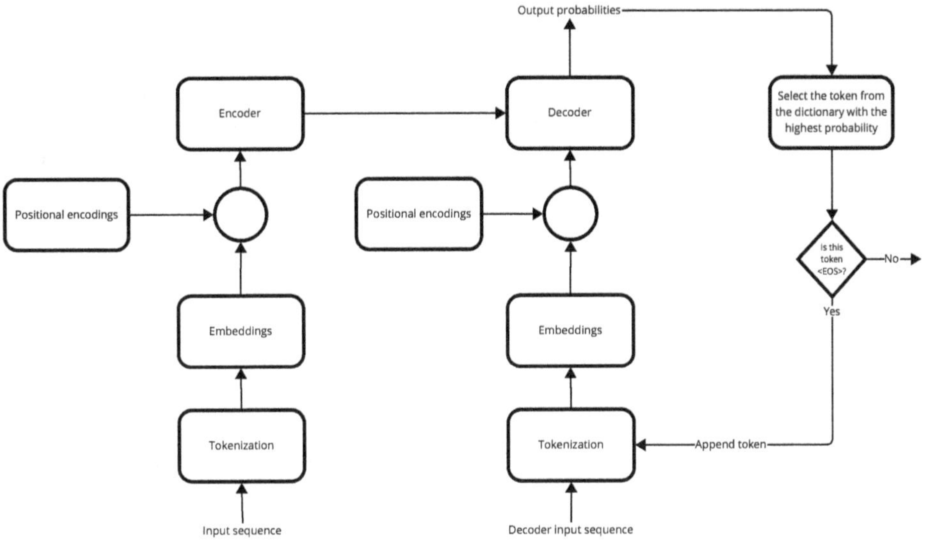

Figure 6.37 Simplified Transformer architecture

Another significant improvement in GenAI was the introduction of the **Generative Pre-Trained Transformer (GPT)**, which is a type of transformer Neural Network designed for natural language understanding and generation (Radford et al. 2018). Developed by OpenAI engineers, GPT models are particularly powerful because they are pre-trained on large amounts of text data, allowing them to generate coherent and contextually relevant text given a prompt. GPT is based on a transformer decoder-only architecture, unlike the original transformer that has both encoder and decoder stacks. This setup allows GPT to generate text in an autoregressive fashion, meaning it predicts each word (or token) one at a time, given the previous context. The extensive pre-training helps GPT learn a general understanding of language, syntax, and even factual information, which it can apply to a wide range of prompts. GPT's architecture scales effectively with increased parameters (more layers, larger hidden dimensions). Models like GPT-3 and GPT-4, with billions of parameters, exhibit far greater fluency and coherence in language generation (Kalyan 2024). In comparison to the transformer architecture, GPT takes only decoder part of the model, and the output token is concatenated directly to the input sentence as it is presented in Figure 6.38.

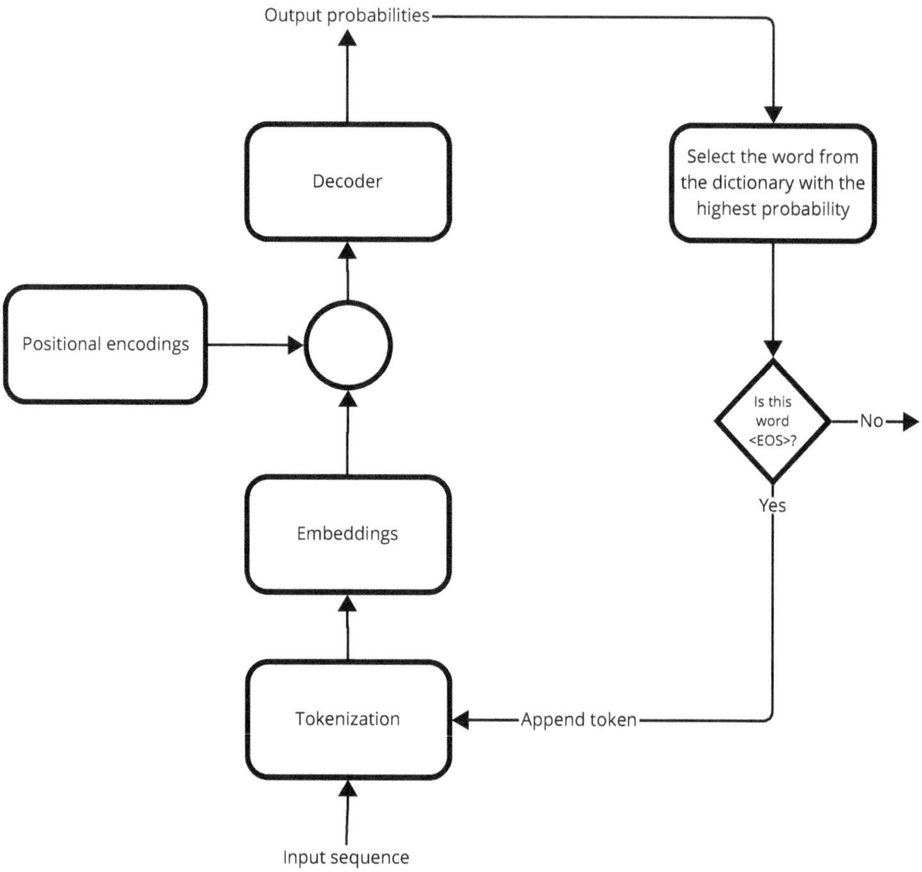

Figure 6.38 Simplified architecture of the of the generative pre-trained transformer which is used in the very popular ChatGPT

GPT models have shown versatility across many language tasks such as text generation, questions answering, text translation or generating summaries of longer texts. Transformers have become the backbone of many state-of-the-art Large Language Models. Their ability to handle extensive datasets and capture intricate linguistic relationships has led to substantial progress in various NLP applications, such as text classification, sentiment analysis, and question answering. Moreover, the parallelizable nature of transformers allows for efficient training on large volumes of data, facilitating the development of highly capable language models that generate coherent and contextually appropriate text. As research continues to advance, transformer models remain central to innovations in Artificial Intelligence and natural language understanding.

Unfortunately, these models suffer from so-called **hallucinations**. A hallucination occurs when a LLM model generates a response that is seemingly realistic yet is nonfactual, nonsensical, or inconsistent with the given prompt. Another challenge is the lack of publicly available documents from safety-critical projects on which such models can be fine-tuned. This is a real problem, when LLM-based tools will be used to automatically generate work products for safety-critical item functions in road vehicles which includes requirements, architecture, test specifications or safety analyses. It could be expected that engineers will pay less attention to the automatically generated content, therefore the whole idea of the V-model could be significantly biased. Using LLMs in safety-critical systems requires more research to better understand potential risks that can be introduced into the final system, when such LLM-based tools are incorporated into the development lifecycle.

6.4.2 Explainable AI

The latest trends in AI postulate also to focus more on legal and ethical aspects of AI. It is also highlighted that potential limitations are not only technical and computational power issues. These concerns are partially addressed in **Explainable Artificial Intelligence (XAI)**. XAI postulates that every decision made by AI must be explainable to the end user (Saranya/Subhashini 2023). It refers to a collection of methods and techniques aimed at making AI systems more transparent, interpretable, and understandable to humans. As AI systems, particularly those utilizing complex models like Deep Learning, are increasingly applied in critical areas such as healthcare, finance, and autonomous systems, the need for users, developers, and stakeholders to comprehend how these systems reach their decisions has become more pressing.

Traditional AI models, especially those based on Neural Networks, are often seen as "black boxes" due to their inability to provide easily interpretable explanations for their decisions. XAI addresses this challenge by offering insights into the internal workings of these models, highlighting the factors that influence their outputs, and helping users assess whether the system's decisions are reliable or biased.

The primary objectives of XAI include promoting transparency, where AI models are made more accessible for humans to understand how decisions are formed, and interpretability, which involves developing tools that explain the model's reasoning in a clear and meaningful way. Furthermore, XAI seeks to enhance trust by enabling users to verify that the system's outputs are fair, unbiased, and grounded in valid reasoning. It also ensures accountability by allowing AI systems to be audited, which is particularly important in high-risk applications such as medical diagnosis or autonomous driving.

XAI employs various approaches, such as visualizing model behaviours, developing simplified models for explanation, and generating natural language explanations to describe the decision-making process of AI systems. By promoting transparency and fostering trust, XAI is essential for the ethical and responsible deployment of AI technologies, ensuring that these systems can be safely integrated into society. One of the most popular methods in this field is Local Interpretable Model-Agnostic Explanations (LIME) (Ribeiro et al. 2016). LIME is an algorithm designed to provide accurate explanations for individual predictions made by any classifier or regressor. The term "local" refers to local fidelity, meaning that the explanation should accurately represent the behavior of the classifier in the vicinity of the specific instance being predicted. For the explanation to be meaningful, it must also be interpretable, allowing humans to understand it. Since LIME can be used with any Machine Learning model, it is considered model-agnostic.

The XAI models may play a crucial role in the development and deployment of autonomous driving vehicles by making the decision-making process of AI systems transparent, understandable, and trustworthy. Such models may also provide traceable justifications for actions taken before and during incidents, aiding in liability determination. XAI can also help to build trust in autonomous systems if society can understand their decisions in critical situations.

6.4.3 Bionics

Artificial Intelligence (AI) and bionics are two distinct fields of study. AI is a branch of computer science that focuses on creating intelligent machines that can perform tasks that typically require human intelligence. **Bionics**, on the other hand, is the study of biological systems and their application to engineering design. While there is no direct relationship between AI and bionics, there are some areas where they overlap. For example, researchers are exploring the use of bionics to create prosthetic limbs that can be controlled by the user's thoughts. AI can also be used to improve the functionality of these prosthetics by enabling them to learn and adapt to the user's movements. AI and bionics are two distinct fields of study with different applications and goals. However, there are some areas where they can complement each other to create innovative solutions.

Bionics and Cyber-Physical Systems (CPSs) are two distinct fields of study. However, researchers are exploring the use of bionics to create Cyber-Physical Systems that can interface artificial systems with biological organisms towards the next generation of bionic Cyber-Physical Systems. For example, the iBionicS Lab at North Carolina State University is researching the development and use of implantable and wearable biomedical microsystems for various in vivo and in vitro applications at different biocomplexity levels. They are particularly interested in biophotonics-based brain-machine interfaces, physiological and behavioral sensing through body area networks, remote control of locomotion in insects, and neuro-electronic interfaces with developing tissues and organisms.

Bionics can be used to create innovative solutions that can support Cyber-Physical Systems. However, more research is needed to explore the full potential of this field. Bionics is a field of study that focuses on the application of biological systems to engineering design. Researchers are exploring the use of bionics to create Cyber-Physical Systems that can interface artificial systems with biological organisms towards the next generation of bionic Cyber-Physical Systems. The human nervous system consists of the brain, spinal cord, and neurons. The Central Nervous System (CNS) consists of the brain and spinal cord, both derived from the embryonic neural tube. Both are surrounded by protective membranes called the meninges, and both float in a crystal-clear cerebrospinal fluid. The Peripheral Nervous System (PNS) consists of the nerves and ganglia outside the brain and spinal cord.

The brain is encased in a bony vault, the neurocranium, while the cylindrical and elongated spinal cord lies in the vertebral canal, which is formed by successive vertebrae connected by dense ligaments. The CNS is the processing center of the body and consists of the brain and the spinal cord. Both of these are protected by three layers of membranes known as meninges. For further protection, the brain is encased within the hard bones of the skull, while the spinal cord is protected with the bony vertebrae of our backbones. The neurons are the basic functional units of the nervous system, and they transmit information throughout the body. The PNS is responsible for transmitting information between the CNS and the rest of the body. The PNS is further divided into the Somatic Nervous System (SNS) and the Autonomic Nervous System (ANS). The SNS is responsible for voluntary movements, while the ANS is responsible for involuntary movements. Muscles and tendons are part of the musculoskeletal system, which is responsible for movement and support of the body. The musculoskeletal system works in conjunction with the nervous system to produce movement.

In the evolving landscape of AI, research in bionics can have great potential to inspire the evolution of AI and ML algorithms. The human control system is a complex network of organs, tissues, and cells that work together to produce movement. It is still unclear how it works, but researchers can explore the use of bionics to create innovative solutions that can interface artificial systems with biological organisms towards the next generation of bionic Cyber-Physical Systems and develop better algorithms closer to Strong AI.

6.4.4 How AI Can Support Processes for Mobility Solutions

Historically, there have been various attempts to automate processes. Over time, these efforts have aimed to improve efficiency and reduce human error. AI has big potential to improve many industries including speed up in product development and mitigation of the existing hazards and risks, especially these, which are safety critical in mobility solutions. It has been proved that processes which are well-defined allow to deploy safe and high-quality products to the market. These automotive processes defined in the standards such as ISO 26262, ISO 21448 or ASPICE require to define several work products including requirement specification, architectural design, implementation, test case specification, etc. Development and maintenance of these work products is a repetitive task, which becomes monotonous and prone to errors. These errors require additional process iterations for correction. If many individuals are involved, managing and coordinating these activities becomes a significant challenge. One of the potential solutions can be the application of Large Language Models (LLM) and Generative AI to support automated work products generation. The potential of these models to produce content, automate processes, and enhance work efficiency can result in substantial productivity improvements and innovation. By comprehending and harnessing their capabilities while addressing their limitations, organizations can open up new opportunities and propel technological advancements. Exemplary tasks which potentially can be resolved by LLMs could be automated extraction of the stakeholder requirements based on the provided regulations, automated architecture generation based on the requirements or support in performing hazard and risk analysis. Diemert and Weber showed that LLMs, when used for System Theoretic Process Analysis (STPA) on simple systems, is a feasible and moderately useful method, but the performance of LLMs in these tasks degrades as the system complexity is increased (Diemert/Weber 2023). This only shows the potential opportunities to build better tools which will probably support engineers in work product generation, and finally more and more documentation will be generated by fully automated tools. Unfortunately, an automated or semi-automated work products generation may bring new problems into the industry. Engineers can put too much trust into automatically generated content, and that will require to develop novel mitigation methods and other challenges for the assessors responsible for the final assessment of the product. It seems that the final assessment always will be performed by humans.

The application of Generative AI for the scenario generation for mobility solutions is another opportunity in the development, testing, and validation of these systems. By creating a wide range of realistic and challenging driving scenarios, AI can help to ensure that AD systems are prepared to handle the complexities of real-world driving. AI can contribute in the simulation of diverse driving conditions and generate a variety of driving scenarios that encompass different adverse weather conditions, lighting conditions, and traffic density. AI methods can be involved into the generation of edge case scenarios that include rare objects and obstacles, unexpected road participants behavior and emergency situations such as road closures, accidents or road works. These models can generate data for all relevant sensors helping to build digital twins. Several companies on the market currently offer digital twin environments, however, the latest research criticizes that the current simulation engines, especially for the camera data, does not reflect the real world. Therefore, the simulation may help in verification and validation of the AD systems, but it seems that they are not mature enough to fully replace a real world testing and build sufficient confidence (Cabrero-Daniel et al. 2024). The synthetic data can be used not only for testing purposes, but they could be also used to train the models. While the synthetic data are fully controlled, then they do not require very costly data labelling processes.

The use of AI for static code analysis and automated error or security vulnerabilities detection is the next opportunity in software development for mobility solutions. Static code analysis involves examining the source code without executing it to identify potential errors, code smells, and security vulnerabilities. Traditionally, static code analysis tools use a set of predefined rules and patterns to detect issues. However, AI-powered static code analysis takes this a step further by leveraging Machine Learning and Natural Language Processing to improve the accuracy and scope of code analysis.

As it was presented before, AI-based models play a crucial role in the perception systems of autonomous driving vehicles, enabling them to interpret and understand their environment accurately. Perception involves the collection and analysis of data from various sensors to identify and track objects, predict their behavior, and make informed decisions. Autonomous vehicles are equipped with multiple sensors, including cameras, lidar, radar, and ultrasonic sensors. AI algorithms are essential for sensor fusion in order to combine data from various sources and perform the tasks such as image recognition, object detection, semantic segmentation, and object tracking.

AI algorithms can be also used for planning application in mobility systems, enabling these systems to navigate complex environments, make real-time decisions, and execute safe and efficient driving strategies. Planning involves determining the best possible path from a starting point to a destination while considering various constraints and dynamic factors. AI algorithms may be used for path planning and determining optimal routes for autonomous vehicles. AI can optimize routes by considering real-time data and various constraints. In scenarios involving multiple autonomous vehicles, AI facilitates coordination to ensure efficient and collision-free navigation. AI algorithms can manage uncertainties in the environment, such as unpredictable behavior from other road users or changes in road conditions.

Apart from direct and indirect applications of AI in vehicle development, the AI algorithms can also support automotive industry in other tasks which are not safety critical such as analysis of the consumer behavior, prediction of the sale volume, automated or semi-automated generation of new advertisements, preparation chat bots which can support customer, etc. The application of AI-based solutions can optimize the costs of the production, but may also bring new risk into the development lifecycle.

6.5 Handling Malfunctions of AI-/ML-Based Systems

AI/ML models have become an integral part of our daily lives, powering everything from our personal devices to complex systems in various industries. However, like any technology, AI/ML-based components have their own limitations and challenges. It is crucial to understand malfunctions that can affect AI/ML-based systems and to identify, similarly as in classical functional safety (FuSa) systems, how we can avoid, detect and control them. These malfunctions can occur due to a variety of reasons, ranging from errors in data processing to random hardware failures. Understanding these failure conditions is crucial for the development of robust, safe and reliable AI systems for mobility solutions.

In this section, we analyze the different types of AI/ML malfunctions, their causes, and their impact on the performance of AI/ML systems. This will provide a comprehensive understanding of the challenges faced in the field of AI and pave the way for future advancements. The first challenge is that the failure modes of AI/ML-based components can be the same as classical ISO 26262 components. In the first iteration, we should deal with systematic and random hardware (HW) faults and systematic software (SW) or HW faults, which can occur in AI/ML-based components. These failure modes can be cover adopting the requirements from ISO 26262.

Additionally, the AI/ML-based components can be impacted by sensor or AI/ML algorithms performance limitations. For example, a system that uses cameras in night driving is unable to correctly detect pedestrians on the road due to sensor limitations even if the system is free from SW and HW faults. Another limitation could be limited performance of the algorithms. For example, a 3D object detection algorithm can have specific requirements related to the number of points in the point cloud to be detected. The main source of the potential malfunction in DL-based systems can be low-quality data. A typical dataset lifecycle describes the set of data-related activities carried out during the entire lifecycle of DL-based system development, including their deployment.

Another source of the potential system failures could be an insufficient specification of the AI/ML-based functions or components. In case of AI/ML systems, the problem may begin with an unclear definition of the system's objectives or target environment in which the AI/ML component should be deployed. For instance, vague system requirements like "collision avoidance system shall detect potential collision with animals by analyzing the collision path between the vehicle and a detected animal" does not provide sufficient details whether a bird flying close to the ego vehicle shall activate the braking system or not. Similarly, insufficient details about data requirements, such as the type, quality, volume, and sources of data, can negatively impact the performance of the developed system and result in incorrect design decision on the sensor selection or algorithmic decisions. Neglecting to specify how data should be preprocessed, labeled, or updated adds further complications.

Another common issue arises from insufficient specification of the non-functional requirements for AI/ML-based systems. This includes the absence of well-defined performance metrics such as accuracy, precision, recall, or F1 score, as well as ambiguity about acceptable trade-offs, like those between precision and recall. The lack of attention to system constraints, such as computational or hardware limitations, latency requirements, scalability, or cost considerations, can also hinder the development and deployment of the system. Moreover, a lack of planning for the iterative nature of AI/ML systems—such as retraining, updates, and field monitoring after deployment—can cause the system to degrade over time.

Such challenges are highlighted in the currently released safety standards such as ISO 21448:2022 (SOTIF) and ISO 8800:2024. In the systematic identification of the potential failures of AI/ML systems, classical safety analysis techniques such as Failure Mode and Effects Analysis (FMEA), Fault Tree Analysis (FTA), and System Theoretic Process Analysis (STPA) can be applied on various levels of abstraction during product development: vehicle level, system level, hardware, software, data collection and labeling. Recently, researchers have been enhanced the basic safety analyses to model the AI/ML systems, demonstrating that the conventional methods used for safety-critical systems must be specifically tailored to address the unique challenges of AI/ML-based systems. We recommend that these safety analyses be conducted collaboratively by safety engineers and Machine Learning engineers, leveraging the technical expertise of Machine Learning engineers to ensure a thorough understanding of the chosen AI/ML models.

Comparison with Linux

The Linux kernel, the heart of the Linux operating system, is a complex piece of software that manages the system's resources and mediates interactions between the hardware and software. It maintains a multitude of system states, reportedly over 10,000, each representing a unique configuration of the system's resources and processes. These states, depending on their context, can lead to multiple system calls. However, these calls can be erroneous due to various factors such as context, timing, sequencing, and more. For instance, a process might make a system call expecting a certain state, but if the state has changed due to another process's activity, it could lead to unexpected behavior or even application failure. Consider a scenario where multiple processes are accessing a shared resource. If the kernel's scheduling state changes and a process is prematurely interrupted, it could leave the shared resource in an inconsistent state. When another process accesses this resource, it could lead to a failure.

Now, let's draw a parallel with AI/ML algorithms. These algorithms operate in a high-dimensional state space, often with uncountably many states due to the continuous nature of many ML parameters. Each state represents a specific configuration of the algorithm's parameters. The decisions that an AI or ML algorithm makes when leading to a process output are based on these parameters. For instance, in a Neural Network, the weights and biases (parameters) determine the output for a given input. The algorithm "learns" by adjusting these parameters based on the feedback it receives (the loss function). However, the complexity of these algorithms, combined with the vastness of the state space, can lead to unpredictable behavior. For example, small changes in input data or parameters can sometimes lead to significantly different outputs, a phenomenon known as "adversarial attacks." Moreover, the presence of millions of "if-then-else" constructs adds another layer of complexity and potential points of failure.

In conclusion, both Linux kernel states and AI/ML algorithm states have the potential to cause system failures, albeit in different ways. While Linux state failures are often due to resource mismanagement, AI/ML failures are typically due to the inherent complexity and unpredictability of the algorithms. Therefore, robust error handling and system design are crucial in both domains to ensure reliability and performance.

System Failures

System failures can significantly impact the operation of AI and ML algorithms, leading to erroneous outputs or even complete system crashes. One common failure condition could be systematic errors in memory management. Memory management is crucial for the efficient operation of AI and ML algorithms, as they often require large amounts of data to be stored and retrieved. However, systematic errors can lead to issues such as memory leaks, where memory that is no longer needed is not properly released, or segmentation faults, where a process tries to access memory that it doesn't have permission to access. For example, an ML model training process could crash if it tries to access more memory than is available due to a memory leak.

Another failure condition is the corruption of tables. Many AI and ML algorithms rely on tables to store data and intermediate results. If these tables get corrupted, for instance, due to a disk failure or a bug in the system, the algorithm might produce incorrect results or fail to run altogether. A typical example would be a lookup table used in a decision tree algorithm getting corrupted, leading to wrong decisions being made.

Arbitration and scheduling failures can also lead to system failures. In a multi-tasking system, the scheduler decides which process gets to use the processor at any given time. If the scheduler fails, it could lead to some processes not getting enough CPU time, causing them to run slowly or not at all. For instance, a real-time object detection algorithm might fail to detect objects in a timely manner if it doesn't get scheduled to run frequently enough.

Runtime and real-time failures of the host processor can also cause system failures. If the processor fails or doesn't meet the real-time requirements of the AI or ML algorithm, it can lead to the algorithm failing to produce results in a timely manner. For example, a speech recognition algorithm might fail to transcribe speech in real-time if the processor is too slow or fails.

Lastly, random hardware faults in the processor can lead to system failures. These faults can be due to various reasons, such as manufacturing defects, aging, or environmental conditions, and can cause the processor to produce incorrect results or fail altogether. For example, a bit flip due to a hardware fault could cause an ML algorithm to make incorrect predictions.

Issues with Training Data

The performance of AI/ML algorithms are heavily dependent on the quality of the training data. Issues with training data can lead to failures, producing incorrect or biased results. One such issue is human errors with training data. AI and ML algorithms learn from the data they are trained on. If this data is incorrect or biased, the resulting AI system will also be incorrect or biased. For example, if a facial recognition system is trained primarily on images of people from a particular demographic, it may perform poorly when presented with faces from other demographics. This bias in training data can lead to a biased AI system, which can have serious implications, such as unfair treatment or discrimination.

Another issue is the inadequate context for training data. AI and ML algorithms need diverse and representative training data to function effectively in all situations. If the training data doesn't represent the full range of possible input data, the AI system might not work well in all situations. For instance, an AI model trained to detect spam emails based on a dataset of English language emails might fail to accurately classify spam emails written in other languages.

Sequence and temporal issues with training data can also lead to failures. Some AI systems, like those used for Natural Language Processing or time series prediction, require data to be in a specific sequence or format. If the training data doesn't meet these requirements, the AI system might not work correctly. For example, a time series prediction model like those used in stock price prediction or weather forecasting requires data in chronological order. If the data is out of order or timestamps are incorrect, the model might make inaccurate predictions.

In conclusion, while AI and ML algorithms offer immense benefits, they are heavily reliant on the quality of their training data. Issues with training data, such as human errors, inadequate context, and sequence or temporal issues, can lead to system failures. Therefore, it's crucial to ensure the quality and diversity of training data to build reliable and fair AI and ML systems.

Adversarial attacks

Malicious data, in the context of Machine Learning, is data that is specifically designed to introduce an error or deceive a model of Machine Learning. Malicious data can be used for so-called adversarial attacks. Such attacks are attempts to deceive a Machine Learning model by providing it with data input that is specifically designed to cause the model to make a mistake or act incorrectly. Malicious data can be added to the dataset used to train the model, or can be used as a stand-alone input to the model. Such data pose a serious threat to UM models, as they can significantly reduce the accuracy and reliability of the model and are difficult to detect and prevent. The alteration of a small number of pixels in the original image caused the model, which had originally correctly recognized that there was an "English Spaniel" in the image, when a few pixels are added to the image, the model changed the classification to the class "Siamese Cat."

We can divide adversarial attacks into two basic groups related to whether or not the attacker has any knowledge of the model. The first group are white box adversarial attacks. In this type of attacks, the attacker has full knowledge of the model and the training set. Many algorithms have now been proposed in literature for generating this type of attack. One of the first methods for adversarial attack generation was the FGSM algorithm (Fast Gradient Sign Method). The purpose of FGSM algorithm is to generate noise (not random) whose direction is the same as the gradient of the cost function with respect to the data, but with opposite returns. In this case, we are trying to achieve the opposite goal to the propagation algorithm backward, namely to increase the value of the loss function. Then, the noise generated in this way is added to the original image, causing a reclassification of the original image. The second group are black box adversarial attacks. In this case, the attacker does not have direct access to the model, and after modifying the input, all he can do is observe whether the model's output has changed its response. One of the more interesting examples of attacks in this category are one-pixel adversarial attacks, which can be generated using various types of heuristic algorithms, for example, evolutionary algorithms.

In literature, adversarial attacks are also divided according to the target of the attack into targeted adversarial attacks and untargeted adversarial attacks. The former refers to a scenario in which the attacker intends to force a change in the model, e. g., in a classification problem to a specific class. Untargeted attacks change the original decision of the model to any other. Intensive work is currently underway to develop techniques for generating hostile attacks and, at the same time, defense strategies against malicious data. Existing defend methods can be divided into the following ones:

- modifying the network weights by generating successful adversarial attacks and then adding these samples to the original training set
- changing the network learning strategy, e. g., through a network distillation technique (network distillation), in which first the network is trained with a large number of parameters and then extracting from them some of the weights
- using low-level input processing methods in order to remove attacks from them, such as the removal of adversarial attacks by noise filtering methods or by color quantization methods
- use of additional classifiers to determine whether or not the data contains an adversarial attack before the actual inference step, which is only applied if the sample is undisturbed

The primary concern with adversarial attacks from a safety and security perspective is the unpredictability of adversarial data modifications. Unfortunately, existing Deep Learning models remain vulnerable to such attacks. Even when a new defense strategy proves effective against known threats, there is a high likelihood that it will fail against emerging attack types.

6.6 Regulations and Standards of Artificial Intelligence

6.6.1 Automotive Standards

The industry is witnessing a paradigm shift with the adoption of AI-based methods also including safety critical domains such as automotive, railway, avionics, medicine, etc. However, the deployment of AI algorithms in mobility systems demands thorough evaluation and mitigation of potential risks.

Given the critical safety implications of such systems, where malfunctions could result in fatalities or injuries, it's imperative to adhere to a rigorous development process aligned with current best practices and relevant safety regulations. However, existing automotive safety standards like ISO 26262 and Safety of the Intended Function (SOTIF) lack comprehensive coverage for the deployment of AI/ML algorithms. Safety standards such as ISO 26262, IEC 61508 still wait to be updated for AI and ML problems. Moreover, the current version of the IEC 61508 standard explicitly advises against the use of Artificial Intelligence algorithms. Consequently, conventional safety standards and their associated verification/validation methodologies are adopted for ML algorithms, failing to adequately address their unique attributes such as non-determinism, opacity, and instability. Other dedicated standards such as ISO/DIS 21448, UL 4600, ISO/PAS 8800, ISO/IEC TR 5469 and ISO/TS 5083 are not mature enough or are still under development. Therefore, they do not address all possible risks that may affect the modern mobility systems such as autonomous driving, ADAS or remotely controlled systems.

Discussing ML and AI algorithms, it is also worth mentioning the fourth generation of the **ASPICE (Automotive SPICE v4.0)** standard, which was published in December 2023 and introduces novel processes dedicated to Machine Learning engineering. Automotive SPICE (Software Process Improvement and Capability dEtermination) is a standard tailored for the automotive industry to evaluate and improve software development processes. It was adapted from the general SPICE (ISO/IEC 15504) standard to address the specific needs of automotive software, including the integration of complex systems and ensuring high safety and quality standards. ASPICE 4.0 introduces specific processes for AI/ML software. It defines the supporting process "SUP.11 Machine Learning Data Development" and four primary lifecycle processes: "MLE.1 Machine Learning Requirements analysis," "MLE.2 Machine Learning Architecture," "MLE.3 Machine Learning Training," and "MLE.4 Machine Learning Model Test." MLE.1 defines and analyzes the Machine Learning requirements to ensure they are clear, complete, and aligned with the system and software level requirements. MLE.2 establishes an ML architecture that supports both training and deployment. It ensures the architecture is consistent with the ML requirements and evaluates it against defined criteria. This includes developing the ML architecture, determining hyperparameter ranges and initial values for training, defining interfaces for the architectural elements, and establishing consistency and bidirectional traceability between the ML architectural elements and the ML requirements. MLE.3 focuses on training and validating Machine Learning models, ensuring they meet specified performance and reliability criteria aligned with the defined ML requirements. MLE.4 addresses the deployment of Machine Learning models into production environments. It includes the continuous monitoring of model performance to ensure they operate reliably and safely over time. SUP.11 focuses on the collection, preparation, and management of data for training and testing Machine Learning models, with a strong focus on ensuring data quality and relevance. This includes defining and aligning the data with ML data requirements, maintaining the integrity and quality of the data, and making it accessible to relevant stakeholders. ML data management is designated as a supporting process, reflecting the assumption that a dedicated unit within the organization handles data collection and preparation. This approach also facilitates the shared use of the same data across various organizational functions.

ISO 21448:2022 provides a general argument framework and guidance on measures to ensure the Safety of the Intended Functionality (SOTIF), which is the absence of unreasonable risk due to a hazard caused by functional insufficiencies, i. e.,

- the specification insufficiencies of the intended functionality at the vehicle level, or
- the specification insufficiencies or performance insufficiencies in the implementation of electric and/or electronic (E/E) elements in the system.

The SOTIF standard provides guidance on the applicable design, verification and validation measures as well as activities during the operation phase that are needed to achieve and maintain the SOTIF. This standard applies to intended functionalities where accurate situational awareness is critical for safety. This situational awareness relies on complex sensors and processing algorithms, making SOTIF particularly relevant for emergency intervention systems and driving automation systems spanning SAE levels 1 to 5. Unlike ISO 26262, this standard acknowledges the possibility of hazards occurring even when a system is free of hardware or software faults. For instance, a component may fail due to inherent sensor limitations rather than random failures or systemic faults. An example of this is a camera-based collision avoidance system that fails to detect a pedestrian at night—not because of a malfunction, but due to the inherent limitations of the camera sensor in low-light conditions. The vehicle should decelerate, but in this case lack of braking when required is caused by a weakness of a sensor.

The SOTIF standard mandates additional risk analysis for both known and unknown scenarios that could lead to hazardous behavior. These risks may arise from sensor limitations, processing algorithm constraints, or improper use of the functionality by the driver. Functional insufficiencies can arise also from insufficiencies of specifications. For example, a pedestrian could be not sufficiently defined in the requirements and a child in a stroller could be not defined as an example of a pedestrian and therefore ignored by the ADAS or autonomous driving feature.

When it is not possible to completely eliminate risk, the final approach is to accept the residual risk. Establishing criteria for risk acceptance involves input from multiple stakeholders, including regulators, society, OEMs, and suppliers. While SOTIF does not define a specific threshold for acceptable residual risk, it outlines several guiding principles, including the following:

- **GAMAB/GAME (Globalement au moins aussi bon/Globalement au moins équivalent):** It means "globally at least as good/globally at least equivalent" associating the newly developed system against the risk associated with a comparable system already in the market.
- **Positive Risk Balance (PRB):** The goal of this criterion is to ensure that the benefits outweigh the risks compared with human driving performance.
- **ALARP (As Low As Reasonably Practicable):** This criterion assumes that we can minimize risks as far as reasonably achievable.
- **MEM (Minimal Endogenous Mortality):** This criterion ensures that the death risk related to the introduction of a new product shall not increase the death rate in the society.

These criteria are very important because based on them an organization can define the global acceptance criteria for the product deployment and on the lowest level the specific acceptance criteria for the sensors and AI/ML algorithms can be defined.

ISO/IEC TR 5469:2024 aims to guide developers of safety-related systems in effectively applying AI technologies within safety functions by enhancing awareness of AI's characteristics, safety risks (specific to functional safety), applicable methods, and potential limitations. This standard is intended to help developers of safety-related systems effectively integrate AI technologies into safety functions by raising awareness of AI's properties, functional safety risks, applicable safety methods, and possible constraints. It also provides insights into the challenges and solution concepts relevant to AI system safety. It contains 11 clauses and four annexes. The ISO/IEC TR 5469 provides a detailed framework for integrating AI technologies into safety-related systems, addressing their unique challenges and requirements.

Clause 5 introduces the concept of functional safety and its importance in the context of AI technologies and systems. Clause 6 categorizes different AI technologies to illustrate how they can achieve compliance with functional safety standards when incorporated into safety functions. It further defines levels of AI usage based on their impact on system outcomes and offers a qualitative assessment of safety risks associated with various combinations of AI technology classes and usage levels. The standard defines the so-called usage levels A1 (AI is used in E/E system for control), A2 (AI is used in E/E system for diagnostic), B1 (AI is used in E/E system as a validation tool), B2 (AI is used in E/E system as a support tool), C (AI is not safety-relevant but can have indirect impact) and D (AI is not safety-relevant and cannot have even indirect effect) and AI technology classes I, II and III. Class I represents AI technologies that are fully developed and reviewed in accordance with functional safety (FS) standards. Class II refers to AI technologies that cannot be entirely developed or reviewed under existing FS standards but where additional complementary requirements, methods, and techniques for design, verification, and validation can be identified to ensure the desired safety properties and achieve the necessary risk reduction. Class III includes AI technologies for which compliance with existing FS standards is not feasible, and no additional properties, methods, or techniques are currently available to address the associated risks or provide guidance.

To address scenarios where a direct adherence to existing functional safety standards is not feasible, Clause 7 presents a three-stage realization approach derived from ISO/IEC 22989. Clause 8 examines the properties of AI systems and their associated safety risks, identifying the challenges and considerations necessary for managing these risks effectively. Clauses 9, 10, and 11 propose comprehensive solutions to these challenges, focusing on verification, validation, control measures, risk mitigation strategies, processes, and methodologies.

The annexes provide additional insights and practical examples. Annex A explains the application of IEC 61508-3 to AI technology components. Annex B offers examples of how the three-stage realization principles can be applied and how essential system properties are defined. Annex C expands on the processes outlined in Clause 9.3, while Annex D maps the safety lifecycle defined in IEC 61508-3 to the AI system lifecycle described in ISO/IEC 5338. Together, these elements form a cohesive framework for ensuring the safe integration of AI technologies into functional safety systems.

The authors of the ISO/IEC TR 5469:2024 agree that AI/ML components used in safety critical systems should be supervised by classical and deterministic components and with classical backup decision systems as presented in Figure 6.39.

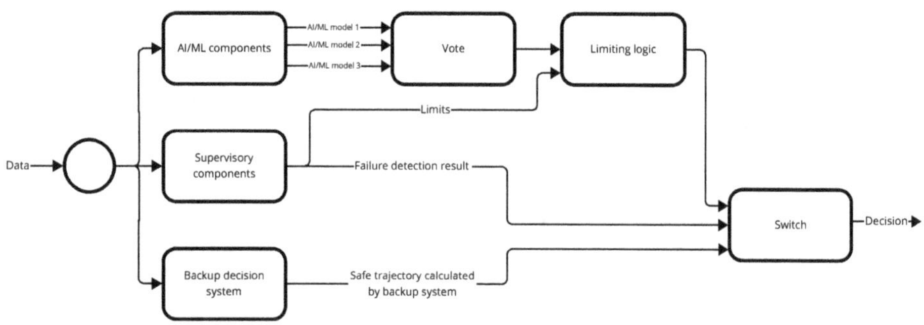

Figure 6.39 Proposed concept of AI- or ML-based system which uses classical components to supervise AI- or ML-based components

ISO/PAS 8800 was released in December 2024. It defines safety-related properties and risk factors impacting the insufficient performance and malfunctioning behavior of Artificial Intelligence (AI) within a road vehicle context. It describes a framework that addresses all phases of the development and deployment lifecycle. This includes the derivation of suitable safety requirements on the function, considerations related to data quality and completeness, architectural measures for the control and mitigation of failures, tools used to support AI, verification and validation techniques as well as the evidence required to support an assurance argument for the overall safety of the system. It is common in AVs to have an incomplete picture, or specification, of the complete operational domain. In those cases, AI/ML is often utilized. ISO 8800 evaluates the impact on AI due to performance insufficiencies such as errors due to bias and lack of robustness, insufficiencies in training data, and many other aspects. It introduces an AI safety lifecycle specific to automotive, while aligning with the safety lifecycle prescribed by the ISO 26262 standard. ISO 8800 will take in requirements derived from ISO 26262 and ISO 21448 but extend the requirements to properties specific to AI such as bias, prediction, robustness and generalization. There are methods prescribed allowing one to reduce the ML performance requirements by compensating using other methods, such as non-AI monitors or supervisors. We can also expect to have safety requirements and safety analysis associated with data sets, since the quality of data plays a significant role in ML.

Another interesting document is **Assurance of Machine Learning for Use in Autonomous Systems (AMLAS)** (Hawkins et al. 2021), a framework that supports the development of a corresponding safety case for components which use ML/AI algorithms. AMLAS provides a structured process designed to produce 34 safety evidence artifacts and recurring safety case patterns for Machine Learning (ML) components. These patterns are represented using the Goal Structuring Notation (GSN), a graphical format for safety cases. The AMLAS process consists of six stages, beginning with the definition of system safety requirements.

The first stage, ML Safety Assurance Scoping, operates at the systems engineering level to define the scope of the safety assurance process for the ML component and its associated safety case. This stage emphasizes the critical interplay between ML-specific and non-ML safety engineering efforts. The subsequent five stages focus on assurance activities related to various aspects of ML development and operation. Each stage concludes with an assurance argument, which, when combined and supported by evidence artifacts, forms the overall ML safety case. In Stage 2, ML Safety Requirements Assurance, requirements engineering is applied to elicit, analyze, specify, and validate ML safety requirements concerning the software architecture and the Operational Design Domain (ODD). Stage 3, Data Management Assurance, begins by using requirements engineering to define data requirements aligned with the ML safety requirements. This stage involves generating data sets, including development data, internal test data, and verification data, and performing quality assurance activities to ensure data integrity. In Stage 4, Model Learning Assurance, the ML model is trained using the development data. The internal test data is then used to assess the model's compliance with the ML safety requirements. Stage 5, Model Verification Assurance, involves testing or formal verification methods to confirm that the ML model meets the defined safety requirements. Crucially, verification data, which has not influenced the training process, is used to validate the model's performance. Finally, Stage 6, Model Deployment Assurance, focuses on integrating the ML model into the overall system and verifying that the system meets its safety requirements. Integration testing is conducted within the specified ODD to ensure compliance. The outcome of the AMLAS process is a comprehensive safety case for the system, providing assurance that the ML components and their integration adhere to the required safety standards.

6.6.2 US Publications

USA was one of the first countries where legislation for autonomous vehicles and AI was initiated. Nevada became the first state to authorize the use of autonomous vehicles in 2011. In 2018, the California Department of Motor Vehicles (DMV) approved the deployment of fully automated vehicles without human drivers, leading to the issuance of numerous permits for self-driving cars across various companies.

The Self-Drive Act (H. R. 3388), officially titled Safely Ensuring Lives Future Deployment and Research in Vehicle Evolution Act, is a federal legislative proposal introduced in 2017. It aims to establish a regulatory framework for the testing and deployment of Autonomous Vehicles (AVs), addressing safety concerns while fostering innovation in the self-driving industry.

The National Highway Traffic Safety Administration (NHTSA) has issued federal guidelines for Automated Driving Systems (ADS) to promote the safe design, testing, and deployment of Autonomous Vehicles. These guidelines are presented in key documents such as the Federal Automated Vehicles Policy (2016), "Automated Driving Systems: A Vision for Safety 2.0" (2017), and subsequent updates including "Preparing for the Future of Transportation: Automated Vehicles 3.0" (2018) and "Ensuring American Leadership in Automated Vehicle Technologies: Automated Vehicles 4.0" (2020).

These documents focus on SAE International automation levels 3 to 5 and emphasize that entities are not required to delay testing or deploying their ADS. They revise safety self-assessment design elements, align federal guidance with current advancements and terminology, and clarify the distinct responsibilities of federal and state governments. The guidance remains voluntary, with no mandatory compliance or enforcement mechanisms, reinforcing its role as a framework rather than a regulatory mandate.

In recent years, the United States has introduced various AI-related bills, acts, and guiding principles at both federal and state levels. The National Artificial Intelligence

Initiative Act, enacted in 2020, was designed to advance U. S. leadership in AI by encouraging innovation, coordinating federal AI efforts, and addressing societal, ethical, and workforce challenges posed by AI technologies. Its primary goal is to ensure that the United States remains a global leader in AI while fostering safe, ethical, and inclusive development and deployment.

The AI Bill of Rights, introduced in 2022 by the White House Office of Science and Technology Policy, established a set of non-binding principles to guide the secure and responsible development of AI. The blueprint focuses on five key principles to mitigate risks such as algorithmic discrimination, to improve access to critical services like healthcare and financial systems, and to ensure AI supports civil rights and democratic values.

In 2024, the Bipartisan House Task Force Report on AI was released, presenting key findings, principles, and proposals for AI policy. It explores AI's impact on education, healthcare, intellectual property, and other areas. The report also introduced several overarching principles, including the following:

- Identify AI issue novelty.
- Promote AI innovation.
- Protect against AI risks and harms.
- Empower government with AI.
- Affirm the use of a sectoral regulatory structure.
- Take an incremental approach.
- Keep humans at the center of AI policy.

6.6.3 Legal Framework of the European Union

The European Union (EU) introduced a new legal framework that aims to significantly bolster regulations on the development and use of AI in Europe. The proposed legislation, the Artificial Intelligence (AI) Act, focuses primarily on strengthening rules around data quality, transparency, human oversight and accountability. The AI Act proposes a risk-based approach to regulating AI, meaning that different types of AI systems will be subject to different levels of requirements and obligations. The AI Act also establishes a governance structure at European and national level, as well as a conformity assessment procedure for high-risk AI systems.

Some of the key features of the AI Act are the following:

- It includes a list of 35 high-risk AI applications that are considered to pose a significant risk to human rights or safety, such as biometric identification, social scoring, recruitment tools, credit scoring, law enforcement tools, and health care tools.

These applications will have to comply with strict requirements before they can be placed on the market or used in public services.

- It includes a ban on the use of AI for mass surveillance purposes, such as facial recognition in public spaces or automated decision-making by law enforcement agencies.
- It includes a requirement for high-risk AI systems to provide clear and comprehensive information about their purpose, functionality, data sources, limitations, and potential impacts. This information will have to be accessible to users and authorities upon request.
- It includes a requirement for high-risk AI systems to ensure human oversight and intervention at all stages of their operation. This could involve appointing a representative or an auditor who can monitor and verify the system's performance and compliance.
- It includes a requirement for high-risk AI systems to implement appropriate technical measures to ensure data quality and security. This could include data minimization, encryption, anonymization, pseudonymisation, or deletion.
- It includes a provision for users and providers of high-risk AI systems to conduct regular audits and tests of their systems' compliance with the rules. They will also have to report any incidents or breaches involving their systems within 72 hours.
- It includes a provision for users and providers of high-risk AI systems to cooperate with authorities in case of investigations or complaints. They will also have to notify authorities if they become aware of any risks or violations related to their systems.
- It includes a provision for imposing administrative fines ranging from 2% to 6% of the annual turnover of the infringing entity. In addition, there could be additional fines up to 4% per day until compliance is achieved.

The EU's ambition is not only to regulate but also to foster innovation and competitiveness in the field of AI across Europe.

6.6.4 AI and AV Regulation in Germany

In Germany, SAE Level 1 (assisted driving) and Level 2 (semi-automated driving) are integrated with national road traffic laws, permitting the use of assistance systems as long as the driver remains attentive, monitors the traffic, and oversees the vehicle. In 2017, an amendment to the German Road Traffic Act introduced regulations for highly automated driving at SAE Level 3 (conditional automation). At this level, the driver is not required to constantly supervise the system but must be prepared to take control within a designated response time when prompted.

The Autonomous Driving Act of 2021 and the Autonomous Vehicle Approval and Operation Ordinance (Autonome Fahrzeuge Genehmigungs- und Betriebs-Verordnung) of 2022 provided a legal framework for fully automated driving at SAE Level 4. Here, the system takes complete control over the vehicle within defined operational boundaries, such as on highways, for prolonged periods without requiring driver monitoring or intervention.

Germany was among the first countries to adopt a national Artificial Intelligence (AI) strategy in 2018. The Federal Government has since prioritized human-centered AI, aiming to strengthen national and European competitiveness while ensuring that AI development benefits workers and society.

The objectives of Germany's AI strategy include fostering growth and competitiveness, ensuring responsible and trustworthy AI development, and embedding AI in society through ethical, legal, and cultural integration, supported by broad societal dialogue. Its implementation is overseen by three federal ministries: the Federal Ministry of Education and Research (BMBF), the Federal Ministry for Economic Affairs and Climate Action (BMWK), and the Federal Ministry of Labor and Social Affairs (BMAS).

In 2019, the German Federal Government published an interim report detailing the strategy's first-year results. In 2020, the strategy was updated to address new developments, including environmental and climate protection priorities. From 2023 to 2024, the strategy underwent further evaluation by the three federal ministries, with support from the OECD through an international benchmarking of Germany's AI ecosystem.

In 2023, Germany introduced the Mobility Data Act (Mobilitätsdatengesetz, BMDV) to improve the accessibility and interoperability of mobility data across transportation modes. The Act establishes standardized frameworks for data sharing to enhance traffic management, foster innovative mobility services, and support intelligent transportation system development.

Germany aligns its AI efforts with OECD AI Principles, focusing on human-centered AI and initiatives for societal good, environmental protection, and responsible public-sector AI systems. The country's approach is also guided by EU regulations such as the General Data Protection Regulation (GDPR) and the European Union Artificial Intelligence Act (EU AI Act), emphasizing ethical AI deployment, transparency in workplace AI, and sustainability.

In the document by BMVI which was introduced in 2023, AI systems must function in a robust, secure and safe way throughout their lifetimes, and potential risks should be continually assessed and managed (BMVI 2023).

- AI systems should be robust, secure and safe throughout their entire lifecycle so that, in conditions of normal use, foreseeable use or misuse, or other adverse conditions, they function appropriately and do not pose an unreasonable safety risk.

- To this end, AI actors should ensure traceability, including in relation to datasets, processes and decisions made during the AI system lifecycle, to enable analysis of the AI system's outcomes and responses to inquiry, appropriate to the context and consistent with the state of the art.
- AI actors should, based on their roles, the context, and their ability to act, apply a systematic risk management approach to each phase of the AI system lifecycle on a continuous basis to address risks related to AI systems, including privacy, digital security, safety and bias.

6.6.5 German Ethics Council

In 2023, the German Ethics Council published its statement "Humans and Machines – Challenges posed by Artificial Intelligence," in which it comprehensively examines the effects of digital technologies on human self-image and interaction. "The use of AI must expand human development and not diminish it. AI must not replace humans. These are fundamental rules for ethical assessment," was said by Alena Buyx, chairwoman of the German Ethics Council. AI systems have now found their way into almost all areas of public and private life and range from cancer diagnostics in medicine and intelligent tutor systems in schools to recommendation systems on online platforms and software that supports decisions in the social and judicial systems or in the police. "AI applications cannot replace human intelligence, responsibility and evaluation," emphasizes Julian Nida-Rümelin, deputy chairman of the German Ethics Council and deputy spokesman for the responsible working group. This conclusion arises in the statement from a consideration of central philosophical and anthropological concepts that are important for the relationship between humans and machines: intelligence, reason, action, and responsibility. This is important for the ethical assessment of AI because it is not enough to simply understand the technologies. The complex interactions between people and technology as well as social effects must also be considered. Therefore, for the Ethics Council, the central key question for ethical assessment is the following: Are human authorship and the conditions for responsible action expanded or diminished through the use of AI?

In its statement, the German Ethics Council addresses this question using examples in four areas of application—medicine, school education, public communication and opinion formation, and public administration. This shows that the assessment of AI must always be context-, application- and person-specific. "When human activities are delegated to machines, this can have very different effects for different groups of people, actors and those affected," says Judith Simon, spokeswoman for the working group. "It is therefore important to take a close look at for whom this is associated with increased scope for action and whose options for action are rather reduced."

This concern is also reflected in the recommendations formulated by the German Ethics Council on the use of AI in each of the four areas of application examined. For the medical sector, recommendations focus, among other things, on quality assurance in the development and use of AI products, on avoiding loss of medical competence and on the goal of reconciling the privacy of patients with intensive data use in medical research. It is important to protect the relationship of trust between all people involved and to avoid the complete replacement of medical professionals.

According to the Ethics Council's recommendations, the use of AI in school education should not be controlled by technological visions but should be based on basic educational ideas and limited to elements that demonstrably expand the learners' skills and social interactions, protect their privacy and promote personality development.

In the area of public communication and opinion formation, the Ethics Council recommends, among other things, further developments in the rules for online platforms about the selection and moderation of content as well as personalized advertising and data trading. It also calls for better access to platform data for research and recommends considering the development of a digital communications infrastructure under public law responsibility.

When using AI in public administration, the Ethics Council recommends approaches that protect against discrimination and prevent blindly following machine recommendations. It also demands that individual case considerations as well as the rights of inspection and objection of those affected be guaranteed. When using AI in the work of security authorities, social negotiation processes should be conducted about an appropriate relationship between the risks and opportunities of such approaches.

In addition, the German Ethics Council identifies ten cross-sectional topics and recommendations that are of cross-sectional importance for the ethical classification of AI applications. This includes, among other things, using AI to support decisions and not to replace them, to prevent the diffusion of responsibility, not to impair human control options and to ensure access to the basis for decision-making, especially in areas with a high level of intervention. Further demands aim to avoid distortions, dependencies, and misuse of technology as well as unwanted losses of human skills. Across all areas of application, it is important to focus on the interests of the people from whom the data used in AI applications come, to prevent excessive invasions of privacy using effective legal and technical precautions, and at the same time to enable data use for the common good.

References

Badrinarayanan, V./Kendall, A./Cipolla., R.: SegNet. A Deep Convolutional Encoder-Decoder Architecture for Image Segmentation. In: IEEE Transactions on Pattern Analysis and Machine Intelligence, 2017, pp. 2481–2495

BMVI – Bundesministerium für Verkehr und digitale Infrastruktur: Policies, data and analysis for trustworthy artificial intelligence. 2023. *https://oecd.ai/en* (accessed 2023)

Bock, J./Krajewski, R./Eckstein, L./Klimke, J./Sauerbier, J./Zlocki, A.: Data basis for scenario-based validation of HAD on highways. 27th Aachen Colloquium Automobile and Engine. 2018

Cabrero-Daniel, B./Abdelkarim, A./Broberg, A.: Digital Twins for Early Verification and Validation of Autonomous Driving Features. Open-source Tools and Standard Formats. In: 2024 IEEE Intelligent Vehicles Symposium (IV), 2024, pp. 2477–2482

Chachula, K./Łyskawa, J./Olber, B./Frątczak, P./Radlak, K.: Combating noisy labels in object detection datasets. Arxiv 2023

Chen, X./Ma, H./Wan, J./Li, B./Tian, X.: Multi-view 3D Object Detection Network for Autonomous Driving. IEEE Conference on Computer Vision and Pattern Recognition (CVPR). 2017

Chollet, F.: Deep Learning with Python. Manning Publications Co. 2018

Dalal, N./Triggs, B.: Histograms of oriented gradients for human detection. 2005 IEEE Computer Society Conference on Computer Vision and Pattern Recognition (CVPR '05), San Diego, CA, USA, 2005

Devlin, J./Chang, M.-W./Lee, K./Toutanova, K.: BERT. Pre-training of Deep Bidirectional Transformers for Language. In: Proceedings of the 2019 Conference of the North American Chapter of the Association for Computational Linguistics: Human Language Technologies, 2019, pp. 4171–4186

Diemert, S./Weber, J.: Can Large Language Models Assist in Hazard Analysis? In: Computer Safety, Reliability, and Security. SAFECOMP 2023 Workshops: ASSURE, DECSoS, SASSUR, SENSEI, SRToITS, and WAISE. 2023. pp. 410–422

Dosovitskiy, A./Beyer, L./Kolesnikov, A./Weissenborn, D./Zhai, X./Unterthiner, T./Dehghani, M./Minderer, M. Heigold, G./Gelly, S./Uszkoreit, J./Houlsby, N.: An Image is Worth 16x16 Words: Transformers for Image Recognition. 9th International Conference on Learning Representations (ICLR). 2021

Fagnant, D. J./Kockelman, K.: Preparing a nation for autonomous vehicles: opportunities, barriers and policy recommendations. In: Transportation Research Part A: Policy and Practice, 2015, pp. 167–181

Gamba, J.: Radar Signal Processing for Autonomous Driving. Springer, Singapore 2020

Gerdes, C.: The Virtues of Automated Vehicle Safety. Mapping Vehicle Safety Approaches to Their Underlying Ethical Frameworks. In: 2020 IEEE Intelligent Vehicles Symposium (IV), 2020, pp. 107–113

German Ethics Council (Deutscher Ethikrat): Ethikrat: Künstliche Intelligenz darf menschliche Entfaltung nicht vermindern March 20, 2023. *https://www.ethikrat.org/mitteilungen/mitteilungen/2023/ethikrat-kuenstliche-intelligenz-darf-menschliche-entfaltung-nicht-vermindern* (accessed 2023)

Goodfellow, I./Bengio, Y./Courville, A.: Deep Learning. MIT Press 2016

Goodfellow, I. F./Pouget-Abadie, J./Mirza, M./Xu, B./Warde-Farley, D./Ozair, S./Courville, A./Bengio, Y.: Generative Adversarial Networks. In: Proceedings of the 28th International Conference on Neural Information Processing Systems, Volume 2, 2014, pp. 2672–2680

Goodfellow, I. F./Pouget-Abadie, J./Mirza, M./Xu, B./Warde-Farley, D./Ozair, S./Courville, A./Bengio, Y.: Generative Adversarial Networks. In: Commun. ACM, 2020, pp. 139–144

Harari, Y. N.: Reboot for the AI revolution. In: Nature, 2017, pp. 324–327

Hastie, T./Tibshirani, R./Friedman, J.: The Elements of Statistical Learning. Springer, New York 2009

Hawkins, R./Paterson, C./Picardi, C./Jia, Y./Calinescu, R./Habli, I.: Guidance on the Assurance of Machine Learning in Autonomous Systems (AMLAS). Technical Report, Version 1.1. University of York 2021

Ilkou, E./Koutraki, M.: Symbolic Vs Sub-symbolic AI Methods: Friends or Enemies? Proc. of International Conference on Information and Knowledge Management. 2020

International Organization for Standardization. Geneva 2011

Kalyan, K. S.: A survey of GPT-3 family large language models including ChatGPT and GPT-4. In: Natural Language Processing Journal, 2024, 100048

Kingma, D. P./Welling, M.: Auto-Encoding Variational Bayes. International Conference on Learning Representations (ICLR). 2014

Krizhevsky, A./Sutskever, I./Hinton, G.: ImageNet classification with deep convolutional neural networks. In: Proceedings of the 26th International Conference on Neural Information Processing Systems, Volume 1, 2012, pp. 1097–1105

Ku, J./Mozifian, M./Lee, J./Harakeh, A./Waslander, S.: Joint 3D Proposal Generation and Object Detection from View Aggregation. IEEE/RSJ International Conference on Intelligent Robots and Systems (IROS). 2018

Laconte, J./Kasmi, A./Aufrere, R./Vaidis, M./Chapuis, R.: A Survey of Localization Methods for Autonomous Vehicles in Highway Scenarios. In: Sensors 2022

Lang, A./Vora, S./Caesar, H./Zhou, L./Yang, J./Beijbom, O.: PointPillars. Fast Encoders for Object Detection From Point Clouds. In: IEEE/CVF Conference on Computer Vision and Pattern Recognition (CVPR), 2019, pp. 12689–12697

LeCun, Y./Boser, B./Denker, J./Henderson, D./Howard, R./Hubbard, W./Jackel, L.: Backpropagation Applied to Handwritten Zip Code Recognition. In: Neural Computation, 1989, pp. 541–551

Li, C.-T.: Source Camera Identification Using Enhanced Sensor Pattern Noise. In: IEEE Transactions on Information Forensics and Security, 2010, pp. 280–287

Li, Y./Ibanez-Guzman, J.: Lidar for Autonomous Driving. The Principles, Challenges, and Trends for Automotive Lidar and Perception Systems. In: IEEE Signal Processing Magazine, 2020, pp. 50–61

Liu, B.: "Weak AI" is Likely to Never Become "Strong AI", So What is its Greatest Value for us? 2021

Luo, W./Xing, J./Milan, A./Zhang, X./Liu, W./Kim, T.-K.: Multiple object tracking. A literature review. In: Artificial Intelligence 2021

Mannings, R.: Ubiquitous Positioning. Artech House 2008.

McCarthy, J.: What is artificial intelligence? 2007. *http://www-formal.stanford.edu/jmc/whatisai*

McCulloch, W./Pitts, W.: A logical calculus of the ideas immanent in nervous activity. In: The bulletin of mathematical biophysics 1943

Minsky, M. L.: Semantic information processing. MIT Press 1968

Mnih, V./Kavukcuoglu, K./Silver, D.: Human-level control through deep reinforcement learning. In: Nature, 2015, pp. 529–533

Nair, V./Hinton, G.: Rectified Linear Units Improve Restricted Boltzmann Machines. In: Proceedings of the 27th International Conference on International Conference on Machine Learning, 2010, pp. 807–814

Pomerleau, D. A.: Advances in Neural Information Processing Systems. Morgan-Kaufmann 1989

Radford, A./Narasimhan, K./Salimans, T./Sutskever, I.: Improving Language Understanding by Generative Pre-Training. Open AI 2018

Radlak, K./Szczepankiewicz, M./Jones, T./Serwa, P.: Organization of machine learning based product development as per ISO 26262 and ISO/PAS 21448. In: 2020 IEEE 25th Pacific Rim International Symposium on Dependable Computing (PRDC), 2020, pp. 110–119

Rashid, A. B./Kausik, A. K.: AI revolutionizing industries worldwide: A comprehensive overview of its diverse applications. In: Hybrid Advances, 2024, 100277

Rauch, H./Striebel, C./Tung, F.: Maximum likelihood estimates of linear dynamic systems. In: AIAA Journal 1965

Redmon, J./Santosh, D./Girshick, R./Farhadi, A.: You Only Look Once. Unified, Real-Time Object Detection. IEEE Conference on Computer Vision and Pattern Recognition (CVPR). 2016

Reed, S./Akata, Z./Yan, X./Logeswaran, L./Schiele, B./Honglak L.: Generative adversarial text to image synthesis. In: Proceedings of the 33rd International Conference on International Conference on Machine Learning, 2016, pp. 1060–1069

Ren, S./He, K./Girshick, R.: Faster R-CNN. Towards real-time object detection with region proposal networks. In: Proceedings of the 29th International Conference on Neural Information Processing Systems, Volume 1, 2015, pp. 91–99

Ribeiro, M./Sameer Singh, S./Guestrin, C.: Why Should I Trust You? Explaining the Predictions of Any Classifier. In: Proceedings of the 22nd ACM SIGKDD International Conference on Knowledge Discovery and Data Mining, 2016, pp. 1135–1144

Richards, M.: Fundamentals of Radar Signal Processing. McGraw Hill 2005

Russell, S./Norvig, P.: Artificial Intelligence. A Modern Approach. 4th Global Edition. Pearson Education 2022

Saranya, A./Subhashini, R.: A systematic review of Explainable Artificial Intelligence models and applications: Recent developments and future trends. In: Decision Analytics Journal 2023

Scarselli, F./Gori, M./Tsoi, A. C./Hagenbuchner, M./Monfardini, G.: The Graph Neural Network Model. In: IEEE Transactions on Neural Networks, 2009, pp. 61–80

Schulman, J./Wolski, F./Dhariwal, P./Radford, A./Klimov, O.: Proximal Policy Optimization Algorithms. 2015

Singh, A.: Transformer-Based Sensor Fusion for Autonomous Driving: A Survey. IEEE/CVF International Conference on Computer Vision Workshops (ICCVW). 2023

Soleimanitaleb, Z./Keyvanrad, M. A.: Single Object Tracking. A Survey of Methods, Datasets, and Evaluation Metrics. 2022

Turing, A. M.: Computing and machinery Intelligence. In: Mind, 1950, pp. 433–460

Vaswani, A./Shazeer, N./Parmar, N./Uszkoreit, J./Jones, L./Gomez, A./Kaiser, L./Polosukhin, I.: Attention is all you need. In: Proceedings of the 31st International Conference on Neural Information Processing Systems. 2017

Viola, P./Jones, M.: Rapid object detection using a boosted cascade of simple features. In: Proceedings of the 2001 IEEE Computer Society Conference on Computer Vision and Pattern Recognition (CVPR). Kauai, HI, USA, 2001

Wang, Y./Chen, X./You, Y./Li, L. E./Hariharan, B./Campbell, M./Weinberger, K. Q./Chao, W.-L.: Train in Germany, Test in the USA: Making 3D Object Detectors Generalize. In: 2020 IEEE/CVF Conference on Computer Vision and Pattern Recognition (CVPR), 2020, pp. 11710–11720

Wang, Y./Pan, Y./Yan, M./Su, Z./Luan, T. H.: A Survey on ChatGPT: AI–Generated Contents, Challenges, and Solutions. In: IEEE Open Journal of the Computer Society, 2023, pp. 280–302

World Health Organization: Global Status Report. 2015

Zendel, O./Murschitz, M./Humenberger, M./Herzner, W.: CV-HAZOP. Introducing Test Data Validation for Computer Vision. In: 2015 IEEE International Conference on Computer Vision (ICCV), 2015, pp. 2066–2074

7 Systems Engineering

Systems engineering involves managing and optimizing a system within its specific context and interactions. This approach can be understood through the lens of Niklas Luhmann's systems theory, which emphasizes the sociological aspects of systems. On the other hand, systems engineering in product development focuses more on the practical application and systematic development of a product within its intended use. This involves detailed work on specific components or technologies to ensure they function optimally within the overall system. While systems engineering from a sociological perspective considers the broader context and interactions of systems within society, the technical perspective in product development emphasizes the detailed, methodical creation and optimization of specific products for their intended applications.

Sociological Perspective: Luhmann's Systems Theory

Niklas Luhmann's systems theory views society as composed of various self-referential systems, each with its own structures, functions, and codes of communication. These systems include the economy, law, politics, and more. Luhmann's theory emphasizes how these systems interact and influence human behavior, system membership, and context (Luhmann 2001).

Example in Automotive

Acceptance of new technologies: Luhmann's theory can be applied to understand how social systems influence the acceptance of new propulsion technologies like electric mobility (e-mobility). For instance, the environmental awareness system might promote the adoption of Electric Vehicles due to their lower emissions, while the economic system evaluates their cost-effectiveness.

Autonomous transport systems: The acceptance of automated or autonomous transport systems can also be analyzed through Luhmann's lens. Social systems such as legal and ethical frameworks play a crucial role in shaping public perception and regulatory acceptance of these technologies.

Technical Perspective: Systems engineering

From a technical standpoint, systems engineering is a holistic methodology that focuses on the design, integration, and management of complex systems throughout their lifecycles. This approach ensures that all components of a system work together efficiently and effectively.

Example in the automotive industry

Holistic vehicle development: Systems engineering in the automotive industry involves integrating various subsystems, such as powertrains, safety systems, and infotainment, to create a cohesive and functional vehicle. This holistic approach ensures that all parts of the vehicle operate harmoniously.

System Engineering

System engineering, in a more specific sense, refers to the systematic development and optimization of individual systems or components within a larger system.

Interdisciplinary nature: Systems engineering requires a multidisciplinary approach understanding the product and the application, the intended use and the typical domain handling and it also requires coordinating efforts across various technology disciplines (e. g., mechanical, electrical, software), whereas system engineering might be more focused on a particular domain for a specific boundary and application.

Example in the automotive industry

Development of specific Components: A system engineer might focus on developing a particular component, such as the Battery Management System (BMS) for electric vehicles. This involves detailed work on the hardware and software to ensure the BMS operates efficiently within the overall vehicle system but excluding chemical and physical processes or the aspects of energy management.

Key Differences

Scope: Luhmann's systems theory provides a broad sociological framework for understanding how different social systems interact and influence technology acceptance. In contrast, the technical perspective of systems engineering focuses on the integration and lifecycle management of complex systems.

Focus: Luhmann's approach is more about understanding societal structures and behaviors, while the technical perspective is about the practical engineering and development of systems and components.

While Luhmann's systems theory offers valuable insights into the sociological aspects of system engineering, the technical perspective emphasizes the practical, methodical development of systems and components within the automotive industry.

System Engineering generally refers to the engineering of individual systems:

- A system engineer is a person who deals with the overall management of engineering projects during their lifecycle, focusing more on physical aspects.
- They follow an interdisciplinary approach governing the total technical and managerial effort required to transform requirements into solutions.
- They are generally concerned with all aspects of computer-based system development, including hardware, software, and process engineering.

On the other hand, systems engineering is a broader field that focuses on the design, integration, and management of complex systems over their lifecycles. It involves the integration of different engineering disciplines and systems thinking to ensure that all parts of a system work together to achieve the desired outcomes. Systems engineering often requires considerations beyond those usually associated with engineering to include socio-technical and sometimes socio-economic phenomena.

7.1 Cyber-Physical System

Cyberphysical Systems (CPSs) are integrations of computation, networking, and physical processes. Embedded computers and networks monitor control the physical processes, with feedback loops where physical processes affect computations and vice versa.

Control engineering: CPSs involve control systems that manage the behavior of other systems using control loops. These systems are often designed using feedback and are central to industrial automation and robotics.

Communication technology: CPSs rely on communication technologies to connect the physical and cyber components. This includes wired and wireless networking technologies that enable data exchange between sensors, actuators, controllers, and user interfaces.

The basis of both engineering approaches derives from the cybernetics. CPSs can be viewed as a form of cybernetics as they involve the study of regulatory systems. They are concerned with systems of "control and communication" in machines and living beings. CPSs often operate under real-time constraints. The correct functioning of a CPS not only depends on the logical correctness of computations but also on the time at which the results are produced.

CPSs often have spatial aspects, as they interact with the physical world. This is particularly important in areas such as transportation, where the geographical location of elements is crucial. While the theory of relativity is not typically a central concern in most CPSs, the principles of information propagation at finite speed and the effects of motion on the observation of events can be relevant in certain high-speed or large-scale CPSs.

Cyber-Physical Systems address very different contexts in many industries, such as the following:

- **medicine:** CPSs play a significant role in healthcare, such as in medical monitoring systems, robotic surgery, and smart health devices.
- **bionics:** CPS principles are applied in the design of advanced prosthetics and bionic devices.
- **automated driving:** CPSs are fundamental to the operation of autonomous vehicles, where onboard systems continuously monitor and interact with the physical environment.
- **avionics:** In the aerospace industry, CPSs involve the integration of physical and computational processes in aircraft systems.
- **remote control:** CPSs enable remote control of systems, such as drones or industrial control systems, where a user interacts with a physical process via a network.
- **space technology:** CPSs play a crucial role in space technology, from the control systems used in spacecraft to the ground-based systems used for tracking and communication.

In all these applications, factors like gravity, weightless, the dualism of light, the speed of light, and the time shift over long distances can come into play, especially in high-speed or large-scale systems such as in avionics and space technology. The design of CPSs must consider these physical phenomena to ensure correct and safe operation. However, the specific impact of these factors can vary greatly depending on the particular CPS and its operating environment. For example, the effects of gravity and weightlessness are much more relevant in the design of a spacecraft's CPS than in a CPS controlling a home heating system. Similarly, the speed of light becomes relevant in CPSs involving long-distance communication, such as satellite systems, where the delay in signal transmission must be taken into account. In the automotive industry, gravity and weightlessness are not key aspects that need to be considered, but especially vehicle dynamics require detailed physical impacts on the necessary safety and security measures to assure safe mobility solutions.

In future mobility solutions for people and goods, Cyber-Physical Systems will play an increasingly important role in the interaction of people with the various means of transport and the means of transport with each other.

Actions and activities must be planned in advance so that the necessary resources are available in sufficient time at the time of the action or activity. There is nothing more dangerous than when a car turns too late into a curve, but the energy required for steering cannot be applied at the right time, or the information for the steering activity is too late or too early, or the necessary forces are too low or too intense. Leaving the specified or intended driving corridor leads to leaving the lane, which in extreme cases can lead to an accident with oncoming traffic. The delay in the flow of information due to hacker attacks can also lead to such an accident scenario, though.

This means that the characteristics of Cyber-Physical Systems, such as control technology and communications technology, considering the physical conditions, are essential for safe driving behavior. Until now, the driver has been held responsible for such correct behavior, but with increasing automation, the demands on the systems and the engineers who are supposed to develop such systems are also growing.

7.1.1 Automated versus Automation

In the automotive industry, the introduction of electronic systems, particularly after the turn of the millennium, has significantly integrated software-based systems into vehicles. Among these are several safety systems that rely on software algorithms. One of the earliest examples is the Anti-Lock Braking System (ABS), which uses software algorithms to ensure the vehicle remains steerable even under high hydraulic pressure gradients in the braking system that could otherwise cause the wheels to lock.

The ABS is designed to prevent wheel lock-up by modulating brake pressure, thereby maintaining traction and allowing the driver to steer during emergency braking situations. However, it is important to note that ABS is a localized system that only affects the hydraulic braking system. It does not provide protection against axle lock-up caused by excessive regenerative braking from the drive motor.

Historically, safety systems have been categorized into active and passive systems. Passive safety systems are those where electronics or software do not directly engage in the protective function. Examples include seat belts, which were initially considered passive safety systems. In modern vehicles, however, automated restraint systems like seat belt pretensioners have been developed. These systems tighten the seat belts in the event of a crash, securing the occupant firmly in the seat. Active safety systems, on the other hand, involve electronics playing a crucial role in their operation. A prime example of an active safety system is the airbag, which deploys in response to a collision to protect the occupants. The evolution of safety systems in the automotive industry has seen a shift from purely mechanical systems to sophisticated electronic and software-based systems, enhancing both active and passive safety measures in vehicles.

With the continued integration of electronics and software into vehicles, more and more driver assistance systems have been introduced. These systems are designed to support the driver in operating the vehicle, enhancing both safety and convenience. Driver assistance systems encompass a wide range of technologies, from basic features like cruise control to more advanced systems such as adaptive cruise control, lane-keeping assistance, and automated parking. These systems use sensors, cameras, and sophisticated algorithms to monitor the vehicle's surroundings and assist the driver in making safer and more efficient driving decisions.

For example, adaptive cruise control adjusts the vehicle's speed to maintain a safe following distance from the car ahead, while lane-keeping assistance helps the driver stay within their lane by providing gentle steering inputs. Automated parking systems can take over the steering and braking to park the vehicle in tight spaces. Lane departure warning systems have evolved into lane-keeping systems. Initially, lane departure warning systems alerted the driver if the vehicle unintentionally drifted out of its lane. Over time, these systems have been enhanced to include lane-keeping assistance, which not only warns the driver but also provides corrective steering inputs to help keep the vehicle centered in its lane.

Today several driver assistant systems are required by the general safety regulations, even for a basic road vehicle.

7.1.1.1 Evolution of Automated Driving Functions

The evolution of driver assistance systems has driven the development of automation systems, with the ultimate goal being the creation of autonomous vehicles. Initially, driver assistance systems were designed to support the driver in specific tasks, such as maintaining a safe following distance or staying within a lane. As technology advanced, these systems became more sophisticated, incorporating more sensors, cameras, and complex algorithms. This progression led to the development of more advanced automation systems capable of handling a broader range of driving tasks.

The ultimate aim of these advancements is to achieve fully autonomous vehicles, which can operate without any human intervention. Autonomous vehicles are designed to navigate, make decisions, and control the vehicle entirely on their own, using a combination of technologies such as artificial intelligence, Machine Learning, and real-time data processing.

Classification Driving Automation Systems

The German VDA and also SAE published levels of automation from Driver Assistant System up to "Full Driving Automation." SAE J3016, "Taxonomy and Definitions for Terms Related to Driving Automation Systems for On-Road Motor Vehicles," defines the typical classification in our days (SAE J3016:2014).

Driver Assistance Systems

Driver assistance systems are designed to assist the human driver with certain aspects of driving, but the driver remains responsible for the overall control of the vehicle. These systems are categorized under lower levels of automation:

- **SAE Level 1 (Driver Assistance):** The system can assist with either steering or acceleration/deceleration, but the driver handles the remaining tasks.
- **SAE Level 2 (Partial Automation):** The system can assist with both steering and acceleration/deceleration, but the driver must monitor the driving environment and be ready to take control at all times.

Automated Driving Functions

Automated driving functions refer to systems that can perform the entire Dynamic Driving Task (DDT) under specific conditions without human intervention. SAE J3016 defines these functions across different levels of automation:

- **SAE Level 3 (Conditional Automation):** The Automated Driving System (ADS) can handle all aspects of driving, but the human driver must be ready to take over when requested.
- **SAE Level 4 (High Automation):** The ADS can perform all driving tasks within certain conditions or environments (e. g., urban areas), and human intervention is not required within these conditions.
- **SAE Level 5 (Full Automation):** The ADS can handle all driving tasks in all conditions, with no need for human intervention at any time.

Role of the driver:

- **Level 1:** The driver is responsible for most driving tasks and must always be engaged.
- **Level 2:** The driver must continuously monitor the driving environment and be prepared to take over immediately if needed.
- **Level 3:** The driver must be available to take control when the system requests.
- **Level 4:** The driver is not required to intervene within the Operational Design Domain (ODD) of the system.
- **Level 5:** No driver intervention is needed at all.

Key Differences

Scope of automation: Automated driving functions (Levels 3–5) can handle the entire driving task under specific conditions, whereas driver assistance systems (Levels 1–2) only assist with certain tasks and require constant driver engagement.

Driver's role: In automated driving functions, the driver's role diminishes as the level of automation increases, eventually becoming non-existent at Level 5. In contrast, driver assistance systems require the driver to remain actively involved and responsible for the vehicle's operation.

Automated driving functions aim to fully automate the driving process, reducing or eliminating the need for human intervention, while driver assistance systems are designed to support the driver, who remains in control of the vehicle.

The strict classification of automated driving systems is not always adhered to by the systems currently available. Specifically, the stringent requirements of UNECE R157 for an Automated Driving System (ADS), comparable to SAE J3016 Level 3, pose significant challenges for vehicle manufacturers, particularly regarding product liability.

For an Automated Lane Keeping System (ALKS) on highways, the driver is no longer responsible for vehicle control. Consequently, the manufacturer assumes responsibility for vehicle control when the active lane-keeping function is engaged, as per UNECE R157. To circumvent this, a hands-free system for highways has been introduced, supported by the legal framework of UNECE R171.

Regulation No. 171 defines Driver Control Assistance Systems (DCAS) as systems that assist the driver in controlling the longitudinal and lateral motion of the vehicle on a sustained basis, without taking over the entire driving task. DCAS are categorized as Automated Driving Systems corresponding to SAE Level 2. This means that while using such systems, the driver retains responsibility for vehicle control and must continuously monitor the surroundings and the vehicle's performance to intervene if necessary.

The function often referred to as Level 2 hands-free is also controversial because the driver must immediately take over vehicle control when needed. This presents a technical challenge, as braking and steering systems often rely on the driver as a fallback for certain failure scenarios.

The extent to which the driver can maintain the necessary attention to detect system errors and whether the warning systems can alert the driver in time for them to initiate the correct steering and braking maneuvers for the given driving situation is a complex issue. The appropriate corrective action by the driver requires a suitable response, which can vary depending on weather conditions, road conditions, and the driving situation.

For instance, a system failure during a lane change may require different corrective actions from the driver compared to a straight drive at constant speed with low traffic. This variability makes it challenging to ensure that the driver can always respond appropriately to system errors, highlighting the importance of robust and reliable driver assistance and alert systems.

7.1.1.2 Automation versus Autonomous Driving

Autonomous and automated driving represent two distinct approaches to revolutionizing transportation. While both aim to enhance safety and efficiency, they differ significantly in terms of the level of human involvement and decision-making capabilities.

Automated Driving: Assisted Driving

Automated driving systems provide assistance to human drivers, taking over certain tasks such as steering, braking, and acceleration. These systems are designed to reduce driver fatigue and improve safety. Examples include Adaptive Cruise Control (ACC), lane-keeping assist, and automatic emergency braking. While Automated Driving Systems can significantly enhance vehicle safety, they still require human supervision and intervention.

Autonomous Driving: Decision-Making Machines

Autonomous vehicles, on the other hand, can operate without human intervention. They rely on advanced sensors, cameras, and Artificial Intelligence (AI) to perceive their surroundings, make decisions, and navigate complex environments. The decision-making processes of Autonomous Vehicles (AVs) are often based on Machine Learning algorithms trained on vast amounts of data.

Ethical and Social Implications

The development and deployment of autonomous vehicles raise numerous ethical and social questions. One of the most pressing concerns is how Autonomous Vehicles should make life-or-death decisions in unforeseen circumstances. For instance, if an Autonomous Vehicle must choose between hitting a pedestrian or swerving into oncoming traffic, how should it prioritize different values such as the preservation of human life, property damage, and the safety of occupants?

Moreover, the societal acceptance of Autonomous Vehicles varies across different cultures and regions. Factors such as trust in technology, concerns about job displacement, and the perceived impact on urban planning can influence public opinion.

The Limitations of AI in Complex Decision-Making

While AI has made significant strides in recent years, it still faces challenges in making complex moral judgments. Theory of Mind (ToM) tests, which assess an AI's ability to understand the mental states of others, have consistently shown that Large Language Models (LLMs) struggle to comprehend and respond to nuanced social situations.

Furthermore, AI systems may be biased based on the data they are trained on. This can lead to discriminatory outcomes, particularly in areas such as traffic enforcement and insurance underwriting.

Alternative Approaches

Given the limitations of AI in making complex ethical decisions, it is essential to explore alternative approaches to autonomous transportation. These may include the following:

- **Human-in-the-loop systems:** These systems combine the strengths of human judgment with the capabilities of AI (in various roles from remote driver approach to supervisor inside or outside of the vehicle).
- **Robust safety systems:** By focusing on redundancy and fail-safe mechanisms, engineers can mitigate the risks associated with Autonomous Vehicle failures.
- **Ethical guidelines:** Developing clear and comprehensive ethical guidelines for the development and deployment of Autonomous Vehicles can help ensure that these technologies are used responsibly.

Autonomous driving sounds like a revolution in transportation, it is important to proceed with caution, though. The ethical and social implications of this technology are complex and require careful consideration. By carefully weighing the benefits and risks, policymakers, engineers, and the public can work together to ensure that Autonomous Vehicles are developed and deployed in a safe and responsible manner.

Additional points to consider are the following:

- **Infrastructure requirements:** The widespread adoption of autonomous vehicles will require significant investments in infrastructure, such as smart roads and connected vehicle technology.
- **Legal and regulatory frameworks:** Developing appropriate legal and regulatory frameworks is essential to address issues such as liability, privacy, and cybersecurity.
- **Public education:** Educating the public about the benefits and limitations of Autonomous Vehicles is crucial to building trust and acceptance.

These subjective aspects form several requirements, which cannot just verify and could not simply argument as fulfilled.

7.1.1.3 Appearances of a System

The Soul of a System: A Philosophical Perspective on Engineering Automated Vehicles

Aristotle's assertion that "the whole is greater than the sum of its parts" encapsulates a fundamental truth about complex systems. When applied to engineering, this principle highlights the interplay between a system's individual components and its overall function. While systems engineering has evolved significantly since Aristotle's time, the philosophical underpinnings remain relevant, particularly in the context of developing highly complex systems such as Automated Vehicles.

The Triad: Body, Mind, and Soul in Systems Engineering

In the context of systems engineering, we can draw parallels between the Aristotelian triad of body, mind, and soul and the components of a system:

- **Body:** This represents the physical components of a system, such as the hardware, sensors, and actuators of an automated vehicle.
- **Mind:** This corresponds to the system's functional behavior, its logic, and its ability to process information. It is defined by algorithms, software, and the system's overall architecture.

- **Soul:** This is the ethical dimension of the system, reflecting the values and principles that guide its design and operation. It is the intangible element that gives a system its character and its ability to respond appropriately to complex and unpredictable situations.

The Challenge of Integrating Ethics into Systems Engineering

While systems engineering has developed robust methodologies for designing and building complex systems, the integration of ethics into the engineering process remains a significant challenge. While thinkers like Luhmann have extended systems engineering principles to sociology, the ethical dimension of systems is often overlooked.

The Role of Induction and Deduction in Systems Engineering

Aristotle emphasized the importance of inductive reasoning in moving from specific observations to general principles. Systems engineering, on the other hand, often relies on a deductive approach, breaking down complex systems into simpler components. Agile development methodologies, while effective for rapid prototyping and iteration, can sometimes neglect the broader systems perspective.

The Need for a Holistic Approach

To create systems that are both technically sound and ethically sound, engineers must adopt a more holistic approach that integrates inductive and deductive reasoning.

This requires the following:

- **A strong foundation in ethics:** Engineers must have a deep understanding of the ethical implications of their work.
- **A systems perspective:** Engineers must consider the system as a whole, rather than focusing solely on individual components.
- **Cultural change:** Organizations must create a culture that values ethical considerations and encourages engineers to think critically about the social impact of their work.

The Case of Automated Vehicles

Automated Vehicles present a particularly challenging case. These systems must be able to make life-or-death decisions in complex and uncertain situations. To ensure that these decisions are made ethically, engineers must consider the following:

- **Consider a wide range of scenarios:** Engineers must develop algorithms that can handle a wide range of possible scenarios, including those that may be difficult to anticipate.

- **Involve diverse stakeholders:** Engineers must involve a diverse range of stakeholders in the development process, including ethicists, sociologists, and representatives from vulnerable populations.
- **Be transparent about the system's limitations:** Engineers must be transparent about the limitations of Automated Vehicles and the potential for human error.

The development of Automated Vehicles highlights the need for a more holistic approach to systems engineering. By integrating ethics into the engineering process, we can create systems that are not only technically sound but also align with our values and principles. As Aristotle observed, the whole is greater than the sum of its parts. To create truly intelligent systems, we must consider the system as a whole, including its ethical dimensions.

Ethical considerations are more about the intensity and frequency of their impact. On a small scale, society may not be perturbed, but as performance increases, diverse societal currents may resist innovations. Beneficiaries are often perceived as societal enemies. Hence, a balanced approach and an evolutionary introduction of new technologies, especially in mass mobility, is crucial. Few willingly relinquish their accustomed freedoms for the benefit of others. Thus, safety, comfort, and societal acceptance must be cultivated.

7.1.1.4 Timing Dimension

To develop intelligent systems, it is essential not only to consider the availability of information regarding time, location, characteristics, and intensity but also to make forecasts and estimations. In weather analysis, this concept is known as nowcasting. It is necessary to estimate energy consumption, resource requirements, and consider traffic and weather conditions in both the near and distant surroundings. Various methods can be employed for this purpose.

Vector Graphics

Trend analysis: Vector graphics can be used to visualize past trends transparently, allowing for estimations of future developments based on historical data.

Markov models – state transitions: Markov models help estimate states and the probabilities of state transitions or transients, providing a framework for predicting future states based on current and past information.

Dead Reckoning

Cultural evolution: Tracking the evolution of cultural norms and values over time can be likened to dead reckoning, where the current state is determined based on past trends and known influences.

Social forecasting: Predicting future social trends based on historical data and current patterns can be seen as a form of social dead reckoning.

It is crucial not only to analyze the results but also to communicate these findings transparently to different societal groups. Ensuring that the information is accessible and understandable to various stakeholders is vital for the effective implementation and acceptance of intelligent systems.

By applying methods like these techniques, we can do the following:

- **visualize complex social systems:** in order to gain a deeper understanding of the dynamics at play
- **metrics about states and conditions:** in order to provide statistical estimation about future states and systematic transitions and adequate safeguarding measures for maintaining safe states and their transitions
- **identify potential risks and opportunities:** in order to anticipate and mitigating societal backlash
- **design more inclusive and equitable technologies:** in order to ensure that innovations benefit a wide range of stakeholders

While vector graphics, Markov models, and dead reckoning are primarily used in technical fields, their underlying principles can be adapted to model and analyze complex social systems. By combining these methods with traditional sociological approaches, we can develop a more comprehensive understanding of how technology shapes society and how society shapes technology.

These are some additional considerations:

- **Ethical implications:** The use of these techniques in a sociological context raises ethical questions about data privacy, surveillance, and the potential for manipulation.
- **Interdisciplinary collaboration:** Successful application requires collaboration between engineers, sociologists, and ethicists.
- **Continuous evaluation:** As societal norms and technological capabilities evolve, so too must our models and methods.

By carefully considering these factors, we can harness the power of systems engineering to create technologies that benefit society as a whole.

7.1.2 Risk Management in the Context of Systems Engineering

Risk management and systems engineering are two complementary processes. The question is how deep risk management and systems engineering need to be integrated and what kind of criterion for independence are recommended.

Separating systems engineering from risk management helps to avoid bias and ensure a balanced assessment of alternatives:

- The primary objective of systems engineering is to define and refine system requirements, while risk management focuses on identifying, analyzing, and mitigating potential risks associated with those requirements.
- Combining the two processes could lead to a bias towards fulfilling requirements at the expense of adequately addressing potential risks.
- For example, if safety risks associated with a specific design choice are downplayed to meet performance requirements, this could result in an unsafe system despite fulfilling its intended function.

Independent assessments from both perspectives enable informed decision-making:

- Requirements need to be traceable to the relevant risk and engineering parties. Especially in case of safety relevant products, the decisions about design and architecture as well as the change management process require transparent communication. The consequence of decisions during the development requires traceable records due to product liability.
- This implies that risk assessments conducted independently of systems engineering provide vital input to the requirement definition process.
- Similarly, risk mitigation strategies, developed through the risk management process, can influence system design choices and lead to more robust and resilient systems.

Separate processes also allow for specialization and focused expertise:

- Systems engineers can concentrate on optimizing system performance and functionality, while risk managers can dedicate their expertise to identifying and mitigating potential threats.
- This separation allows for a more comprehensive and thorough assessment of both aspects, leading to better-informed decisions.

Separating systems engineering from risk management is crucial for comprehensive risk assessment, unbiased evaluation of alternatives, and robust and traceable decision-making in complex projects. This separation fosters specialization, facilitates risk consideration during requirements development, traceable decisions and ultimately contributes to the development of safer, more reliable, and successful systems.

NASA developed a handbook that addresses the decision-making process and the adequate communication of decisions within program, projects and product groups (Figure 7.1). The *NASA Risk-Informed Decision Making Handbook* provides insights into the top-down breakdown of objectives and the handling of inherent risks, especially when balancing ambitious goals in technology, cost, and schedule with safety (NASA 2010).

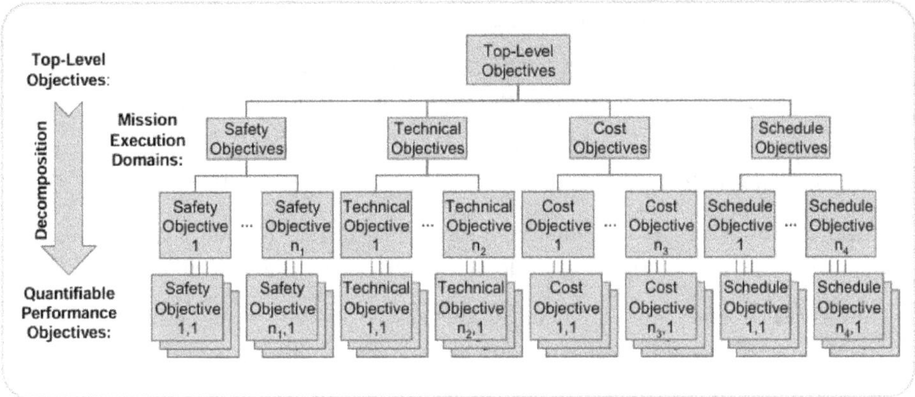

Figure 7.1 Notional objectives hierarchy (source: NASA 2010)

Objectives Breakdown and Risk Management

- High-level objectives, like NASA's strategic goals or goals for a vehicle platform, are decomposed into progressively detailed performance requirements that cascade down the organizational hierarchy.
- Each organizational unit negotiates specific objectives, deliverables, performance measures, baseline requirements, resources, and schedules with the unit below it.

This hierarchical breakdown ensures alignment with overarching goals while enabling each unit to manage its risks within agreed-upon specifications and thresholds.

Addressing Interdependencies and Risks

- The sources emphasize that objectives across safety, technology, cost, and schedule are inherently interdependent. Decisions in one domain inevitably impact the others.
- Risk-Informed Decision Making (RIDM), as described in the sources, is crucial to addressing these interdependencies. It involves the following:
  - identifying and analyzing potential risks associated with each decision alternative
  - quantifying these risks and their potential impact on performance measures
  - deliberating on the trade-offs between performance and risk across all objectives
  - selecting the alternative that best balances stakeholder expectations and risk tolerance

Balancing Stakeholder Requirements

- The challenge lies in balancing the desire for cutting-edge technology, cost efficiency, and rapid development with the paramount importance of safety.
- The sources stress that risk acceptance should be a conscious and informed decision, not an accidental outcome of pursuing ambitious objectives.
- Clear communication and documentation of risks and trade-offs are essential for ensuring that all stakeholders understand and accept the chosen path.
- Iterative processes, involving continuous risk assessment and feedback loops, are crucial for adapting to evolving knowledge and unforeseen circumstances.

Achieving the Necessary Balance

- The sources offer no simple formula for achieving the perfect balance. It is a complex and context-dependent challenge.
- However, a structured approach like RIDM, combined with a culture of open communication and continuous risk awareness, can help organizations navigate these complex trade-offs and make informed decisions that balance stakeholder expectations with acceptable risk levels.

The NASA handbook distinguishes between fundamental objectives and means objectives, highlighting the importance of using fundamental objectives for decision making (Figure 7.2, see also NASA 2010).

Challenges with Means Objectives

The handbook points out that focusing on means objectives can lead to suboptimal decisions. Means objectives prescribe specific methods for achieving a higher-level objective, potentially limiting the exploration of alternative solutions. For instance, mandating a particular safety feature as a means objective might preclude the consideration of a more innovative design that achieves the same level of safety through a different approach. This reliance on predetermined methods can stifle creativity and innovation.

Benefits of Fundamental Objectives

Fundamental objectives describe the desired end state without specifying how to achieve it. This approach encourages a performance-based approach to decision-making, focusing on what needs to be accomplished rather than dictating how to do it. The handbook emphasizes that using fundamental objectives promotes early risk identification and allows for the exploration of diverse design solutions that could be more efficient, safer, or both.

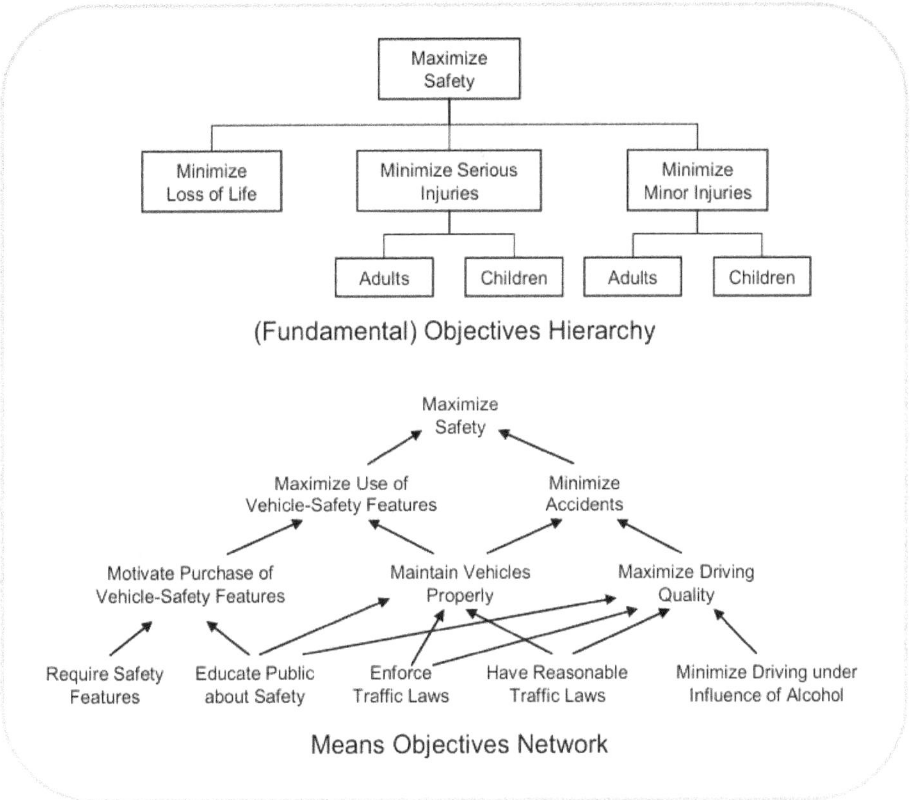

Figure 7.2 Fundamentals versus means objectives (source: NASA 2010)

Example from the Handbook

The handbook uses the example of "Maximize safety" to illustrate the difference between the two types of objectives:

- **Hierarchical Fundamental Objectives:** "Maximize safety" is broken down into fundamental objectives like "Minimize loss of life" and "Minimize injuries." These objectives clearly define the desired outcome without prescribing specific methods.
- **Means Objectives:** The handbook contrasts this with a means objectives network, where "Maximize safety" might lead to objectives like "Maximize use of vehicle-safety features," "Educate public about safety," or "Enforce traffic laws." These objectives focus on specific methods rather than the desired end state of safety.

The NASA handbook advocates the use of fundamental objectives in the decision-making process. This approach fosters a focus on desired outcomes, encourages the exploration of diverse solutions, and promotes early risk identification. By concentrating on the "what" rather than the "how," organizations can achieve more innovative, efficient, and robust solutions.

According to the NASA handbook, risk and safety objectives are intricately woven into the system development process. This integration necessitates a concrete product development process that continuously synchronizes stakeholders and risk analysts, ensuring regular evaluations of the outcomes. In such a milestone-based concept, the maturity of various objectives can be assessed, providing the necessary transparency for consistent development.

7.2 Extended Lifecycle of Automotive Vehicle Development

The safety lifecycle is generally only a subset of the overall lifecycle of the target product. In the context of automobiles, this lifecycle spans from the initial concept of the vehicle to its decommissioning. ISO 26262 focuses on a specific subsystem of the vehicle system (referred to as the "system under Consideration," SuC in the standard). However, it is essential to consider the entire vehicle lifecycle. In literature, this is also referred to as the "System of Interest" (SoI).

The "System under Consideration" is a term defined in ISO 26262 and refers to the specific part of the vehicle system being analyzed for functional safety. This could be a particular component or subsystem, such as the braking system or the airbag control unit.

On the other hand, the "System of Interest" is a broader term used in systems engineering to describe the entire system being studied, which includes all its components and interactions.

The "System of Interest" is for example defined in the SEBoK glossary as follows:

- the system whose lifecycle is under consideration (ISO/IEC/IEEE 15288:2015)
- the System of Interest to an observer (Bertalanffy 1968)

The first definition is the common systems engineering definition, referring to a collective set of all elements of any system considered by a lifecycle, including both operational and enabling systems. The second definition applies to all systems, irrespective of whether they have a purpose or lifecycle in an engineering sense.

Particularly, maintainability through over-the-air updates (Software Update Management System, SUMS) and remote diagnostics to support workshops means a significant role in vehicle development and must be appropriately considered. This leads to additional architectural and design requirements, as well as significant constraints on the solution space for technical implementation.

The ISO 26262 standard, which addresses the safety lifecycle of automotive systems, is undergoing a significant paradigm shift. Traditionally, the focus has been on the safety of individual systems, or "items," within a vehicle. However, there is a growing recognition that this approach needs to be expanded to encompass the entire lifecycle of the vehicle itself, as well as the environment in which it operates.

This broader perspective necessitates the consideration of various factors throughout the vehicle's life cycle, from its initial application and purpose, through refueling or charging, goods and/or passenger transport, maintenance and repair, modifications during its lifetime (including over-the-air updates), to eventual recycling or resale in other regions, or repurposing for other uses.

The safety lifecycle as defined by ISO 26262 must therefore be extended to include these additional development steps and periods in the vehicle's lifecycle. This involves not only the design and manufacturing phases but also the operational phase, which includes the use of the vehicle in its intended environment, and the end-of-life phase, which deals with the decommissioning, disposal, or repurposing of the vehicle.

In this context, the safety lifecycle becomes a continuous process that spans the entire existence of the vehicle, from conception to decommissioning. This expanded view of the safety lifecycle is essential to ensure the safety of increasingly complex and interconnected automotive systems, particularly with the advent of Autonomous Vehicles and Advanced Driver-Assistance Systems (ADAS).

In conclusion, the extension of the safety lifecycle as per ISO 26262 is a necessary evolution in automotive safety engineering, reflecting the increasing complexity of modern vehicles and the environments in which they operate. This holistic approach ensures that safety considerations are integrated into every stage of a vehicle's lifecycle, ultimately contributing to the safer operation of vehicles on our roads.

Implementing an expanded safety lifecycle as per ISO 26262 presents several challenges:

- **Awareness:** There is often a lack of awareness about the importance of functional safety (FuSa), cybersecurity and other sources for safety requirements and their compliance or conformity, especially in the context of the entire vehicle lifecycle.
- **Complexity:** The inherent complexity of automotive systems, coupled with the need to consider a wide range of factors throughout the vehicle's lifecycle, adds to the challenge. This includes everything from the initial application and purpose, through refueling or charging, goods and/or passenger transport, maintenance and repair, modifications during its lifetime, to eventual recycling or resale.
- **Project development time and cost:** The implementation of safety and security processes can increase project development time and subsequent cost. This is due to the additional development steps and periods in the vehicle's lifecycle that need to be considered.
- **Managing safety-critical projects:** Managing safety relevant projects requires not only deep awareness of the stakeholder requirements, but also on the applied technology and their physical aspects within the physical, situational, temporal,

and operational context. This includes the managing of systems, software and hardware teams, and addressing challenges that span across the entire safety lifecycle.

- **Technological challenges:** There are also technological challenges. These systems depend on sophisticated models and advanced sensor technology, which can introduce additional complexities for the developers and the usability of drivers and other people involved with the transportation means.
- **Regulatory compliance:** Ensuring compliance with regulatory standards is another significant challenge. This includes not only ISO 26262 but also other relevant standards and regulations.

These challenges highlight the need for a comprehensive approach to safety lifecycle management, one that considers the entire vehicle lifecycle and the complex interplay of various factors that can impact vehicle safety. For the vehicle level, the homologation dominates the source of requirements. On the operation level, we see all the regulations under the umbrella of behavior law. The influences from product liability and product safety is fuzzier and could not generally considered.

7.2.1 Safety and Risk as a Condition

Safety and risk can influence the context of product development or the entire vehicle platform in different ways. While product safety and risk focus on individual components or subsystems, platform safety and risk take a more comprehensive view, considering the entire vehicle and its interactions. Effective systems engineering requires addressing both levels to ensure a safe and reliable vehicle.

Here are some high-level definitions summarizing the focus of vehicle systems engineering:

- **Safety** is a condition in which the risk of harm or damage is reduced to an acceptable level. In different conditions, there are varying rules for safety tolerance and the necessary risk-reducing measures.
- **Risk** is the combination of the probability of an event occurring and the severity of its consequences. In the automotive industry, this involves assessing the likelihood of accidents and the potential harm they could cause to occupants and other road users.
- A **safe state** is a condition where the vehicle or system is in a state that poses no unacceptable risk. This could mean the vehicle is parked securely, or all systems are functioning within safe parameters. For example, during maintenance, ensuring the vehicle is in a safe state might involve disconnecting the battery to prevent accidental starts.

- A **Minimal Risk Condition** (MRC) is a state where the vehicle is brought to a stable and safe stop in the event of a system failure. For example, if an autonomous vehicle detects a critical failure, it might pull over to the side of the road and stop to minimize risk.
- The **tolerable risk level** refers to the maximum level of risk that is acceptable in a given context. In the automotive industry, this is often determined by regulations and standards such as ISO 26262. For example, the risk of a vehicle's braking system failing must be reduced to a level that is considered acceptable by industry standards.
- **Necessary risk reduction** involves implementing measures to lower the risk to an acceptable level. This can include design changes, adding safety features, or implementing stricter quality control processes. For instance, adding Electronic Stability Control (ESC) helps reduce the risk of skidding and loss of control.

Perceived Safety for Drivers

Perceived safety for drivers is crucial for their confidence and comfort while driving. Drivers feel safe when

- the vehicle behaves and reacts as expected to their inputs,
- there are no critical errors or warnings communicated to them, and
- the vehicle provides supportive and predictable driving experience.

For example, features like blind-spot monitoring, lane departure warnings, and adaptive cruise control help drivers feel more secure by providing additional awareness and control.

Perceived Safety for Vulnerable Road Users

Vulnerable Road Users (VRUs) such as pedestrians, cyclists, and motorcyclists feel safe when the following applies:

- They can walk on sidewalks or cycle in bike lanes without significant risk.
- Vehicles are driving smoothly within road markings and adhering to traffic rules.
- There are clear and visible signals from vehicles, such as turn signals and brake lights.

Advanced systems like Vehicle-to-Everything (V2X) communication could enhance VRU safety by providing pre-emptive information to vehicles about the presence of VRUs, especially in poor visibility conditions.

Relation to Different Conditions

Parking: It needs to be ensured that the vehicle is securely parked with the parking brake engaged and systems powered down to prevent unintended movement.

Standstill: When the vehicle is stationary but not parked, such as at a traffic light, systems like Auto Hold can keep the vehicle from rolling.

Driving: Active safety systems like Adaptive Cruise Control (ACC) and Lane Keeping Assist (LKA) help maintain safe driving conditions.

Maintenance in the garage: It needs to be ensured that the vehicle is in a safe state by using proper lifting equipment, disconnecting power sources, and following safety protocols to prevent accidents.

7.2.2 Regional Constraints

Vehicles are designed to be versatile and adaptable to a wide range of applications, environments, and operational scenarios across various regions. Each region presents its own unique set of challenges due to differing road conditions, weather patterns, temperatures, humidity levels, altitudes, traffic regulations, laws, and signage. Despite these variations, the goal is to operate the same vehicle effectively in all these conditions. This complexity extends beyond the concerns of holiday travel, rental cars, type of cargo, and the number of passengers. It also encompasses the diverse legal regulations that exist within cities and countries, as well as the broader differences between markets such as North America, Europe, and China. Adapting a single vehicle to meet all these requirements is a significant engineering challenge that involves careful consideration of global standards and local adaptations.

When designing a vehicle, it is imperative to consider the necessary operating environment and its influencing parameters. This holistic approach ensures that the vehicle can perform optimally under various conditions, such as different climates, terrains, and legal frameworks. By accounting for these factors, manufacturers can create vehicles that are not only functional but also compliant and adaptable to the diverse requirements of global markets.

In the realm of vehicle design, it is understood that not all influences can be addressed with a single solution. The diversity of regions, transport tasks, and uses necessitates the development of configurable vehicle platforms. These platforms must be designed with the flexibility to be optionally implemented across different regions, catering to the specific needs of each area. For a particular vehicle, this means that features may be added optionally, for an additional cost, or activated during use, depending on the application.

The challenge lies in ensuring that drivers, based on their driver's license training, can competently operate the vehicle with all its functions and/or features. This requires a design that is intuitive and user-friendly, while also being sophisticated enough to handle the various optional features that might be activated depending on the vehicle's use and the region it operates in. The goal is to strike a balance between versatility, compliance, and ease of use, allowing drivers to safely and effectively utilize their vehicles in a wide range of scenarios.

7.2.3 Vehicle Purpose

The design of a vehicle is significantly derived from its intended use:

- A long-distance vehicle will feature essential comfort attributes for a comfortable ride. For instance, it will likely include adjustable seating, climate control, and advanced infotainment systems to ensure passenger comfort over extended periods.
- A city car, on the other hand, does not necessarily require a large battery range or, as in the past, a large fuel tank. Its design prioritizes maneuverability and efficiency in urban environments, often with smaller dimensions and a focus on energy conservation.
- An SUV is designed to transport sports equipment and to possess a certain suitability for rough roads. It typically features higher ground clearance, robust suspension systems, and ample cargo space to accommodate various outdoor activities.
- A station wagon should offer a large cargo volume, catering to those who need extra space for luggage, shopping, or business goods. Its design often includes a long roofline and a rear hatch for easy access and loading.
- A family car needs to provide ample space for children and entertainment options. This could mean having multiple seating configurations, durable interior materials, and technology such as rear-seat entertainment systems to keep passengers engaged on journeys.
- A convertible is designed to deliver outdoor fun, with features like a retractable roof, stylish design, and often a focus on performance to enhance the driving experience.

However, all these vehicles must meet safety and security requirements, which are often mandated by laws. This includes the integration of airbags, Anti-Lock Braking Systems (ABS), Electronic Stability Control (ESC), and Advanced Driver-Assistance Systems (ADAS) to ensure the well-being of passengers and compliance with regulatory standards.

In essence, while the primary function shapes the vehicle's design, the underlying safety and security features remain a constant necessity across all vehicle types. These elements are non-negotiable and form the bedrock of vehicle design principles, ensuring that regardless of the vehicle's purpose, it adheres to the highest standards of safety and security.

Vehicles possess numerous performance characteristics that must be evaluated using matrices. One matrix dimension includes the registration regions, such as UNECE, FMVSS (USA), or GB (China). The second dimension of the matrix is the vehicle type, whether it is a station wagon, convertible, sedan, SUV, or off-road vehicle (Figure 7.3). Additionally, there are many regional requirements influenced by social structures and market trends in different regions. For example, northern countries versus southern regions, hot regions versus cold regions, with varying impacts from temperature and other weather conditions such as ice, snow, cold start behavior, or battery preheating. Furthermore, the reaction of electronics to salt near the sea presents a challenge. These characteristics must be adaptable through configuration in vehicle and product platforms; otherwise, each new component would need to be developed from scratch.

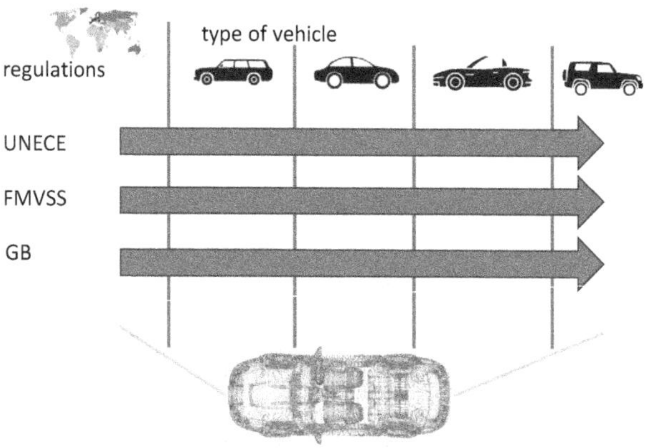

Figure 7.3
Regions and type of vehicles

Therefore, a multidimensional database is necessary to meet the requirements of a specific vehicle platform. This database should transparently combine requirements and solution possibilities, ensuring that vehicles meet the desired performance characteristics before delivery to customers. Additionally, it is crucial to ensure that vehicles comply with local homologation and registration requirements while meeting the promised performance characteristics to customers.

7.2.4 Current Safety Lifecycle of ISO 26262

The current safety lifecycle of ISO 26262 starts with the "Item Definition." The item is considered as an element of a vehicle. The standard defines the item as follows:

3.84, item: *"system (3.163) or combination of systems (3.163), to which ISO 26262 is applied, that implements a function or part of the function at the vehicle level"* (ISO 26262:2018, Chapter 1, 3.84)

From a process point of view the safety lifecycle is defined as follows:

3.86, lifecycle: *"entirety of phases (3.110) from concept through decommissioning of the item (3.84)"*

Based on these definitions of the standard it starts with a concept development.

The ISO 26262 standard, a cornerstone in automotive functional safety, provides a robust framework for ensuring the safety of electrical and electronic (E/E) systems in vehicles. However, its safety lifecycle and the scope of the standard is limited in several key areas (Figure 7.4).

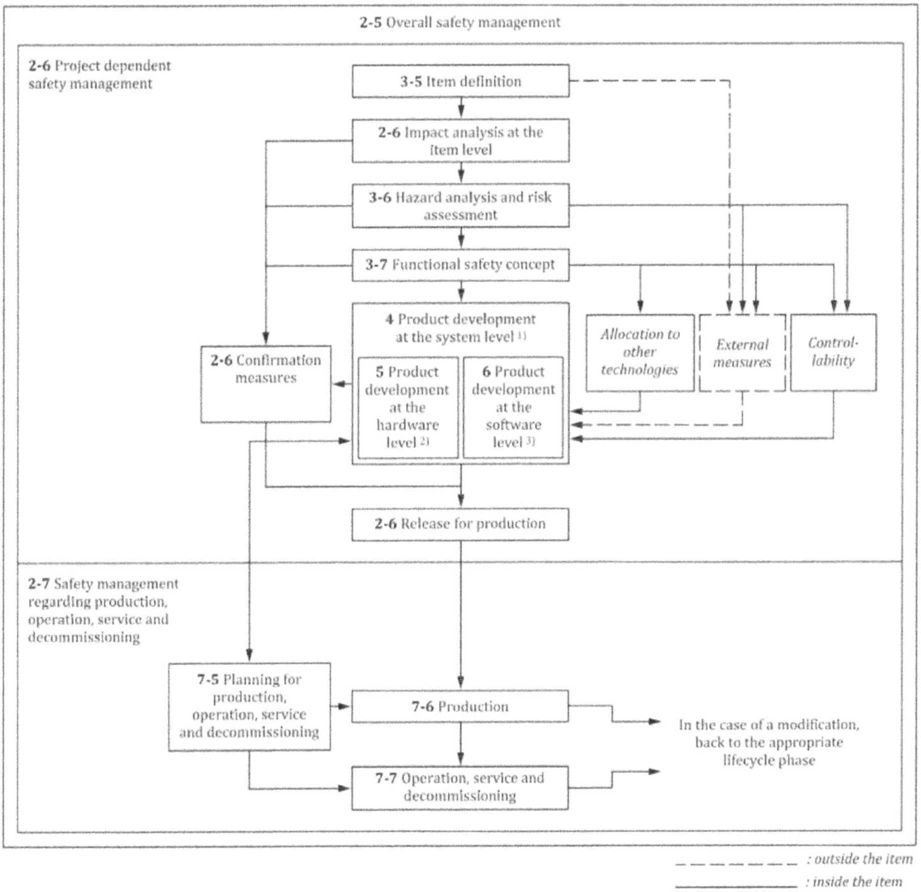

Figure 7.4 Safety lifecycle according to ISO 26262 (source: ISO 26262:2018, Chapter 2)

Narrow Focus on E/E Systems

One of the primary limitations of ISO 26262 is its exclusive focus on malfunctioning behavior E/E system or their failures. This means that the standard does not address a broader range of potential hazards, such as the following:

- **Physical phenomena:** These include issues related to sensor measurement principles, the physical impact of vehicle functions on the road, and failures in mechanical or hydraulic systems.
- **Nominal performance:** The standard does not delve into the design and validation processes that ensure a vehicle performs as intended under normal operating conditions.
- **Human factors:** While the concept of "Safety in Use" is mentioned, the standard primarily focuses on the driver's ability to control failures rather than on the overall human-machine interaction.

SOTIF and Broader System Risks

The introduction of the term "Safety of the Intended Functionality" (SOTIF) has helped to address some of the limitations of ISO 26262, particularly in the context of Advanced Driver Assistance Systems (ADAS). However, SOTIF still does not fully capture the range of risks associated with vehicle design and operation. For instance, it does not account for the risks arising from the following:

- **Vehicle design:** Flaws in the overall vehicle design can contribute to accidents, even if the E/E systems are functioning correctly. This includes the entire vehicle dynamics, measures related to crash impacts, and occupancy protection by means of passive safety.
- **Vehicle usage:** The way a vehicle is used can also introduce risks, such as driving at excessive speeds or under adverse conditions.

Lack of Comprehensive Risk Mitigation

While ISO 26262 recognizes the need for risk mitigation measures beyond E/E systems, it does not provide specific requirements for these measures. This means that organizations may struggle to ensure that all potential risks are adequately addressed.

Verification and Validation Challenges

The limited scope of ISO 26262 poses significant challenges for verification and validation. Specifically, it is difficult to do the following:

- **Verify negative requirements:** It is challenging to prove that a system will not perform an unintended function.
- **Assess conditions outside the scope:** Factors that are explicitly excluded from the standard's scope cannot be easily evaluated.
- **Manage change requests:** Incorporating changes into the design iteratively can lead to complex impact analyses and increase the risk of introducing new failures.

Implications for Safety Argumentation

The initial idea of safety integrity was that the standard would serve as a basis for risk-reducing measures using EE functions. For such risk reduction measures by means of EE functions, very reliable and robust design principles are required. However, this is no longer clearly evident in the standard. It does not address the control and reduction of misbehavior, incorrect design, or the suitability of functions and systems in terms of performance characteristics through electronic and software measures, as seen in other industries.

Since due to applying several lifecycle approaches and management systems for safety and security, many process iterations are required. In case of detected inconsistencies or lack of completeness any of the iterations lead to significant change requests. The risk associated with change requests is highlighted as a significant risk in the Toyota Production System (TPS). The traceability through all the risk analysis and the potential change impact leads to exponentially growing complexity, which cannot be controlled by a limited organization.

The limitations of ISO 26262 have implications for the overall safety argumentation process. A safety argument based solely on this standard may not be comprehensive or convincing. Additionally, the lack of clear guidance on addressing broader system risks can make it difficult to demonstrate compliance with safety regulations.

While the standard and the provided safety Lifecycle is a valuable tool for ensuring the functional safety of E/E systems in vehicles, it is essential to recognize its limitations and adopt a more holistic approach to safety. By considering a wider range of factors, including physical phenomena, human factors, and vehicle-level risks, the automotive industry can develop vehicles that are even safer and more reliable.

7.2.5 Lifecycle Extension

The safety lifecycle from ISO 26262 is proposed to be extended with two additional layers:

- **Operational Layer:** This is the highest layer. In this layer, the context, including the operational domain, is defined (Figure 7.5). It also involves defining the stakeholder requirements. At this level, the vehicle is considered as a "black box," meaning its internal workings are not taken into account, only its inputs and outputs are considered.
- **Vehicle Layer:** This is the second layer. Here, the lifecycle requirements of the system of interest are addressed. This spans from the conception of the vehicle up to the end of the vehicle's life or the end of the vehicle platform's life. This layer includes considerations for maintenance and legal obligations.

These two layers provide a more comprehensive approach to safety lifecycle management, extending the scope beyond the existing ISO 26262 framework to consider broader operational and vehicle-specific factors. This ensures a more robust and thorough approach to safety in the automotive industry.

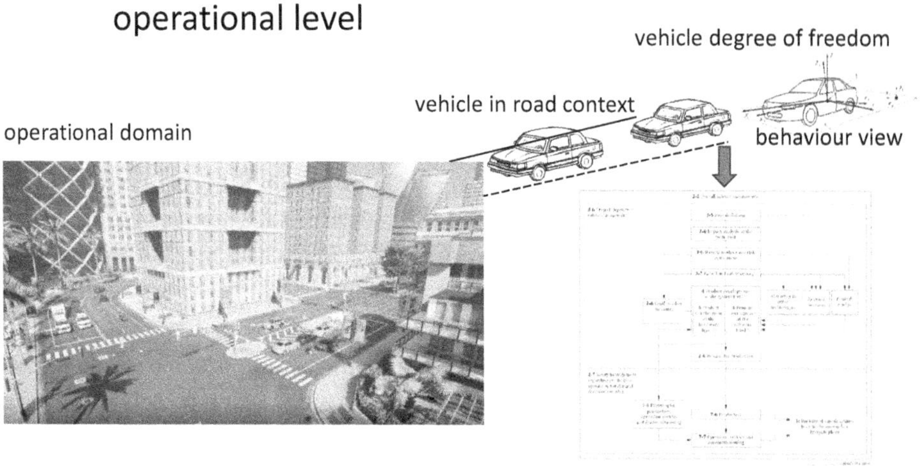

Figure 7.5 Operational Layer

The operational layer is the highest tier where the behavior of all actors in public road traffic is described to extend the basis of the safety lifecycle according to ISO 26262. It provides the foundation for developing requirements that can then be derived for an ego vehicle with automated driving functions.

The operational domain offers the framework for the scenery (the backdrop) for various actors in road traffic. The scenery itself is dynamic, including elements such as weather conditions, road conditions, construction sites, broken-down vehicles, pedestrians, cyclists, scooter riders, and traffic signs. These elements form ever-changing combinations and rules at intersections or impose restrictions on speed, parking, and stopping maneuvers, as well as traffic lights, all of which can affect the ego vehicle in any combination.

The scenery is dynamic, and the behavior of different actors in road traffic affects the vehicle, which is also dynamic. This results in multi-complex dependencies and influences, to which a driver learns to respond appropriately during driving school.

These are some examples of key terms:

- **Scenery:** This includes static and dynamic elements such as lane information, traffic lights, weather conditions, pedestrians, other vehicles, and road signs.
- **Scenario:** A scenario is a sequence of scenes that describe a specific situation in road traffic, for example, a vehicle driving through a city with varying traffic densities and pedestrian crossings.

- **Scene:** A scene is a snapshot of the environment at a particular moment, including the scenery and dynamic elements. For instance, a scene could depict a busy intersection with pedestrians waiting to cross and vehicles stopping at a red light.

The term Operational Design Domain (ODD) is interpreted differently in various sources, such as the Pegasus project, UNECE regulations, FMVSS, and Nevada Law. However, the key point is that the dynamic interaction of all traffic participants in a dynamic operational environment must be described as a scenario. Traffic rules provide the basis for correct behavior in each traffic situation, which should be described as different scenes.

The legal requirements for behavior in public road traffic are derived from the respective national road traffic laws. The Geneva Convention on Road Traffic (1949) provides the following basis in its General Provisions: Every driver, pedestrian, or other road user shall conduct themselves in such a way as not to endanger or obstruct traffic. They should avoid all behavior that might cause damage to persons, or public or private property. This fundamental rule has been incorporated into many national road traffic laws, such as Paragraph 1 of the German Road Traffic Regulations (StVO). It serves as an essential foundation for behavior in road traffic, which all road users must consider, even when vehicle control is performed by a system.

Since the amendment in 2016 of the Vienna Convention on Road Traffic (1968), vehicle control can now also be performed by a system and is no longer restricted to a driver in the car or a guide for animal-drawn vehicles. However, there are still very different national interpretations of this provision.

The rules for breaking down requirements and the different levels of abstraction not only formally define the methodology derived from Axiomatic Design (AD), but there is also a good template for model-based development in automotive engineering, such as SAE J3049 (Model Architecture and Interfaces Recommended Practice for Ground Vehicle System and Subsystem Dynamical Simulation). Additionally, ISO 26262 recommends rules to reduce complexity across multiple system levels.

We have the vehicle level, which is fundamentally at the center of development for traditional vehicle manufacturers, and the system development, for example, of steering systems. In mechanics and common design tools, an exploded view is a common method that visually shows the individual elements. However, a picture of a resistor or microcontroller does not really convey the function, nor is a picture of software meaningful, regardless of how it is presented.

SAE J3049 starts with a Ground Vehicle System (GVS) and breaks it down into environment, driver/passenger, and the vehicle itself. The vehicle is further decomposed into Vehicle Supervisory Control (VSC), power, chassis, and trailer (Figure 7.6). While further clustering can certainly be discussed, the advantage of SAE J3049 is that it provides continuous decomposition for an entire vehicle, down to the effect chains consisting of sensors, actuators, and ECUs, including the structuring and classification of the various interfaces. These interface definitions are interesting to regard since they allow also dynamic simulation of the different subsystems.

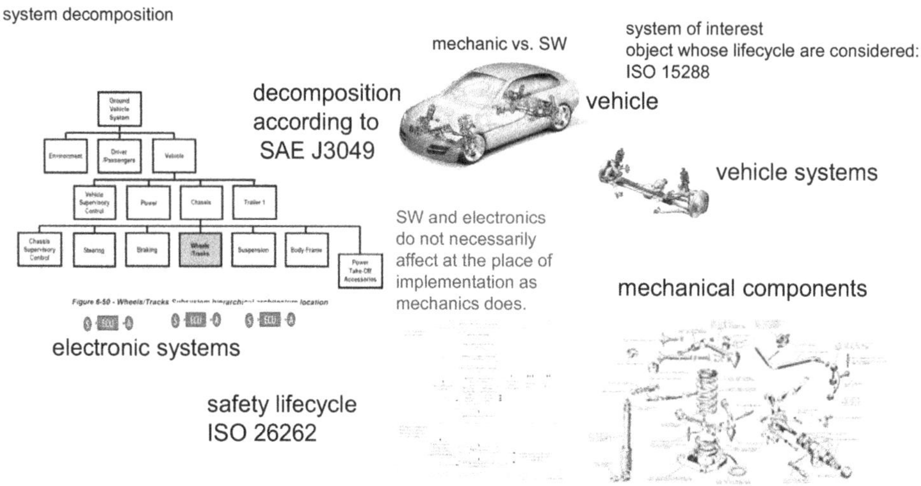

Figure 7.6 Vehicle decomposition

7.2.6 Structuring and Modularization in Avionics Industry

In aviation, the clustering of functions and decomposition are even defined in regulations and documented in the Air Transport Association of America (ATA). In the realm of automation, the establishment of a standardized basis is imperative for fostering a systematic development approach. This standardization would serve as the cornerstone for ensuring the safety and reliability of distributed development processes. Just as the ATA regulations have streamlined the categorization and management of aircraft systems, a similar framework in automation could significantly enhance the coordination and integration of various components and subsystems. A standardized basis would facilitate clear communication among stakeholders, promote best practices, and ensure consistency across different projects and teams. It would also provide a structured method for addressing the complexities inherent in distributed development, such as version control, change management, and traceability of requirements.

7.2.6.1 Modularization for Civil Aircrafts

The evolution from ATA 100 to ATA iSpec 2200 represents a significant shift in the standardization of technical documentation within the aviation industry. Here's a brief overview of the development:

- **ATA Spec 100:** Established in 1956, ATA Spec 100 introduced an industry-wide numbering scheme to organize aviation technical documentation. It provided guidelines for content and formatting, primarily for printed distribution.
- **ATA Spec 2100:** Focused on electronic data exchange, implemented in SGML (Standard Generalized Markup Language), ATA Spec 2100 was developed to address the growing need for digital documentation and data interchange.
- **ATA iSpec 2200:** First published in 2000, ATA iSpec 2200 combined the previous ATA Spec 100 and ATA Spec 2100. It marked a transition to a more modern approach, incorporating improvements such as a new numbering system and gender-neutral terms. The structure of iSpec 2200 is designed to facilitate the digital representation and exchange of technical data, aiming to minimize costs, improve information quality, and meet airline operational needs.

Over the years, the structure of ATA iSpec 2200 has been continuously updated to adapt to technological advancements and industry requirements. For instance, it has embraced a modular approach to documentation, allowing for more efficient data management and distribution across various platforms. This evolution reflects the industry's move towards digitalization and the need for more flexible and interoperable documentation standards.

The adoption of a systematic framework akin to the ATA regulations in aviation is crucial for the automation industry. It would not only streamline the development process but also bolster safety and efficiency in the creation of automated systems.

In their book *Aircraft Systems,* Ian Moir and Allan Seabridge provide an example of how multiple control devices could integrate (Figure 7.7, see also Moir/Seabridge 2001).

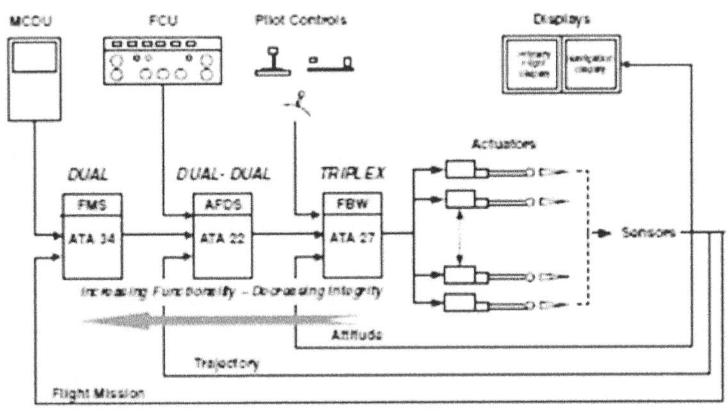

Figure 7.7 Example for an ATA clustering for the flight pilot (source: Moir/Seabridge 2001)

Dual Flight Management System (FMS)—ATA 34

The Flight Management System (FMS) is a specialized computer system used in aircrafts to automate a wide variety of in-flight tasks, reducing the workload on the flight crew. According to ATA Chapter 34, which covers navigation systems, the FMS integrates various subsystems to manage flight planning, navigation, performance management, and aircraft guidance. It uses inputs from navigation sensors, such as GPS and inertial reference systems, to determine the aircraft's position and provide guidance along the planned route.

Dual-Dual Autoflight System (AFOS)—ATA 22

The Autoflight System (AFOS), covered under ATA Chapter 22, is responsible for the automatic control of the aircraft's flight path. This includes autopilot functions, autothrottle, and flight director systems. The AFOS works in conjunction with the FMS to maintain the desired flight path, altitude, speed, and other parameters, ensuring efficient and safe operation.

Triplex Flight Control System (Fly-By-Wire system, FBW)—ATA 27

The Flight Control System (FCS), particularly the Fly-By-Wire (FBW) system, is detailed in ATA Chapter 27. FBW technology replaces traditional manual flight controls with an electronic interface. The pilot's inputs are converted into electronic signals, which are interpreted by flight control computers to move the control surfaces (e.g., ailerons, elevators, rudder). This system enhances aircraft performance, safety, and efficiency by providing precise control and reducing the weight and complexity of mechanical linkages.

Integration and Application

In modern aircraft, these systems are highly integrated to ensure seamless operation. The FMS provides the flight plan and navigation data to the AFOS, which then automates the flight path control. The FBW system ensures that the pilot's commands and the automated inputs from the AFOS are accurately translated into control surface movements, maintaining the desired flight trajectory. This integration allows for advanced features such as automatic landing, optimized flight paths for fuel efficiency, and enhanced safety through redundancy and fault-tolerant design. Please use this more than 25 years old example just as a reference, how to deal with complex cascaded control systems.

7.2.6.2 Analysis of Architectural Approaches by Ray Hudson

Ray Hudson, a distinguished academic from Durham University, has critically examined the weaknesses of current architectural approaches in his various publications. In his academic workgroup, ARO 4510, he explores multiple dimensions of architecture. He wants to overcome with his work the static view on current architectural approaches.

A fundamental issue, e. g., within the SysML (Systems Modeling Language) effort in the MBSE community, is that SysML is not genuinely a modeling language but rather Just Another Drawing Standard (JADS). Ray Hudson stated (Hudson 2024):

"My assessment of SysML is that it is 'necessary but not sufficient' for effective MBSE. In subsequent discussions, I intend to elucidate why this is a fundamental problem and advocate for a language-based approach, termed the 'Schema First Approach,' which must precede any drawing standard like SysML. This approach ensures that the barriers to understanding effective MBSE are sufficiently low for average engineers."

He describes an 3D MBSA approach under ARO 4510 (Figure 7.8). The entry in the first dimension he called a "thread trace." A thread trace is a detailed, chronological record of how a specific action or event propagates through a system. It's like following a single thread through a tangled tapestry to see where it starts and where it ends. In the context of a complex system, this "thread" represents the flow of information, energy, or matter as it triggers a series of interconnected physical processes.

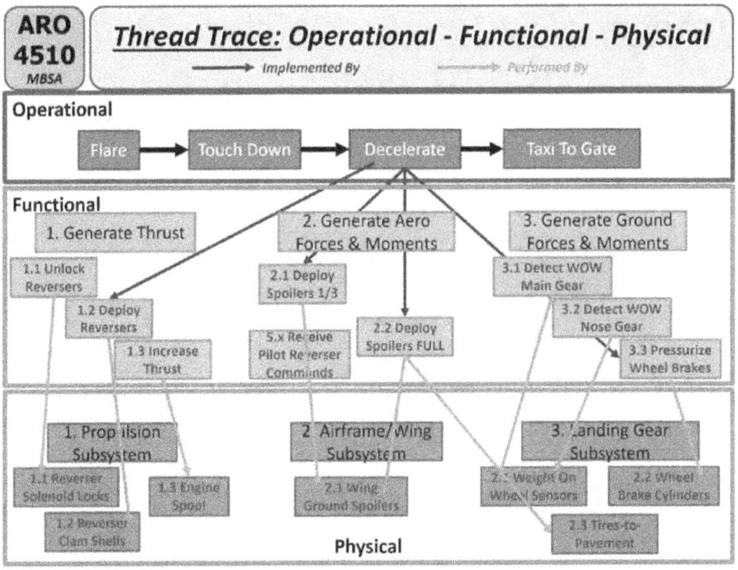

Figure 7.8 ARO 4510 MBSA (source: Hudson 2024)

One dimension delineates three horizontal layers within the aircraft domain: the Operational Level, the Functional Level, and the Physical Level (Figure 7.9).

Operational Level

At the Operational Level, Hudson identifies the sequence of actions involved in aircraft operations. The focus question is "How one uses the system." Characteristics are activities and duration.

The following examples could be seen:

- flare
- touchdown
- decelerate
- taxi to gate

These actions are interconnected through the relation "implemented by," indicating that each step is executed by specific functions.

The focus question is "What the system does." Characteristics are transformers and algorithms.

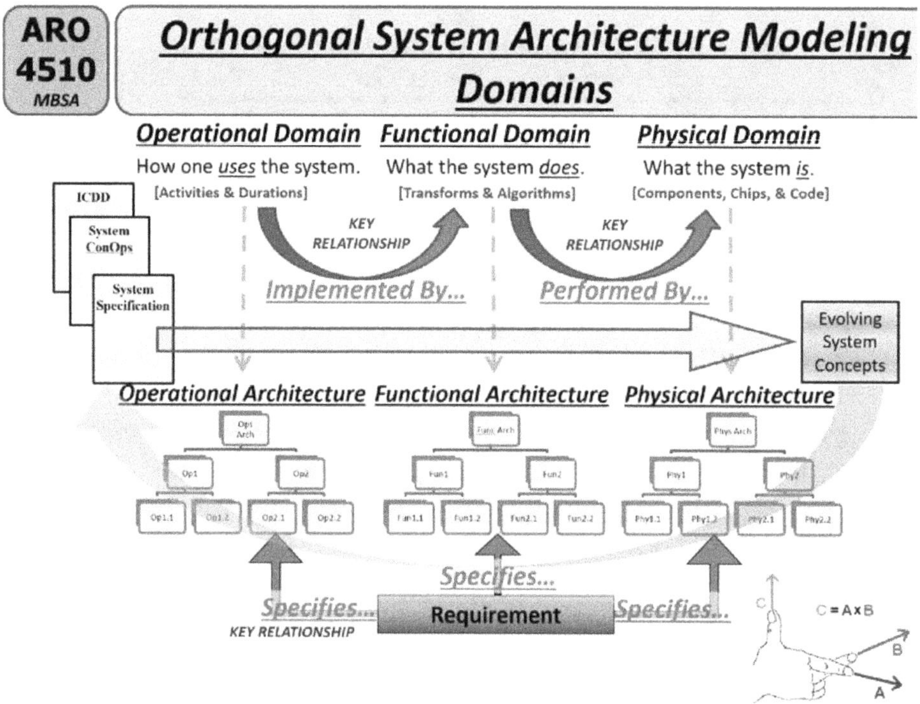

Figure 7.9 Three dimensions of an airplane level architecture (source: Hudson 2024)

Functional Level

The Functional Level breaks down the operational actions into detailed tasks:

- **1:** generate thrust
- **1.1:** unlock reversers
- **1.2:** deploy reversers
- **1.3:** increase thrust
- **2:** generate aero forces & moments
- **2.1:** deploy spoilers 1/3
- **2.2:** deploy spoilers full
- **3:** generate ground forces & moments
- **3.1:** detect WoW (Weight on Wheels) main gear
- **3.2:** detect WoW nose gear
- **3.3:** pressurize wheel brakes

The relation to the Physical Level is described as "performed by," indicating that the physical components are responsible for the technical realization of these functions.

Physical Level

The focus question is "What the system is." Characteristics are components, chips and code. The Physical Level encompasses the actual hardware and systems that perform the functions:

- **1:** propulsion subsystem
- **1.1:** reverser solenoid locks
- **1.2:** reverser clam shells
- **1.3:** engine spool
- **2:** airframe/wing subsystem
- **2.1:** wing ground spoilers
- **3:** landing gear subsystem
- **3.1:** Weight on Wheels (WoW) sensors
- **3.2:** wheel brake cylinders
- **3.3:** tires to pavement

Interrelation of Levels

Hudson emphasizes that the Operational and Functional Levels generate requirements that define what needs to be done. These requirements are then fulfilled by the Physical Level, which focuses on the technical realization. This hierarchical structure ensures that design specifications are clearly defined and systematically implemented.

Requirement Development

Another critical dimension in Ray Hudson's architectural analysis is Requirement Development (Figure 7.9). This process involves developing requirements both top-down and along the operational chains. The top-down approach ensures that high-level objectives are broken down into specific, actionable requirements. Simultaneously, requirements are developed along the operational chains to define the relationships between sensors and actuators and to specify the interfaces.

System and Functional Breakdown

The third dimension is the System and Functional Breakdown. This involves decomposing the system from higher-level components or functional groups down to individual parts in both software (SW) and hardware (HW). This method ensures that each component and function is clearly defined and understood, facilitating effective implementation and integration.

This approach is similar to the methodologies described by Nancy C. Leveson in her works on STAMP (Systems-Theoretic Accident Model and Processes) and Intent Specifications. Both approaches emphasize the importance of clear, hierarchical breakdowns to ensure traceability, safety and functionality in complex systems.

By incorporating these additional dimensions—Requirement Development and System and Functional Breakdown—Ray Hudson's framework provides a comprehensive and structured approach to architectural analysis. This ensures that all aspects of the system are thoroughly considered, from high-level requirements to detailed technical specifications, promoting both clarity and precision in design and implementation.

Ray Hudson's analysis provides a comprehensive framework for understanding the complexities of aircraft architecture. By dissecting the process into operational, functional, and physical dimensions, he highlights the importance of clear requirements and precise technical execution. This approach not only identifies weaknesses in current methodologies but also offers a structured pathway for improvement.

7.2.7 Basic Structure of Automotive Development

The basic structure for motor vehicles is also based on the road traffic regulations. The development of automotive technology and its regulations is a complex process that begins with international conventions and is refined through regional and local regulations. Here's how this process unfolds:

- **Vienna Convention:** The 1968 Vienna Convention on Road Traffic is a fundamental framework that sets out the basic rules of the road and driving requirements. It has been the core of road safety and automated driving development efforts.[1]
- **UNECE Regulations:** The United Nations Economic Commission for Europe (UNECE) plays a crucial role in setting regulatory standards for vehicle certification, road traffic safety, and environmental concerns. UNECE regulations are instrumental in shaping the development of automated driving technologies, ensuring they conform to safety standards and can be integrated into traffic legally.[2]
- **EU Regulations:** The European Union adopts and enforces UNECE regulations, further developing them to suit the specific needs of its member states. The EU has its own set of directives and regulations that govern vehicle construction, safety, environmental performance, and approval processes. This includes the harmonization of vehicle regulations to facilitate international trade and ensure a high level of safety and environmental protection.[3]
- **Local Regulations:** National laws, such as the German Road Traffic Act, incorporate these international and EU regulations into their legal framework. Germany, for example, has amended its Road Traffic Act to allow for the operation of autonomous vehicles under specific conditions, reflecting the advancements in automotive technology.[4]
- **Vehicle Classification:** Within this regulatory framework, vehicles are classified into different categories. For passenger cars, the classification "M1" is used, which refers to vehicles designed for the carriage of passengers and comprising not more than eight seats in addition to the driver's seat.

This regulatory cascade from international to local levels ensures that automotive development aligns with global safety standards while also meeting regional and national requirements. It provides a structured approach to the safe and efficient integration of new automotive technologies into society.

---

1 *https://clepa.eu/mediaroom/vienna-conventions-from-1968-are-still-at-the-core-of-automated-driving-development-and-road-safety-efforts*

2 *https://unece.org/automated-driving*

3 *https://clepa.eu/mediaroom/vienna-conventions-from-1968-are-still-at-the-core-of-automated-driving-development-and-road-safety-efforts*

4 *https://clepa.eu/mediaroom/vienna-conventions-from-1968-are-still-at-the-core-of-automated-driving-development-and-road-safety-efforts*

In many motor companies you find a typical structure like chassis, drive, and body. It was detailed in the following organizational units:

- chassis, brake, steering, and chassis control
- powertrain, energy management, and propulsion
- interior, body electronics, and HMI
- ADAS

Due to the development of a more software dominated vehicle architecture, such organizational interfaces are difficult to identify.

Most new vehicle platforms have a unit EE architecture, which needs to be basic input for all the concepts of vehicle development. Unfortunately, the EE architecture shows no link to the observable behavior of the vehicle, so that traces between vehicle architecture and design, vehicle behavior and the systems breakdown are not linked.

7.2.7.1 Example from SAE J3049

SAE published a modelling guideline for dynamic interfaces, the SAE J3049 (Figure 7.10). In that standard they gave recommendations for a typical breakdown of vehicle structure and provide requirements for dynamic interfaces.

The primary intent of the 2015 version of SAE J3049 is to establish a standardized framework for the architecture and interfaces of the dynamical simulation models of ground vehicle systems and subsystems (SAE J3049:2015). This standardization aims to achieve the following:

- **Promote interoperability and reusability:** This remains a core objective. The standard seeks to enable the seamless exchange and reuse of vehicle dynamics models between different simulation tools, organizations, and projects. This reduces redundant model development and facilitates collaborative engineering.
- **Facilitate system-level simulation:** By defining clear interfaces between subsystems, J3049 (2015) enables the creation of complex, full-vehicle simulations by integrating individual component models. This allows for a more holistic understanding of vehicle behavior.
- **Improve model integration and coupling:** The standard provides guidelines for how different models should be coupled together, ensuring consistency in data exchange and interaction. This is crucial for accurate simulation results.
- **Support Model-Based Development (MBD):** J3049 (2015) aligns with the principles of MBD by providing a structured approach to model development and integration. This supports the use of simulation throughout the vehicle development process.

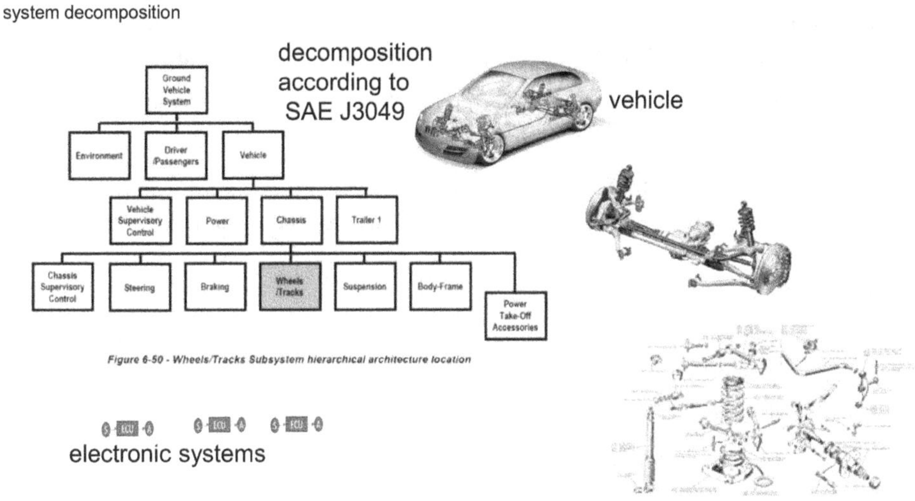

Figure 7.10 Example from SAE J3049 (source: SAE J3049:2015)

Scope of SAE J3049 (2015)

The scope of the 2015 version of SAE J3049 encompasses the following key aspects related to the dynamical simulation models of ground vehicles:

Model architecture: The standard defines a recommended architecture for vehicle models, including a hierarchical decomposition of the vehicle into subsystems. This decomposition typically includes the following:

- vehicle body (including chassis)
- powertrain (engine, transmission, etc.)
- drivetrain (driveshafts, differentials, etc.)
- suspension (springs, dampers, linkages, etc.)
- steering (steering wheel, linkages, etc.)
- brakes (brake actuators, friction elements, etc.)
- tires
- aerodynamics

Interface definitions: A central focus is the specification of standardized interfaces between these subsystems. This includes the following:

- **Signal definitions:** defining the types of signals exchanged between subsystems (e. g., forces, torques, displacements, velocities, accelerations)
- **Units of measurement:** specifying the units for each signal (e. g., N, Nm, m, m/s, m/s^2)

- **Coordinate systems:** defining the coordinate systems used to represent the signals, ensuring consistent interpretation and transformation of data
- **Data types and formats:** specifying the data types (e. g., scalar, vector, matrix) and formats used for signal transmission
- **Model fidelity and complexity:** J3049 (2015) recognizes that different simulation applications require different levels of model fidelity. It does not prescribe specific modelling methods or levels of detail but provides a framework that can accommodate models of varying complexity.
- **Application to ground vehicles:** The standard is primarily intended for on-road vehicles (passenger cars, trucks, buses, etc.) but can also be applied to off-road vehicles and other ground vehicle types.

SAE J3049 did not provide detailed internal modelling of individual components (e. g., detailed combustion models within the engine) such as the following:

- vehicle dynamic aspects
- mechanical design modelling
- specific simulation software tools or platforms
- control system modelling (although the interfaces defined in J3049 can be used to connect vehicle dynamics models with control system models)
- driver models or traffic simulation

By adhering to the guidelines outlined in SAE J3049 (2015), developers can create vehicle dynamics models that are more easily integrated, reused, and validated, leading to more efficient and effective vehicle development processes. It's important to note that newer versions of the standard might exist with updates and additions, so always refer to the specific version you are working with.

7.2.7.2 4+1 View from Kruchten

The 4+1 architectural view model was introduced by Philippe Kruchten to describe the architecture of software-intensive systems from different perspectives, catering to various stakeholders' concerns (Kruchten 1995). Here's an explanation of the original four views, the extension with the scenario view, and their relation to UML (Unified Modeling Language):

The original 4 views (Figure 7.11):

- **Logical view:** It focuses on the functionality the system provides to end-users. UML diagrams like class diagrams and state diagrams are used to represent this view.
- **Development view:** It illustrates the system from a programmer's perspective, concerned with software management. This view is represented using UML package diagrams and component diagrams.

- **Process view:** It deals with the dynamic aspects of the system, such as processes, communication, and concurrency. It uses UML diagrams like sequence diagrams, communication diagrams, and activity diagrams.
- **Physical view:** It depicts the system from a system engineer's perspective, showing the topology of software components on the physical layer. The UML deployment diagram is used for this view.

The "+1" scenario view:

Scenarios (use case view): This view is used to illustrate and validate the architecture design through a small set of use cases or scenarios. These scenarios describe sequences of interactions between objects and processes. UML use case diagrams are typically used here.

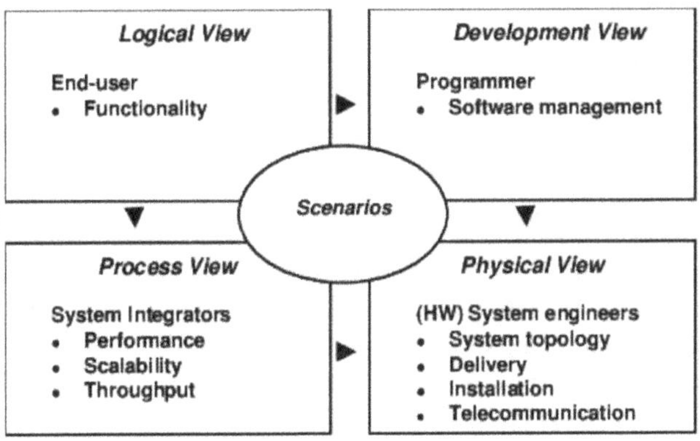

Figure 7.11 4+1 views derived from Kruchten: In the original model, four architectural views are defined (source: Kruchten 1995).

The 4+1 view model is generic and not restricted to any specific notation, tool, or design method. However, UML has been successfully used to represent different views. UML provides a standardized way to visualize the system's design, and each view in the 4+1 model can be expressed using one or more UML diagrams. This model helps in addressing various concerns of the stakeholders, such as end-users needing to understand the system's functionality, developers requiring insight into the system's structure and management, and system engineers needing to know the system's physical deployment.

Kruchten's model has been influential in the field of software architecture, providing a structured approach to capturing the complexities of software systems and ensuring that different stakeholder perspectives are considered during the design process. The use of UML in this context serves as a bridge between the abstract architectural views and the concrete implementation details, facilitating communication and understanding among all parties involved in the development of a software system.

7.2.7.3 Views from SPES

The SPES 2020 framework, developed at the Technical University of Munich (TUM), offers a structured methodology for Model-Based Systems Engineering (MBSE) specifically tailored for the complexities inherent in embedded systems. This framework, often referred to as the SPES matrix, provides a systematic approach to managing system complexity through the delineation of distinct views and levels of abstraction. The following explanations elaborate on the core components of the SPES matrix.

SPES Views (Perspectives on the System)

The SPES matrix defines four key views, each offering a unique perspective on the system under development (Figure 7.12):

- **Requirement view:** This view serves as the foundation of the system, capturing the needs and expectations of various stakeholders. It focuses explicitly on what the system is intended to achieve, articulating the desired functionalities and performance characteristics without delving into implementation details. This view is crucial for establishing a shared understanding of the system's purpose and scope among all involved parties. It employs natural language, use cases, and other elicitation techniques to document requirements comprehensively.
- **Functional view:** This view describes the system's behavior from an external, observer-centric perspective. It delineates the system's functions and their interactions with the external environment and other systems. Critically, this view abstracts away from the internal structure or implementation of the system, focusing solely on its observable behavior. This "black box" perspective is essential for understanding the system's overall functionality and its interactions within its operational context.
- **Logical view:** This view transitions from the external perspective of the functional view to an internal representation of the system's architecture. It defines the logical components that comprise the system and their respective interfaces, specifying how these components interact to realize the functions defined in the functional view. This view remains abstract regarding specific implementation technologies, focusing instead on the logical organization and interaction of system elements. It often employs block diagrams and other structural representations to depict the system's internal architecture.
- **Technical view:** This view represents the concrete realization of the system, detailing the specific hardware and software technologies employed to implement the logical components defined in the logical view. It addresses aspects such as hardware platforms, software frameworks, communication protocols, and other implementation-specific details. This view bridges the gap between the abstract logical architecture and the tangible physical implementation, providing a complete specification for system construction.

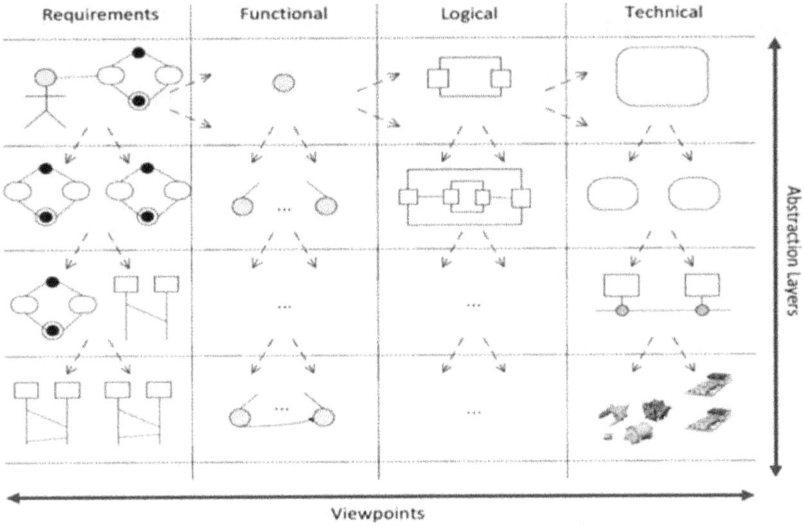

Figure 7.12 SPES 2020: Viewpoints and abstraction layers (source: Daun et al. 2016)

Levels of Abstraction (Hierarchical Decomposition)

SPES calls them "abstraction layers", in this book we use the term "level of abstraction." Complementary to the views, the SPES matrix employs a hierarchical decomposition of the system into distinct levels of abstraction:

- **Vehicle level:** This represents the highest level of abstraction, encapsulating the entire system as a holistic entity. For example, in the context of automotive systems, this level would represent the complete vehicle, encompassing all its subsystems and components. This level provides a high-level overview of the system's purpose and overall scope.
- **System layer:** This mid-level abstraction decomposes the vehicle level into its major constituent systems. In the automotive example, this could include systems such as the powertrain, brake system, navigation system, or infotainment system. This level allows for a more detailed examination of the system's constituent parts and their interrelationships.
- **Component layer:** This represents the lowest level of abstraction, detailing the individual components that comprise each system layer. For example, within the braking system, components such as sensors (e.g., wheel speed sensors), actuators (e.g., brake calipers), and control units would be represented at this level. This level provides the most granular view of the system, enabling detailed design and implementation.

Interplay of Views and Levels

The power of the SPES matrix lies in the interplay between these views and levels of abstraction. By navigating between these different perspectives and levels of detail, engineers can effectively manage the complexity of embedded systems. The matrix facilitates a top-down design approach, commencing with high-level requirements at the vehicle level and progressively refining them through the system layers and component layers, while simultaneously considering the system from the perspectives of requirements, functional, logical, and technical views. This structured approach ensures a consistent and comprehensive system development process, from initial conceptualization to final implementation. This structured approach, while robust, requires careful consideration and application to avoid unnecessary complexity.

7.2.7.4 Breakdown of Architecture and Design Constraints

In the intricate process of requirements decomposition and architectural design, the transparent documentation of design decisions, along with the resulting design constraints they impose on the solution space, is not merely advisable but absolutely essential. Without this meticulous record-keeping, the crucial link between initial requirements and the final system architecture becomes fractured, hindering traceability, verification, and ultimately, the success of the project. Missing design decisions or undocumented design decisions could lead to systematic issues during requirement development.

Aspects and Impacts of Design Decisions and Constraints

Design decisions and design constraints are inextricably linked. A design decision, by its very nature, narrows the available solution space by imposing constraints on subsequent design choices. These constraints can manifest in various forms:

Performance constraints: Decisions about performance targets (e. g., speed, fuel efficiency, processing power) impose constraints on the selection of components and technologies.

Cost constraints: Budgetary limitations restrict the use of certain materials, components, or manufacturing processes.

Technical constraints: The availability of technology, compatibility with existing systems, and physical limitations (e. g., size, weight, power consumption) constrain design choices.

Regulatory constraints: Legal and regulatory requirements (e. g., safety standards, environmental regulations) impose strict limits on design parameters.

Schedule constraints: Time limitations dictate the feasibility of certain design approaches and the selection of readily available solutions.

The Consequences of Neglecting Design Decisions and Constraints

Failing to document design decisions and the resulting constraints leads to a cascade of negative consequences:

Loss of design rationale: The reasons why a particular design was chosen, including the trade-offs made between competing requirements and constraints, become obscured. This makes it difficult to understand the design's underlying logic and makes future modifications or adaptations risky.

Impaired traceability: The ability to trace requirements through the architecture and back to the original stakeholder needs is severely compromised. This makes it challenging to verify that the system fulfills its intended purpose.

Constrained solution space exploration: Without explicit documentation of constraints, designers may inadvertently explore solutions that are ultimately infeasible. This wastes time and resources on dead-end paths.

Difficult impact analysis: Assessing the impact of changes becomes significantly more complex. A change in a requirement or constraint can have ripple effects throughout the design, but without clear documentation, it's difficult to identify all the affected elements. This can lead to unforeseen consequences and integration problems.

Impeded verification and validation: Demonstrating compliance with requirements and constraints becomes nearly impossible without a clear record of design decisions and their rationale. This undermines confidence in the system's correctness and reliability.

Increased risk of design flaws: Undocumented constraints can lead to design flaws that are only discovered late in the development process, resulting in costly rework and delays.

Ensuring Traceability and Managing Constraints

To mitigate these risks, a robust approach to documenting design decisions and managing constraints is essential:

Explicitly document design decisions: Every significant design choice should be documented, explaining the rationale behind it, the alternatives considered, and the trade-offs made.

Clearly define and document design constraints: Constraints should be explicitly stated and documented, including their source, rationale, and impact on the design.

Establish traceability links: Traceability links should be established between requirements, design decisions, architectural elements, and constraints. This can be achieved using traceability matrices, dedicated requirements management tools, or MBSE tools.

Use a formalized approach to constraint management: This includes identifying, documenting, analyzing, and tracking constraints throughout the development lifecycle.

Regularly review and update documentation: As the design evolves, the documentation of design decisions and constraints should be kept up-to-date to reflect the current state of the system.

Example Revisited

Mathematical analogy: This can be likened to a mathematical function where the domain is restricted by a higher-level decision. If we define the function

$$f(x) = \sqrt{x}$$

the decision to take the square root of x means that the domain of f is constrained to $x \geq 0$ as square roots of negative numbers are not defined in the set of real numbers.

Practical example

Consider the requirement "The vehicle shall have a braking system capable of stopping the vehicle within a specified distance." Decomposing this, we might have "The braking system shall provide a maximum deceleration of X m/s^2." A design decision might be to use a specific type of brake caliper. Critically, this decision might be driven by the constraint of the maximum allowable wheel weight (a technical constraint) or a cost target (a cost constraint). The documentation should explicitly state: "The selection of [specific brake caliper] was driven by the constraint of a maximum wheel weight of Y kg, which is necessary to meet the overall vehicle weight target and maintain desired handling characteristics. Alternative calipers were considered but rejected due to exceeding this weight constraint."

By explicitly linking the design decision to the weight constraint, the rationale becomes clear, and the impact of potential changes (e. g., a change in the weight target) can be readily assessed.

By meticulously documenting design decisions and their associated constraints, organizations can create a clear and auditable record of the design process, ensuring traceability, facilitating verification and validation, and ultimately improving the quality and success of their systems.

High-Level Decisions Shape the Solution Space

When decisions are made at a high level of abstraction, they set the direction for the entire project. For instance, choosing a specific architectural style, like service-oriented networks or classical CAN-communication, will inherently limit the types of technologies and patterns that can be used subsequently.

Constraints on Lower Levels

These high-level decisions act as constraints for the lower levels. If a services-oriented network or microservices architecture is chosen, it constrains the system to use distributed patterns and possibly specific communication protocols. It might also necessitate the use of containerization and orchestration tools like Docker and Kubernetes.

Impact on Design Freedom

As a result, designers and architects working on the lower levels have a reduced set of options. Their choices must align with the high-level decisions to ensure consistency and feasibility of the overall system. High-level design and architectural decisions are like setting the rules of a game; they define what moves are possible, and this in turn limits the actions that can be taken as one moves forward with the design and implementation process. These constraints ensure that the system develops in a controlled and predictable manner, adhering to the initial vision and requirements.

Negative Requirements

Consider a typical non-requirement for an automobile, for example, "The car should not be noisy." This statement is problematic for several reasons:

- **Lack of specificity:** The term "noisy" is subjective. What is considered noisy to one person may not be to another. Without a specific decibel level or a clear standard to measure against, it's impossible to verify whether the requirement has been met.
- **Infinite possibilities:** By stating what should not be, there are infinite conditions that could satisfy the requirement. The car could be silent, but it could also make any number of sounds below the threshold of "noisy," making the requirement too broad to be useful.
- **Negative framing:** It's difficult to design and test against a negative. It's more constructive to frame requirements positively, specifying what the car should do, such as "The car should operate at a noise level below 70 decibels under normal driving conditions."
- **Verification and validation:** Negative requirements are not conducive to the verification and validation process. Testing typically involves checking that the vehicle meets certain criteria, not that it avoids an undefined set of characteristics.

To address these issues, it's best to convert non-requirements into positive, testable requirements. Instead of saying what the car should not be, define what it should be. This approach simplifies the design process, testing, and quality assurance, ensuring that the vehicle meets the intended standards and customer expectations.

7.2.7.5 ISO 26262 Breakdown of Information

In its first edition, ISO 26262 proposed a structured information flow that outlines how the requirements, design, and test phases can be integrated into the product development process for system hardware and software. This standard also illustrates the relationship between the Functional Safety Concept (FSC) and the Technical Safety Concept (TSC).

The information flow provided by ISO 26262 helps to understand the dependencies among the various work products required by the standard. Essentially, it shows how each phase and its outputs are interconnected, ensuring that all safety-related aspects are addressed systematically (Figure 7.13). However, the diagram from the 2011 edition of the standard only references the standard itself. It does not consider the intended function of the system but focuses solely on the functions that are attributed to an Automotive Safety Integrity Level (ASIL). This means that the standard emphasizes the safety functions that have been assigned to a specific ASIL rating, rather than the broader intended functionality of the system.

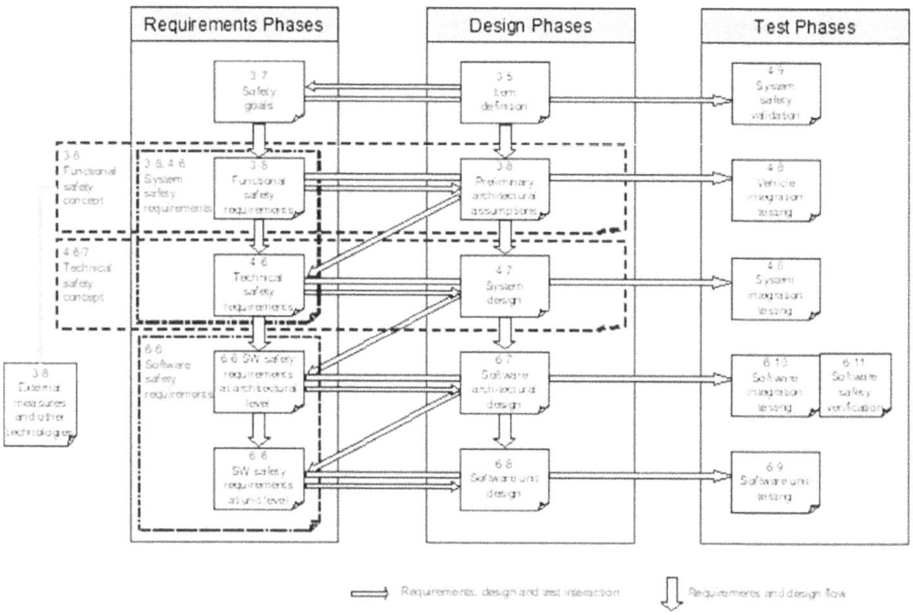

Figure 7.13 Information flow as proposed by ISO 26262:2011 (source: ISO 26262:2011, Part 10, Figure 8)

7.2.7.6 Full Matrix Breakdown

In the ISO 26262:2011 standard, there was no distinction made between architecture and design. Consequently, there was no reference to how design changes in the intended function would impact on the safety measures attributed with an Automotive

Safety Integrity Level (ASIL). Furthermore, the analysis and the impact of changes in the intended function were not considered.

A change in the intended function, if not properly modularized and without systematic barriers to separate functions, can lead to the need for a complete revision and review of the safety case. To address these issues, I introduced several phases in earlier books.

The following phases were considered (Figure 7.14):

- **requirement phase:** development of requirements
- **architecture phase:** modularization and allocation of requirements
- **analysis phase:** including change impact analysis
- **design phase:** documenting design decisions and technical feasibility after analysis
- **verification phase:** including design verification and checking for completeness, correctness, and consistency
- **integration phase:** planning for integration capability and additional requirements
- **validation phase:** validating the design and development goals against the respective baseline, and providing the basis for assessments, approvals and release, as well as supporting the safety case with evidence

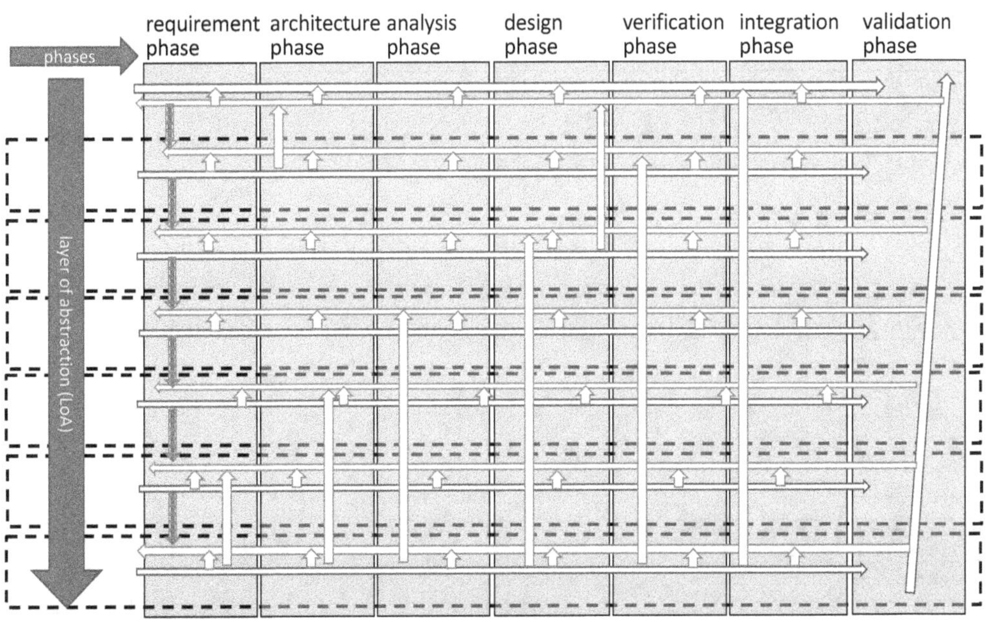

Figure 7.14 Matrix for systems safety engineering (source: Ross 2021)

The necessary levels of abstraction have been expanded to include the operational level and the various system and component levels. This allows the seven phases to be considered at each level of abstraction using appropriate methods and their respective implementations. This upward expansion aligns with the proposed extension of the ISO 26262 safety lifecycle to include an operational level and a vehicle level. At these levels, the corresponding phases can also be executed to ensure that architectural and design requirements are traceable, and that relevant analyses, verifications, and validations are thoroughly documented and conducted.

Specifically, for Safety in Use and the Safety of the Intended Functionality (SOTIF according to ISO 21448), the design and validation phases at the operational and vehicle levels are crucial additions to the safety case. These phases ensure that the design and validation processes are comprehensive and that all safety-related aspects are addressed systematically, providing a robust foundation for assessments and approvals.

Safety in Use and SOTIF require safety considerations at the operational and vehicle levels. Especially the homologation standards define several requirements on that level of abstraction, which need to be considered during the development process in all further phases. The matrix can then be adapted at these levels, allowing for the development of necessary safety concepts at each level of abstraction. This includes the operational safety concept, vehicle safety concept, and system safety concepts at various levels, down to the hardware safety concepts and the software safety concepts (Figure 7.15). This approach ensures that safety measures are consistently applied and managed across all levels of the system.

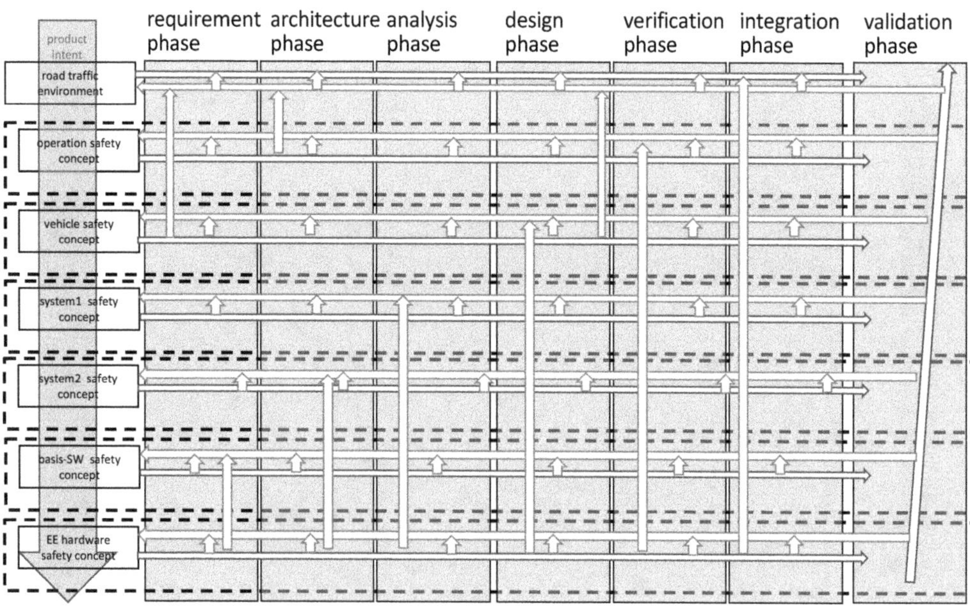

Figure 7.15 Product intent breakdown

7.2.7.7 Effect of Emergence

In such a matrix with phases and abstraction levels, the effects at the lower levels can also be traceably considered through a bottom-up analysis. Here, the effect of emergence is particularly important to note. Emergence refers to the phenomenon where a system exhibits properties or behaviors that its individual components do not possess on their own. These properties or behaviors arise only when the components interact within the system as a whole. Emergence is a key concept in understanding complex systems, as it highlights how new and unexpected properties can arise from the interactions of simpler elements. For example, a snowflake has a symmetrical and fractal pattern that is not found in the water molecules that form it.

Emergence of software is the concept that software systems can exhibit properties or behaviors that are not explicitly programmed or intended by the developers but emerge from the interactions of the software components and the environment.[5] Emergence of software can be seen as a form of self-organization, adaptation, or evolution of software systems.[6]

Examples of Emergence in Electronics Hardware and Software

Electronics Hardware

Neural networks: In hardware implementations of neural networks (e. g., hardware acceleration modules, or special hardware for AI modules), individual neurons (simple electronic components) interact to form a network capable of complex tasks like image recognition or language processing. The ability to recognize patterns is an emergent property that does not exist in individual neurons.

Microprocessors: A microprocessor consists of millions of transistors. While a single transistor can only perform basic operations, their collective interaction within the microprocessor enables complex computations and processing capabilities.

Software

Operating systems: An operating system (OS) manages hardware resources and provides services for application software. The emergent property here is the ability to run multiple applications simultaneously, manage memory, and handle input/output operations, which individual software components cannot achieve alone.

Distributed systems: In a distributed computing environment, multiple software agents work together to solve a problem. Emergent properties are the system's ability to perform tasks more efficiently and reliably than any single agent could, such as in cloud computing or blockchain networks.

---

5 *https://en.wikipedia.org/wiki/History_of_software*

6 *https://ecomputernotes.com/software-engineering/emergence-of-software-engineering*

Understanding and managing emergence is crucial for ensuring that complex systems function as intended, especially in safety-critical applications like automotive and aerospace industries.

Emergence has also well-known negative impacts, though. Here are some examples of the emergence of software:

- the emergence of bugs or errors that are not detected by testing or debugging but only manifest in certain situations or inputs
- the emergence of new features or functionalities that are not planned or designed but result from the combination or reuse of existing software modules or libraries
- the emergence of patterns or structures that are not specified or expected but arise from the dynamics or feedback of the software system
- the emergence of intelligence or creativity that are not inherent or predefined but emerge from the learning or problem-solving capabilities of the software system

---

The emergence of software can be beneficial or detrimental, depending on the context and the perspective of the stakeholders. The emergence of hardware and software can lead to innovation, diversity, robustness, or efficiency, but also to unpredictability, complexity, instability, or vulnerability.

The emergence of passive electronic parts at material interfaces is the phenomenon of new properties or behaviors arising from the interactions of passive electronic components, such as resistors, capacitors, and inductors, with the materials they are made of or connected to. For example, the capacitance of a capacitor may depend on the dielectric material, the temperature, the frequency, and the voltage applied to it.

Electromagnetic Compatibility (EMS) is the ability of electrical equipment and systems to function acceptably in their electromagnetic environment, by limiting the unintentional generation, propagation and reception of electromagnetic energy that may cause unwanted effects such as Electromagnetic Interference (EMI) or even physical damage to operational equipment.

The functionality of electronics can be affected by various factors, such as temperature, vibration, material, wear, aging, and physical stress. Some of the possible impacts are the following:

- **Temperature:** Temperature can affect the electrical resistance, capacitance, inductance, and power dissipation of electronic components. High or low temperatures can also cause thermal expansion or contraction, which can lead to mechanical stress, cracking, or deformation of the components. Temperature cycling can induce fatigue and failure of solder joints, wires, and connectors.
- **Vibration:** Vibration can cause mechanical stress, strain, or displacement of electronic components, which can alter their electrical properties or damage their

structure. Vibration can also induce noise, interference, or resonance in electronic signals. Vibration can be caused by external sources, such as engines, fans, speakers, or by internal sources, such as oscillators, relays, or switches.

- **Material:** Material can affect the electrical, thermal, mechanical, and chemical properties of electronic components. Different materials have different resistivity, conductivity, permittivity, permeability, coefficient of thermal expansion, modulus of elasticity, and corrosion resistance. Material selection can influence the performance, reliability, and durability of electronic components.
- **Wear:** Wear can refer to the wear and tear of electronic components due to friction, abrasion, erosion, or corrosion. Wear can reduce the functionality of electronic components by increasing their electrical resistance, decreasing their capacitance or inductance, altering their shape or size, or breaking their connections. Wear can be caused by mechanical, thermal, electrical, or chemical factors.
- **Aging:** Aging can refer to the gradual degradation or deterioration of electronic components over time. Aging can affect the functionality of electronic components by changing their electrical, thermal, mechanical, or chemical properties. Aging can be caused by various factors, such as temperature, humidity, voltage, current, radiation, or oxidation. Aging can result in reduced performance, increased failure rate, or shortened lifespan of electronic components.
- **Physical stress:** Physical stress can refer to the force or pressure applied to electronic components, which can cause deformation, strain, or fracture of the components. Physical stress can affect the functionality of electronic components by changing their electrical resistance, capacitance, inductance, or power dissipation. Physical stress can be caused by various factors, such as bending, twisting, stretching, compressing, or impacting of the components.

Physical effects such as temperature, vibrations, wear, aging, or other physical stress can significantly influence the behavior of electronics, particularly in high-frequency microcontrollers and microprocessors. These physical stresses can cause various emergent effects in electronic hardware (E-HW) and silicon, leading to unpredictable and often difficult-to-trace impacts on software and overall system behavior. Emergent effects occur when these physical stresses interact with the electronic hardware and silicon, creating new and often unexpected behaviors that were not present in the individual components. These emergent effects can propagate through the system, affecting software and overall system performance in unpredictable ways.

Understanding and mitigating these physical and emergent effects is crucial for ensuring the reliability and stability of electronic systems, especially in safety-critical applications.

7.2.8 Interface Breakdown

The hierarchical structuring of architectures are fundamental principles in systems engineering and design. This approach ensures that complex systems are broken down into manageable components, facilitating better understanding, implementation, and maintenance. The process involves decomposing functional, logical, and technical elements according to the axioms of Axiomatic Design. However, it is equally important to systematically break down interfaces, recognizing that physical, functional/logical, and technical interfaces have distinct characteristics and properties.

Functional, Logical, and Technical Elements

In the context of Axiomatic Design, the decomposition of functional, logical, and technical elements follows two primary axioms: the Independence Axiom and the Information Axiom. The Independence Axiom states that a design should maintain the independence of functional requirements, meaning that each function should be achieved without affecting others. The Information Axiom focuses on minimizing the information content of the design, aiming for simplicity and clarity.

Functional elements refer to the specific actions or operations that a system must perform. Logical elements represent the abstract, conceptual relationships between these functions, often depicted in models or diagrams. Technical elements are the tangible components and technologies that realize the functional and logical elements.

Importance of Interface Structuring

While decomposing these elements is crucial, it is equally important to systematically break down interfaces. Interfaces are the points of interaction between different components or systems, and play a critical role in ensuring seamless integration and operation. There are three main types of interfaces: physical, functional/logical, and technical.

Physical interfaces: These are the tangible connections between hardware components. They include mechanical joints, electrical connectors, and other physical linkages. Physical interfaces are often straightforward to identify and define, as they involve direct, observable interactions.

Functional/logical interfaces: These interfaces represent the interactions between different functions or logical elements. They are more abstract and can be challenging to define because they involve the flow of information, control signals, or other non-physical interactions. Functional and logical interfaces are dynamic and context-dependent, often requiring careful analysis to ensure they are correctly specified.

Technical interfaces: These interfaces involve the interaction between different technical components or systems. They include software APIs, communication protocols, and other technical linkages. Technical interfaces must be precisely defined to ensure compatibility and interoperability between different systems or components.

Characteristics and Properties of Interfaces

Each type of interface has unique characteristics and properties that must be considered during the design process. Physical interfaces are typically static and can be directly observed and measured. Functional and logical interfaces, on the other hand, are dynamic and context-dependent, often requiring simulation or modelling to fully understand their behavior. Technical interfaces must be rigorously specified to ensure that different systems or components can work together seamlessly.

Functional and logical interfaces, in particular, cannot be inherited in the same way as physical interfaces. They are often emergent properties of the interactions between physical or technical elements. This means that they must be carefully analyzed and specified to ensure that the overall system functions as intended.

The hierarchical structuring of architecture is a critical process in systems engineering and design. It involves decomposing functional, logical, and technical elements according to the principles of Axiomatic Design. Equally important is the systematic breakdown of interfaces, recognizing the distinct characteristics and properties of physical, functional/logical, and technical interfaces. By carefully analyzing and specifying these interfaces, designers can ensure that complex systems are integrated and operate seamlessly, achieving the desired functionality and performance.

In typical mobility solutions, we see different levels of interfaces (Figure 7.16). Unless we consider only static interfaces, we can see them very transparent when we disassemble our vehicle or if we see an airplane in a repair workshop. If you decompose the various system levels into components, a pure view on static interfaces could be observable and could be described by the observed characteristics. On the component level, you could identify hardware parts and SW units and specify their interfaces.

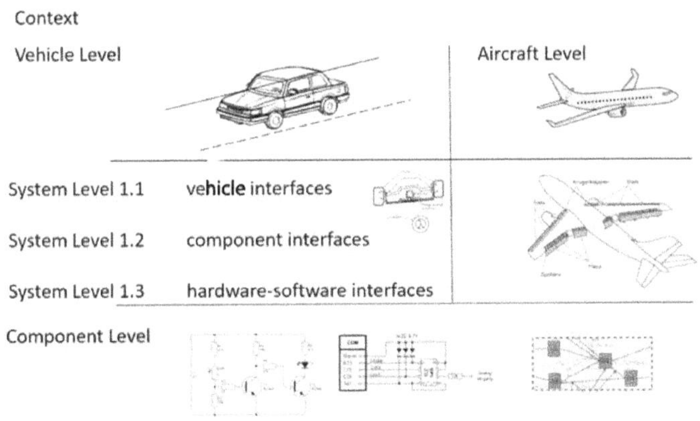

Figure 7.16 Hierarchy of mobility solutions

For a vehicle made of Lego bricks the interfaces are standardized (Figure 7.17). The interfaces of Lego bricks are characterized by their simplicity, generality, and compatibility. Each Lego brick has a standard size and shape, with studs on the top and tubes on the bottom. The studs and tubes allow the bricks to connect with each other in various ways, without requiring any additional tools or parts. The bricks also come in different colors and types, such as basic bricks, plates, tiles, slopes, etc. These features make Lego bricks easy to use, versatile, and adaptable for creating different models and structures. Even temperature and material impacts do not impact their compatibility.

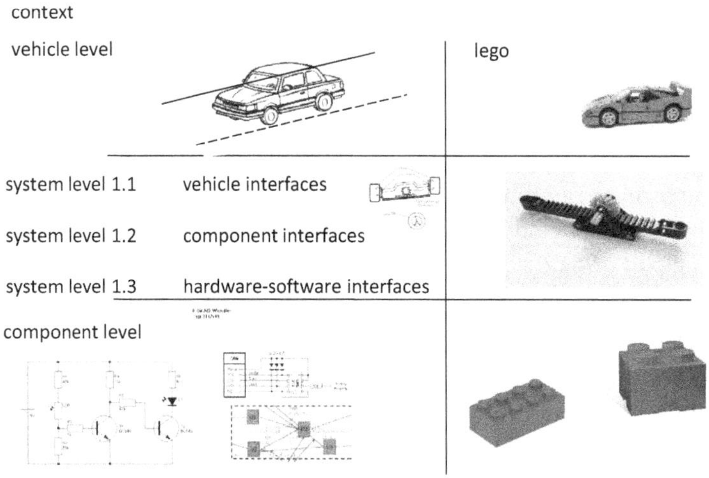

Figure 7.17 Vehicle interfaces versus Lego interfaces

To understand the effect and impact of emergences you see that the characteristics of single Lego bricks have no relation to the target of the intended Lego construction. Nobody would think of studs and tubes when you want to build a Ferrari out of your Lego bricks.

On the system level, a subsystem from the Lego Ferrari could be a steering system. This requires already a basic idea of the target Lego vehicle. If not, you would not be able to integrate the steering system into the target Lego vehicle.

7.2.9 Dynamic Interfaces

Lego bricks and dynamic SW interfaces are both examples of systems that can exhibit emergence. Lego bricks are physical components that can be assembled in various ways to create different structures and models. Dynamic SW interfaces are software elements that can interact with each other and with external inputs to produce different outputs and behaviors.

One way to compare Lego bricks and dynamic SW interfaces is to look at their levels of abstraction and modularity. Abstraction is the process of simplifying a system by hiding its details and focusing on its essential features. Modularity is the degree to which a system is composed of independent and interchangeable parts. Both abstraction and modularity can facilitate emergence by allowing for more flexibility and creativity in designing and building complex systems.

Lego bricks have a high level of abstraction and modularity. They are designed to be simple, generic, and compatible with each other. They can be used to create a variety of models, from buildings and vehicles to animals and characters. The models can have different functions and aesthetics, depending on the imagination and skill of the builder. The models can also be modified, disassembled, and reassembled easily, allowing for experimentation and evolution.

Dynamic SW interfaces have a lower level of abstraction and modularity than Lego bricks. They are more specific, complex, and dependent on each other. They can be used to create a variety of applications, from games and simulations to databases and web pages. The applications can have different inputs and outputs, depending on the requirements and design of the developer. The applications can also be modified, updated, and integrated with other software, but this may require more effort and expertise.

Effects of Dynamic Interfaces

Dynamic interfaces play a crucial role in the functionality and performance of complex systems, particularly in software-intensive environments. Unlike mechanical components or Lego bricks, where the type of interface and the degree of compatibility can be visually identified, software interfaces are more abstract and require careful architectural planning to ensure effective operation.

Visual Identification vs. Abstract Interfaces

In mechanical systems, interfaces are often straightforward to identify. The physical connections between components, such as joints or connectors, can be seen and assessed for compatibility. This visual identification simplifies the process of integration and troubleshooting.

In contrast, software interfaces are not as easily identifiable. Software is typically implemented in one location but exerts its effects through an effect chain of operations that ultimately influence actuators. This requires a well-designed architecture to ensure that the software interacts correctly with electronic hardware (E-Hardware) and mechanical components, resulting in the desired physical actions at the actuator level.

Dynamics of Software Interfaces

The dynamic nature of software interfaces means that their interactions are not static but evolve based on the system's state and inputs. These interfaces facilitate the flow of information and control signals between software modules, E-Hardware, and mechanical components. The effectiveness of these interactions depends on the architecture's ability to manage and coordinate these dynamic relationships.

Emergent Effects

Dynamic interfaces can lead to emergent effects, which are behaviors that arise from the interactions between system components. These effects can be both positive and negative. Positive emergent effects enhance system performance and functionality, while negative effects can disrupt the intended operation and degrade performance.

For example, in an aircraft system, the interaction between software controlling the thrust reversers and the mechanical actuators must be precisely coordinated. Any misalignment or delay in these interactions can lead to suboptimal performance or even safety hazards.

Robust Design for Resilience and Fault Tolerance

To manage the potential negative emergent effects, robust design principles must be applied. This involves creating resilient and fault-tolerant systems that can maintain desired behavior even in the presence of faults or unexpected conditions. Robust design ensures that the system can handle variations and disturbances without significant degradation in performance.

Resilience and fault tolerance are achieved through various strategies, such as redundancy, error detection and correction mechanisms, and adaptive control algorithms. These strategies help to mitigate the impact of negative emergent effects and ensure that the system continues to operate safely and effectively.

Functional and Logical Interfaces

Functional and logical interfaces are foundational to the behavior of software-intensive systems. These interfaces define how different software modules interact with each other and with the hardware components. They are dynamic and context-dependent, requiring careful specification and management to ensure correct operation.

Unlike physical interfaces, functional and logical interfaces cannot be inherited directly. They are emergent properties of the interactions between physical and technical elements. This means that their behavior must be analyzed and specified in detail to ensure that the overall system functions as intended.

Dynamic interfaces are a critical aspect of complex systems, particularly in software-intensive environments. Unlike mechanical interfaces, they require careful architectural planning and robust design to manage their dynamic nature and emergent effects. By ensuring resilience and fault tolerance, designers can create systems that maintain desired behavior and performance, even in the presence of faults or unexpected conditions. Functional and logical interfaces, in particular, are essential for defining the interactions within the system, forming the basis for effective and reliable operation.

7.3 Analysis Methods

In the automotive industry, a wide range of different analyses are applied. They are applied in different fields and with different analysis targets, such as the following:

- **Safety analysis:** This involves identifying and evaluating potential hazards that could harm the vehicle or its occupants. It's often used in the design and testing phases to ensure that safety-critical systems function correctly.
- **Risk analysis:** This involves identifying and assessing potential risks that could impact the vehicle or its operations. It's often used in the planning and decision-making processes to mitigate potential risks.
- **Formal analysis:** This involves using mathematical and logical methods to verify the correctness of system designs. It's often used in the design and testing phases to ensure that systems meet their specifications.
- **Informal analysis:** This involves using less rigorous methods, such as expert judgment and inspection, to evaluate system designs. It's often used when formal methods are not feasible or practical.
- **Design analysis:** This involves evaluating the design of the vehicle and its components to ensure they meet their intended functions. It's often used in the design and testing phases to optimize the design and performance of the vehicle.
- **Requirement analysis:** This involves identifying and defining the requirements of the vehicle and its systems. It's often used in the planning and design phases to ensure that the vehicle meets the needs of its users.
- **Stress analysis:** This involves evaluating the stress and strain on vehicle components under various loads and conditions. It's often used in the design and testing phases to ensure that components can withstand the expected loads.
- **Robustness analysis:** This involves evaluating the ability of the vehicle and its systems to maintain their performance under variations in inputs or conditions. It's often used in the design and testing phases to ensure that systems are robust and reliable.
- **Feasibility analysis:** This involves assessing the technical and economic viability of a proposed automotive project. It's often used in the early stages of a project to guide decision-making and planning.

Each of these analysis methods has its own strengths and weaknesses, and the choice of method may depend on the specific requirements and constraints of the automotive system being analyzed.

7.3.1 Risk Analysis

There are many different risk analysis methods for different fields of application. These are typical failure analyses:

- **Failure Modes and Effects Analysis (FMEA):** used in automotive, avionics, machinery, and space industries for analyzing potential reliability problems early in the development cycle
- **Fault Tree Analysis (FTA):** commonly used in the avionics and space industries to quantify system safety and reliability
- **Event Tree Analysis (ETA):** used in the avionics and space industries to analyse the effects of a failure
- **Root Cause Analysis (RCA):** identifies and eliminate root causes to solve problems

On a higher level, the hazards are in the focus of the analysis, such as in the following analyses:

- **Hazard Analysis and Risk Assessment (HARA)**: This is the basis for ISO 26262 and today it is often extended for hazards which derive from insufficiencies or inadequacies, e. g., in the intended function (e. g., Sotif-HARA).
- **Hazard and Operability Study (HAZOP):** used in the machinery industry to identify hazards in a system
- **Hazard Identification Study (HAZID):** used in the machinery industry to identify hazards in the early design stage
- **Preliminary Hazard Analysis (PHA):** used in the avionics industry to evaluate potential hazards

The risk is also matched with business impact, such as in the following analyses:

- **Risk Benefit Analysis (RBA) & Cost Benefit Analysis (CBA):** used in the machinery industry to weigh the pros and cons of an action
- **Business Impact Analysis (BIA):** used in the machinery industry to plan for operational disruptions

Design analysis methods are often used in various ranges of applications, such as the following:

- **PSPICE technology:** This is based on models of discrete electronic parts and provides simulations for stress, robustness, or failure of electronic hardware.
- **Multicriteria Decision-Making Methods (MCDM):** These methods are efficiently used as a support to the engineering design in automotive design. They help solve

engineering design problems related to the selection of materials and manufacturing processes in the detail design phase.

- **Formal and semi-formal verification techniques:** These techniques are applied to embedded, safety-critical code. Despite certain successes, the introduction of formal techniques on a larger scale has been challenging.
- **Artificial Intelligence (AI) models and methods:** AI models and methods dominate current research and automotive applications, with Machine Learning methods and neural networks being the most used.
- **AI-driven computer modelling:** This helps identify problems and play out scenarios for every aspect of a vehicle's design and performance.

Especially due to multiple new ideas of using artificial intelligence solutions, numerous new analysis methods are being used.

Many of the analysis methods are using quantitative models. Here are some examples to combine them:

- **Statistical analysis:** Statistical analysis is the foundation for transformation in the automotive industry. It leverages Artificial Intelligence (AI), Machine Learning (ML), and other advanced technologies to enhance vehicle performance, quality, safety, efficiency, and more.[7)]
- **Markov analysis:** Markov analysis is a state-based analysis technique based on a stochastic model. In the automotive industry, Markov Chain-based reliability analysis is used for automotive fail-operational systems.[8)] However, the use of Markov chains for failure analysis of complex safety-critical systems is limited due to exponential explosion of the size of the model.[9)]
- **Monte Carlo simulation:** The Monte Carlo simulation is a computational technique that uses random sampling to model and analyze complex systems or processes.[10)] In the automotive industry, Monte Carlo simulations are used to study the probability of confidence level during production under the PFMEA's Action Priority.[11)]

---

7 *https://www.bing.com/aclk?ld=e8wro0cdVEa3_ERAGafY892TVUCUwaUDS3BT6MFpzdanAJq3m6xot0n_b7GKL_XUuzH2Rg-SFHjUBXtY25RzEJG5Ho7nPhm789HKpHe-kSxLAmiMP0dQvTVZQugsDZYMclP5Z4DEMUVx-ca1Q-vpFdAfO6wasgUEZvOvLWPsxDOcYUwg7A&u=aHR0cHMlM2ElMmYlMmZ3d3cuY2Fzc2luaS5kZSUyZmxlaXN0dW5nZW4lMmZkYXRhLXNjaWVuY2UtYW5hbHl0aWNzLWFp&rlid=d9f801b1d9a81ff0513a67938bdec07c*

8 *https://www.jstor.org/stable/pdf/26169319.pdf*

9 *https://arxiv.org/pdf/2106.00247*

10 *https://arxiv.org/pdf/2206.11021*

11 *https://www.mdpi.com/2227-7390/10/15/2596*

7.3.1.1 Zwicky Box or Morphological Box Analysis for Road Traffic Scenarios

The Morphological Box, also known as the Zwicky Box, is a method used for multi-dimensional, non-quantified complex problems. Named after its inventor, Fritz Zwicky, it's a tool used in the field of product development, design, and problem-solving. In the context of road traffic scenarios, this method can be particularly useful in analyzing and predicting traffic patterns, and in developing effective traffic management strategies.

Methodology

The Morphological Box analysis involves identifying the parameters that define the problem space, and then systematically exploring all possible combinations of these parameters. Each parameter is represented as a row in the box, and each possible value that the parameter can take is represented as a column. The different combinations of these values represent different scenarios or solutions.

For instance, in a road traffic scenario, the parameters could include time of day, weather conditions, type of vehicles on the road, and road conditions. Each of these parameters can take several values. For example, the time of day could be morning, afternoon, evening, or night; weather conditions could be sunny, rainy, or snowy; and so on. By exploring all possible combinations of these parameters, we can analyze a wide range of traffic scenarios.

Real-World Application

Consider a scenario where a city's traffic management authority wants to optimize traffic flow during peak hours. Using the Morphological Box analysis, they could identify the key parameters affecting traffic flow during these hours, such as the number of vehicles, traffic light timings, and road conditions. By exploring all possible combinations of these parameters, they could identify potential solutions to improve traffic flow, such as adjusting traffic light timings or implementing congestion charges.

These are the benefits:

- **Comprehensive analysis:** The Morphological Box allows for a comprehensive exploration of all possible traffic scenarios, helping to identify novel solutions to traffic problems.
- **Structured approach:** It provides a structured approach to problem-solving, making it easier to understand and tackle complex problems.
- **Versatility:** The method is versatile and can be applied to a wide range of problems, not just traffic scenarios.

These are the weaknesses:

- **Time-consuming:** The method can be time-consuming, especially for problems with a large number of parameters and possible values.
- **Complexity:** While the method helps in dealing with complex problems, the analysis itself can become complex and difficult to manage with an increase in the number of parameters.
- **Subjectivity:** The effectiveness of the analysis depends on the accurate identification of relevant parameters and their possible values, which can be subjective.

The Morphological Box or Zwicky Box analysis can be a powerful tool for analyzing road traffic scenarios. However, like any method, it has its strengths and weaknesses, and its effectiveness largely depends on how well it is applied. Therefore, it should be used as part of a broader traffic analysis and management strategy, complemented by other data analysis and prediction tools.

7.3.1.2 Scientific Analysis Methods as a Basis for AI Solutions

Many known principles are also used in the context of AI methods and tools. Some examples are listed in the following.

Game Theory

Game Theory is a mathematical framework designed for understanding the behavior of players in strategic situations. It involves the study of mathematical models of strategic interaction among rational decision-makers. It has applications in all fields of social science, as well as in logic, systems science and computer science.

In the context of statistical analysis, Game Theory can be used to model and analyze various scenarios where entities (or "players") interact. Each player has a set of possible actions or "strategies" and their outcomes depend on the strategies chosen by all players. The aim is to determine the optimal strategies that maximize a player's expected payoff.

Game Theory has been showing its multi-dimensional applications in the last few decades. The application of Game Theory to Deep Learning includes another dimension in research. Game Theory helps to model or solve various Deep Learning-based problems.

The theory is being used in adversary training in GANs, multi-agent systems, and imitation and Reinforcement Learning. In the case of perfect information and symmetric games, many Machine Learning and Deep Learning techniques are applicable.

Toy Theory

The distinction between toys and games is built into grammar itself: One plays games but plays with toys. Although some thinkers have recognized the importance of distinction, their insights are often contradictory and vague, and the word toy is used unsystematically to refer to a wide range of objects and associated play-activities.

Toy Theory could be interpreted to support a dichotomy of playability and usability, saying that playability is related only to a game's "capability to provide enjoyment for a player over an extended period of time."

Since measurable factors of gameplay are being discovered for objective quantitative measurement, these need to be incorporated in future playability definitions. The assumption is made in these games that all players will act as rational decision-makers, that is, that they will act to maximize their own return.

Game Theory and Toy Theory have their unique applications and potential for statistical analysis. They can be used in conjunction with AI and ML techniques to solve complex problems and enhance usability concepts. Many theories could be applied to bring it in the context of AI solutions. Hereafter, some aspects are listed, as far as the methods could be applied in the context of safety. The application of Game Theory and Toy Theory as a framework for Machine Learning-based analysis in mobility solutions presents a promising approach to enhance interoperability with human users.

Game Theory, a mathematical model of strategic interaction among rational decision-makers, can be a powerful tool in the realm of mobility solutions. It can model the interactions between different agents (vehicles, pedestrians, cyclists, etc.) in a transportation network as a game, where each agent's strategy affects the others. This can help in predicting and optimizing traffic flow, reducing congestion, and improving overall efficiency.

For instance, consider a scenario where two vehicles are approaching an intersection. Using Game Theory, each vehicle can be modeled as a player in a game, where the strategies could be to either continue at the same speed, slow down, or stop. The payoff for each player would be determined by factors such as safety, time taken to cross the intersection, fuel efficiency, etc. An ML model can learn these games' outcomes over time and make optimal decisions that benefit all players.

On the other hand, Toy Theory, which involves the use of simplified models to understand complex systems, can be particularly useful in ML for mobility solutions. It can help in creating simplified models of the transportation network, which can be used to train ML algorithms more efficiently. These "toy" models can capture the essential features of the system without the complexity, making them ideal for initial stages of ML model development and testing.

Moreover, Toy Theory can also aid in making ML models more interpretable and thus more interoperable with human users. By using simplified models, it becomes easier for human users to understand how the ML model makes its decisions, increasing trust and acceptance. The use of Game Theory and Toy Theory in ML for mobility solutions can offer significant benefits. Game Theory can help model and optimize interactions between different agents in the transportation network, while Toy Theory can simplify complex systems for efficient ML model training and improved interpretability. Both methods can enhance the interoperability with human users, leading to more effective and user-friendly mobility solutions. However, it's important to note that these are theoretical assumptions, and their practical implementation would require rigorous testing and validation.

For safety applications, both methods are not a specific solution for the implementation of algorithms or even safety mechanisms. They could be a strong tool to help humans in the role of a developer or assessor to understand operational dependencies and help them to communicate complex solutions to other members of the development team.

7.3.1.3 Analysis of the Operational Design Domain (ODD)

The Operational Design Domain (ODD) is a fundamental concept in the development of automated driving solutions. It refers to the specific conditions under which an automated driving system is designed to function. These conditions can include geographical location, weather conditions, time of day, traffic scenarios, and road types, among others. Understanding and accurately defining the ODD is crucial for the safe and efficient operation of automated driving systems. It helps in setting the boundaries within which the system can operate safely and reliably. Moreover, it aids in the development of fallback strategies for scenarios that fall outside the defined ODD.

Empirical Morphological Boxes and Advanced AI Solutions

To analyze the ODD, a combination of empirical morphological boxes and advanced AI solutions can be employed. Empirical Morphological Boxes, a method used to explore all possible solutions in a complex problem space, can be used to identify and categorize the different parameters that define the ODD. For instance, parameters such as road type, weather conditions, and traffic density can be systematically explored using this method.

On the other hand, advanced AI solutions can be used to model and predict the behavior of the Automated Driving System within the defined ODD. Machine Learning (ML) algorithms can be trained on data collected from various sensors and systems of the vehicle to learn the patterns and make accurate predictions.

Game Theory and Toy Theory for Model Validation

Game Theory and Toy Theory can be used for validating the models developed using the above methods. Game Theory, which involves mathematical models of strategic interaction, can be used to simulate different traffic scenarios within the defined ODD. The performance of the Automated Driving System in these simulated scenarios can provide valuable insights into its reliability and safety.

Toy Theory, on the other hand, involves the use of simplified models to understand complex systems. It can be used to create simplified versions of traffic scenarios, which can be used for initial stages of model testing and validation. These "toy" models can capture the essential features of the system without the complexity, making them ideal for initial testing and validation.

The analysis of the Operational Design Domain is crucial for the development of safe and efficient automated driving solutions. A combination of empirical Morphological Boxes and advanced AI solutions, validated by using Game Theory and Toy Theory, can provide a comprehensive approach to defining and analyzing the ODD. However, it's important to note that the effectiveness of these methods depends on the quality and availability of data, as well as the complexity of the traffic scenarios being analyzed. Therefore, continuous monitoring, validation and updating of the ODD and the models are necessary to ensure the safe operation of automated driving systems.

Comparative Analysis of Empirical and AI/ML Methods for Operational Design Domain (ODD) Scenarios

The Operational Design Domain (ODD) is a critical concept in the development of Automated Driving Systems (ADS). It defines the specific conditions under which an ADS is designed to function. Analyzing ODD scenarios is a complex task that can be approached using empirical methods like Morphological Boxes or AI- and Machine Learning-based techniques. Both methods have their pros and cons, and their effectiveness can vary depending on the specific real-world scenario.

Empirical Methods: Morphological Boxes

Morphological Box analysis is an empirical method that involves identifying and categorizing different parameters that define the problem space. In the context of ODD scenarios, these parameters could include factors like road type, weather conditions, traffic density, etc.

These are the pros for the ODD analysis:

- **Comprehensive:** Morphological Box analysis allows for a comprehensive exploration of all possible traffic scenarios.
- **Structured:** It provides a structured approach to problem-solving, making it easier to understand and tackle complex problems.

These are the cons for the ODD analysis:

- **Time-consuming:** The method can be time-consuming, especially for problems with a large number of parameters and possible values.
- **Subjectivity:** The effectiveness of the analysis depends on the accurate identification of relevant parameters and their possible values, which can be subjective.

AI/ML Based Techniques

AI -and ML-based techniques involve using algorithms to learn patterns from data and make predictions. In the context of ODD scenarios, these techniques could be used to model and predict the behavior of the ADS within the defined ODD.

These are pros for the ODD analysis:

- **Scalable:** AI/ML techniques can handle large amounts of data and complex scenarios.
- **Predictive:** These techniques can make accurate predictions about future traffic scenarios based on past data.

These are the cons for the ODD analysis:

- **Data dependency:** The effectiveness of AI/ML techniques largely depends on the quality and availability of data.
- **Complexity:** These techniques can be complex and require a high level of expertise to implement and interpret.

Real-World Scenarios and Validation

In real-world scenarios, a combination of both methods can often yield the best results. Empirical methods can be used to define the problem space and identify potential solutions, while AI/ML techniques can be used to validate these solutions and make predictions about future scenarios.

However, the validation of these methods in real-world scenarios can be challenging. It requires rigorous testing and continuous monitoring to ensure that the ADS operates safely and effectively within the defined ODD. Moreover, the breakdown of design requirements for ADS is a complex task that requires a deep understanding of both the technical aspects of ADS and the specific traffic scenarios it will encounter.

Proposal to Combine Both Approaches

A promising approach to analyzing ODD scenarios could be to combine empirical methods and AI/ML techniques. Empirical methods like the Morphological Box analysis could be used to set up the scenario database. This would involve identifying and categorizing the different parameters that define the ODD and systematically exploring all possible combinations of these parameters to create a comprehensive database of potential traffic scenarios.

Once the scenario database is set up, AI and ML solutions could be used for simulation and validation. ML algorithms could be trained on the data from the scenario database to learn the patterns and make accurate predictions about future traffic scenarios. These predictions could then be validated by human observers, who could provide feedback to further refine the ML models.

This combined approach could offer several benefits. It would allow for a comprehensive exploration of all possible traffic scenarios, while also leveraging the predictive power of AI and ML techniques. Moreover, it would provide a structured approach to problem-solving, making it easier to understand and tackle complex problems. Finally, it would allow for continuous monitoring and updating of the ODD and the models, ensuring the safe operation of ADS.

7.3.2 Failure Analysis

ISO 26262 considers the following in their requirements to safety analysis approaches:

- In a **top-down approach (deductive analysis)**, the analysis starts at a high level and works down to the details. This is often associated with deductive analysis, which starts with a theory or hypothesis and tests it with observations. It's a kind of "top-down" approach to data analysis.[12] This approach is often used when there is a lot of existing knowledge or theory about the subject.[13]
- In a **bottom-up approach (inductive analysis)**, the analysis starts with specific observations or data and works its way up to general conclusions or theories.[14] This is often associated with inductive analysis, which starts with observations and uses them to derive a theory.[15] It's a kind of "bottom-up" approach to data analysis.[16] This approach is often used when there is little to no existing literature on a topic.[17]

In the context of safety analysis, both approaches can be valuable depending on the situation. A top-down approach might start with a broad view of overall system safety and work down to the details of individual components or events. A bottom-up approach might start with an analysis of individual failures or hazards and work up to an understanding of how they impact overall system safety.

ISO 26262 requires for all ASIL-classified functions a bottom-up approach and for ASIL C and ASIL D functions an additional top-down approach.

[12] *https://link.springer.com/content/pdf/10.1007/s10669-014-9488-3.pdf*

[13] *https://www.scribbr.com/methodology/inductive-deductive-reasoning*

[14] *https://link.springer.com/content/pdf/10.1007/s10669-014-9488-3.pdf*

[15] *https://www.andreajbingham.com/resources-tips-and-tricks/deductive-and-inductive-approaches-to-qualitative-analysis*

[16] *https://link.springer.com/content/pdf/10.1007/s10669-014-9488-3.pdf*

[17] *https://www.scribbr.com/methodology/inductive-deductive-reasoning*

7.3.2.1 Basic Bottom-Up Approach

In the realm of safety analysis, the bottom-up approach, applied as inductive analysis, plays a crucial role. This approach begins at the granular level, examining individual components or events, and works its way up to an understanding of how they impact the overall system safety. The bottom-up approach is particularly useful when dealing with complex systems where the interaction of individual components can lead to emergent behaviors that are not immediately obvious. By starting with a detailed analysis of individual failures or hazards, we can build a comprehensive understanding of the system's overall safety profile. This method is often employed in Failure Modes and Effects Analysis (FMEA), where each component's failure modes are analyzed for their effects on the system. The bottom-up approach's strength lies in its ability to uncover seemingly minor issues that can cascade or interact, leading to significant system-level effects.

A classical and deeper view into methodologies is a hardware safety analysis.

ISO 26262 did consider in their examples the typical **fault—error—failure propagation** approach. In an FMEA approach you could convert it to **failure cause—failure—failure effect** (Figure 7.18). Since the name bottom-up could be always mapped to a propagation from a lower level of abstraction into a higher level of abstraction, most of the analysis is based on these principles.

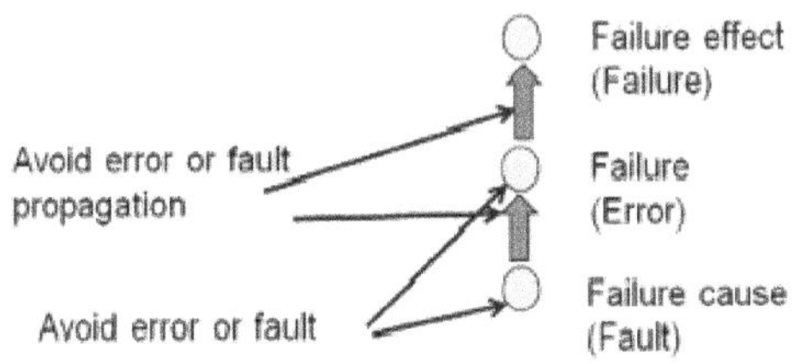

Figure 7.18
Classical bottom-up failure propagation

Figure 7.19 shows a typical example for a classical bottom-up analysis. The following steps are applied:

- Determine all possible failure conditions on a specific level of abstraction.
- Consider a functional and a physical model from the lower level of abstraction to the higher level of abstraction.
- Evaluate for any lower level of failure condition their possibility to propagate to the higher level of abstraction.

For a typical hazard analysis, the failure effect could be analyzed to their potential propagation to hazardous scenarios, or potential hazards or danger.

In hazard analyses and safety analyses such as the FMEA, metrics about reliability, probabilities, occurrence, detectability, severity, extent of hazard, or scaling about likelihood, controllability etc. are assigned systematically in order to achieve quantitative approaches.

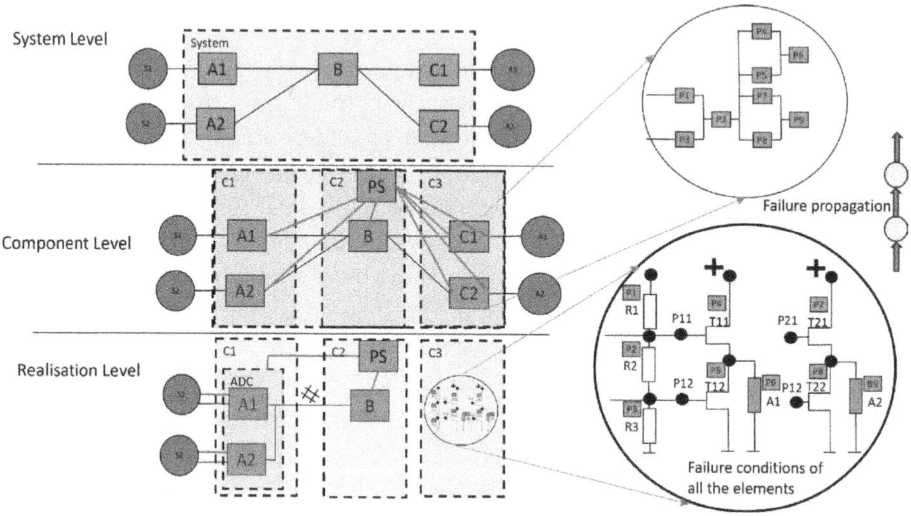

Figure 7.19 HW failure propagation

7.3.2.2 Deductive Top-down Approach

Deductive analysis asa top-down approach is a dissimilar alternative to the bottom-up inductive analysis.

Methods of Deductive Analysis

Several methods can be employed for deductive analysis in EE systems. One such method is Logical Reasoning, which involves applying general principles to specific instances. Another method is Mathematical Proof, which uses mathematical techniques and formulas to prove or disprove a hypothesis.

Advantages of Deductive Analysis for EE Systems

Deductive analysis offers several advantages for EE systems. Firstly, it allows for the prediction of outcomes. By starting with a general principle and applying it to specific instances, engineers can predict how a system will behave under certain conditions. Secondly, deductive analysis can lead to a better understanding of the system. By breaking down a complex system into its constituent parts, engineers can gain a deeper understanding of how the system works as a whole.

Inductive Analysis and the Emergence Effect

While deductive analysis offers numerous benefits, inductive analysis also has its advantages. Inductive analysis considers the emergence effect, which refers to the phenomenon where the whole is greater than the sum of its parts. This effect is particularly relevant in EE systems, where the interaction between components can lead to emergent properties that would not be apparent when considering the components in isolation.

Inductive analysis also allows for the consideration of hidden interfaces. These are interfaces that may not be immediately apparent but can have a significant impact on the system's performance. By forming a hypothesis and testing it, engineers can identify and address these hidden interfaces, leading to a more robust and reliable system.

Deductive Top-down Approach for Requirements Development

The deductive top-down approach is beneficial for requirements development because it allows for a systematic and structured development process. Starting from a high-level system requirement, the system is broken down into subsystems and components. This approach ensures that all requirements are derived from the top-level system requirement, ensuring consistency and traceability. It also allows for early identification of interfaces and dependencies between components, which can be crucial for the system's overall performance and reliability. This early identification aids in the development of a comprehensive and cohesive system, where all components are well integrated and function harmoniously.

Inductive Bottom-up Approach for Requirements Verification

On the other hand, the inductive bottom-up approach is beneficial for requirements verification. This approach involves verifying individual components or subsystems and then integrating them to verify the system as a whole. This approach allows for early detection of issues at the component or subsystem level, reducing the risk of system-level failures. It also allows for more thorough testing, as each component or subsystem can be tested in isolation before integration.

7.3.3 Failure Analysis in the Automotive Industry

In the automotive industry, both deductive and inductive analysis methods are used for failure analysis:

- **Fault Tree Analysis (FTA):** FTA is a deductive failure analysis method. It starts with a top-level event (such as a system failure) and breaks it down into its contributing events using a tree-like structure. This method allows for the identification of all possible failure paths leading to the top-level event, helping engineers design systems that are robust against these failures.
- **Failure Modes and Effects Analysis (FMEA):** FMEA is an inductive failure analysis method. It starts with individual components or sub-systems and identifies all possible failure modes, their effects on the system, and their causes. This method

allows for the identification of critical components and failure modes, helping engineers design systems that are robust against these failures and develop appropriate mitigation strategies.

Both deductive and inductive analysis offer unique advantages in the development of EE systems. While deductive analysis allows for the systematic specification of interfaces and the prediction of outcomes, inductive analysis considers the emergence effect and the ability to identify hidden interfaces. By employing both methods, engineers can develop robust, reliable, and efficient EE systems.

These are more elaborated deductive analyses:

- **Hypothesis testing in statistics:** In statistics, deductive reasoning is often used in the form of hypothesis testing. A general hypothesis is formed, and specific data is then collected to test the hypothesis. If the data supports the hypothesis, it is accepted; otherwise, it is rejected.
- **Physics laws:** In physics, laws such as Newton's laws of motion or the laws of thermodynamics are general principles derived from a large number of specific observations. These laws can then be applied to specific systems to predict their behavior.
- **Circuit analysis:** In electrical engineering, circuit analysis often involves applying Ohm's law and Kirchhoff's laws (general principles) to specific circuits to determine the current, voltage, and power in various parts of the circuit.
- **Algorithm design:** In computer science, algorithms are often designed using deductive reasoning. The general problem is broken down into smaller, more manageable problems, and solutions to these smaller problems are then combined to solve the original problem.
- **Database queries:** In databases, SQL queries often use deductive reasoning. A general query is written, and the database management system applies this query to the specific data in the database to return the requested information.

7.3.3.1 Application of Deductive and Inductive Analysis

Inductive and deductive analyses are two fundamental approaches used in various fields, including the automotive industry, to ensure safety and reliability. These methods are particularly crucial when complying with standards like ISO 26262, which is an international standard for functional safety in road vehicles. The following example mixed inductive and deductive analysis methods. ISO 26262 recommends the use of both inductive and deductive analyses for different aspects of the safety lifecycle. For instance, deductive analysis can be used in the hazard analysis and risk assessment phase to identify potential hazards based on the intended functionality of the system. Inductive analysis, on the other hand, can be used in the functional safety concept phase to derive safety goals based on the identified hazards.

One of the challenges in complying with standards like ISO 26262 is defining different work products for both inductive and deductive analyses. Each analysis requires different types of data, tools, and techniques, and the results need to be documented in different formats. Moreover, these work products need to be independently verified and validated to ensure their correctness and completeness.

For deductive analysis, the work products might include system requirements, design specifications, and verification plans. For inductive analysis, the work products might include test cases, test results, and validation reports.

The planning of tests and the development of test cases would be recommended to be aligned with the development of requirements and follows a deductive approach, since various test methods like

- fault injection tests or
- over-limit tests

are more considered as an inductive analysis since you could develop hypotheses from the observed test result for the entire system under development.

In aviation industry, particular test methods are recommended for integration tests such as the following:

Highly Accelerated Life Testing (HALT) is a more inductive method used during the product design and integration phase. In HALT, the product is subjected to progressively harsher conditions until it fails. The goal is not to simulate the product's life cycle but to uncover its weaknesses. The information gathered is then used to make improvements, enhancing the product's reliability and robustness. This process is inductive because it starts with specific observations (the product's response to stress) and uses them to derive a general conclusion (the product's limits).

Highly Accelerated Stress Screening (HASS) is a more deductive method used during the production and integration phase. It involves testing the finished products under accelerated stress conditions to detect any defects or failures before they reach the customer. The stress levels used in HASS are derived from the HALT results but are usually less severe to prevent over-stressing the product. This process is deductive because it starts with a general principle (the product's robustness determined by HALT) and applies it to specific instances (each product unit). In HASS you are using hypothesis about the stress behavior of the products based on the experienced observation in HALT.

Top-down Requirements Development

For a systematic breakdown of requirements, an architectural structure shall be drafted, so that the considered elements of their interfaces could be identified. A typical function has by nature no limits and interfaces, the limits and interfaces of a function need to be defined and specified.

In Figure 7.20, we consider two requirements A1 and A2 as an input for two derived functions. The two functions shall be assigned to four subelements with defined interfaces between the four elements. The elements could be functional of technical elements. Any physical effects like heat or noise should be neglected.

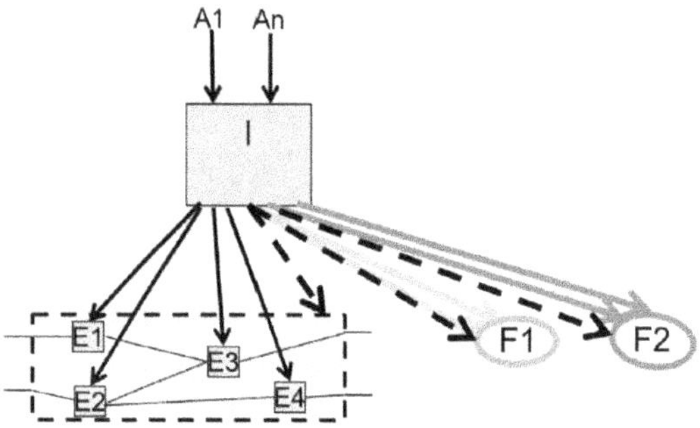

Figure 7.20
Breakdown of requirements

In this example we could consider the following allocation (Figure 7.21):

- F2 shall be allocated to E1 and E3.
- F1 shall be allocated to E2, E3, and E4.
- E3 is a common element of F1 and F2.

Due to this allocation, the possible propagation of failure will be influenced. In case of another allocation, the fault propagation could be different.

Considering any noise, heat, other technical needs, or physical impacts are neglected in the example.

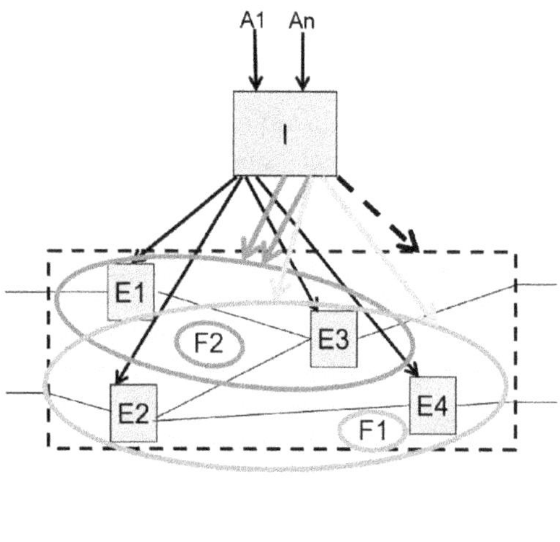

Figure 7.21
Allocation of requirements

In this example we would consider E2, E3 and E4 as component 1 and we would consider that E1 is another component. Maybe you would consider a camera system as component 2, you would add a radar to improve the object detection, and you would implement the sensor fusion into the camera (Figure 7.22).

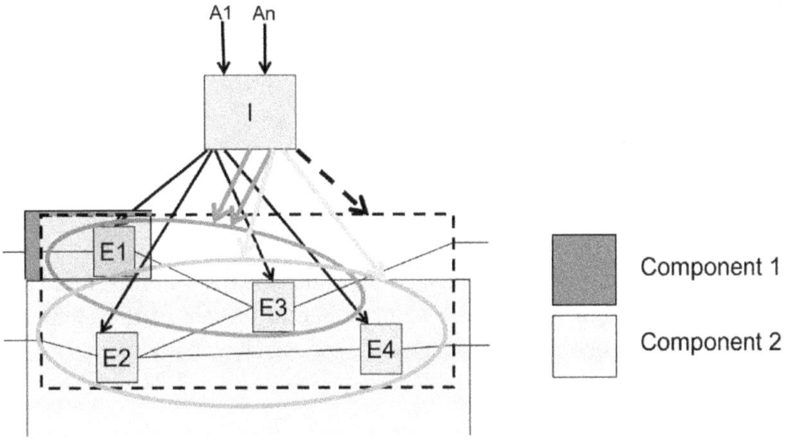

Figure 7.22 Introducing component interfaces

In case that you introduce two failure conditions like

- too high and
- too low

to the function and their characteristics, any of the characteristics have the potential to be too high or too low. It leads to a multiplication of the characteristics of the function with the number of considered failure conditions of the function (Figure 7.24).

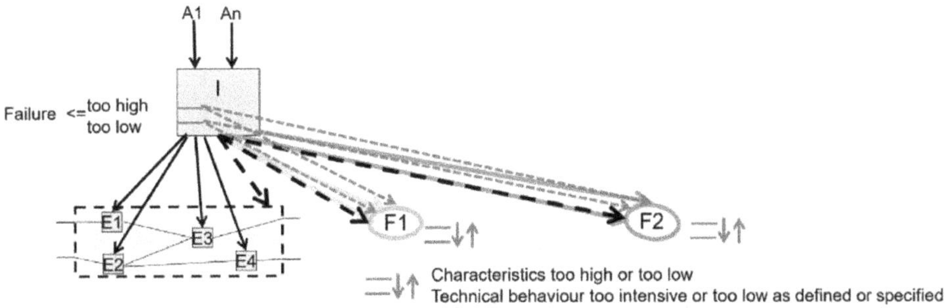

Figure 7.23 Failure conditions

In case of considering functional elements implemented in different technical components an additional increase of interfaces and failure conditions occurs (Figure 7.24). These are the technical interfaces:

- environment and technical elements
- between technical elements

These are the functional interfaces:

- environment and logical elements
- between logical elements
- between technical elements
- between logical and technical elements

In addition to functional and technical interfaces not only the desired interfaces are potential causes of failure, but also all the environmental impact of the elements could not be neglected anymore since EE parts need, e. g., electrical energy.

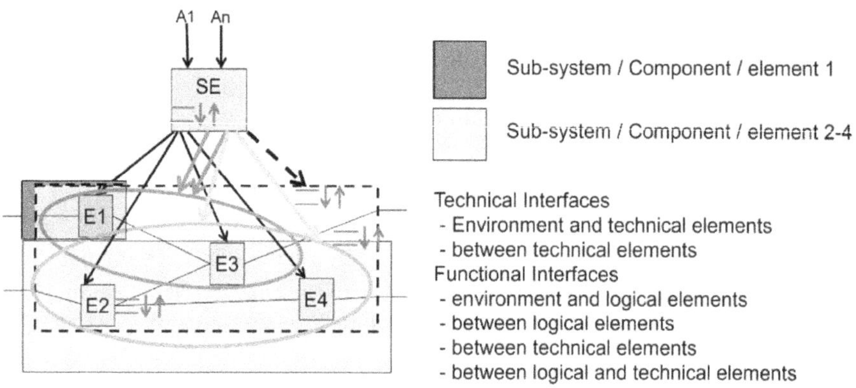

Figure 7.24 Failure conditions for technical and functional interfaces

Failure Analysis versus safety analysis

In order to perform a complete failure analysis, a lot of information needs to be collected just to initiate a basic analysis. In order to transform a failure analysis to a safety analysis, the measures during development and during runtime of the product need also to be considered. The safety could be only assessed if the effectiveness of the measures during development and measures during runtime of the product are documented by adequate evidence.

ISO 26262 provides in Chapter 9 sets of requirements for the additional supplements for a safety analysis to comply with the standard.

Such an inductive approach also needs to consider other failure propagation approaches, such as the following:

- horizontal failure propagation, e. g., from sensor to processing to output
- failure impact due to failure in the feedback or readback with close loop control systems (Figure 7.25)
- typical common cause effect derived from outer effects like heat or noise on the function or any other physical effect due to the spatial environment of the considered elements
- emergence effects due to hidden interfaces which are not considered in the propagation model.

In case of an inductive analysis, the pure failure propagation from the lower level of abstraction to the higher level of abstraction is not sufficient. It is very important, that the model also includes all the functional effects, and the potential parasitic effect.

In case of unauthoritative data access, e. g., hacker attacks, of course also human intelligence could be considered to manipulate the function.

The use of P-Diagrams (Parameter Diagrams) in the systematic analysis of technical systems, such as electronic hardware, is a well-established practice in engineering. P-Diagrams provide a structured framework for identifying and understanding the interactions between various components of a system, including inputs, outputs, and potential sources of variation.

In the context of electronic hardware, P-Diagrams can be used to map out the functional relationships between different components and subsystems. This can help engineers identify critical parameters that affect system performance, as well as potential failure modes.

Moreover, the integration of P-Diagrams with simulation tools like PSPICE can further enhance the analysis process. PSPICE is a popular tool for simulating and analyzing electronic circuits. By using PSPICE, engineers can model the behavior of their hardware under a variety of conditions and inputs, as defined by the P-Diagram.

This combination of P-Diagrams and PSPICE simulations allows for a more comprehensive and accurate analysis of electronic hardware. It enables engineers to validate their designs, optimize performance, and mitigate potential risks, thereby contributing to the development of safer and more reliable hardware systems.

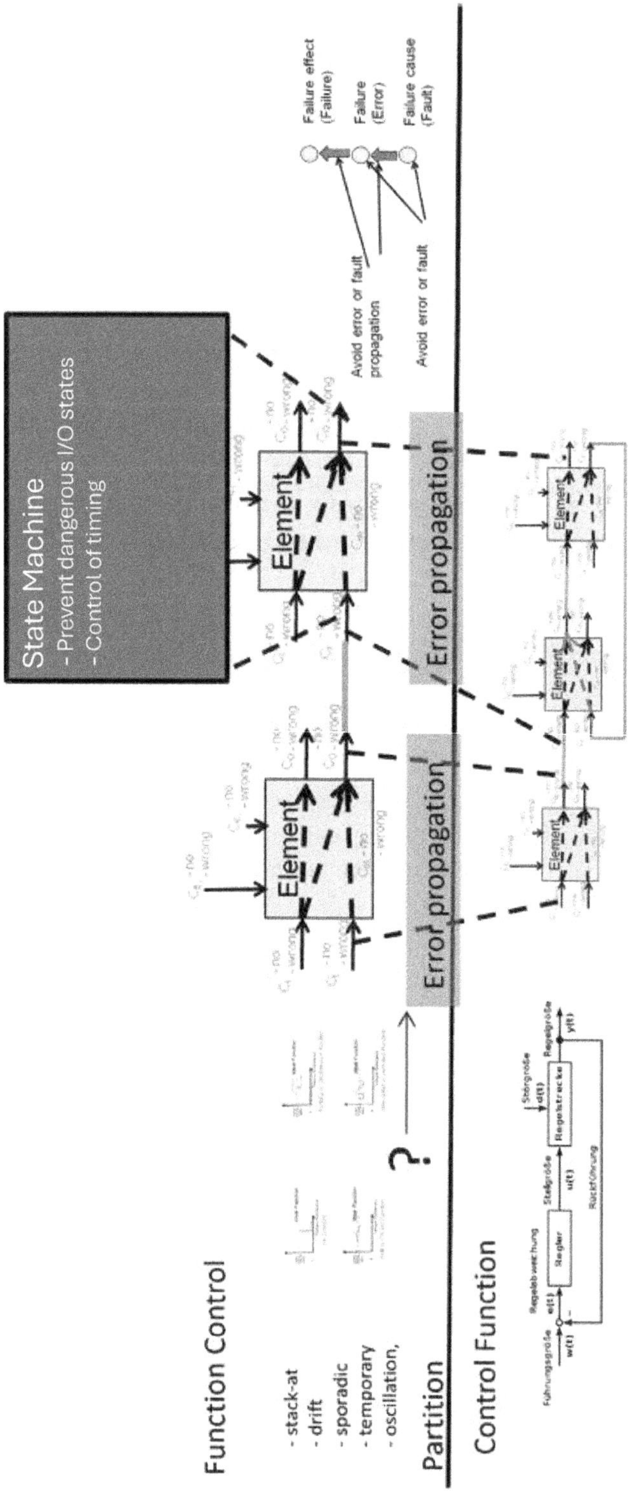

Figure 7.25
Failure propagation in close loop control circuits

7.3.3.2 Software Safety Analysis

A software safety analysis could be also considered by means of an inductive analysis.

Of course, you could consider any line of code or a C function as a potential source of failure, but you can imagine that this leads to a very heterogeneous complex analysis. Typical in the lower level, you apply software unit tests and statical code analysis, so that you can detect the further failure during integration of the software into higher level of abstraction.

Due to the transparency of the models of the SW units or any higher level of abstraction SW elements, it is recommended to develop a model, considering measurable software interfaces, so that the demanded correctness or potential failure could be detected during tests (Figure 7.26):

- Most of the SW errors are systematic failures, which means they derive from failures of the engineers and the programmers or the applied tools.
- Other important sources of failure derive from the hardware of the microcontroller, or any other physical effects form the computing device.
- Even in microprocessors with virtual memory management, stochastic failures not only derive from HW failure but also derive from interrupts scheduling issues or other timing impacts.

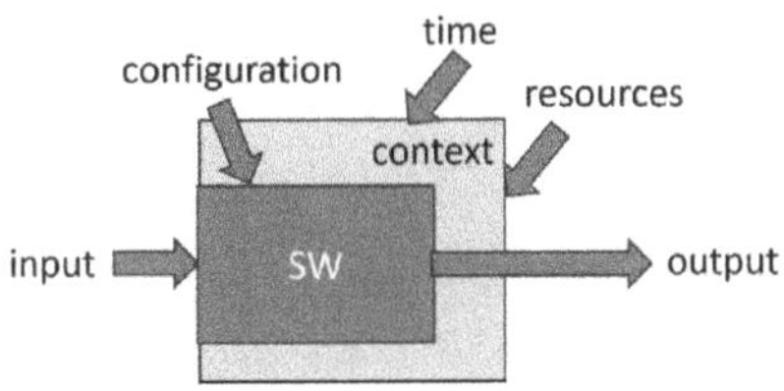

Figure 7.26
Basic SW model

For all the possible failure impact and their potential propagation to a higher level of abstraction you could develop a model and an inductive hypothesis how and with what probability of failure could propagate within the considered system (Figure 7.27).

Based on such a model you also could consider

- hardware impact,
- timing issues,
- interface inconsistency, or
- scheduling, interrupt and sequencing issues etc.

as a potential cause of failure.

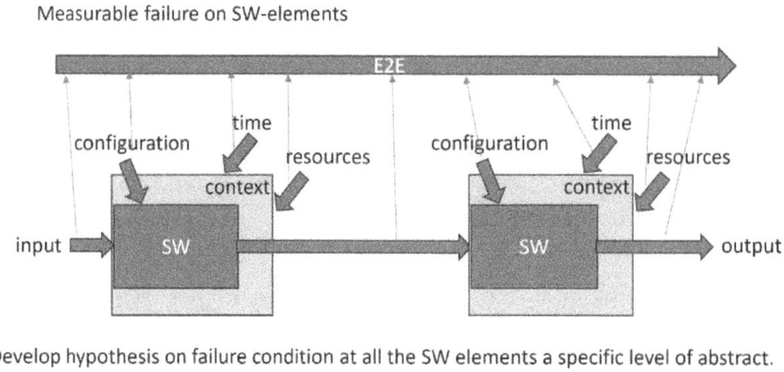

Figure 7.27 SW failure model

7.3.4 Signal Chain Analysis

The topic of emergence with its hidden interfaces makes a multi-level analysis necessary. In particular, to assess functional safety, it is important to understand how the elements that are necessary to implement the intended function relate to one another in the system context and in the network or interconnectivity with their internal and external interfaces. For a systematic analysis of the signal chains, a top-down approach must also be chosen so that the interaction of the sensors and the actuators is analyzed individually.

7.3.4.1 Functional Signal Chain

A signal chain analysis starts with a hypothesis about how the intended function affects the end-to-end (E2E) function to the actuator (Figure 7.28). This hypothesis is based on the system's architecture and design and the known properties of its components. For example, in an audio system, the hypothesis might be that increasing the input signal amplitude will increase the loudspeaker's sound level.

The first approach is to develop a functional model that implements the consolidated stakeholder requirements in an observable functional signal chain based on the intended E2E functions. Important is, that the highest-level function is defined in a functional and logical structure, which shows the physical stimuli from events outside of the vehicle or the driver's physical impact up to the physical behavior of the vehicle.

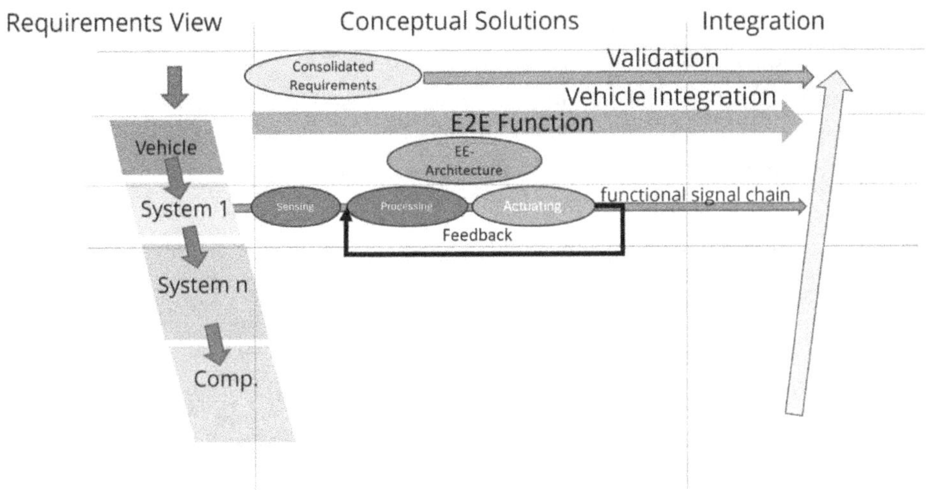

Figure 7.28 Top-down development of a functional signal chain

Such a functional signal chain is based on the following:

- sensing elements as the receiving element for physical effects, which are into for the intended E2E function
- processing units that convert the input to the intended output and control the actuation for the intended E2E functionality
- an actuating device that converts the output into the intended observable behavior of the E2E function
- In case of a close loop concept, feedback elements have to be considered that monitor the initiated actuator behavior and provide input to the processing unit that corrects the output in a way that the intended behavior of the E2E function meets the given requirements

The system is then analyzed from the input to the output, applying the known properties of each component to the signal. This is a top-down approach (Figure 7.28), starting with a general principle (the hypothesis) and applying it to specific instances (the signal at each stage of the chain). It could be considered as a deductive part of the analysis.

The inductive part of the analysis comes in when the actual behavior of the system is observed and compared to the predictions made by the deductive analysis. For example, the sound level of the loudspeaker might be measured while varying the input signal amplitude.

During test planning and verification, various failure scenarios, stress conditions can be developed based on the know-how about the product architecture and design. Based on simulation or real product testing in its operation environment, the behavior in case of stress or fault-condition etc. could be observed and hypothesis about the lifecycle behavior could be determined.

Based on the functional signal chains a complete testable and verifiable functional and logical model can be developed so that the requirements could be improved and detailed based on the defined functional elements and their inner and outer interfaces.

In the functional signal chain, you could identify the requirements for

- the intended E2E function,
- the individually considered functional and logical elements,
- the functional and logical interfaces,
- the resulting E2E functions and their intended behavior as well as the behavior of the integrated E2E functions.

For all elements and all interfaces, also the dysfunctional behavior and the individual possible malfunctions of the elements and the interfaces identified.

7.3.4.2 Technical Signal Chain

A technical signal chain based on technical elements and all the physical effects on the elements of the signal chain could be analyzed (Figure 7.29). Based on that information from the functional elements, the technical signal chain could be developed. In order to avoid a too increased complexity, the interfaces considered of the functional interfaces shall be selected in a way so that they are also valid for the technical interfaces. The most recommended approach is that technical elements only implement one functional element from the functional signal chain.

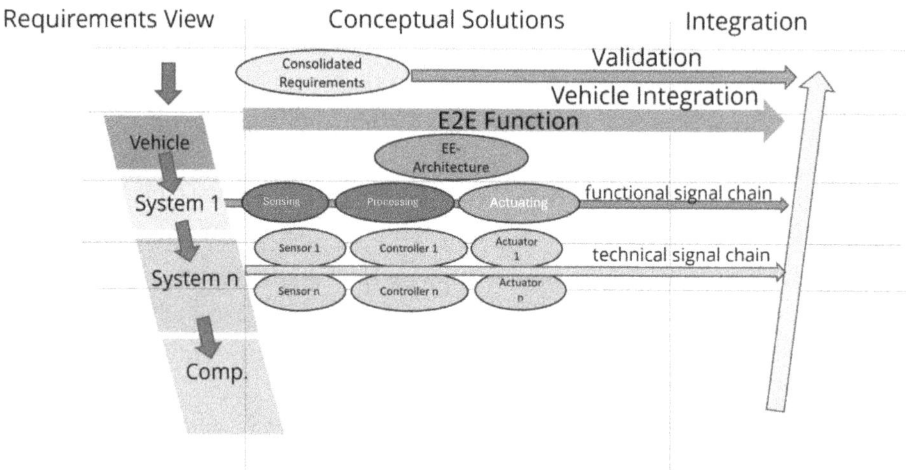

Figure 7.29 Development of a technical signal chain

Due to introducing the technical elements to the signal chain, the following further information could be determined: The technical requirements for any of the implemented physical elements and their utilities in terms of

- energy consumption, heat dissipation, and space consumption,
- reactivity, aging effects, noise impact etc., as well as
- physical interface characteristics.

Based on this information, the failure analysis could be extended so that all the new potential faults, dysfunctional and malfunctional behavior could be analyzed. In this step, the systematic errors in the top-down approach to requirements development become transparent (Figure 7.30). From the bottom-up you can see

- whether the lower technical level reflects the expected functional behavior under the specified boundary conditions,
- that the EE architecture can support the expected functional requirements, and

that the E2E functions meet the consolidated stakeholder requirements.

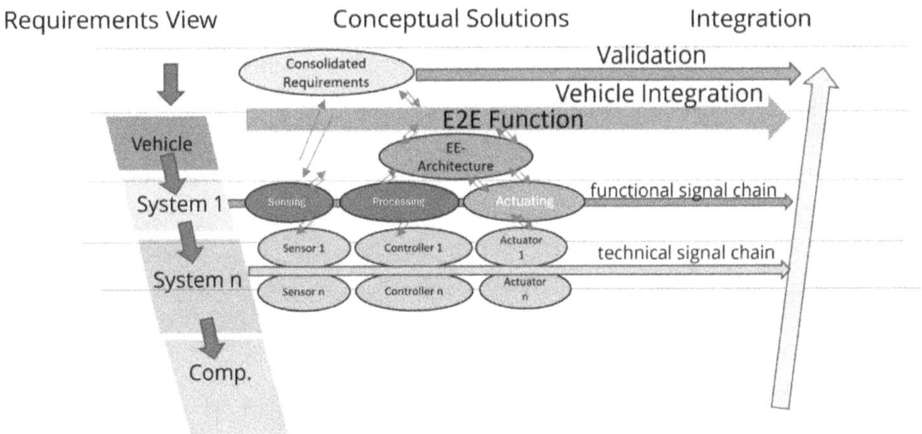

Figure 7.30 Systematic failure in top-down requirement development process

When analyzing errors, you not only see the error propagation horizontally and how the errors propagate from the context to the different levels. Of course, you can also see how the errors propagate horizontally along the functional and technical signal chain (Figure 7.31).

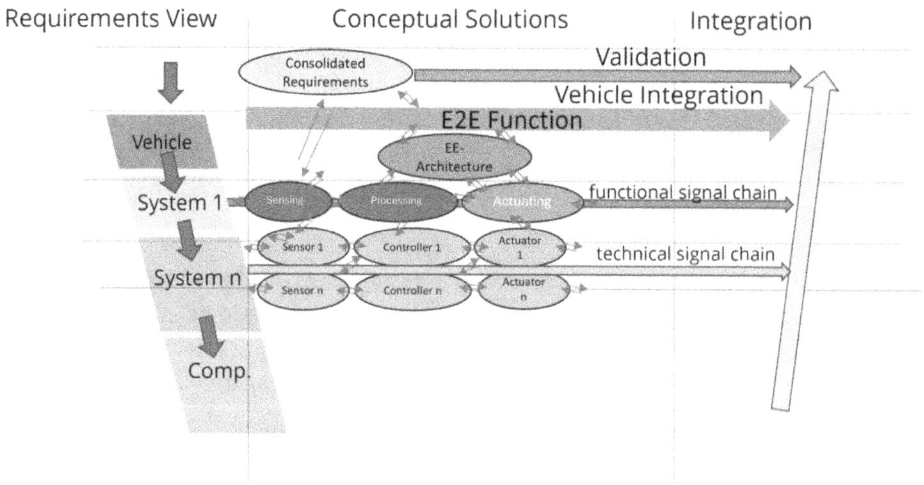

Figure 7.31 Failure propagation along the signal chain

In typical development projects you don't just have one signal chain and a functional and technical safety concept that you have to integrate into different ECUs, but many (Figure 7.32). Without such a hierarchically structured functional and implementation model, it will not be possible to assess the requirements and their correct implementation. Often it is only when it is integrated into the vehicle during validation that you realize that the development goals based on stakeholder requirements cannot be met.

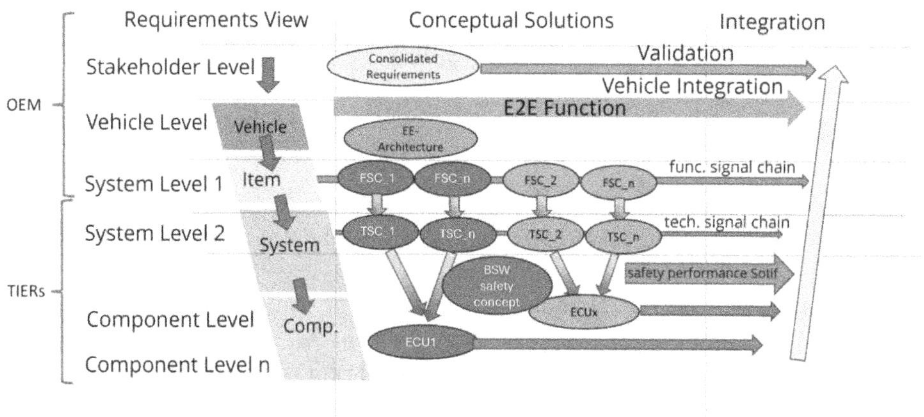

Figure 7.32 Multiple safety concepts

7.3.4.3 Time Chain Analysis

Specific engineering measures are necessary to ensure timely operation in software intensive systems. Software operates in a unique manner, distinct from the physical location where it is implemented. The primary objective of a software solution is to produce a physical effect on an actuator or generate outputs such as optical information on a display, acoustic effects, or other sensory outputs like temperature, pressure, torque, or haptic feedback.

Latency in the Control Chain

One critical aspect of software operation is the presence of latencies along the control chain leading to the actuator. This means that the software's physical effects are delayed relative to the information generation within the microcontroller or microprocessor. When information is captured by a sensor, it already has a certain age before it can be processed. This inherent delay must be accounted for in the system's design and operation.

Von Neumann Architecture and Cyclic Operation

Typical controllers that operate based on the "Von Neumann" architecture function in a cyclic manner. The cycle time can vary depending on several factors, including program memory size, clock frequency, program size, and the number of program iterations. This variability introduces challenges in achieving deterministic behavior in data processing units.

Ensuring Deterministic Behavior

To ensure deterministic behavior, additional measures must be implemented. These measures are necessary to manage the inherent variability and latency in the system, ensuring that the software can reliably produce the desired physical effects in a timely manner.

Time chain analysis is a crucial process in the design and operation of real-time systems. It involves the summation of individual time budgets that constitute a correctly functioning signal chain. This section discusses the time chain analysis in the context of best-case and worst-case, and firm real-time systems.

A signal chain can be analyzed on different levels of abstraction, but functional and technical views also provide different information (Figure 7.33). On different levels of abstraction, a function requires different time spans. On a higher level, to the timing in the effect chain, effects can adhere as a ramp but also in different curve characteristics. On the implementation level it is maybe just an operation of microseconds in the microcontroller.

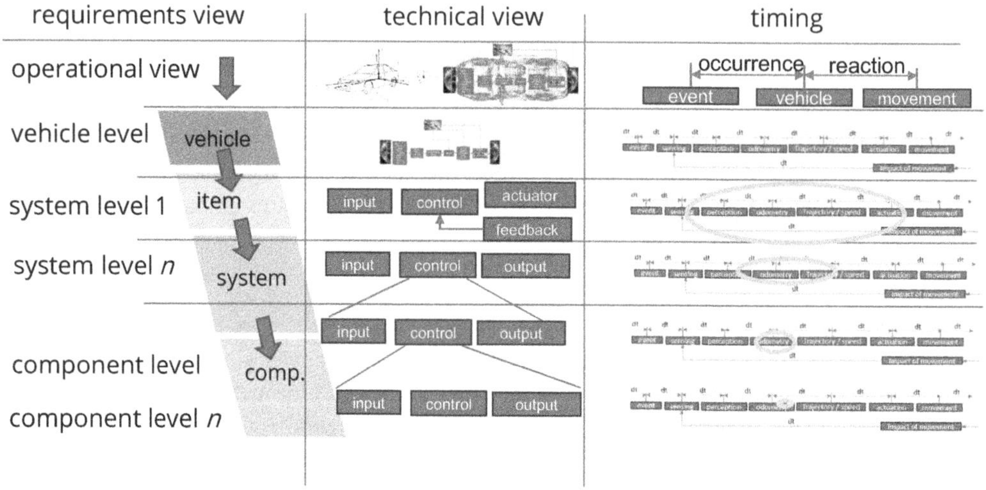

Figure 7.33 Timing chain on different levels of abstraction

Functional Signal Chain

The functional signal chain, also known as the requirements, view or problem space, refers to the sequence of processes or stages through which a signal passes from source to destination. In order to assure a correctly functioning time-critical signal chain, each stage operates within its specified time budget, ensuring that the overall system meets its real-time requirements.

Technical Signal Chain

The technical signal chain, also referred to as the solution space, is derived from the functional signal chain. It represents the actual implementation of the functional signal chain in hardware and software. The technical signal chain's performance is crucial in determining the system's real-time performance and is the basis for verifiable evidence needed for design assurance, such as part of the safety case for hard real-time safety applications.

Firm Real-Time Systems

In a best-case scenario, all stages of the signal chain operate at their maximum efficiency, resulting in the fastest possible signal processing time. This scenario is often used for system optimization. In a worst-case scenario, all stages of the signal chain operate at their minimum efficiency. This scenario is critical for ensuring that the system can meet its real-time requirements even under the most challenging conditions.

Firm real-time systems are designed in a way that any element of the signal chain produces a valid output based on an input change in a fixed time interval. Especially in safety applications with hard real-time requirements, it is important that even under stress and other worst-case scenarios, the fixed response of any of the elements could be assured. The time chain analysis for these systems focuses on ensuring that the system can meet its deadlines during any safety-related functionality, while also considering the impact of missed deadlines lead to an acceptable degradation of the system.

Time chain analysis is a vital tool for designing and operating real-time systems. By considering the best-case, worst-case, and firm real-time scenarios, designers can ensure that their systems can meet their real-time requirements under a wide range of conditions. The functional and technical signal chains play a crucial role in this analysis, determining the system's overall performance and reliability.

7.4 Hierarchical System Development

The vehicle matrix is a complex framework used to categorize and manage various aspects of vehicle design and compliance. It consists of multiple dimensions, each representing different facets of vehicle characteristics and requirements.

There are different dimensions to be considered, they could be classified as follows (Figure 7.34):

Vehicle type and usage: This dimension includes the type of vehicle and its intended use, such as sedan, convertible, station wagon, SUV, or off-road vehicle.

Registration regions and regulations: This dimension covers the regions where the vehicle will be registered and the applicable regulations in those regions, such as UNECE, FMVSS (USA), or GB/GB/T (China).

Main vehicle systems: This dimension includes the various primary systems of the vehicle, which are further subdivided into specific components and subsystems:

- **chassis:** brake systems, steering systems, damper systems, and crash design
- **powertrain:** drive systems, power and communication networks
- **interior:** access systems, theft protection, body systems, occupancy safety, airbags, convenience features, and human-machine interface (HMI)
- **ADAS/AD (Advanced Driver Assistance Systems/Autonomous Driving):** lane keeping, and collision warning
- **exterior:** fueling/charging systems, connectivity (e. g., V2x)
- **diagnostics and maintenance:** diagnostic systems, maintenance, and repair
- **pre and post-crash systems:** systems designed to manage vehicle safety before and after a crash

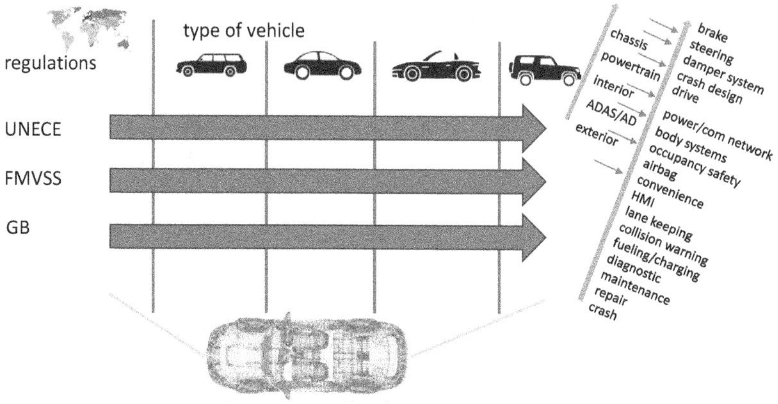

Figure 7.34 Develop matrix

These dimensions create a multi-faceted matrix that must be managed and coordinated to ensure that all interfaces and requirements are met. The complexity of this matrix varies depending on the organizational structure and the specific vehicle platform. Effective management of this matrix is crucial to ensure that all components and systems work together seamlessly and comply with the relevant regulations and performance standards.

7.4.1 Cube Perspective

In her book *Engineering a Safer World*, Nancy C. Leveson describes a multi-dimensional structure of specification using a cube model (Leveson 2012). This model includes the dimension of "Part as a Whole," where the system and its derived subsystems represent only a small dimension. The refinement process extends beyond the system itself to include the environment, operator (Human-Machine Interface), verification, and validation (Figure 7.35).

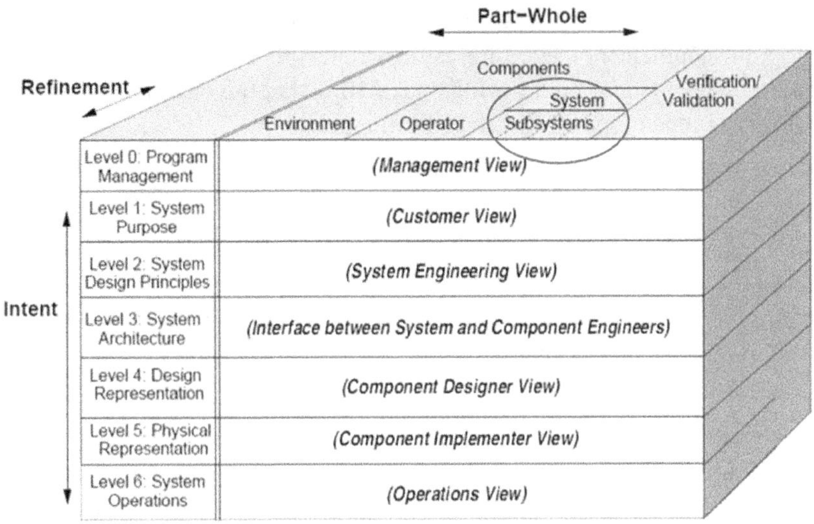

Figure 7.35 Cube perspective (source: Leveson 2012, Figure 10.1, "The structure of an intent specification")

Leveson outlines the intent across various levels (Figure 7.35):

- Level 0: program management
- Level 1: system purpose
- Level 2: system design principles
- Level 3: system architecture
- Level 4: design representation
- Level 5: physical representation
- Level 6: system operation

These levels correspond to different views:

- management view
- customer view
- system engineering view
- interface between system and component engineers
- component designer view
- component implementer view
- operations view

Each level and view highlight the roles and competencies of the different groups involved in the development process. Interestingly, Leveson addresses "system engineering" rather than "systems engineering," suggesting that the methodology is considered within the scope of specific product development.

Translated into the automotive systems engineering, the levels could be considered as follows:

- **Level 0—Program Management:** Level 0 oversees the development of a new electric vehicle (EV) model, ensuring that all project milestones are met, budgets are adhered to, and resources are allocated efficiently. This includes coordinating between different departments such as design, engineering, marketing, and production.
- **Level 1—System Purpose:** Level 1 defines the primary objectives of the EV, such as achieving a specific range on a single charge, meeting safety standards, and providing a high level of driver comfort and convenience. This involves setting clear goals for the vehicle's performance, safety, and user experience.
- **Level 2—System Design Principles:** Level 2 applies principles such as modularity and scalability in the design of the EV's powertrain. This could involve using a modular battery pack design that allows for easy upgrades or replacements and ensuring that the vehicle's software can be updated over-the-air to improve performance or add new features.
- **Level 3—System Architecture:** Level 3 develops the overall architecture of the EV, including the integration of various subsystems such as the battery management system,the electric motor, and the infotainment system. This involves creating detailed diagrams and models to ensure that all components work together seamlessly.
- **Level 4—Design Representation:** Level 4 defines the design of the EV, for subsystems such as the battery management system, electric motor, and infotainment system. This involves defining detailed design patterns, characteristics and their verification.
- **Level 5—Physical Representation:** Level 5 creates physical prototypes of the EV, including full-scale models and working prototypes. This stage involves rigorous testing of the vehicle's physical components, such as crash testing, durability testing, and performance testing under various conditions.
- **Level 6—System Operation:** Level 6 ensures that the EV operates efficiently and safely in real-world conditions. This includes developing and testing the vehicle's autonomous driving features, optimizing energy consumption, and ensuring that all systems function correctly during normal operation and in emergency situations.

These examples illustrate how each level contributes to the successful development and operation of an automotive system. All the people involved in the program need their own views and perspectives, so that a successful product could be developed. Such a systematic approach assures that the stakeholders could find their requirements fulfilled and that the final product could fulfil their purpose as expected.

7.4.2 Axiomatic Design

Axiomatic Design is a systems design methodology developed to systematically transform customer needs into functional requirements, design parameters, and process variables using matrix methods. This approach was pioneered by Dr. Nam Pyo Suh at the Massachusetts Institute of Technology (MIT) in the 1990s.

Origin of Axiomatic Design

The methodology was introduced by Dr. Suh in 1977, aiming to establish a scientific approach to design. Dr. Suh's work was inspired by the need for a structured method that could analyze, and determine the merits of a design early in the process, rather than relying on trial and error. He drew analogies between design and thermodynamics, proposing that a set of fundamental principles could guide good design practices.

The Two Axioms of Axiomatic Design

Axiomatic Design is built on two primary axioms:

- **Independence Axiom:** This axiom states that the independence of functional requirements (FRs) should be maintained. In other words, a design should be structured so that each functional requirement is satisfied without affecting others. This helps in creating a robust design where changes in one aspect do not lead to unintended consequences in another.
- **Information Axiom:** This axiom emphasizes minimizing the information content of the design. A design with less information content is preferred because it implies higher reliability and simplicity. This axiom helps in selecting the best design among alternatives by evaluating the probability of achieving the desired functional performance.

Motivation Behind the Axioms

The motivation for these axioms is rooted in the desire to create a systematic and scientific approach to design. The Independence Axiom ensures that designs are modular, and changes can be made without widespread impact, which is crucial for maintaining functionality and reducing complexity. The Information Axiom aims to simplify designs, making them more reliable and easier to manage.

By adhering to these axioms, designers can achieve high-quality, efficient, and reliable designs that meet customer needs effectively. The structured approach provided by Axiomatic Design helps in reducing the trial-and-error aspect of design, leading to more predictable and successful outcomes.

Axiomatic Design utilizes four primary design domains to systematically transform customer needs into a final product or solution (Figure 7.36). These domains help in organizing and structuring the design process.

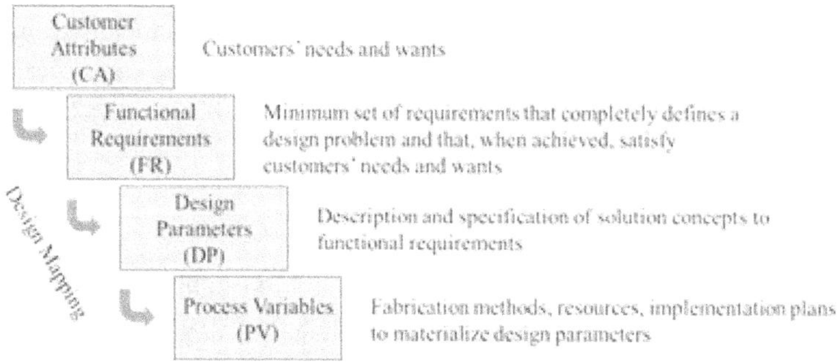

Figure 7.36 Design domains (source: Nordlund et al. 2015, Figure 1, Design domains and mapping)

The four design domains are listed in the following (Figure 7.37).

Customer Domain

What it represents: This domain captures the needs and desires of the customer.

Purpose: It defines what the customer wants to achieve, often referred to as Customer Attributes (CAs).

Example: If a customer or the user demands certain features or functionality or even doesn't want a certain feature or functionality, this need is captured in the customer domain.

Functional Domain

What it represents: This domain translates customer needs into Functional Requirements (FRs).

Purpose: It specifies what functions the design must perform to meet the customer's needs.

Example: One example is the way the customer experiences their requirements, or how a neutral observer will experience this function or feature, or the function of an air conditioning system to maintain a certain temperature.

Physical Domain

What it represents: This domain involves the physical Design Parameters (DPs) that will fulfil the functional requirements.

Purpose: It defines how the design will achieve the required functions.

Example: In the case of automotive systems, the physical domain might include the design of a refrigerator with specific cooling mechanisms.

Process Domain

What it represents: This domain outlines the Process Variables (PVs) necessary to produce the design.

Purpose: It specifies the processes and methods needed to manufacture and implement the design.

Example: For an air condition system, the process domain would detail the manufacturing steps, materials, and assembly processes required to build the air condition system so that it could be integrated into the targeted vehicle.

The Zigzagging Process

In Axiomatic Design, designers often move back and forth between these domains in a process known as "zigzagging." This iterative approach ensures that each domain is adequately addressed and aligned with the others, leading to a coherent and effective design.

Practical impact on systems engineering: This zigzagging causes significant issues due to changes in the various other modules. Here transparent modularization is compulsory in order to perform a change impact analysis.

By systematically working through these domains, designers can ensure that the final product meets customer needs, performs the required functions, is physically realizable, and can be efficiently produced.

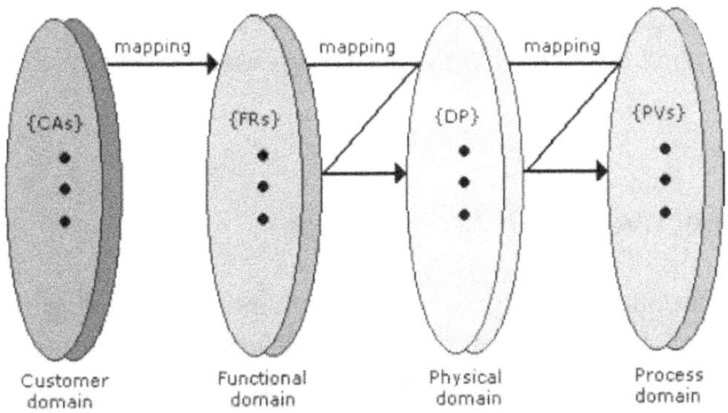

Figure 7.37 Mapping of domains (source: Functional Specs 2024)

Key Principles in Axiomatic Design for Complex Systems

In complex systems, it can be challenging to apply principles consistently, but the axioms of Axiomatic Design are crucial aspects. These axioms form the essential foundations for structured systems engineering.

Here are some important principles:

- **When deriving from a higher level to a lower level, only clearly countable elements should be considered.**

  This principle ensures that each element at a lower level is distinct and measurable, facilitating precise design and analysis.

- **The interfaces of the elements should be clearly described.**

  Clear descriptions of interfaces ensure that interactions between elements are well understood and managed, reducing the risk of integration issues.

- **The elements and interfaces must also be clearly specified regarding their independence, connections, and dependencies.**

  Specifying the independence and dependencies of elements and interfaces helps maintain the integrity of the design, ensuring that changes in one part do not adversely affect others.

- **All domains, including the process domain for technical feasibility, should be considered.**

  Considering all domains ensures that the design is not only functional but also manufacturable and implementable, covering customer needs, functional requirements, physical design, and production processes.

- **The horizontal levels must be defined and specified by the organization.**

  Defining horizontal levels within the organization ensures that there is a clear structure and hierarchy in the design process, facilitating better communication and coordination.

By adhering to these principles, designers can create robust, efficient, and reliable systems that meet the required specifications and customer needs. These principles help in managing complexity and ensuring that the design process is systematic and structured.

7.4.3 Views and Perspectives of an Architecture

In the automotive industry, various architecture models are employed to manage the complexity of modern vehicles, which are essentially sophisticated systems of interconnected hardware and software components. These models serve as blueprints, guiding the development, integration, and maintenance of these complex systems.

They focus on decompositions, views, perspectives, structural notations, and behavioral descriptions which are described as follows.

Decompositions of Structures

Hierarchical Decomposition: This involves breaking down the vehicle system into smaller, manageable subsystems and components. For example, a vehicle can be decomposed into powertrain, body, chassis, and electronics systems. Each of these can be further decomposed. The electronics system, for instance, can be divided into body control, infotainment, driver assistance, etc. This hierarchical structure allows engineers to focus on specific areas and manage complexity.

Functional Decomposition: This focuses on the functions the vehicle performs. For example, braking, steering, acceleration, and communication are functions. These functions are then mapped to the components and systems that realize them. This helps in understanding how different parts of the vehicle contribute to overall functionality.

Representations of Views and Perspectives

Logical view: This describes the abstract organization of the system, focusing on the logical relationships between components and functions, independent of their physical implementation. It answers the question “what components are needed and how do they interact logically?”

Physical view: This describes the physical layout and interconnection of hardware components within the vehicle. It considers aspects like wiring harnesses, ECU locations, and communication buses. It answers the question “where are the components located and how are they connected physically?”

Development view: This focuses on the software and hardware development process, including modules, interfaces, and dependencies. It supports concurrent engineering and management of software development.

Deployment view: This describes how software components are deployed onto the hardware (ECUs) in the vehicle. It considers resource constraints, communication bandwidth, and real-time requirements.

Each of these views provides a different perspective on the vehicle architecture, allowing stakeholders with different roles (e. g., software engineers, hardware engineers, system architects) to focus on relevant aspects.

Structural Notations

Various notations are used to represent the architecture models.

UML (Unified Modelling Language): This is a widely used standard for modelling software systems, including class diagrams, component diagrams, sequence diagrams, and state diagrams. It provides a rich set of notations for describing static structure and dynamic behavior.

SysML (Systems Modelling Language): This is an extension of UML specifically designed for systems engineering. It adds specific diagrams and constructs for modelling system requirements, architecture, and behavior.

AUTOSAR (Automotive Open System Architecture) Methodology: This is a standardized software architecture for automotive ECUs. It defines a layered architecture and standard interfaces, facilitating software reuse and interoperability. AUTOSAR methodology uses specific notations and templates for describing software components and their interactions.

Behavioral Descriptions

Architecture models also capture the dynamic behavior of the vehicle system.

State machines: These describe the different states a system or component can be in and the transitions between these states triggered by events. This is useful for modelling control logic and error handling.

Sequence diagrams: These illustrate the interactions between different components over time, showing the sequence of messages exchanged. This is useful for understanding how different parts of the system collaborate to perform a function.

Activity diagrams: These depict the flow of activities within a process or system, including parallel and conditional execution. This is useful for modelling complex workflows and control logic.

Architecture models in the automotive industry provide a structured and comprehensive way to represent and manage the complexity of vehicle systems. They assure that engineers with different competences and skills could work together and could optimize their work efficiency. By using decompositions, different views and perspectives, standardized notations, and behavioral descriptions, engineers can design, develop, and integrate complex automotive systems effectively. These models are essential for ensuring functionality, safety, and reliability in modern vehicles.

7.4.4 Hierarchical Level of Abstraction

The V-model is one of the core principles of systems engineering. In order to manage the complexity of large systems, decomposition and modularization are essential. These techniques break down the system into smaller, more manageable parts, making it easier to understand, design, implement, and test. However, simply breaking a system into parts isn't enough. To ensure these parts work together seamlessly, the integration requirements must be considered during the decomposition process. This is the fundamental idea behind the V-model.

The V-Model and Hierarchical Development

The V-model is a graphical representation of a structured development process. It emphasizes the relationship between development phases (left side descending branch of the "V") and testing/validation phases (right side ascending branch of the "V").

Top-down requirements development (left side): This side represents the decomposition process. It starts with high-level system requirements and progressively refines them into more detailed subsystem, component, and module-level requirements. This hierarchical breakdown continues until the requirements are specific enough to be directly implemented in hardware or software. This top-down approach ensures that all lower-level elements contribute to the overall system goals.

Bottom-up integration and verification/validation (right side): This side represents the integration and testing process. It mirrors the decomposition process, but in reverse. Individual modules or components are first tested in isolation (unit testing). Then, they are integrated into larger subsystems, which are then tested (integration testing). This process continues until all subsystems are integrated into the complete system, which is then tested against the original high-level requirements (system testing and acceptance testing).

Ensuring Consistency and Integrability

To guarantee the consistency and integrability of modules, several key practices are crucial.

Clear hierarchy levels: Defining distinct hierarchy levels (e. g., system, sub-system, component, module) is essential. Each level represents a specific level of abstraction, ensuring that all elements at that level are described with similar granularity. This prevents inconsistencies and misunderstandings about the scope and function of different parts of the system.

Consistent abstraction levels (horizontal consistency): At each horizontal level of abstraction within the V-model, the same system is described with varying levels of detail. This horizontal consistency is vital for verification and validation.

On a higher system level, you have high-level requirements like "The vehicle shall accelerate from 0 to 60 mph in under 6 seconds." At a lower level (e. g., powertrain subsystem), this requirement is decomposed into more specific requirements related to engine power, transmission efficiency, and traction control. The key is that the lower-level requirements must trace back to the higher-level requirements, ensuring that the system as a whole fulfils its intended purpose.

Verification and validation at each level: The V-model emphasizes verification and validation at each level of integration.

Verification: Verification ensures that the system or component is built right (i. e., it meets its specified requirements). This is often done through testing, reviews, and analysis.

Validation: Validation ensures that the system or component is the right thing (i. e., it meets the user's needs and the overall system objectives). This is often done through user testing, simulations or other evidence to assure achievement of specified targets or behavioral, logical or functional intent.

By adhering to these principles, the V-model provides a structured framework for developing complex systems. It helps ensure that all parts of the system are consistent, complete, and work together as intended, ultimately leading to a higher quality and more reliable final product. The emphasis on integration requirements from the initial stages of decomposition is the key to minimizing integration issues later in the development cycle, saving time and resources.

Such a modular breakdown and systematic integration along the V-model is essential for model-based development. Figure 7.38 shows how the models and their abstracts lead to detailed logical scenarios.

Since the system boundary for the different models requires a different context, different test setups also need to be specified so that the models could be verified and validated by adequate tests.

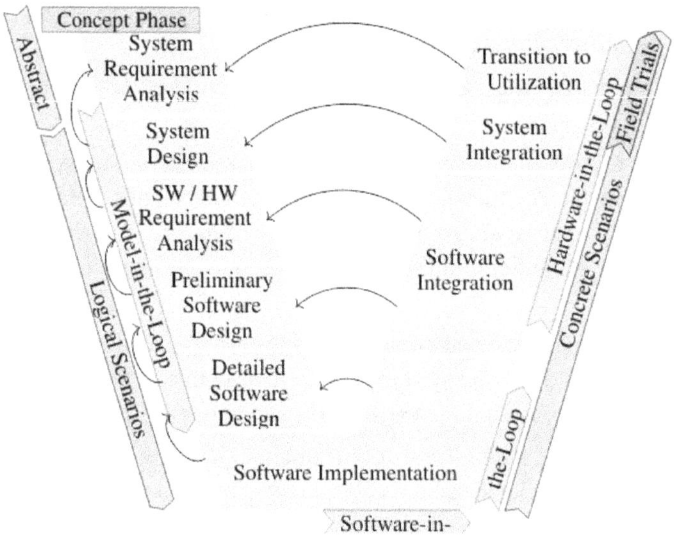

Figure 7.38 Integration testing along a V-model (source: Schütt et al. 2022, Figure 1)

7.4.4.1 Introduction of Different Levels of Abstraction

The idea is to structure the development of a car in a way similar to how aircraft are designed and built. Aircraft engineering is known for its rigorous systems engineering approach, with well-defined levels of abstraction and strict adherence to requirements and testing. This approach is becoming increasingly important in the automotive industry due to the rising complexity of modern vehicles, driven by electrification, automation, and connectivity.

Aircraft Level vs. Vehicle Leve—the Highest Level of Abstraction

Aircraft: This represents the complete aircraft as a single entity, black box or behavior model, with overall performance characteristics (e. g., range, speed, payload capacity), mission requirements, and certification standards.

Vehicle (car): This represents the complete car as a single entity, with overall performance characteristics (e. g., fuel efficiency, acceleration, top speed), intended use cases (e. g., city driving, off-road), and regulatory requirements.

System level (similar in both): This level decomposes the overall vehicle or aircraft into major functional systems.

Aircraft examples: propulsion system (engines, fuel system), flight control system (wings, control surfaces, actuators), avionics (navigation, communication, sensors), landing gear, etc.

Vehicle examples: powertrain (engine/motor, transmission, drivetrain), chassis (suspension, steering, brakes), body (structure, exterior, interior), electrical/electronic architecture (ECUs, sensors, actuators, communication buses), ADAS (Advanced Driver Assistance Systems)

Component level (similar in both): This level further decomposes the systems into individual components or subsystems.

Aircraft examples

Within the propulsion system: engine components (turbines, compressors, combustion chamber), fuel pumps, fuel tanks, control valves

Within the flight control system: ailerons, elevators, rudder, hydraulic actuators, flight computers

Vehicle examples

Within the powertrain: engine components (pistons, cylinders, valves), transmission gears, differentials, drive shafts

Within the chassis: shock absorbers, springs, steering rack, brake calipers

Within the electrical/electronic architecture: individual ECUs (Engine Control Unit, Brake Control Unit, BCU), sensors (wheel speed sensors, temperature sensors), actuators (valves, motors)

Software/module mevel (more pronounced difference): This level becomes particularly relevant in the automotive industry due to the increasing software content in vehicles.

Software Modularization

Aircraft: Software is often deeply embedded within specific components (e. g., flight control computers, engine controllers). The focus is on real-time performance, safety criticality, and deterministic behavior.

Vehicle: Modern vehicles have a distributed software architecture with numerous ECUs running complex software. This includes operating systems, application software for various functions (e. g., engine control, braking, infotainment), and communication protocols. This level can be further broken down into software modules, functions, and even individual lines of code.

There are several benefits of applying aircraft-style abstraction also to automotive industry.

Complexity management: Breaking down the vehicle into manageable levels makes it easier to understand, design, and integrate complex systems, especially with the increasing software content.

Improved traceability: A clear hierarchy allows for better traceability between requirements, design, implementation, and testing. This is crucial for ensuring safety and compliance with regulations.

Enhanced reusability: Modular design and well-defined interfaces promote the reuse of components and software across different vehicle models, reducing development costs and time.

Facilitated collaboration: Clear levels of abstraction and well-defined interfaces facilitate collaboration between different engineering teams and suppliers.

Better safety and reliability: A rigorous systems engineering approach, like the one used in aerospace, helps to identify and mitigate potential safety risks early in the development process, leading to more reliable vehicles.

By adopting a structured, hierarchical approach with well-defined levels of abstraction, the automotive industry can better manage the increasing complexity of modern vehicles, improve development efficiency, and enhance safety and reliability. The analogy to aircraft engineering provides a valuable framework for achieving these goals (Figure 7.39):

- The top level is the **operational level**, where the means of transport are shown in their application context.
- The **vehicle or aircraft level** considers the transport system in the context of the higher level of abstraction in which case the operational level is the higher level of abstraction that forms the context for the vehicle or aircraft level.
- At the **system level**, the subsystems that provide the basic functionalities so that the means of transport can function are considered. The system level often consists of several sublevels that serve to structure and identify the internal and external interfaces.
- The component level primarily serves as the basis for the realization of the product; the technical production aspects and the physical influences on the individual components and their internal and external interfaces are then considered here.

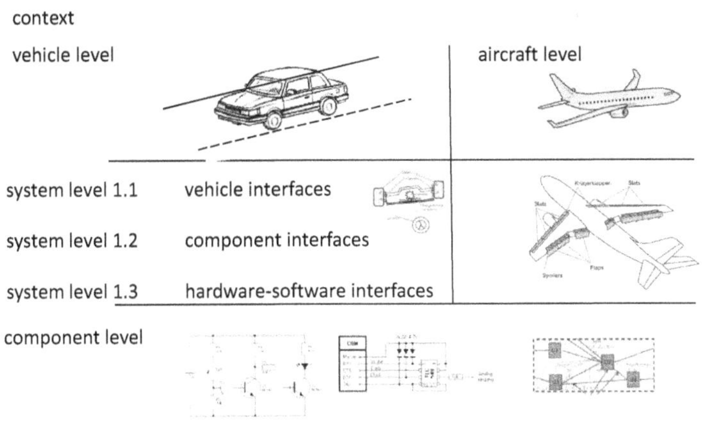

Figure 7.39 Hierarchy for means of transportation

In the different levels, an observer takes on different views (Figure 7.40):

- He sees the vehicle in the traffic context from a **bird's eye view**.
- He sees the vehicle itself from the so-called **normal perspective**. You can imagine a car salesman who wants to sell the car to a customer without going into too many technical details.
- The observer sees the various systems, such as the steering, from a **frog's eye view**. Here you can imagine the workshop foreman driving the car up on the lifting platform in order to be able to point out a possible error in the steering gear from below.

It is essential that the upper abstraction level always forms the context for the lower, derived abstraction level.

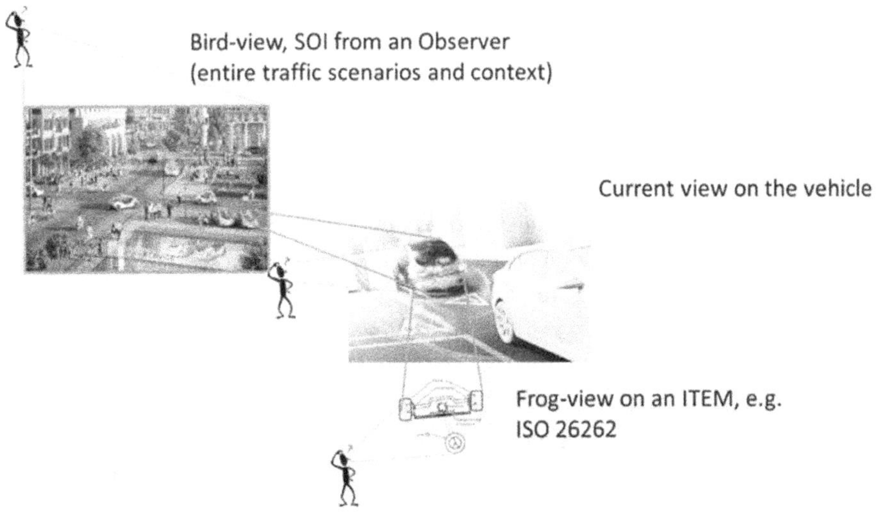

Figure 7.40 Views from a neutral observer

In order to reduce the complexity and structure of the integration level for suppliers and other development partners, multiple levels of system should be introduced. In these different levels of abstractions, the interfaces are different and need to be specified specifically for this level of abstraction. Especially for different technologies like hydraulic, pneumatic, electric and electronic systems, a separation of the system levels is recommended.

The interface characteristics depend on system, product or component properties and their needs like geometrical, energy and dynamic aspects (Figure 7.41). The interfaces are derived from their function, task and the respective context. For example, the individual components of the chassis have a significant influence on the driving dynamics and driving behavior and provide limits for the vehicle stability, which could not only be compensated by software algorithms.

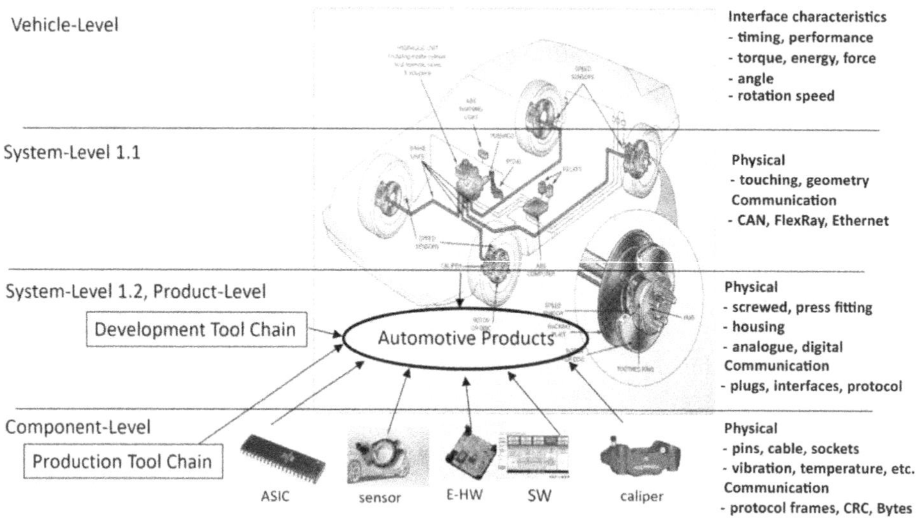

Figure 7.41 Interface characteristics

7.4.4.2 Operational Level

The highest level of abstraction is the stakeholder requirements level. The first task here is of course to identify the stakeholders. When it comes to mobility solutions, we should mention the legislators who provide the basis for social acceptance of mobility solutions. This not only includes the approval guidelines, but also the requirements for the state of the art arise from product liability, or for new technologies the state of science and technology. In this way, not only the legal regulations come into focus but also the immense landscape of standards.

Of course, such a means of transport should also be sold to customers. Therefore, usability, ergonomics and manageability come with many requirements that drive the architecture and the accepted design solutions. If you look at the vehicle down to its smallest parts and try to reflect the stakeholder requirements, this leads to an explosion of complexity in the functions, the system elements, and their interfaces. However, it makes no sense to only look at stakeholder requirements in isolation. Therefore, the vehicle in its operating environment must be considered at this level. The operating environment and the relevant stakeholder requirements form the context for this level of abstraction, also known as the operational domain.

All vehicles are considered as black boxes, with their degree of freedom (Figure 7.42). Not only the ego vehicle is considered as black box, but also other vehicles as well as any other road users are black boxes within their possible behavior space. Occupancies including the driver are not visible.

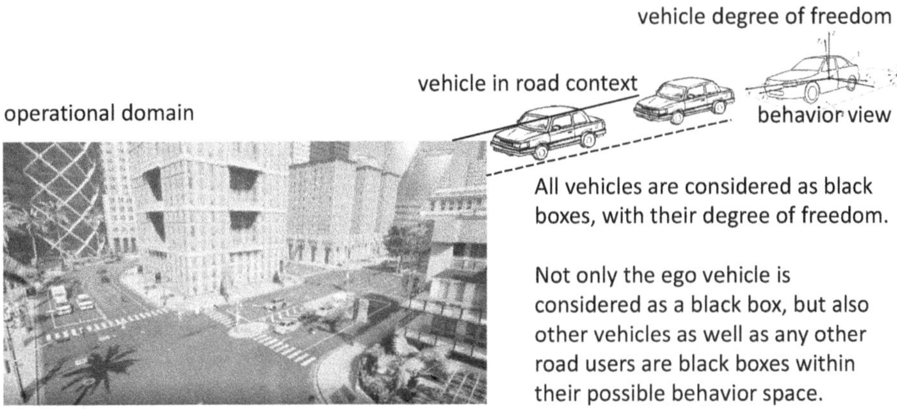

Figure 7.42 Operational Level

7.4.4.3 Simulation and Testing on Operational Level

The number of real road traffic scenarios cannot be tested in real world scenarios, therefore model-based testing is unambiguous. Even at the vehicle level, one cannot simply observe the vehicle; instead, one must systematically examine how deeply to delve into the architecture of the vehicle. This level must also be standardized so that the results can be comparable and assessable.

In the example, four levels of abstraction are considered (Figure 7.43):

- **Macroscopic simulation:** This is the simulation of traffic flow, e. g., on highways. Traffic is seen as a fluid.
- **Mesoscopic simulation:** Traffic is modeled as discrete particles (in contrast to fluid) without mass and personality.

- **Microscopic simulation:** Traffic is modeled as the behavior and interactions of each simulated entity individually with specific state, variables such as position, speed and acceleration.
- **Nanoscopic or sub-microscopic simulation:** Each vehicle is divided into several subunits to achieve a higher level of detail.

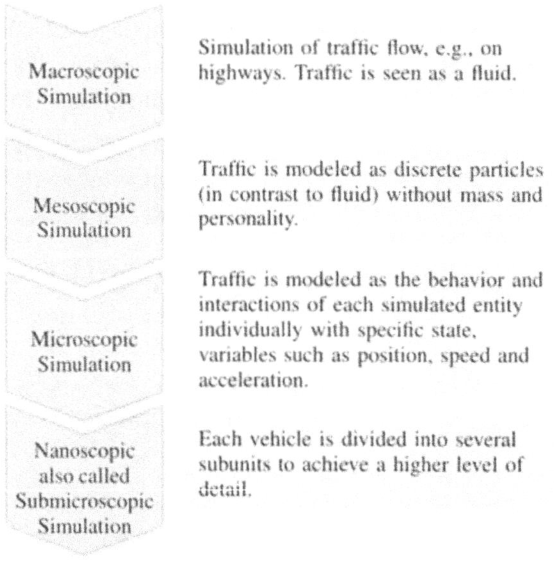

Figure 7.43
Levels of traffic simulation (source: Schütt et al. 2022, Figure 2)

In this proposal the first three levels of abstractions are very meaningful and should be assigned to the operational level. The third level, the microscopic simulation, could be already considered as a vehicle level simulation, since the vehicle would be considered as a black box device and only the behavior would be considered. The fourth level, the nanoscopic simulation should already be considered a system level, since you need to introduce different kinds of interfaces of the different elements or systems of the vehicle. White box or at least grey box approaches should be considered.

7.4.4.4 System Level

The system level serves to systematically structure the requirements for the development object and prepare these requirements in such a way that the various discipline engineers receive the necessary views and perspectives. This ensures they have the information needed to realize the individual components and specify the software so that it can be implemented effectively.

The views must be prepared in such a way that stakeholders, project managers, assessors, controllers, and others receive the necessary information to monitor the development's maturity at any time. It is essential to evaluate the change impacts for any modifications and assess the corresponding changes and the change impacts and their implementation.

It is important to consider that mechanics, electronics, and software have different realization and implementation processes (Figure 7.44). The design process and the development of requirements must align with a product-structure-oriented architecture, even the organization to be responsible for the realization of the components. For example, mechanics and electronics are produced in a production process in a factory environment, since software would be realized in a more laboratory environment.

A unique aspect of mechanical products is that their function can often be inferred from their appearance. For example, a suspension spring is recognizable as such, and its design indicates that it stabilizes the chassis and absorbs shocks from road irregularities in conjunction with the shock absorbers. In contrast, this is not the case with electronics. A microcontroller is identifiable as such, but its functions for customer benefit or observable functions are not apparent from its appearance.

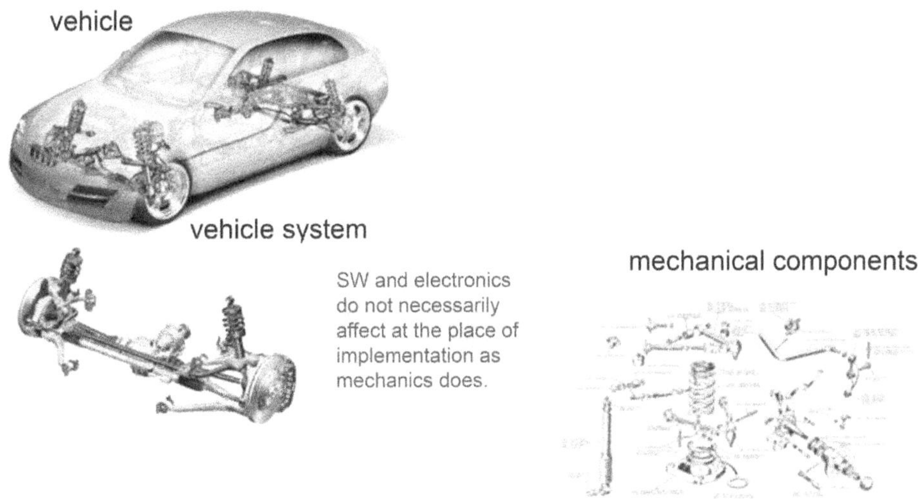

Figure 7.44 Vehicle decomposition/mechanics versus software

The essential aspect of software is that it does not act where it is implemented. Software operates through electronics, communication systems or busses and their physical effect chains, generally affecting mechanical components to produce physical effects such as heating, force or torque transmission, movement, or steering effects.

It is also crucial to understand that when steering a typical vehicle axle knuckle, the knuckles move through the trapezoidal structure depending on the moments, but the vehicle does not move around the lateral axis when stationary. Only when the vehicle moves along the longitudinal axis does the movement of the knuckles lead to lateral movement (Figure 7.45).

Since forces depend on internal and external influences, sensors must be installed to transfer the necessary forces as feedback or disturbance variables to a controller, following the principle of a closed control loop, ensuring uniform and linear movements.

> Where software is implemented, there is always a latency to where it has an effect. This is also true for sensors; all the information a sensor captures or measures already has a certain age before the software in the controller or processor can process the information.

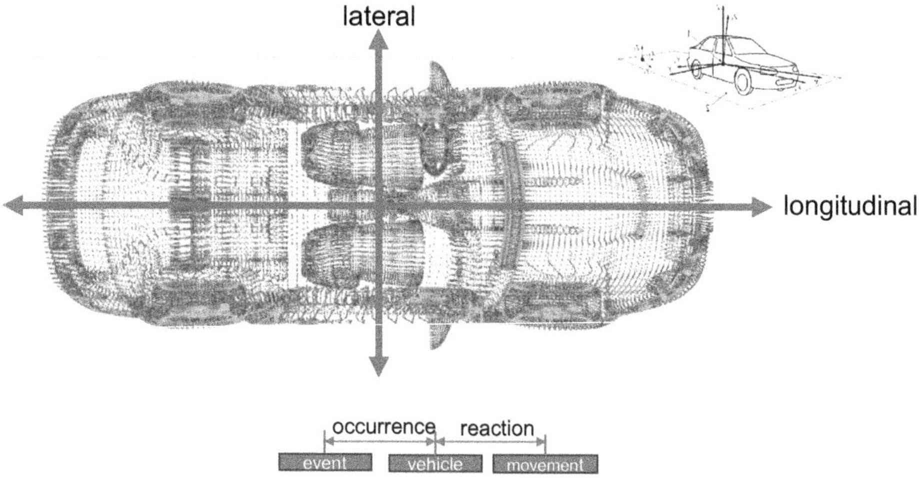

Figure 7.45 Vehicle degree of freedom and their movement

A dynamic system is always in interaction with its environment. For instance, a car requires more engine power to ascend a hill. The higher the friction coefficient of the road, the greater the rolling resistance. Additionally, the shape of the vehicle significantly influences its sensitivity to wind. The drag coefficient (CW) indicates the resistance the vehicle's shape poses to airflow. Every event in the operational environment affects the car, leading to changes in its movement. For example, without continuous power supply, the car would gradually slow down and eventually stop. These effects are typically managed by the engine control system.

Since the information used by a controller is already somewhat outdated, the latencies along the chain of effects have a crucial impact on the behavior initiated by the software. Using the example of a lane-keeping function, the following latency chain can be outlined (Figure 7.46):

- **Occurrence of the physical event:** The vehicle system needs to respond to this event.
- **Sensing the information:** Sensors detect the relevant data.
- **Perception of the information:** The system processes and interprets the sensed data.
- **Odometry adaptation:** planning and predicting the latency for the necessary movement of the vehicle, considering the vehicle's geometry, current dynamic properties, and speed
- **Trajectory planning:** The system calculates the vehicle's position and plans the trajectory, taking into account the current vehicle speed (usually as a feedforward control).
- **Actuation of the steering:** The system sends commands to the steering actuator and the associated mechanism.
- **Physical steering movement:** The vehicle executes the steering movement.
- **Feedback:** This movement is measured by a steering angle sensor, which provides feedback on the actual steering movement. However, this measured steering movement is also transmitted back to the controller with a latency, creating a closed-loop control system. Before this value leads to a correction of the steering movement, the entire chain must be traversed again. The more deterministically these time chains are executed, the more precisely the vehicle's steering movement can be implemented.

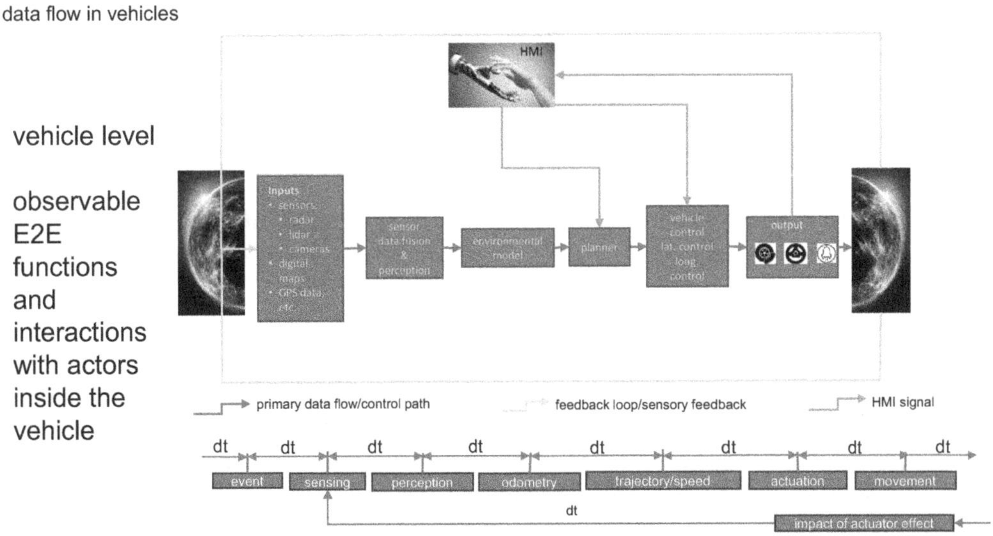

Figure 7.46 Effect chain and consequence on timing

7.4.4.5 Component Level

In the realm of component-level design, elements must be realized in a manner that aligns with the requirements derived from the system level. This necessitates a systematic and structured approach to ensure traceability, integrability, configurability, and adaptability to both system and application requirements. According to Axiomatic Design principles, system requirements should be derived to countable elements.

Literature often suggests treating mechanics, pneumatics, and hydraulics as separate components, alongside electronics and software (Figure 7.47). However, this separation can be challenging due to the intricate interfaces between electronics and mechanics, as well as the hardware-software interface. Modern groupings integrate more intelligent units as components, particularly in actuators within the drive and chassis domains. These upper integration levels present significant challenges, leading to heterogeneous interface structures during requirement derivation.

For instance, a hydraulic brake must be electronically controlled, as purely hydraulic braking systems no longer meet current standards and homologation requirements. Regulations for homologation such as UNECE, FMVSS, EU regulations etc., require electronic control functions like ABS, AEB, and ESC to assure safe and stabile braking of motor vehicles. To electronically control a hydraulic brake, sensors and software-based algorithms must be integrated into hardware controllers, transforming the brake into a complex, software-driven component with various physical interfaces.

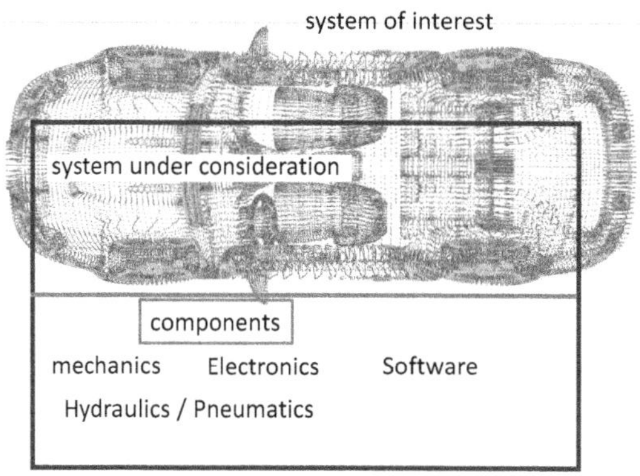

Figure 7.47
Components made of mechanics, electronics and software (source: NASA 2010)

In contemporary electronic steering systems, hydraulics are often absent, yet safety functions previously ensured by hydraulics must now be achieved through electrical and electronic measures. This necessitates defining and systematically specifying interface classes within the system. Boundary diagrams are effective tools for describing mechatronic systems, illustrating what lies within the component and the static and dynamic effects acting upon it (Figure 7.48).

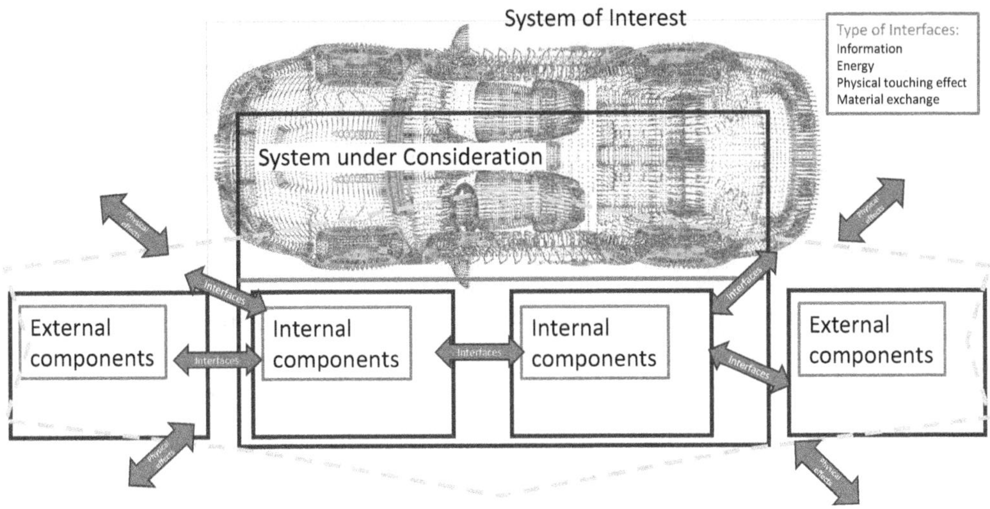

Figure 7.48 Simplified boundary diagram

Statically, interfaces can be categorized into information, energy, physical touching or effect, and material transfer types. Dynamically, transfer functions must be described, comprising various classes. For example, the driver's foot pressure is a physical movement converted into hydraulic pressure by a brake booster in a hydraulic brake system. This hydraulic pressure facilitates information transfer and provides the energy for the foundation brake's operation. The hydraulic fluid also enables material transfer, increasing pressure in the brake caliper. In a purely hydraulic brake, the brake pressure in the caliper piston is regulated by valve openings and throttle orifices, leading to the vehicle's deceleration.

To implement a by-wire function, all hydraulic functions must be replicated by electronics and software. This includes capturing physical requirements, transmitting necessary information, providing required energy, and control the braking force to ensure stable and user-friendly vehicle deceleration.

In the context of mechatronic systems, the term "component" refers to a grouping that is even more complex and heterogeneous at a lower level (Figure 7.49). While the functions of mechanical and hydraulic parts in mechatronic systems can still be inferred, it is not immediately apparent what customer benefit a microcontroller provides in a given context, nor can this be discerned from the software implemented within it. Due to the intricate interactions in software-intensive systems, an isolated examination of the software alone will not yield any results.

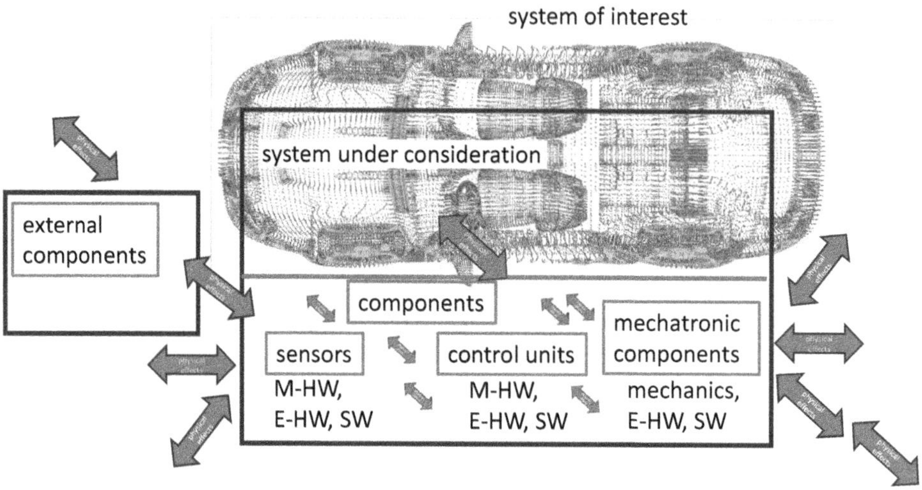

Figure 7.49 Mechatronic components

Within the components that fall under the system of consideration, there are inherent physical effects, such as the intended current flow. However, this current flow also leads to unintended internal physical effects, including heat generation, Electromagnetic Interference (EMI) and other noises, stress, and voltage fluctuations etc. Unintended disturbances also impact the system and its components from the outside. External components can as well introduce unexpected disruptive effects that negatively influence the system and component behavior.

Heat dissipation and EMI protection are crucial tasks to build stable electronic systems. Networks and bus communications, particularly those operating at high frequencies, are especially sensitive to such effects. Ensuring proper thermal management and shielding against electromagnetic interference is essential to maintain the reliability and performance of these systems.

7.4.5 Systems Breakdown

Systems engineering in the automotive domain starts with a thorough understanding of stakeholder requirements (Figure 7.50). At its core lies the System of Interest, in this case, the vehicle. This vehicle needs to be defined not just by its physical components, but also by its intended use cases and behavior within the context of its operational environment, which is the real world with its complex traffic situations. This necessitates a detailed description of relevant operational scenarios within this domain. This analysis, often referred to as Operational Design Domain definition, directly influences the design criteria for the vehicle.

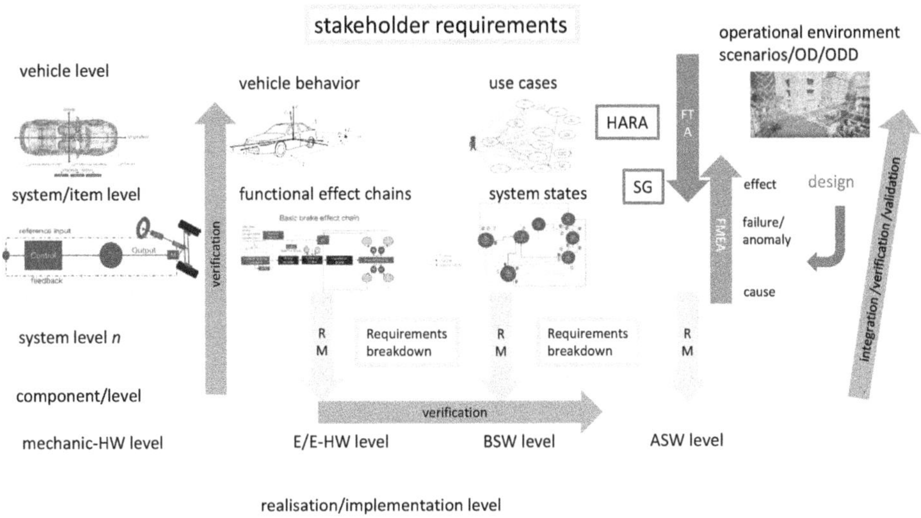

Figure 7.50 Systems breakdown as a top-down approach

Safety is paramount, and a comprehensive Hazard Analysis and Risk Assessment (HARA) is crucial. This analysis should focus on the vehicle's behavior in its various use cases, considering all potential actors: the driver, passengers, and vulnerable road users. Safety in Use is particularly critical for those within the vehicle.

Performance requirements are largely derived from the identified traffic and operational scenarios. To identify potential failures and anomalies, deductive analyses like Fault Tree Analysis (FTA) are employed.

Vehicle design is heavily influenced by its historical evolution. Features like four wheels and Ackermann steering are deeply ingrained. Functional effect chains, system state machines, sequence diagrams, and timing diagrams are used to describe the system's behavior. Anomalies and errors identified in these behavioral descriptions can be further analyzed using techniques like Failure Mode and Effects Analysis (FMEA). This allows for a systematic breakdown of the system to understand the root causes of potential failures. At the vehicle level, the FMEA should address not only dysfunctional or malfunctioning behavior but also how the system impacts the defined safety goals. This analysis cascades down through various system levels, influencing design and architecture decisions, and ultimately guiding the implementation at the component level.

Verification activities occur horizontally and vertically throughout this development matrix. Integration processes are intertwined with further verification and validation efforts. All work products, including the safety argumentation, culminate in a comprehensive Safety Case (SC). This structured approach ensures that the final vehicle design meets all requirements, performs as intended, and prioritizes the safety of all road users.

7.5 Automotive Control Systems

In today's automotive electronics, the fundamental principle of the Von Neumann architecture still prevails. This means that the sequence of operations is controlled by a program stored in the memory. Essentially, the system fetches instructions from the memory, decodes them, and then executes them sequentially. This architecture ensures that the vehicle's Electronic Control Units (ECUs) can efficiently manage various functions, from engine control to infotainment systems, by following a predefined set of instructions stored in the program memory.

A microcontroller or microprocessor must be enabled by an operating system to organize the entire periphery, various memory units, and the Arithmetic Logic Unit (ALU) into a structure that makes the controller applicable for adapting to application requirements. At the hardware-software interface, so-called low-level drivers are required to establish the connection and access between registers and other functional units. Much of the ALU and low-level memory, cache, and so on, is often implemented in controller-specific firmware.

For microprocessors, the POSIX architecture (Portable Operating System Interface) is used with virtual memory units (VMUs) or general virtual memory management to adapt the hardware to the application task. In embedded controllers and microprocessor applications, a so-called middleware is developed today to provide the framework for application software. Within the framework of Classic AUTOSAR, the overall architecture is referred to as the basic software, which systematically prepares the controller for the application software. To extend this classic approach to microprocessors with POSIX architectures and networks, the AUTOSAR standard has been expanded with Adaptive AUTOSAR.

Classic AUTOSAR

Classic AUTOSAR is a standardized software architecture designed for automotive Electronic Control Units (ECUs). It is structured into three main software layers (Figure 7.51): the application layer, the Runtime Environment (RTE), and the basic software (BSW). The application layer contains the application functionality, which should be abstracted from the hardware. The RTE manages communication between the application software and the BSW, providing a clear separation between component-based and layered approaches. The BSW includes services, ECU abstraction, and microcontroller abstraction, offering functionalities such as operating system services, network communication, memory management, and diagnostics.

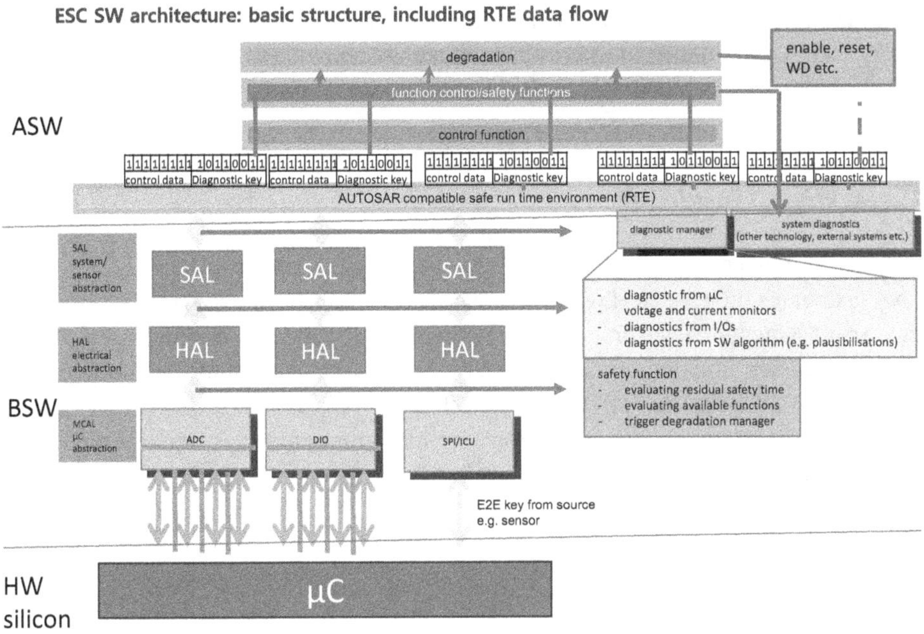

Figure 7.51 Typical AUTOSAR application

Adaptive AUTOSAR

Adaptive AUTOSAR, on the other hand, is designed to address the needs of modern, connected, and highly automated vehicles. It supports dynamic deployment of applications and is suitable for high-performance computing platforms. Adaptive AUTOSAR provides a flexible and scalable architecture that can handle complex software functions, such as those required for autonomous driving and Advanced Driver Assistance Systems (ADAS). It allows for the integration of new technologies and communication standards, ensuring that the software can be updated and extended over the vehicle's lifetime.

By following such common frameworks, one can appreciate the systematic preparation and adaptability required for modern automotive software systems.

7.5.1 Control Theory

Control theory is the science of designing and analyzing systems that modify or manipulate variables to achieve desired outcomes. In the context of automotive engineering, control theory plays a pivotal role in ensuring safety, efficiency, and optimal performance. With the beginning of electrification in the automotive industry, more and more electrical control systems were integrated into the vehicles. Automotive control systems aim to maintain stability, improve safety, enhance fuel efficiency, and optimize vehicle behavior. Whether it's maintaining a steady speed on the highway, preventing skidding during braking, or adjusting engine parameters for optimal fuel consumption, control theory provides the framework.

7.5.2 E/E Architecture

The Electrical/Electronic (E/E) architecture of a vehicle is the foundational framework that delineates the electrical power network, communication networks, and the geometric placement of electronic systems within the vehicle. This architecture is pivotal for the vehicle's functionality and efficiency and exerts a considerable influence on the vehicle's design. A fundamental approach to E/E architecture is the Controller Area Network (CAN) bus-based tree approach, which interconnects various vehicle domains such as chassis, powertrain, charging, body, Human-Machine Interface (HMI), and Advanced Driver-Assistance Systems (ADAS). This hierarchical network facilitates efficient communication between control units.

Examples of E/E architectural approaches include the following:

- **Zonal architecture:** Control units are grouped within specific zones of the vehicle, simplifying wiring and reducing complexity.
- **Ring architecture:** A ring structure connects all control units in a closed loop, offering redundant communication paths and enhancing fail-safety.
- **Service-oriented network solutions:** These leverage software definitions to provide services, allowing for greater flexibility and scalability.
- **High-power computing approach:** Instead of multiple gateways, a centralized high-performance computing system is employed, centralizing data processing and communication.

Each of these approaches has its merits and demerits.

CAN bus-based tree approach:

- **Pros:** proven technology, widespread adoption, cost-effective
- **Cons:** Limited bandwidth can lead to bottlenecks in complex systems.

Zonal architecture:

- **Pros:** reduces the number of cables and control units needed; simplifies assembly
- **Cons:** can complicate troubleshooting as faults in one zone can affect multiple functions

Ring architecture:

- **Pros:** increased reliability through redundant paths, improved fail-safety
- **Cons:** higher costs, more complex wiring

Service-oriented network solutions:

- **Pros:** high flexibility facilitates updates and upgrades
- **Cons:** requires advanced software development and maintenance

High-power computing approach:

- **Pros:** centralized management, enhanced performance capabilities
- **Cons:** higher costs, potentially higher risk in case of system failures

These architectures underpin the electrical and electronic systems of a vehicle and are crucial for integrating modern technologies into vehicle design. They must be carefully considered to meet the demands of performance, safety, and user experience.

The Electrical/Electronic (E/E) architecture of a vehicle encapsulates its intrinsic values, reflecting the core attributes that define a vehicle's platform. This architecture is not merely a network of wires and protocols; it is a manifestation of the vehicle's potential and versatility. A robust E/E architecture showcases the breadth of variations and the potential of the vehicle's architecture. A well-designed E/E architecture allows for a range of functionalities to be integrated seamlessly, from basic vehicle operations to advanced driver-assistance systems. It supports the vehicle's scalability, enabling the addition of new features and technologies without extensive redesigns. This adaptability is crucial in today's rapidly evolving automotive landscape, where consumer demands, and technological advancements are in constant flux.

Moreover, the E/E architecture is a significant marker for the vehicle platform because it directly impacts efficiency, safety, and user experience. It determines how power is distributed, how systems communicate, and how well the vehicle adapts to new challenges and requirements. In essence, the E/E architecture is the backbone that supports the vehicle's intelligence, defining its capability to evolve and embrace future innovations. A vehicle with a dynamic and flexible E/E architecture is poised to meet the demands of modern mobility and stand out in the competitive automotive market.

It is essential to highlight that the E/E architecture is the backbone of a car, providing the foundational structure upon which all essential performance characteristics are built. However, it does not necessarily have to significantly influence the vehicle's design. The E/E architecture underpins the critical functionalities of a vehicle, from managing power distribution and electronic communications to supporting advanced features such as autonomous driving and connectivity.

While the E/E architecture is integral to the vehicle's operation, it operates largely behind the scenes. Designers can often work around the E/E architecture when crafting the vehicle's aesthetics, allowing for creative freedom in the vehicle's external and internal appearance. This means that while the E/E architecture is crucial for performance, safety, and functionality, it can be designed in a way that minimizes its impact on the overall design of the vehicle.

The E/E architecture is the unsung hero of automotive engineering, quietly but critically supporting the vehicle's capabilities without dictating its design language. It is the basis for the vehicle's key performance features, ensuring that the car operates smoothly, efficiently, and safely, while leaving room for designers to define the vehicle's visual identity.

7.5.3 Basic Control Systems

Basic control systems consist of the following (Figure 7.52):

- **Sensors:** These devices collect data from the environment (e.g., wheel speeds, accelerometers, temperature sensors).
- **Actuators:** Actuators execute control actions (e.g., adjusting throttle, applying brakes, steering).
- **Controller:** This is the brain of the system. It processes sensor data, computes control signals, and adjusts actuators.

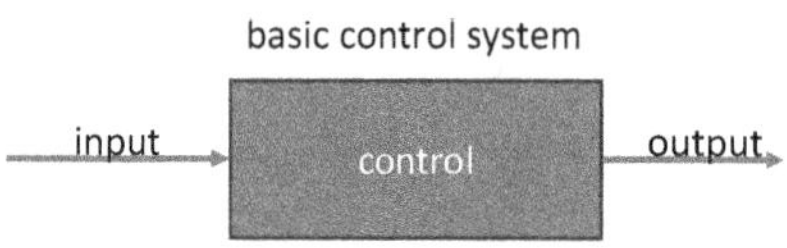

Figure 7.52
Basic control system

Basic control systems are extended. Typical open loop control systems or Model Predictive Control (MPC) do not use feedback during operation. They use predictive models to optimize control actions over a future time horizon. Adaptive Control Systems (ACS) adjusts control parameters based on varying conditions.

Neural Networks and Machine Learning learn from data to enhance control performance. There are two basic control principles for neuronal networks:

- tuned field feedback systems
- trained field feedback systems

Tuned Field Feedback Systems

Definition: Tuned field feedback systems are control systems where the controller parameters (such as gains, time constants, etc.) are manually adjusted or "tuned" based on engineering expertise and experience.

Process:

- Engineers analyze the system dynamics, behavior, and requirements.
- They adjust the controller parameters to achieve desired performance (e. g., stability, response time, robustness).
- Tuning often involves trial and error, simulation, and real-world testing.

Pros:

- **Simplicity:** parameters directly set by engineers
- **Intuitive:** based on practical understanding
- suitable for well-understood systems

Cons:

- **Labor-intensive**: requires manual adjustments
- It may not adapt well to changing conditions.
- limited robustness

Trained Field Feedback Systems

Definition: Trained field feedback systems utilize Machine Learning techniques (such as reinforcement learning) to automatically learn optimal control policies.

Process:

- A simulation environment or real-world system is used.
- An RL agent interacts with the environment, learning from feedback (rewards or penalties).
- The agent adjusts its control actions based on the observed outcomes.
- Training continues until the agent performs well.

Pros:

- **Adaptability:** learns from data and adapts to changes
- **Robustness:** handles uncertainties and disturbances
- **Scalability:** can handle complex systems

Cons:

- **Complexity:** requires understanding of RL algorithms
- **Data requirements:** needs sufficient training data
- exploration vs. exploitation trade-offs

Validation approach for Safety:

The behavior of open-loop systems could be completely verified and validated during development, if the environmental model is complete in the laboratory simulation (Figure 7.53). For any input a possible output could be defined, so that the behavior of the open-loop control function is unambiguous. Safety mechanisms can be implemented that assure the correct input/output relation by checks during runtime:

- the E2E behavior according to a mathematical or logical formular
- reverse calculating of the function
- signal range checks
- change rate checks
- gradient checks

In summary, tuned field feedback systems rely on manual tuning by engineers, while trained field feedback systems leverage Machine Learning to automatically learn control strategies. The choice depends on system complexity, available data, and desired adaptability.

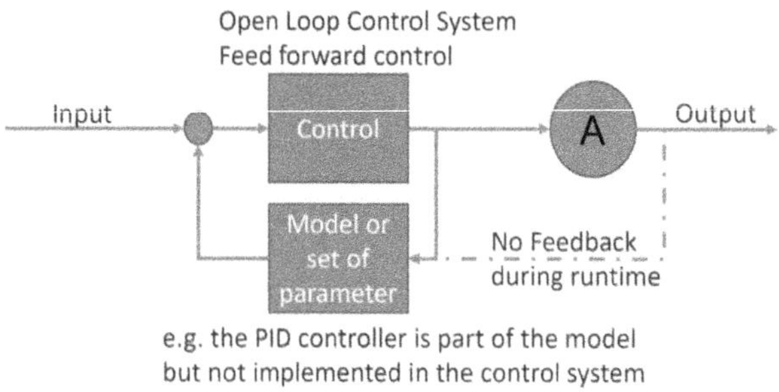

Figure 7.53
Variants of open-loop control systems

Open-Loop Control (Feed Forward Control, FFC)

Input: In an open-loop control system, we start with an input signal (often a desired setpoint or reference value).

Processing: The input signal is processed by a controller to generate a control action.

Output: The control action is directly sent to an actuator (e. g., motor, valve, heater).

No feedback: Importantly, there is no feedback from the system. The controller assumes that the actuator will achieve the desired effect.

Practical example: A microwave oven with a timer setting is one example. You set the time, and the microwave runs for that duration without adjusting based on the actual temperature inside.

Model-Based Design

Before deploying the control system, we create a mathematical or logical model of the process. This model describes how the system responds to changes in input and disturbances. The model includes parameters (such as time constants, gains, etc.) that affect the system's behavior.

Controller design: Based on the model, we design a controller (e. g., PID controller). The controller parameters (proportional gain, integral time, derivative time) are tuned using simulation or offline testing.

Pre-field validation: We validate the controller's performance using the model and simulated scenarios. Adjustments are made to ensure stability, robustness, and desired response characteristics.

Deployment: The tuned controller is deployed in the real system. However, there is no real-time feedback from the system during operation.

Pros:

- improved performance due to pre-tuning
- faster commissioning time
- predictable behavior based on the model
- suitable for systems with minimal disturbances

Closed-Loop Control (Feedback Control), see Figure 7.54

Input: It is similar to open-loop. We start with an input signal (desired setpoint).

Processing: The input signal is processed by a controller.

Feedback: Unlike open-loop, here we have feedback from sensors measuring the actual system output (e. g., temperature, position, pressure).

Comparison: The controller compares the desired setpoint with the actual output (feedback).

Adjustment: Based on the feedback error, the controller adjusts the control action to minimize the difference.

Practical example: A thermostat controlling room temperature is one example. It adjusts the heating or cooling based on the actual room temperature.

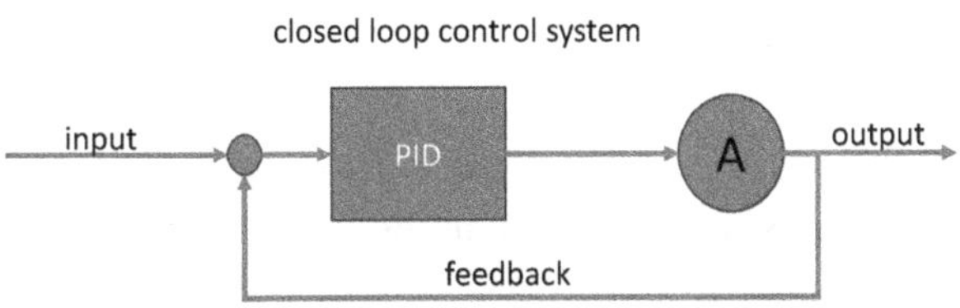

Figure 7.54 Closed-loop control system

Closed-Loop Control with real actuator feedback

Model-based design (optional): Similar to open-loop, we create a mathematical model. However, the model is not strictly required for closed-loop control.

Controller design: We design a controller (e. g., PID controller) based on the desired system response. Initial parameters may be estimated from the model or based on experience.

Deployment and feedback: The controller is deployed in the real system. The system operates, and the controller receives real-time feedback from sensors.

Online tuning and adaptation: The controller continuously adjusts its parameters based on the actual system behavior. If the system changes (due to wear, environmental conditions, etc.), the controller adapts.

Pros:

- **Robustness:** adapts to variations and disturbances
- **Real-world performance:** considers actual system behavior
- suitable for dynamic systems with uncertainties
- improved accuracy
- stability
- responsiveness

Validation approach for safety: The behavior of closed-loop systems, especially with real world feedback, could *not* be completely verified and validated during development. Gradients and states of the environmental conditions could not be assessed completely in the laboratory simulation. Only the control parameter and the gradient of the output could be checked during runtime. In safety applications, a safe control approach shall be implemented to assure that the vehicle does not enter a dangerous state.

7.5.4 Timing Constraints in Closed-Looped Control Systems

In closed-control systems there are many functional and temporal dependencies. This affects the execution of the control algorithm as well as the temporal aspects in the data transport to the controller as a reference variable or as feedback, internally in the controller and from the controller to the actuator.

An automotive power steering system is a prime example of a closed-loop control system (Figure 7.55). The Electric Power Steering (EPS) system uses an electric motor to control the vehicle's steering. The EPS system is based on the sensors listed in the following.

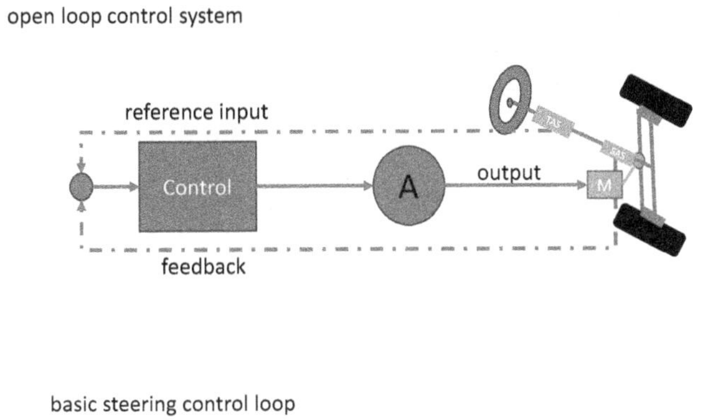

Figure 7.55
Basic Power Steering System

Torsion Angle Sensor (TAS)

The Torsion Angle Sensor (TAS) consists of an angle sensor and a torque sensor element. The TAS measures the angle over a torsion bar and the commanded hand torque from the driver. This measured torque is the demanded torque, which serves as the input or reference value for the closed-loop controller.

Steering Angle Sensor (SAS)

The Steering Angle Sensor (SAS) provides the rotation angle from the steering wheel column. It plays an essential role in modern vehicle safety systems by constantly monitoring the driver's steering input. This sensor enhances safety by aiding in stability control during evasive maneuvers, ensuring precise steering accuracy for peak vehicle control.

Timing Issues in Closed-Loop Systems

In closed-loop control systems, timing is crucial. The control algorithm must execute at the right time to ensure the system responds correctly. Delays or inaccuracies in the control algorithm's execution can lead to system instability or performance degradation. Data transport in these systems also has temporal aspects. The data, whether it's a reference variable or feedback, must be transported to the controller in a timely manner. Delays in data transport can cause the controller to make decisions based on outdated information, leading to system instability.

Actuator in Closed-Loop Control Systems

The actuator in a closed-loop control system is responsible for executing the control action. In an automotive power steering system, the actuator might be the electric motor that turns the steering wheel and the steering knuckles. The actuator must respond quickly and accurately to the controller's commands. Delays or inaccuracies in the actuator can lead to system instability or performance degradation.

Timing issues in closed-loop control systems, such as an automotive power steering system, can significantly impact system performance and stability. These issues can arise in the execution of the control algorithm, the data transport to the controller, and the actuator's response to the controller's commands. Therefore, it's crucial to consider these temporal dependencies when designing and implementing closed-loop control systems.

A Proportional-Integral-Derivative (PID) controller is a control loop mechanism that is widely used in industrial control systems and a variety of other applications requiring continuously modulated control. The mathematical formula for a PID controller is often called the parallel equation.

Here are the typical terms and variables in the formula:

$$m = K_p e + \frac{1}{\tau_i} \int edt + \tau_d \frac{de}{dt} + b$$

m: This is the output of the PID controller.

K_p: This is the proportional gain. It determines how much of the error will be corrected immediately.

e: This is the error, calculated as the difference between the desired setpoint and the measured process variable.

τ_i: This is the integral time constant. It determines how much of the past errors will be corrected.

$\int edt$: This is the integral of the error with respect to time. It represents the accumulation of past errors.

τ_d: This is the derivative time constant. It determines how much of the future trends will be corrected.

de/*dt*: This is the derivative of the error with respect to time. It represents the rate of change of error.

b: This is the bias or offset. It is used to achieve a desired value in the steady state.

The PID controller continuously calculates an error value as the difference between a desired setpoint and a measured process variable and applies a correction based on proportional, integral, and derivative terms. This allows the PID controller to automatically apply accurate and responsive correction to a control function.

Based on the formula it becomes obvious that timing and timing differences have a significant impact on the control function and consequently on the stability of the steering function and resulting vehicle stability.

If you look at an EPS as a PID controller, then in a typical system implementation there are the following latency intervals (Figure 7.56):

- **t_TASi:** time interval to perceive the physical hand torque of the driver and generating the EE-output for the controller
- **t_TASo:** time interval of the torque and angle signal before the signal is available for processing of the control or PID algorithm (if the TAS is a digital signal or a bus signal, the time interval needs to be considered)

- **t_CC:** This is the time interval to acquire the **TAS signals** and the synchronized information from the **SAS feedback**. Based on this synchronized input information, the control output would be generated.
- **t_Co:** time interval to send the controller output to the EE-actuator
- **t_EE_A:** time interval to generate the information and the energy to generate the physical movement of the physical actuation
- **t_p_A:** time interval to generate, based on the electrical input, the physical movement of the actuator and consequently of the vehicle
- **t_SASi:** This is the time interval of the SAS to perceive the reaction of the physical actuation, in this case the steering angle of the vehicle. In current technology, it is a linear function to the steering knuckles, but it is no equivalent to the lateral movement of the vehicle. To acquire the feedback for the lateral movement of the vehicle, other feedback information is necessary.
- **t_SASo:** time interval of the controller to make the steering angle available for the control function

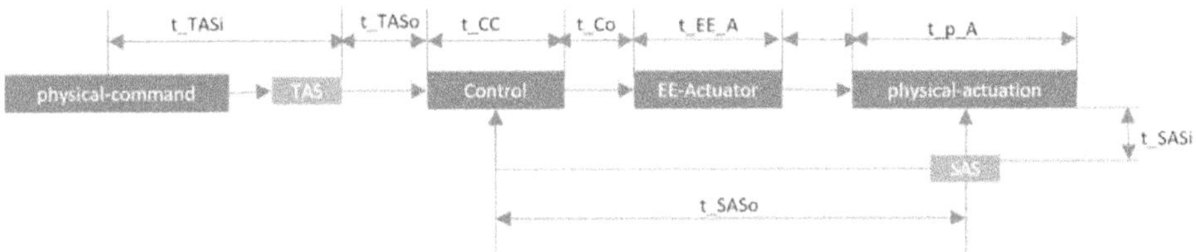

Figure 7.56 Timing in an EPS system

Along a functional effect chain, a temporal distribution of the time budgets can already be described over the physical end-to-end (E2E) function. This means that while you can describe how time is distributed across the entire functional chain, these descriptions shouldn't be treated as strict requirements. When implementing redundant functional chains, it's essential to plan and synchronize the timing to ensure everything works correctly.

7.6 Safe Control Solutions

The most common safety approach is the VDA EGAS Approach, which had been tailored during recent years to an AUTOSAR compatible approach with an QM RTE (run time environment) and a safety RTE (Figure 7.57). A QM RTE could not be applied for safety integrity functions which are typically attributed with an ASIL. Unfortunately, the AUTOSAR standard does not provide mechanisms for temporal and spatial separation. For example, in memory for communication data, the entire caching mechanisms are typically not designed for safety applications.

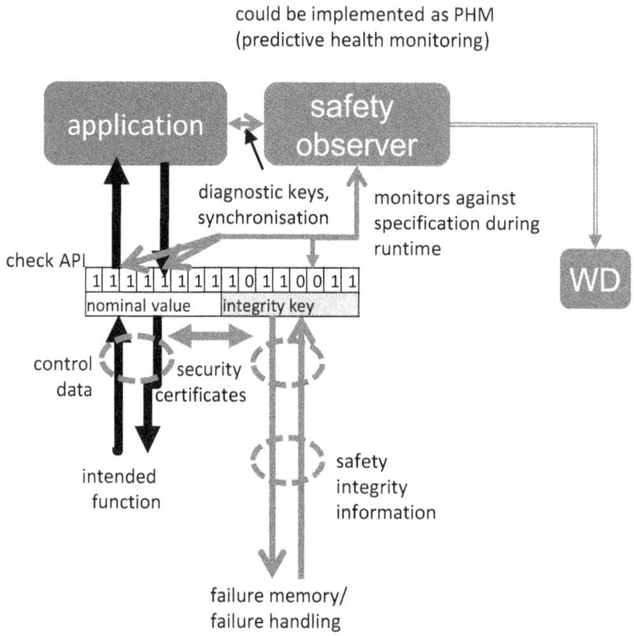

Figure 7.57
Basic EGAS Approach

To implement closed-loop control systems and time-critical systems, special measures are necessary in the software architecture to manage typical temporal dependencies. This often requires redundancies to compensate for stochastic runtime variations.

In essence, ensuring that these systems function correctly involves careful planning and design to handle timing dependencies and variations, often by incorporating redundant pathways to maintain reliability and performance.

To manage typical stochastic runtime variations, a Safety Observer can be implemented. Checkpoints can be integrated into the application software and the data flows of the basic software. When these checkpoints are reached during program execution, they provide a signed checksum to the Safety Observer. The Safety Observer can be implemented as Predictive Health Monitoring (PHM) based on this information. The Safety Observer can also deliver error predictions through trend analysis (AUTOSAR 2022). This approach ensures that the system can monitor its performance in real-time and predict potential issues before they become critical, enhancing overall reliability and safety.

These are the arguments for a Safety Observer:

- In the architecture, you only see static architecture and not the runtime behavior during the operation of the software, but the correct availability of the application data must be made available to an API or the RTE (Safety RTE) in accordance with the specifications.
- The error monitors according to the AUTOSAR solutions only address static errors. Short sporadic errors, runtime violations or sequential errors in runtime behavior are not checked at runtime as is the case with debugging.
- Causes, here bus errors, hardware errors, cable errors and errors in the basic software cannot be determined without observation during runtime.
- In particular, the influence of systematic errors such as aging, EMC, and increased bus load can no longer be traced as a consequence of the error.
- The determinism of control units and their performance quality can be guaranteed by a runtime observer.
- In a hard real-time system, where the violation of the real-time thresholds leads to a violation of the safety goal, the observer must be implemented in accordance with the highest ASIL.
- In case of a single implementation of time-critical closed-loop control system, any deviation of the timing could lead to a violation of a safety goal or the assigned Fault Tolerance Time Interval (FTTi). In case of any timing failure, a shutdown of the control circuit is the only solution. A switch to a independent redundance could be a solution.

The PID controller should be implemented independently and redundantly to compensate for variations in its own timing execution (Figure 7.58). This ensures that the first valid value can be directed to the outputs, avoiding stochastic timing error latencies. However, the effectiveness of these measures depends on the type of microcontroller. Many multicore solutions have dependent program, cache, and application memory, making this approach rarely effective. In such cases, the redundant PID controllers need to be implemented on different hardware microcontrollers. This strategy helps to maintain the reliability and accuracy of the control system by mitigating timing-related issues.

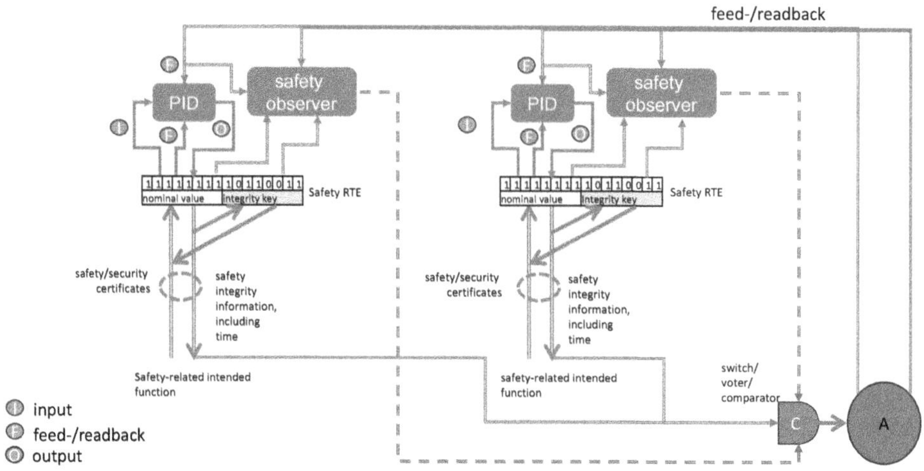

Figure 7.58 Redundant PID controller

To effectively implement such redundant closed-loop systems, it is crucial to consider the typical separation mechanisms for the control loop (Figure 7.59). This means that the setpoint (reference value) must be independent of the feedback, which corresponds to spatial and temporal separation.

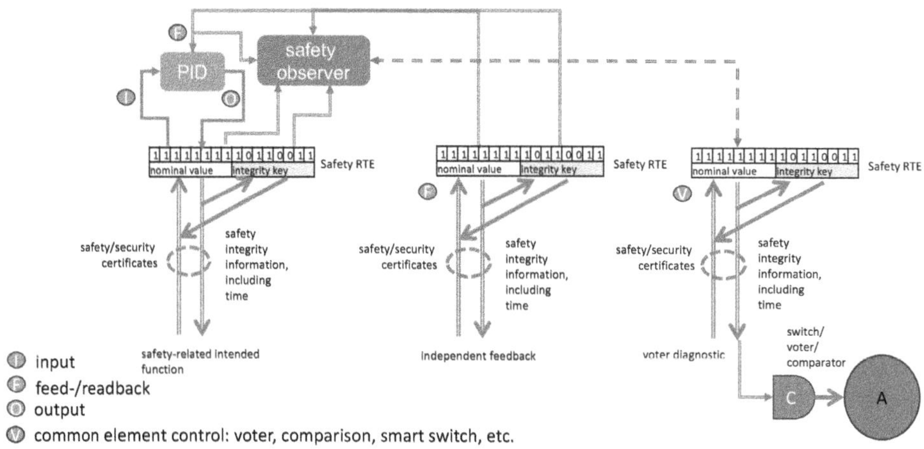

Figure 7.59 Separated RTE

Spatial and temporal separation: The setpoint should be isolated from the feedback to ensure that the control loop operates independently in both space and time. This helps in maintaining the integrity of the control process.

Voting and comparison mechanisms: To realize mechanisms like voting, comparison, or first valid serve, the final elements must be supplied with independent redundant data flows. This ensures that the system can make accurate decisions based on multiple sources of data.

Cybersecurity measures: Each data line should be secured with MACsec (Media Access Control Security) to prevent hackers from intentionally affecting the controller's runtime. This is critical to ensure that the control system remains reliable and secure.

Critical timing considerations: In applications like by-wire steering, even a delay of 50 milliseconds can result in the vehicle leaving its lane or moving into oncoming traffic. Therefore, it is essential to minimize latency and ensure timely responses.

Hardware considerations: The effectiveness of these measures depends on the type of microcontroller used. Many multicore solutions have dependent program, cache, and application memory, making redundancy less effective. In such cases, redundant PID controllers should be implemented on different hardware microcontrollers to ensure true independence.

By following these principles, you can design robust and reliable redundant closed-loop systems that are capable of handling timing variations and maintaining high levels of safety and performance.

7.6.1 Aspects of Realtime

Realtime systems are all about lean, high-precision processes. A distinction is made between, among other things, "hard" and "soft" real time. Real-time systems are systems that must produce correct reactions within the shortest possible defined period of time. If the reactions exceed the time limit, this leads to reduced performance, malfunctions and/or even danger to people and material. DIN 44300 from the German Institute for Standardization describes the term real time as follows:

"Real time refers to the operation of a computing system in which programs for processing data are constantly ready for operation in such a way that the processing results are available within a predetermined period of time. Depending on the application, the data can be generated randomly or at predetermined times."

In contrast, the real-time system is described in the Oxford Dictionary of Computing as follows:

"A real-time system is a system in which the time at which outputs are available is significant. This is usually because the input corresponds to some changes in the physical world and the output must relate to those changes. The delay between the time of input and the time of output must be sufficiently small for acceptable 'timeliness'."

Depending on the requirements, a distinction is made in practice between "hard" and "soft" real-time systems. This distinction is made exclusively based on the nature of the consequences of violating the time limits.

A real-time system is called a **hard real-time system** if exceeding the response time limits can have significant consequences. These consequences include endangering people and damaging machines, i. e., impacts on health and the integrity of the environment. Typical examples of this are some control systems in airplanes or cars, e. g., brake system.

A violation of the execution times in a **soft real-time system** only leads to a reduction in quality, but not to damage or danger. Examples of this are multimedia systems in which the occasional deviation from a playback rate of 30 frames/sec. leads to jerking or similar.

As a note, it should be added that almost always only the upper time limits are listed. The reason for this is that compliance with an upper time limit requires considerable design effort in cases of doubt, while a lower limit, i. e., a minimum time before which no reaction is allowed, is structurally harmless. An example of a system in which both values are important is the control of the ignition timing in the internal combustion engine: This can only occur within a narrow ignition interval.

7.6.1.1 Operating Systems

Classic AUTOSAR used OSEK-OS for most of the early AUTOSAR solutions. OSEK is the abbreviation for: "Offene Systeme und deren Schnittstellen für die Elektronik in Kraftfahrzeugen" (Open System and Interface for Electronic in Motor Vehicles). OSEK is an open standard developed by the German automotive industry to define a common interface for embedded software in motor vehicles. It aims to facilitate the development of automotive control systems by providing a standardized framework. However, OSEK itself is not an RTOS (Real-Time Operating System).

OSEK shows the following characteristics:

- **Purpose:** OSEK defines a set of specifications for automotive software architecture, communication, and task scheduling.
- **Safety-critical focus:** OSEK is primarily used in safety-critical automotive applications, such as engine control, braking, and airbag systems. Consider that OSEK itself provides no safety characteristics, any safety mechanism needs to be implemented additionally.
- **Static memory allocation:** OSEK-compliant systems allocate memory statically during compilation time.

The following characteristics are typical on task levels:

- **basic tasks:** execute to completion without blocking
- **enhanced tasks:** support blocking and sleep states
- **Priority ceilings:** OSEK uses priority ceilings to prevent priority inversion.

While OSEK is widely used in automotive ECUs, it does not inherently guarantee safety or timing aspects. Safety-critical applications often require additional safety mechanisms beyond OSEK compliance.

Here are some examples for OSEK variants:

- **OSEK/VDX:** basis for AUTOSAR Classic and widely used in automotive ECUs
- **OSEK time:** an extension for time-triggered scheduling in OSEK-compliant systems

OSEK is not an RTOS but rather a framework for automotive software development. When safety and real-time guarantees are critical, developers often combine OSEK with an actual RTOS or other safety mechanisms.

7.6.1.2 RTOS and ROS

In the past years, the electrification of vehicle propulsion systems and the increased use of ADAS and AD functions in the automotive applications ROS (Robotics Operating Systems) and RTOS (Real-Time Operating Systems) came in the focus of automotive applications.

What is RTOS?

Purpose: RTOS is designed for applications where timing is critical, ensuring predictable and deterministic behavior.

Real-time guarantees: RTOS provides strict timing guarantees for task execution. It ensures that critical tasks meet their deadlines.

Resource efficiency: RTOS minimizes resource consumption (memory, CPU) while maintaining real-time behavior. It is ideal for embedded systems, microcontrollers, and safety-critical applications.

RTOS shows the following key features:

- task scheduling:
  - prioritizes tasks based on urgency
  - supports preemptive or cooperative scheduling
  - ensures timely execution of critical tasks
- deterministic behavior:
  - predictable response times for interrupts and task execution
  - minimizes jitter (variation in response time)

- resource management:
  - efficient memory allocation (static or dynamic)
  - minimal overhead for context switching

RTOS is used for safety-critical applications in automotive ECUs, industrial control systems, medical devices, etc.

The following advantages of RTOS could be considered:

- **predictable timing:** ensures tasks meet deadlines
- **resource optimization:** efficient use of memory and CPU
- **safety and reliability:** critical for safety-critical systems

What is ROS (Robot Operating System)?

- **Purpose:** ROS is not an actual operating system but a middleware framework for robotics and automation.
- **Modularity and Abstraction:**
  - ROS provides a set of libraries and tools for building robot applications.
  - It focuses on modularity, communication, and sensor-to-actuator abstraction.
- **Platform Independence:** ROS is platform-independent and widely used across various robots and hardware.

ROS shows the following key features:

Node-based architecture: ROS applications consist of nodes (software modules) that communicate via topics (publish-subscribe) or services (client-server) etc.

Typical automotive applications for ROS:

- sensor integration:
  - ROS abstracts sensors (e. g., cameras, lidars) into standardized interfaces
  - allows easy integration of various sensors
- simulation and visualization:
  - Solutions like Gazebo and RViz provide simulation and visualization tools.
  - Developers can test and visualize robot behavior.

In the past years several community and ecosystems established, that offers extensive package repository (ROS packages) and supports rapid prototyping and collaborative development.

ROS2 (Robot Operating System 2)

ROS2 offers the following improvements:

- real-time capabilities:
  - ROS2 introduces real-time features using the Data Distribution Service (DDS) protocol.
  - achieves deterministic communication and timing guarantees
- middleware layer:
  - ROS2 uses DDS as the middleware layer for communication.
  - supports Quality of Service (QoS) settings for reliability, latency, and bandwidth
- embedded systems and microcontrollers:
  - ROS2 targets resource-constrained devices.
  - enhancements for embedded architectures (e. g., ARM Cortex-M)
- POSIX compatibility:
  - ROS2 integrates with POSIX APIs, making it easier to port existing code.
  - supports both ROS2-specific APIs and POSIX APIs

ROS2 interfaces are commonly used for applications like V2X (Vehicle-to-Everything). It enables vehicles to exchange information with infrastructure, other vehicles, and pedestrians.

ROS2 facilitates sensor fusion by integrating data from multiple sensors (e. g., cameras, lidars, radars). You can create comprehensive perception pipelines.

RTOS provides real-time guarantees for critical tasks, while ROS focuses on modularity, abstraction, and sensor integration. ROS2 enhances real-time capabilities, supports embedded systems, and extends to V2X and sensor fusion applications. Preferably it would be great to combine all these advantages in one operating system.

7.6.2 Event Control or Time Control?

The immediate question now arises as to how hard real-time systems can be designed. A distinction can be made here between two different concepts, which of course appear mixed in practice: event-triggered and time-triggered systems.

- **Event-controlled systems** are controlled by interruptions: If there is an event on a sensor (what that is, of course, must be defined), then it can send an interrupt request to the processor and thus draw attention to its request for operation. An asynchronous Interrupt Request (IRQ) is a signal generated by the environment external to the processor that indicates status and/or requests processing by the processor. This signal is not synchronized with the program run. The interrupt is handled as part of the interrupt service routine.

- In **time-controlled systems**, there is no reaction to input events. The interruptions are only triggered by one or possibly several periodic timers (timers). Sensors are then actively queried by the control unit. This method has the great advantage that the behavior of all system activities can be completely planned at compile time. This is a significant advantage, especially for use in real-time systems, as it can be checked a priori whether real-time requirements are being met.

Depending on the consequences, a distinction is sometimes made between hard real-time, soft real-time and firm real-time. Different real-time requirements apply in each case.

- **Hard real-time requirements:** Exceeding the response time is considered a failure. After an exact time recording of the application to be provided, calculations according to the theory of real-time systems are necessary. Real-time systems always deliver the correct result within the specified time limits. This property can be relied upon when using a hard real-time system.
- **Soft real-time requirements:** Such systems typically process all incoming inputs quickly enough. For example, the response time reaches an acceptable mean or another statistical criterion. The time requirements should be seen here as guidelines. Exceeding the time requirement does not have to be considered a failure. On the one hand, the time can often be exceeded as long as it is still within a tolerance range. On the other hand, it can rarely be exceeded significantly.
- **Firm real-time requirements:** With firm real-time requirements, there is no immediate threat of damage. However, after the time requirements expire, the result of the calculation is useless and can be discarded.

Depending on the problem and perspective, the following definition is also used:

- **Soft real-time requirements:** The system must respond in the specified time period but not deliver the complete result. If there is no response, the process is considered failed and is aborted. Alternatively, soft real-time systems can sometimes deliver the result correctly, but too late.
- **Hard real-time requirements:** The system must present the result completely within the defined time window.

Within the soft real-time systems, there are sometimes additional classifications that differentiate the response time exceedances more precisely. Common criteria are the following:

- The result no longer has any value. The calculation is aborted and discarded.
- The usefulness of the result decreases from the end of the response time.
- Exceeding the response time is accepted and the result is accepted when it is available.

Even the definition of soft real-time systems is of a rather colloquial nature, so that a finer division is difficult to give.

The DIN standard 44300 defines real-time operation (there referred to as real-time operation) as the operation of a computer system in which programs for processing the resulting data are constantly ready for operation in such a way that the processing results are available within a predetermined period of time. This terminology standard has not been able to establish itself in practice as the only accepted definition. Terms from the English-speaking world dominate.

According to DIN 44300, the following definition applies to the term "real-time system:" It is an operation of a computing system in which programs for processing the resulting data are constantly ready for operation in such a way that the processing results are available within a specified period of time. Depending on the application, the data can be generated randomly or at predetermined times. This means that a real-time system must guarantee both the functional correctness and the time it takes to determine the results (temporal correctness). A distinction must be made between hard and soft real time. For example, a video conferencing system that records, sends, and displays images and sound within milliseconds only needs to meet soft real-time requirements, as a delay in some images only results in a "jerky" display. In contrast, a car's engine controls must adhere to hard real-time, as failure to adhere correctly will result in mechanical damage or, in the worst-case scenario, an accident.

Real-time requirements: Real-time systems have the property of reacting reliably to an external stimulus within a specified period of time [in real time]. With regard to compliance with the time limit, a distinction is made between hard and soft real-time requirements. A hard real-time requirement exists when all required system responses to an external stimulus must arrive absolutely reliably within a fixed period of time, otherwise there is a risk of significant damage. A soft real-time requirement, on the other hand, exists if exceeding a specified time limit can be tolerated to a certain extent since there are no fears of fatal consequences.

7.6.3 Modelling of Behavior

Markov chains are a mathematical framework used to model systems that transition between different states. They are particularly useful for predicting future behavior when the system's future state depends only on its current state, known as the Markov property.

The process of using Markov chains for prediction involves defining the relevant states, determining the probabilities of transitioning between states, and organizing these probabilities into a transition matrix. By repeatedly applying the transition matrix to an initial state, we can generate a sequence of probability distributions that represent the likelihood of being in different states at future time steps.

Markov chains have found applications in various fields, including weather forecasting, financial modelling, traffic prediction, and customer behavior analysis. However, it's crucial to remember that the accuracy of predictions depends on the validity of the assumptions made about the system's behavior.

Here's a breakdown of how to use Markov chains for prediction:

- **State definition:** Clearly define the distinct states that the system can occupy. These states should capture the essential information relevant to the prediction task.

  For example, in weather prediction, states could be "sunny," "cloudy," and "rainy."

- **Transition probabilities:** They determine the probabilities of moving from one state to another. This can be done by analyzing historical data or using expert knowledge. Each transition is associated with a probability value, representing the likelihood of the system changing from one state to the next.

  For instance, if it's sunny today, the probability of it being rainy tomorrow might be 0.2, while the probability of it remaining sunny might be 0.7.

- **Transition matrix:** It organizes the transition probabilities into a matrix, known as the transition matrix. This matrix provides a clear representation of the probabilities of moving between any two states. Each row and column of the matrix represents a state, and the entry at the intersection of a row and column gives the probability of transitioning from the row state to the column state.

- **Initial state:** It identifies the current state of the system. This serves as the starting point for the prediction. The initial state can be obtained from observations or estimations.

- **Future state prediction:** In order to predict future states, repeatedly apply the transition matrix to the current state's probability distribution. This process generates a sequence of probability distributions, each representing the likelihood of being in different states at future time steps. The predictions are based on the current state and the transition probabilities encoded in the matrix.

- **Practical Applications:** Markov chains are particularly well-suited for systems that exhibit the Markov property, where the future state depends only on the current state and not on the history. Markov chains often rely on assumptions about the stationarity and Markov properties of the system. The accuracy of predictions depends heavily on the validity of these assumptions. If the system's behavior changes over time or if the future depends on past states, the predictions may be less reliable.

7.6.4 Safe State in Highly Available Systems

For highly available systems, a so-called safe state cannot be accepted as we know it from other systems. Most systems have an "ultimate safe state," which lead to the unavailability of the "intended function," in nearly all automotive systems a watchdog enables even the power-supply of the controller, so that the controller cannot recover the latest states and conditions into a buffered or non-volatile memory. The ultimate safe state is a shut down to a de-energized state. With that option you lose control for a dynamic system loss of control is the ultimate safety goal.

Safe State versus Minimal Risk Condition

In the context of IEC 61508, a safe state is a condition where safety is achieved. The fundamental principle of IEC 61508 is that any safety-related system must work correctly or fail in a predictable (safe) way. The standard requires that hazard and risk assessment be carried out for bespoke systems. The residual risk must be reduced to a level which is below the tolerable risk. This basic approach had been taken over by ISO 26262, but the idea of safety integrity in order to introduce a safety integrity function (a function assigned with an ASIL or SIL) to reduce the risk to a tolerable level is not that transparent and not often understood from the standard.

A Minimal Risk Condition (MRC) is a stable condition with an acceptable risk, given the situation when the decision to enter MRC is taken. It is a condition to which the Automated Driving System (ADS) may bring a vehicle after performing the Dynamic Driving Task (DDT). The transition from the DDT to MRC is called "Minimal Risk Maneuver" (MRM).

The difference between a safe state and MRC lies in their application and context. A safe state is a general concept used in safety-related systems where the system is in a state that does not pose any risk. On the other hand, MRC is a specific term used in the context of automated driving systems, where the vehicle is brought to a stable condition that minimizes risk to a tolerable level.

The need for introducing MRC in automated driving functions arises from the challenges of ensuring the safety of Automated Driving Systems (ADSs). One of the main challenges is that a direct validation of the safety of ADSs is both infeasible and potentially unsafe. Therefore, there is a need for alternatives to ensure safety before releasing ADSs on public roads. This is where MRC comes into play. It serves as a fallback function to achieve a stable state, reducing the risk of a crash when a trip cannot or should not be completed or at least the ADS is not able to perform the required functions with the required performance. This is particularly important when the ADS is at risk of exiting its operational design domain (ODD), or when the fallback ready user or the driver is unable or fails to take over the Dynamic Driving Task. In such cases, the ADS initiates a minimal risk maneuver function to provide enhanced safety for all passengers and other road users.

In the context of Automated Driving Systems (ADS), the vehicle is typically under the control of the ADS. However, when the ADS or the vehicle actuator systems are affected by a possible failure, the driver may not be able to take over control of the vehicle in the time interval immediately following the failure. This is due to the fact that the driver may not be prepared to immediately assume control, and the vehicle may not be in a state that allows for safe manual control.

To address this, additional active safety functions need to be implemented. These functions actively provide a transition to a Minimal Risk Condition (MRC) or maintain the Dynamic Driving Task (DDT) on a degraded performance level, allowing the driver time to take over control of the vehicle.

Alternatively, an independent redundant control instance can control the vehicle during the recovery of the failure. This control instance can switch back to the DDT, for example, through a hot standby mechanism, once the failure has been resolved.

In some failure conditions, a speed reduction can already mean an MRC. For instance, if the system is equipped with an auxiliary brake system, it could brake the vehicle, and a further independent steering system would be able to keep the vehicle in the lane. This means that depending on the system design, a reduced speed could already be considered an MRC.

It's important to note that the specific implementation of these safety functions and the definition of MRC can vary depending on the specific design and requirements of the Automated Driving System. The ultimate goal is to ensure the safety of all passengers and other road users.

Fault Tolerant Systems are designed to prevent disruptions arising from a single point of failure, ensuring the availability and business continuity of applications or systems. Fault-tolerant systems use backup components that automatically take the place of failed components, ensuring no loss of service.[18] However, they can tolerate some of their own failures, but they cannot control or tolerate failure of other systems involved of the control task. [19]

High availability refers to a system's ability to operate continuously—without downtime or failure—usually by using built-in failover mechanisms.[20] High availability systems are designed to operate without fail even in the case of unexpected events. The aim of high availability is to limit the duration of the downtime, not to completely eliminate it or control the downtime by an independent redundancy.[21]

---

[18] *https://www.imperva.com/learn/availability/fault-tolerance*

[19] *https://www.mdpi.com/2079-9292/9/9/1513*

[20] *https://www.f5.com/glossary/high-availability*

[21] *https://www.imperva.com/learn/availability/fault-tolerance*

The difference between fault tolerant systems and highly available systems lies in their approach to handling failures. While fault tolerant systems are designed to prevent disruptions and continue operating in the event of a failure, highly available systems are designed to minimize downtime and recover quickly when a failure occurs and using redundant control paths during downtime of the core system.[22)]

In the context of a vehicle, a highly available system aims to maintain the intended function even in case of any failure of the entire control circuit or effect chain. This includes control of failure of the mechanical parts (e. g., mechanical steering system), EE-system, and the related energy and power-supply systems.[23)] The goal is to ensure that the vehicle can continue to operate safely and effectively, even when certain components or systems fail. This is achieved through redundancy, failover mechanisms, and other strategies designed to prevent single points of failure and ensure continuous availability.[24)]

The state diagram represents various states of vehicle operation and the transitions between these states. It based on the following states (Figure 7.60):

- **Safe parked:** The vehicle is safely parked, locked to assure unauthorized access and the immobilizer is active to assure rolling away.
- **Steady driving:** The vehicle is driving in an adequate traffic environment, all systems are working correctly, and several monitoring functions check for external and internal events that require a transition to a degraded state.
- **Minimal risk condition:** The vehicle could be stopped at an adequate place in the traffic environment or is driving, e. g., with a lower speed, and redundant control systems are in case of any event able to control the moving vehicle. Takeover by the human driver could be also a minimal risk condition.
- **Dangerous condition:** The vehicle is in a condition where whether the driver, nor the system could control the vehicle. It usually means lateral or longitudinal control fails in a way that they are not able to control the vehicle adequately.

[22] *https://www.baeldung.com/cs/high-availability-vs-fault-tolerance*

[23] *https://reqi.io/articles/reliability-maintainability-availability-ram*

[24] *https://systemsatscale.io/basics/systems/redundnacy*

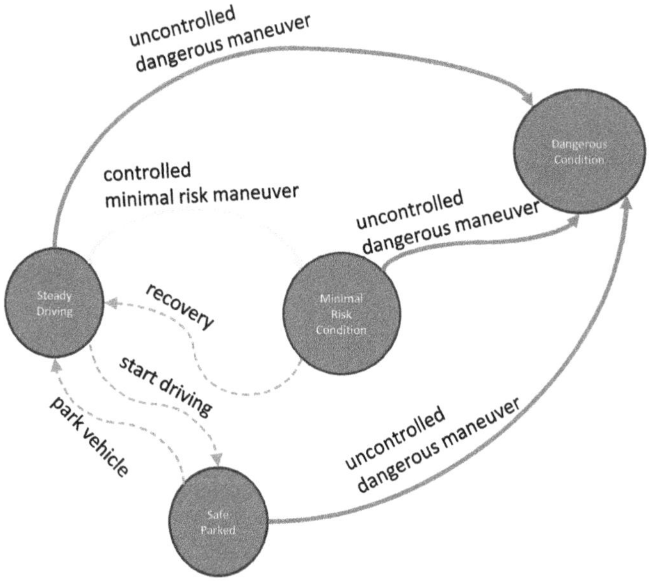

Figure 7.60
Principle of the Minimal Risk Condition similar to the safe state according to IEC 61508

The following transitions are shown as dashed lines (Figure 7.60):

- **From parked vehicle to steady driving:** Usually the driver approaches the vehicle and unlocks the vehicle, ignites the engine and starts driving. For an automated vehicle, the process is similar. The vehicle gets a driving task, the vehicle could be unlocked, e. g., to provide access for passengers and start routines. Pre-drive checks are performed. After successful pre-drive checks, the immobilizer should be deactivated so that the vehicle could start. Usually the park-lock locks the vehicle so that it can not roll away. It is an important condition for the Over-the-Air (OtA) flashing and charging of BEVs. Only in this condition, OtA and charging are permitted.
- **From steady driving to parked vehicle:** The driver stops the vehicle at the intended park field, turns off the engine and applies the park-brake. After an automated driving task, the vehicle stops as well, and the immobilizer would be activated. If necessary, the passengers leave the vehicle, and the vehicle would be locked so that the vehicle is again in a safe parking condition.
- **From MRC to steady driving:** In case of an MRC the vehicle could recover to steady driving, e. g., when the failure of the system is recovered, the external event that causes degradation is not present anymore or the vehicle is back in the intended ODD.

The dashed transition areas are of tolerable risks and also their transitions do not control any failure condition, even the failing of the active transition function would never lead to a dangerous situation (Figure 7.60).

The light-colored transitions are dangerous transitions (Figure 7.60). If they fail the vehicle could enter into dangerous or hazardous situations. The following transitions are in this class:

From steady driving to MRC (controlled minimal risk maneuver): The transition needs to be initiated by adequate detection mechanisms. The energy must be available to control the systems. The active transition function needs to be available to transfer the vehicle into the MRC.

There are three types of uncontrolled dangerous transitions:

- The parked vehicle could roll away. Unauthorized people could get access to the vehicle, drive and cause accidents, or hackers could start the vehicle and drive away.
- During steady driving a dangerous failure could not be detected so that a transition to MRC could not be initiated, or the active control function for the minimal risk maneuver could fail and the MRC could not be achieved.
- The vehicle could be in an MRC and the systems to maintain the MRC fails so that the vehicle loses lateral or longitudinal control. This is very dangerous since the MRC is driving with lower speed and the vehicle could not keep the lane or could not brake anymore. In case the driver takeover is the MRC, for example the steering or brake actuator could fail, and the driver could not control the vehicle anymore.

The state diagram could be converted into a Markov model. Markov models are a mathematical tool often used in safety analysis to represent the transitions between different states in a system. Each state in the model represents a condition of the system (like the states in the diagram), and the transitions between states are represented by probabilities. These probabilities can be used to calculate the likelihood of the system being in a particular state at a given time, which is useful for assessing the overall safety of the system. In the context of IEC 61508 (part 6), Markov models can be used to analyze the safety of a system and ensure that it meets the required safety standards.

7.7 Systems Engineering for Future Mobility Solutions

Modern transportation systems are increasingly complex, integrating numerous technologies and interacting with diverse stakeholders. To ensure safety, efficiency, and user acceptance, a holistic and structured approach is essential. A comprehensive systems engineering (SE) approach provides the necessary framework to manage this complexity, mitigate risks, and deliver successful transportation solutions.

Core Tenets of a Thorough Systems Engineering Approach

There is a clear hierarchy of abstraction levels:

- A clearly defined hierarchy of abstraction levels is integral to the system architecture, allowing for system comprehension at various levels of detail, appropriate to each stakeholder.
- This structured decomposition enables better complexity management and clear communication among development teams.

There are diverse perspectives and views:

- A comprehensive SE approach incorporates multiple perspectives and views to capture both static and dynamic behaviors, ensuring a thorough analysis of the system's behavior.
- Employing diverse perspectives such as use cases, operational and system states, and various scenarios, engineers can thoroughly analyze the system's behavior, identify potential issues, and ensure robust performance.

The behavioral modeling is hierarchical:

- The system's behavior must be structured and modelled at all levels, from the overall transport vehicle down to the lowest implementation level, guaranteeing consistency and traceability throughout the development process.
- This involves defining how components at each level interact to achieve the desired system behavior, enabling predictable and reliable system behavior.

These are the Axiomatic Design principles for risk management and modularization:

- Axiomatic Design provides a framework for structuring the system to manage both internal and external risks, crucial for effective modularization.
- By separating functional requirements from design parameters, engineers can minimize dependencies and improve the system's modularity, facilitating risk management and ensuring system controllability.

Temporal considerations include the following:

- Time-related aspects must be considered throughout the system.
- This involves modeling the timing of events, response times, and synchronization requirements at all levels of the system architecture.
- Ensuring that these temporal aspects are well-defined and validated is critical for predictable and reliable system behavior.

For transportation systems to be commercially successful, their behavior must be understandable and unambiguous to both users and the general public. Acceptance hinges on trust, which is built on clear, predictable, and safe system behavior. If people do not understand how the system works or do not trust its behavior, they are unlikely to accept it, regardless of its technological capabilities.

By integrating these elements, a thorough systems engineering approach ensures that complex transportation systems are not only technologically advanced but also safe, reliable, and user-friendly, thereby increasing their likelihood of commercial success.

References

AUTOSAR GbR: Specification of Health Monitoring. AUTOSAR FO R22-11. 2022. *https://www.autosar.org/fileadmin/standards/R22-11/FO/AUTOSAR_ASWS_HealthMonitoring.pdf*

Bertalanffy, L. von: General System Theory. Foundations, Development, Applications. Revised Edition. George Braziller, Inc., New York, NY, USA 1968

Daun, M./Bohn, P./Brings, J./Weyer, T.: Structured Model-Based Engineering of Long-living Embedded Systems. The SPES Methodological Building Blocks Framework. 2016

Functional Specs, Inc.: Introduction to Axiomatic Design Concepts. 2024. *https://www.axiomaticdesign.com/technology/introduction-to-axiomatic-design-concepts*

Hudson, R.: ARO 4510 – Model-Based Systems Architecture. California State Polytechnic University, Pomona 2024

ISO 26262:2011: Road vehicles – Functional safety. Part 10

ISO 26262:2018: Road vehicles – Functional safety

ISO/IEC/IEEE 15288:2015: Systems and Software Engineering – System Life Cycle Processes. International Organization for Standardization (ISO)/International Electrotechnical Commission (IEC), Institute of Electrical and Electronics Engineers (IEEE) 2015

Kruchten, P.: The 4+1 View Model of architecture. In: IEEE Software. Volume 12, Issue 6, November 1995, pp. 42–50

Leveson, N. G.: Engineering a safer world. Systems thinking applied to safety. MIT Press, Boston, MA, USA 2012

Luhmann, N.: Soziale Systeme. Grundriß einer allgemeinen Theorie. Erstveröffentlichung: 1984. Frankfurt am Main 2001

Moir, I./Seabridge, A.: Aircraft Systems. Mechanical, Electrical, and Avionics Subsystems Integration. 3rd Edition. Aerospace Series. John Wiley & Sons 2001

NASA: NASA Risk-Informed Decision Making Handbook. NASA/SP-2010-576, Version 1.0. April 2010

Nordlund, M./Taesik, L./Sang-Gook, K.: Axiomatic Design. 30 Years After. In: Proceedings of the ASME 2015 International Mechanical Engineering Congress and Exposition (IMECE 2015), Houston, Texas, November 13–19, 2015. *https://dspace.mit.edu/bitstream/handle/1721.1/107378/Kim_Axiomatic%20design.pdf*

Ross, H.-L.: Safety for Future Transport and Mobility. Springer 2021

SAE J3016: Taxonomy and Definitions for Terms Related to Driving Automation Systems for On-Road Motor Vehicles. SAE International 2014

SAE J3049: Model Architecture and Interfaces Recommended Practice for Ground Vehicle System and Subsystem Dynamical Simulation. SAE International 2015

Schütt, B./Steimle, M./Kramer, B./Behnecke, D./Sax, E.: A Taxonomy for Quality in Simulation-Based Development and Testing of Automated Driving Systems. In: IEEE Access, Vol. 10, 2022, pp. 18631–18644

8 Evidence and Confidence in Technology

The successful adoption of new technologies hinges on societal trust and acceptance. Without these, even the most innovative ideas can falter. This chapter explores the critical role of societal trust in technology, the importance of generating enthusiasm and perceived benefits, and the challenges faced in achieving widespread acceptance.

8.1 Building Trust and Acceptance for New Technologies: A Societal Challenge

To gain acceptance for new technologies, society must first trust them. Trust is the foundation upon which acceptance is built. Without trust, skepticism and resistance can hinder the adoption process. Historical examples, such as the Hindenburg disaster, illustrate how a lack of trust can abruptly end technological advancements. The Hindenburg accident in Lakehurst marked the end of the airship era due to the catastrophic loss of trust in the safety of airships.

A business case for new technology can only be established if it generates enthusiasm and demonstrates clear benefits to society. Enthusiasm drives early adoption, while perceived benefits ensure sustained use. For instance, the shift towards electric mobility (e-mobility) is driven by environmental concerns. However, the transition from the familiar internal combustion engines to the electric vehicles faces resistance from both manufacturers and users. Overcoming this resistance requires demonstrating the tangible benefits of e-mobility, such as reduced emissions and lower operating costs.

Trust and acceptance in mobility solutions are significant societal tasks. Acceptance of specific characteristics of transportation mode deeply influences individual decisions. Questions arise: Is an individual willing to pay for a personal vehicle? What features are important to them? Will future generations prefer car ownership, or will they opt for use-per-demand models? Additionally, the acceptance of public transportation plays a crucial role in shaping future mobility landscapes.

Another perspective involves gaining acceptance from authorities and third parties for specific technologies. This requires transparent methods and acceptable technologies. Evidence must be provided to demonstrate compliance with legal regulations, product liability standards, and trade laws. Strategies during the development of technology or mobility systems must focus on creating evidence to ensure product safety, integrity, and usability.

Achieving societal acceptance for new technologies is a multifaceted challenge. It requires building trust, generating enthusiasm, and demonstrating clear benefits. Historical examples remind us of the consequences of failing to gain societal acceptance. As we move towards future mobility solutions, understanding and addressing societal concerns will be crucial in ensuring the successful adoption of new technologies.

Introducing new technology into an organization or even in a public environment is not a singular task. It requires more than just the organization itself to be convinced of the quality and product characteristics. There are multiple levels that need to be persuaded by the product solutions and the design decisions derived.

Internal Evaluation

Firstly, there are internal entities that must evaluate the product or subproduct based on certain quality and safety features. These entities play a crucial role in ensuring that the product meets the organization's standards and can deliver the promised benefits.

Management and Stakeholders

Secondly, the management and stakeholders need to be provided with suitable evidence that the product can deliver its promised benefits. This evidence needs to be presented with sufficient confidence to convince these parties of the product's potential.

Regulatory Bodies and Third Parties

The process continues with providing evidence to regulatory bodies authorities, and third parties like TÜV etc. Consumer organizations, such as NCAP, also need to be convinced of the product's quality and safety.

Public Transport Mobility Solutions

In the case of mobility solutions for public transport, the entire society, including political bodies as representatives of society, must be convinced. This requires building sufficient trust in the new product.

Introducing new technology is a complex process that involves convincing multiple parties of the product's quality, safety, and potential benefits. Each party has different requirements and standards, making this a challenging but necessary part of technological innovation. It is only through this rigorous process that we can ensure the successful and safe implementation of new technologies.

8.2 Providing Evidence

Generating evidence with sufficient confidence is a challenging aspect of introducing new technology, particularly in the automotive and transportation sectors. This process involves various methods, including verification, validation, testing, simulations, and reviews. However, it's important to note that the context of application and societal acceptance also play a significant role in this process.

In a rough consideration, verification and validation can be described as follows:

- Verification is the process of checking that a product, system, or component meets specified requirements. It often involves reviews and non-execution-based methodologies.
- Validation, on the other hand, is the process of evaluating a system during or at the end of the development process to determine whether it satisfies the specified requirements. It typically involves execution-based methodologies such as testing and simulations.

There are other methods for generating evidence with sufficient confidence. There are approaches to provide transparency for assessing confidence in qualitative evidence syntheses findings. There are methods that add value by providing decision makers with additional evidence to improve understanding of intervention complexity, contextual variations, implementation, and stakeholder preferences and experiences.

When generating evidence for a specific product application, it's crucial to consider the context of application and the societal acceptance of the product. For instance, in the case of mobility solutions, societal acceptance can be influenced by factors such as safety, efficiency, environmental impact, and cost. Evidence must be generated to demonstrate that the product meets these societal expectations with sufficient confidence.

Evaluating societal confidence in product features can be challenging. It is important to consider the different target groups. One approach for young technology enthusiasts is to conduct surveys or studies to gauge public opinion. For instance, a study on the effects of celebrity-product/consumer congruence found that the congruence of Internet celebrities with followers and products promotes follower product adoption by enhancing followers' perception that the recommendations are credible, in turn increasing the followers' confidence in, desire for, and motivation to buy the product.

Generating evidence with sufficient confidence is a multifaceted process that involves various methods and considerations. It's not only about meeting technical specifications but also about ensuring that the product meets societal expectations and fits within the context of its intended application. By carefully considering all these factors, we can generate evidence that not only demonstrates the product's capabilities but also builds trust and confidence in its use.

8.2.1 New Regulations and Standards

In response to the evolving landscape of automated driving, new regulations and standards have been developed to ensure systematic safety validation. Notably, UNECE WP.29 has introduced the VMAD (Validation Methods for Automated Driving) framework, which mandates a three-pillar approach based on processes, simulations, and real-world tests to provide the necessary evidence of safety. Additionally, Safety Management Systems (SMS) are now required, as outlined in UNECE R157 and UNECE 171. These SMS frameworks include Cybersecurity Management Systems (CSMS) in accordance with UNECE R155 and Software Update Management Systems (SUMS) as per UNECE R156. These regulations and methods are crucial for maintaining the safety and security of automated driving systems throughout their lifecycle.

8.2.2 The Evolution of Safety Standards in the Age of Automated Driving

The automotive industry is in the midst of a transformative shift, with Automated Driving (AD) and Advanced Driver-Assistance Systems (ADAS) at the forefront of innovation. This evolution has brought about significant challenges to traditional safety standards, necessitating a paradigm shift in how we approach vehicle safety. This section examines the evolution of safety standards, from the foundational ISO 26262 to the more recent SOTIF (ISO 21448), highlighting the key changes and the rationale behind them.

ISO 26262, the cornerstone of automotive functional safety, focuses on mitigating the risks arising from electrical and electronic (E/E) system failures. Its primary objective is to prevent harm to vehicle occupants and other road users by ensuring the dependability of E/E systems. The standard provides a risk-based approach, defining Automotive Safety Integrity Levels (ASILs) to classify the severity of potential hazards and specifying corresponding safety requirements.

Limitations of ISO 26262: While ISO 26262 has been instrumental in improving vehicle safety, it has certain limitations, particularly in the context of ADAS and AD.

Focus on E/E failures: ISO 26262 primarily addresses failures within the E/E architecture. It does not explicitly cover hazards arising from the intended functionality of the system, such as those related to perception or decision-making in automated driving.

Limited consideration of nominal performance: The standard does not explicitly address the validation of nominal performance, which is crucial for ADAS and AD functions that rely on complex algorithms and sensor data.

Item-level focus: ISO 26262 often focuses on individual components or subsystems (items) rather than the entire vehicle as a system. This approach may not adequately capture the interactions and emergent behaviors that can arise in complex automated driving systems.

ISO 21448 (SOTIF): Addressing the Safety of the Intended Functionality

To address the limitations of ISO 26262, particularly concerning the safety of the intended functionality, the industry developed ISO 21448, also known as Safety of the Intended Functionality (SOTIF). SOTIF complements ISO 26262 by focusing on the hazards that can arise from the intended behavior of the system, rather than just random hardware failures (Figure 8.1).

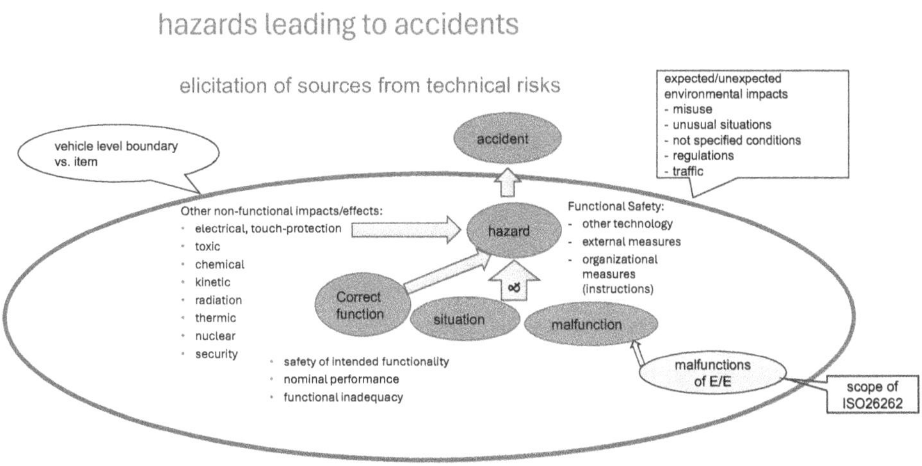

Figure 8.1 Limitations of ISO 26262

Key aspects of SOTIF

Systematic perception: SOTIF emphasizes the validation of the perception system, which is critical for ADAS and AD functions that rely on accurate and reliable environmental sensing. It addresses the challenges of dealing with imperfect sensor data, environmental variations, and complex traffic scenarios.

Human factors: SOTIF considers the interaction between the driver and the automated driving system, recognizing that human behavior can significantly impact safety. It addresses issues such as driver misuse, misunderstanding of system capabilities, and the transition of control between the driver and the automated system.

Operational scenarios: SOTIF requires the identification and analysis of a wide range of operational scenarios, including edge cases and challenging situations, to ensure that the automated driving system can handle diverse real-world conditions.

Expanding the Safety Lifecycle

The evolution of safety standards has also led to an expansion of the safety lifecycle, encompassing aspects beyond the traditional focus on design and development. This expanded lifecycle includes the following:

- **Maintenance and updates:** With the increasing complexity of Software-Defined Vehicles (SDVs), the safety lifecycle now extends to include over-the-air updates and maintenance procedures, ensuring that safety is maintained throughout the vehicle's lifespan.
- **Cybersecurity:** The convergence of safety and security has become increasingly important, as vulnerabilities in cybersecurity can potentially lead to safety hazards. The safety lifecycle now incorporates cybersecurity considerations to protect against malicious attacks and unauthorized modifications.

Validating and verifying the safety of automated driving systems presents significant challenges due to the complexity of the systems and the vast number of possible scenarios. Traditional testing methods are often insufficient to cover the full range of real-world conditions. To address these challenges, the industry is exploring multi-pillar approaches that combine the following:

- **Simulation:** Virtual testing environments allow for the evaluation of automated driving systems in a wide range of scenarios, including those that are difficult or dangerous to reproduce in real-world testing.
- **Real-world testing:** On-road testing is still essential to validate the performance of automated driving systems in real-world conditions and to gather data for further refinement.
- **Formal methods:** Formal verification techniques can be used to mathematically prove certain properties of the system, providing additional confidence in its safety.

Scenario-Based Safety Assessment

Scenario-based safety assessment has emerged as a crucial approach for evaluating the safety of automated driving systems. This approach involves defining a comprehensive set of scenarios that represent the various situations the vehicle might encounter in real-world driving. These scenarios are then used to test the system and assess its performance.

Goal-Structured Notation (GSN) provides a structured way to represent safety arguments, making it easier to communicate and evaluate the evidence supporting the safety of an automated driving system. GSN is often used in safety cases to demonstrate compliance with safety standards and regulations.

Confidence and Situation Awareness

Ensuring safety in automated driving requires not only robust design and validation but also the ability to maintain confidence in the system's performance during operation. This involves the following:

- **Situation awareness:** The automated driving system must have a comprehensive understanding of its environment, including the state of the vehicle, the surrounding traffic, and the road conditions.
- **Confidence monitoring:** The system must be able to assess its own confidence in its perception, decision-making, and control capabilities. If the system's confidence falls below a certain threshold, it should be able to take appropriate action, such as requesting driver intervention.

8.3 Aspects of Verification and Validation

Understanding and managing system risks involves a range of considerations and strategies. These include replicating system accidents or other adverse processes, designing system accidents, adverse processes, and intentional adverse propagations, and defining consistent rules or axioms to enable adverse process design. It also involves thinking in various ways to understand system risks, modelling adverse progressions, and not relying solely on probability and stochastic processes.

In order to find arguments for a safety case not only the static risk condition needs to be understood. As often understood by single-minded safety stands like ISO 26262 which led to so-called "safety goals" the risk scenarios need to be understood in order to communicate the risks to the organization.

Applying temporal logic and understanding system and adverse processes lifecycles, applying multi non-learner propagations, integrating hardware, software, firmware, logic, the human and environment as well as associated adverse interactions or inactions, and applying decision analysis are all important strategies. Noting that decisions will adversely affect the system throughout the lifecycle, as well as applying and integrating many analytical techniques to solve complex risks, are crucial considerations in managing system risks. Noting that designing for early detection, isolation, correction, contingency, and recovery, and thinking in linear and non-linear multi-progressions, forming adverse sequences that initiate via latent or real-time hazards, triggers, contributors, outcomes (harm), detection, isolation, correction, contingency, recovery, re-stabilization, given various system states, are also crucial considerations in managing system risks.

Today the different industries follow different V&V approaches. Some examples are listed in the following.

Automotive Industry

In the automotive industry, V&V activities are essential to ensure that vehicles and their components meet safety, performance, and reliability standards. The ISO 26262 standard for automotive systems requires several V&V activities, which have to be planned and performed jointly by the Original Equipment Manufacturers (OEMs), the vehicle producer, brands and the suppliers. These activities include checking whether the design is properly validated and verified and is in compliance with the project's requirements, the stakeholder requirements and the related goals.

Rail Industry

In the rail industry, V&V activities are integral to ensure that railway projects comply with all the specified railway requirements and standards for Reliability, Availability, Maintainability, and Safety (and security) (RAMS), as well as functionality, and quality. The verification tasks are generally included within each lifecycle phase, and their objective is to demonstrate that the railway requirements of each lifecycle phase have been fulfilled. The validation tasks aim to ensure that the railway project under consideration meets the specified requirements for the intended use or application. Like the machinery industry, the rail industry follows CEN or CENELEC standardization. CENELEC (European Committee for Electrotechnical Standardization) has developed standards for the rail industry that are harmonized based on the Machinery Directive. The Machinery Directive (Directive 2006/42/EC) is a European regulation that ensures a uniform level of safety for machinery and incomplete machinery when placed on the market within the European Economic Area. The CENELEC standards, such as EN 50126, EN 50128, and EN 50129, have been derived from the IEC 61508 standard to meet the specific requirements of the railways. These standards provide a systematic point of reference for all operators and manufacturers, allowing them to work together beyond boundaries. They form the backbone of railway design, manufacturing, and maintenance, ensuring better and safer railways.

In the rail industry, quality management is of utmost importance due to the high safety and reliability requirements. The International Railway Industry Standard (IRIS), which is based on ISO 9001, sets certifiable quality guidelines.

In the context of regulations, verification is an activity that involves examination, inspection, testing, and reviewing a product to establish that it meets the design specifications and/or is fit for the purpose for which it was designed. Validation, on the other hand, is a subprocess of the verification or certification process. It involves testing a product to ensure that it meets a pre-determined specification.

Aviation Industry

In the aviation industry, V&V processes are critical to ensure the safety and functionality of aircraft systems. The Federal Aviation Administration (FAA) develops, maintains, and improves V&V related standards and processes. These processes involve a wide variety of analysis and tests to ensure that control laws will function properly under both nominal and failure conditions

Space Technology

In space technology, V&V activities are of extreme importance due to the high costs and complexities associated with deploying hardware (e. g., robots, satellites) to space. These activities ensure that on-board systems, especially autonomous ones, are verified and validated through a variety of techniques. NASA's Independent Verification and Validation (IV&V) program applies system and software engineering best practices to evaluate the correctness and quality of critical and complex software systems throughout the System Development Life Cycle (SDLC).

Military

In the military, V&V processes are essential for ensuring that military models and simulations meet operational and system-level requirements. The Department of Defense has established templates for the four core products of the Modeling and Simulation (M&S) V&V processes:

- accreditation plan
- V&V plan
- V&V report
- accreditation report

These processes subject the control laws to a wide variety of analysis and tests to ensure that they function properly under both nominal and failure conditions.

8.3.1 The Importance of Validation and Verification

Validation and Verification (V&V) are critical processes in the defence industry to ensure that systems and components meet the required standards and perform reliably under all conditions. These processes help in identifying and mitigating risks, ensuring safety, and maintaining the integrity of defence systems. Let's explore the importance of V&V through the key artifacts used in the defence industry.

Accreditation Plan

The accreditation plan outlines the criteria and procedures for accrediting a system or component. It serves as a roadmap for the entire accreditation process, detailing the standards and requirements that must be met. The plan ensures that all necessary steps are taken to achieve accreditation, providing a structured approach to compliance.

The accreditation plan addresses the following criteria:

- **guidance:** provides clear guidelines and criteria for the accreditation process
- **consistency:** ensures a uniform approach to accreditation across different projects
- **accountability:** establishes roles and responsibilities, ensuring that all stakeholders are aware of their duties

V&V Plan

The V&V plan (verification and validation plan) describes the methods and activities that will be used to verify and validate the system or component. It includes the scope, objectives, resources, and schedule for V&V activities. The plan ensures that all aspects of the system are thoroughly tested and evaluated.

The V&V plan addresses the following criteria:

- **comprehensive testing:** ensures that all system components are tested under various conditions
- **risk mitigation:** identifies potential issues early in the development process, reducing the risk of failures
- **resource allocation:** helps in efficient allocation of resources and scheduling of V&V activities

V&V Report

The V&V report documents the results of the verification and validation activities. It provides evidence that the system meets the specified requirements and performs as intended. The report includes test results, analysis, and any discrepancies found during the V&V process.

The V&V Report addresses the following criteria:

- **evidence of compliance:** provides documented proof that the system meets all requirements
- **decision making:** assists stakeholders in making informed decisions based on the V&V results
- **continuous improvement:** identifies areas for improvement and helps in refining the system

Accreditation Report

The accreditation report is the final document that certifies the system, or component has met all the accreditation criteria. It summarizes the findings of the accreditation process and provides official recognition of compliance.

The accreditation report addresses the following criteria:

- **official certification:** provides formal recognition that the system meets all required standards
- **trust and confidence:** builds trust among stakeholders and users by demonstrating compliance with stringent standards
- **regulatory compliance:** ensures that the system adheres to all relevant regulations and standards

The accreditation plan, V&V plan, V&V report, and accreditation report are essential artifacts in the defence industry. They provide a structured approach to ensure that systems are safe, reliable, and compliant with all necessary standards. These documents play a crucial role in maintaining the integrity and performance of defence systems, ultimately contributing to the safety and security of military operations. It is also the best practice for other transport industries, but they show the importance of the safety argumentation to achieve acceptance.

8.3.2 Verification and Validation Methodology

Verification is fundamentally the process of confirming that a system or its individual components meet the specified requirements. It's about ensuring that the "right thing" is built "right." In critical sectors such as automotive, verification is not merely a best practice but an absolute necessity to guarantee safety.

8.3.2.1 Verification in the Automotive Context

The verification of technical systems is often guided by established standards. Two key standards that are guiding system safety engineering are ISO 15288 and ISO 26262.

ISO 15288: Systems and software engineering—System life cycle processes: This standard offers a comprehensive framework for all stages of a system's lifecycle, including verification. It stresses a structured approach, ensuring that verification is an integral part of the entire development process.

ISO 26262: Road vehicles—Functional safety: Specifically designed for the automotive industry, this standard concentrates on functional safety. It mandates rigorous verification to effectively reduce risks associated with electronic and electrical systems in vehicles.

Verification occurs at various levels throughout the development lifecycle. These levels encompass different aspects of the system and its creation:

- **Process, method, and tool verification:** This critical level involves verifying the development processes themselves, the design methods employed, and the tool chains utilized to create the product. For instance, ISO 26262 mandates tool qualification, which means demonstrating that the software tools used in development are appropriate for their intended purpose, thereby minimizing the risk of errors introduced by the tools.
- **Verification of functions, requirements, and specifications:** Here, the focus is on ensuring that the system's functions, requirements, and specifications are consistent, correct, and complete. An example of this is requirements reviews, where stakeholders meticulously examine the requirements for clarity and completeness.
- **Verification of test-related artifacts:** This includes the verification of test plans, test specifications (along with acceptance criteria), and test reports, for example, ensuring that test plans cover all relevant scenarios and that acceptance criteria are measurable and unambiguous is crucial.
- **Verification of implementation and realization:** This level concentrates on verifying the actual hardware and software implementations of the system. Examples include code reviews, unit testing, and Hardware in the Loop (HIL) testing.
- **Verification of safety case argumentation:** A safety case needs a verifiable structure, and this level involves the overall verification of the complete safety argumentation. This might involve a final review process by an independent group that checks all safety-related documents, test reports, and other relevant artifacts.

A variety of methods are employed to verify technical systems:

- **Reviews:** These are formal evaluations of documents, code, or designs. They are valuable for identifying defects early in the development process. An example are formal technical reviews of software design documents.
- **Simulations:** Simulations allow for the evaluation of system behavior in a virtual environment. For example, software simulations can be used to test the performance of a vehicle's electronic stability control system.

- **Tests:** Testing involves executing the system or its components to verify their behavior. Different types of tests include the following:
- **Unit, component testing:** This means the testing of individual software elements within a specific context. Usually these are static tests.
  - **Integration testing:** This is the testing of combined software or hardware components.
  - **System testing:** This means the testing of the complete system. An example is performing physical tests of a vehicle's braking system on a test track.
  - **Scenario-based testing:** The element will be implemented in the target context and specified testing patterns, which cover the entire application space, are executed. Inputs and context need to be animated within the specified range of use. Additional over-limit tests, fault-injection tests, and penetration tests for cybersecurity are needed as evidence for the integrity and fault-tolerance requirements of the element under test.
- **Consistency, correctness, and completeness checks:** These involve verifying that the system's elements are consistent with each other, meet their intended purpose, and cover all required aspects. An example is cross-referencing requirements with design documents to ensure consistency.

Special Considerations for Automotive Systems

Automotive systems engineering, particularly in the context of ISO 26262 does not define specific test methods. The test methods are addressing over-limit tests, fault injection tests, etc., but in the context of dynamic systems the following aspects should be considered:

- **Verification of dynamic systems:** Automotive systems are often dynamic, meaning their behavior changes over time. Verification must account for this dynamic nature. An example is the dynamic testing of an adaptive cruise control system under varying traffic conditions.
- **Scenario-based verification:** Automotive systems must operate safely in a wide range of scenarios. Scenario-based verification involves testing the system's behavior in representative scenarios. An example is testing an autonomous emergency braking system's response to different pedestrian crossing scenarios.
- **Integration of simulations and dynamic tests:** Combining simulations and dynamic tests enables more comprehensive verification. Simulations can be used to explore a broad spectrum of scenarios, while dynamic tests validate the system's real-world behavior. An example is using HIL simulations to test a vehicle's Electronic Control Units (ECUs) in realistic driving scenarios, followed by physical tests on a test track.

Within the framework of ISO 26262, the term verification is employed in core processes for all manner of checks, whereas validation is utilized solely to confirm the achievement of safety objectives at the vehicle or system level in order to provide arguments for the correctness and achievement of safety goals within the scope of that standard (Figure 8.2). This distinction, however, raises pertinent questions, particularly concerning the dynamic nature of software. Given that the behavior of software can often only be fully apprehended through direct observation and scrutiny in real or simulated environments, a debate emerges as to whether a systematic application of the term validation would not be more appropriate. Traditional verification, which concentrates on confirming compliance with specifications, may fall short in assessing the actual functionality and safety of dynamic systems. This is especially true in the context of design reviews, where mere adherence to requirements is insufficient to thoroughly evaluate a design's suitability for its intended purpose. A methodological shift towards a stronger emphasis on validation, even in early development phases, could aid in the timely detection of potential faults and vulnerabilities, thus enhancing the overall safety of vehicle systems. The necessity for comprehensive validation is further accentuated by the complexity of modern vehicle software, which frequently operates within intricate, networked systems whose behavior under diverse operating conditions is difficult to predict. Therefore, it is of paramount importance that validation not only occurs at the system level but also at the level of individual software components and modules to ensure a holistic assessment of system safety.

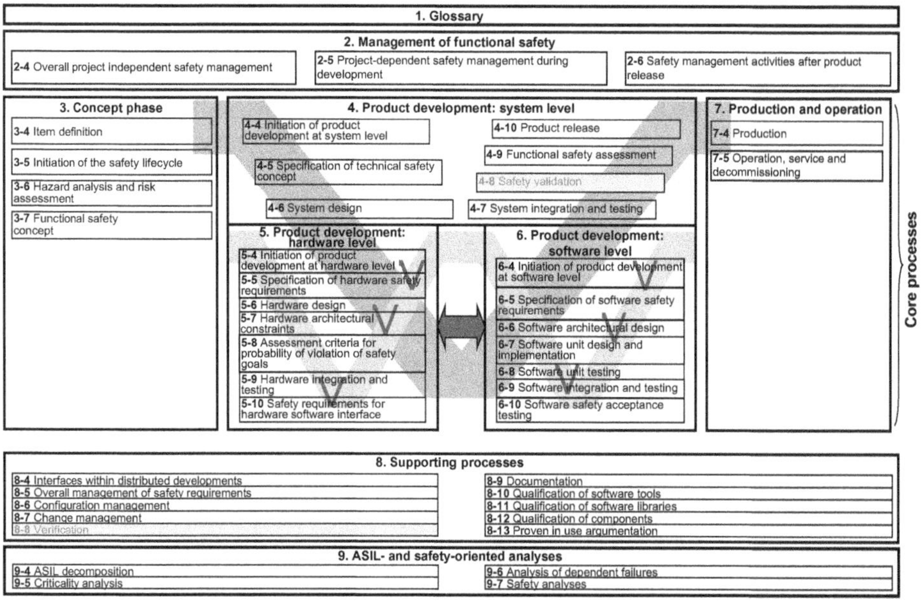

Figure 8.2 Verification and validation in ISO 26262:2011 (source: ISO 26262:2011)

Verification Deliverables

Verification activities result in several key deliverables:

- **Test reports:** These documents provide a detailed record of test activities and their outcomes.
- **Release protocols:** These documents formally authorize the release of the system or its components.
- **Safety cases:** These documents offer a comprehensive argument demonstrating that the system is safe for its intended use.

Effective verification is indispensable for the development of safe and reliable technical systems, especially in safety-critical domains like the automotive industry. By adhering to standards such as ISO 15288 and ISO 26262 and by utilizing a combination of various verification methods, organizations can ensure that their systems meet the highest standards of quality and safety.

8.3.2.2 Verification on the System Level

In order to provide verification arguments a hierarchical structure for the entire mode of operation is necessary. From this top-down approach complete traceability is compulsory, so that arguments for the following attributes could be provided:

- completeness
- consistency
- correctness

The first argument, completeness, becomes a very heterogeneous task due to new top requirements. It is important to consider not only the positive effects on the dependent products but also the negative products. Every new feature, every new function need resources and therefore energy, space and time. These resources must be created additionally in the system, or they must be introduced at the expense of other functions or similar. You can see that this requires considerable knowledge of the other functions which resources, energy, space, and time they require. It is even more complex if you also must consider the negative aspects of the malfunctions, errors, and failure modes of the new function to be introduced and the functions or features that have already been considered.

Traceability can only be achieved if the interfaces are consistent. Functional, technical, organizational interfaces lead to an explosion of interfaces in a disordered system, which become more potent with every additional dimension. In a mobility solution where there are immense integration and coordination tasks, such heterogeneous interfaces become an unmanageable and controllable structure. Therefore, the interfaces must be structured hierarchically, functionally, technically, and organizationally for all parties involved in the development.

Correctness can only be argued for requirements based on their context. In order to argue for correctness, evidence must be provided that shows that the requirements are also met in the respective application context. To do this, the appropriate context must also be created and specified for all requirements so that other people involved in the development and parties who are supposed to assess the work results are aware of the requirement and the respective context.

In addition to the requirements, the permissible or accepted restrictions, constraints or limitations for the solution space must of course also be known. This includes constraints on the nature of the requirements, the functionalities, the architecture, and the design decisions made.

8.3.2.3 Validation and Safety of the Intended Functionality

While ISO 26262 mandates that measures to prevent systematic failures be considered within its processes, it conspicuously lacks concrete solutions for addressing these issues (Figure 8.3). In the realm of automated driving functions, the ISO 21448 standard, concerning Safety of the Intended Functionality (SOTIF), has emerged to provide more specific guidance on the design of Advanced Driver Assistance Systems (ADAS) and Automated Driving (AD) systems. This standard emphasizes the validation of design and the consideration of uncertainties as an effective means to achieve adequate confidence in system safety.

The issue extends far deeper into the development lifecycle. At the highest level, it involves the comprehensive interaction of a vehicle with its operational environment, and how its interaction with the traffic environment must proceed to prevent accidents. Descending further, it encompasses operational and maintenance incidents, as well as faults and anomalies in vehicle behavior. This cascades down to anomalies, faults, and inadequacies at the system levels, across the hardware-software interface, and ultimately to the hardware and software themselves. A crucial aspect is the correct realization, implementation, and production of electronic and mechanical components.

In a conventional development process, one would expect the creation of a fully traceable specification that covers all relevant aspects. However, the heterogeneous nature of external influences poses significant challenges, particularly for software and electronics developers. These external factors range from operational conditions such as weather, traffic, and road conditions, to more subtle influences like the disposition and mindset of production personnel, software engineers, and those involved in the safety case evidence, including test planners, testers, tool developers, and regulatory auditors.

Essentially, ensuring safety requires a holistic approach that acknowledges the intricate interplay of numerous variables, both within and outside the vehicle system.

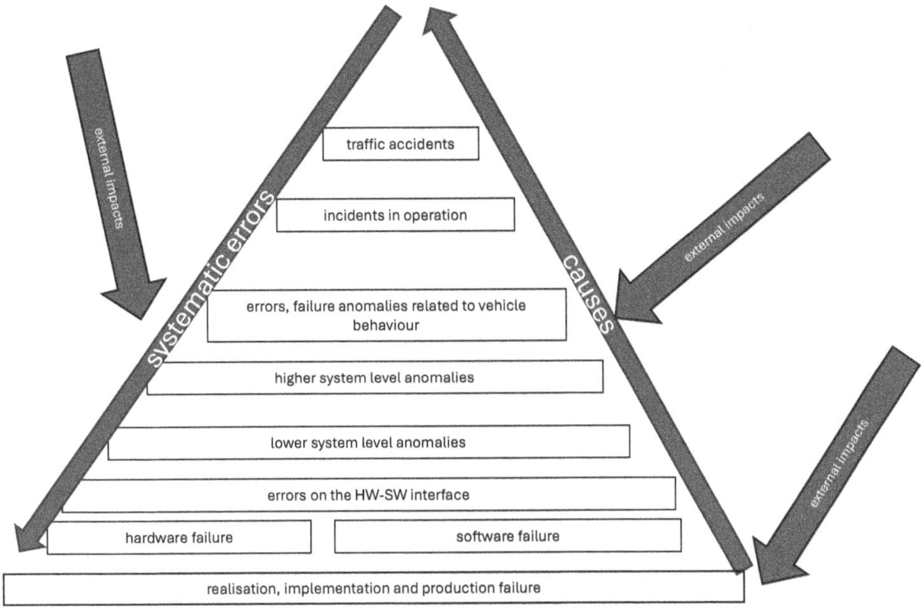

Figure 8.3 Systematic failure and their causes

Validation, in essence, is the process of confirming that a product or system meets the intended needs of its users and stakeholders. This differs from verification, which ensures that the product is built correctly according to its specifications. Validation methodologies vary significantly depending on the level of abstraction, each level requiring a tailored approach to ensure comprehensive safety and functionality.

Vehicle level validation

At the vehicle level, validation focuses on assessing the integrated system's performance in real-world or highly realistic scenarios.

Tests are well-established validation methods; they address similar aspects:

- **Real-world testing:** This includes extensive driving tests under various conditions (e. g., weather, traffic, road types) to evaluate the vehicle's overall behavior and safety. It also encompasses user studies to assess the Human-Machine Interface (HMI) and ensure that the vehicle is safe and intuitive to operate.
- **Simulation and virtual testing:** Advanced simulation tools are used to create virtual environments that replicate real-world scenarios, including edge cases and hazardous situations. This allows for the testing of a wide range of scenarios that would be impractical or dangerous to test in the real world.
- **Scenario-based testing:** Specific scenarios, such as emergency braking, lane changes, and obstacle avoidance, are tested to evaluate the vehicle's response and ensure that it meets safety requirements.

System level validation

At the system level, validation focuses on assessing the interaction between subsystems and ensuring that they function correctly as a whole. This involves the following:

- **Hardware in the Loop (HIL) testing:** HIL testing involves integrating real hardware components with simulated environments to test the system's behavior under realistic conditions. This allows for the testing of complex interactions between hardware and software components.
- **Software in the Loop (SIL) testing:** SIL testing involves running software components in a simulated environment to test their behavior and ensure that they meet system requirements. This is very important for testing ADAS and AD systems.
- **Integration testing:** Integration testing involves testing the interaction between different subsystems to ensure that they function correctly as a whole.

Component level validation (Hardware and Software)

At the component level, validation focuses on assessing the individual hardware and software components to ensure that they meet their specific requirements. This involves the following:

- **Hardware validation:** Hardware validation includes testing the performance, reliability, and durability of individual hardware components. This may involve environmental testing, stress testing, and electrical testing.
- **Software validation:** Software validation includes testing the functionality, performance, and robustness of individual software components. This may involve unit testing, integration testing, and system testing. Dynamic testing of the software is very important.
- **Model in the Loop (MIL) testing:** These testings are used to test the algorithms of a software in a virtual environment.

Key validation principles

Across all abstraction levels, the following principles are essential for effective validation:

- **Traceability:** Validation activities should be traceable to requirements and specifications.
- **Reproducibility:** Validation tests should be reproducible to ensure consistent results.
- **Coverage:** Validation activities should cover a wide range of scenarios and conditions.
- **Objectivity:** Validation results should be objective and based on measurable criteria.

As seen on this example in the context of provision of evidence in a typical lifecycle approach, it is meaningful to consider verification and validation and the related planning aspects as one process. Introducing too many interfaces leads to increased complexity.

8.3.2.4 Receiver Observer Characteristics (ROC)

The accuracy of any diagnostic test, including those used for COVID-19, is crucial for making informed decisions. However, no test is perfect, and understanding the potential for errors is essential. These errors can be categorized into four outcomes: true positives, true negatives, false positives, and false negatives:

- **True Positive (TP):** This occurs when the test correctly identifies the presence of a condition.

  Example: A COVID-19 test correctly identifies that a person has the virus.

- **True Negative (TN):** This occurs when the test correctly identifies the absence of a condition.

  Example: A COVID-19 test correctly indicates that a person does not have the virus.

- **False Positive (FP):** This occurs when the test incorrectly indicates the presence of a condition when it is not actually present.

  Example: A COVID-19 test shows that a person has the virus when they are actually healthy. This could be due to cross-reactivity with other substances or errors in the testing process.

- **False Negative (FN):** This occurs when the test incorrectly indicates the absence of a condition when it is actually present.

  Example: A COVID-19 test shows that a person does not have the virus when they are actually infected. This could be due to low viral load, improper sample collection, or limitations in the test's sensitivity.

We could consider the following impacts of these errors:

- False positives can lead to unnecessary anxiety, isolation, and treatment.
- False negatives can lead to the spread of infection, as infected individuals may believe they are not contagious.

Factors Affecting Test Accuracy

Several factors can influence test accuracy, including the following:

- the sensitivity and specificity of the test
- the quality of the sample

- the skill of the person administering the test
- the prevalence of the condition in the population

Understanding these concepts is crucial for interpreting test results and making informed decisions about health and safety.

In the technology of automated driving Safety of the Intended Functionality (SOTIF) is facing similar issues for perception systems. For implemented effect chains, the distinction between these information's are unambiguous.

ROC means "Receiver Operating Characteristic." It is the visual representation of performance of binary classifier. "False Positive Rate" versus "True Positive Rate" visualize the understanding of the classifier's performance. **AUC** means "Area Under the Curve." It is a numerical representation of the performance of binary classifiers. When using normalized units, the Area Under the Curve (AUC) is equal to the probability that a classifier will rank a randomly chosen positive instance higher than a randomly chosen negative one.

- **True Positive (TP):** It is predicted positive and it is true.
- **True Negative (TN):** It is predicted negative and it is true.
- **False Positive (FP):** It is predicted positive and it is false.
- **False Negative (FN):** It is predicted negative and it is false.

The "predicted values" are described as positive and negative and the "actual values" are described as true and false. A great tool for understanding the real world and how much, what, and how well we need to perceive from the real world is the ROC (Receiver Operating Characteristic) method or the ROC curve (Figure 8.4).

When implementing complex effect chains, particularly in safety-critical systems, achieving runtime correctness is paramount. Often, redundancy serves as the primary mechanism to ensure this correctness. However, the effectiveness of redundancy hinges on the critical concept of independence.

The role of redundancy includes the following:

- Effect chains, which involve a series of sequential processing steps, are susceptible to errors at any stage.
- Redundancy, in this context, involves replicating critical processing steps or dataflow.

By comparing the outputs of redundant paths, the system can detect and potentially correct errors.

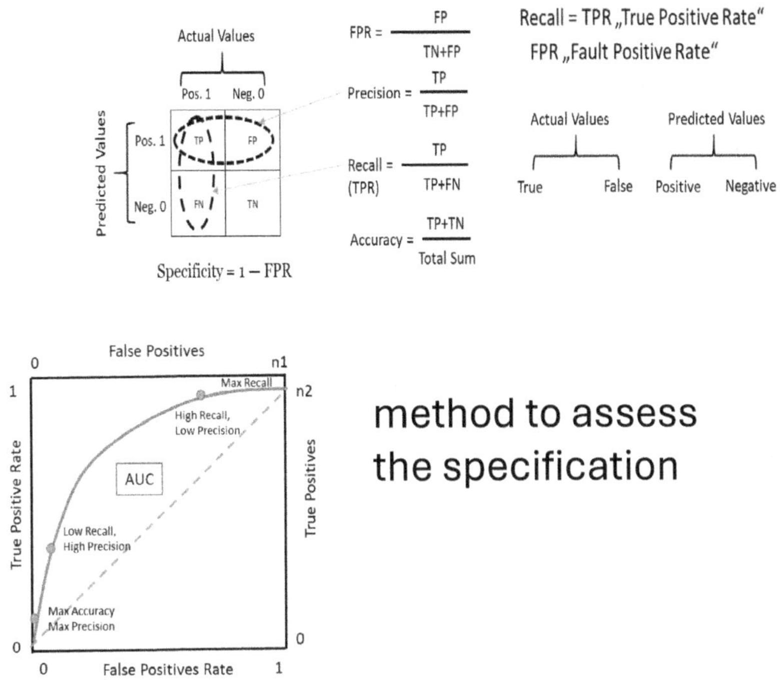

Figure 8.4 Receiver observer characteristics

The Importance of Independence

The true value of redundancy is realized only when the redundant paths are truly independent. Independence means that the redundant paths should be designed and implemented in such a way that they are unlikely to be affected by the same faults. If redundant paths are not independent, they may suffer from "common cause" failures. A common cause failure occurs when a single event or condition affects multiple components simultaneously.

For example, if redundant software modules share a common library with a bug, both modules will produce the same incorrect result. When common cause failures occur, the system loses its ability to distinguish between correct and incorrect information. The comparator will simply see that all redundant paths produce the same result, even if that result is wrong.

Therefore, to maximize the effectiveness of redundancy, it is essential to do the following:

- use diverse hardware and software components,
- implement independent error detection and correction mechanisms,
- ensure that redundant paths are physically separated.

Redundancy is a powerful tool for ensuring runtime correctness in effect chains. However, its effectiveness depends on the independence of the redundant paths. Without independence, common cause failures can undermine the benefits of redundancy.

8.4 Confidence and Uncertainties

It is important to acknowledge your points regarding the limitations of solely relying on the ISO 26262 approach for highly automated driving systems. The standard emphasizes a structured approach to functional safety, including rigorous verification and validation to mitigate risks associated with electronic and electrical systems in road vehicles.

However, the approach with safety goals can lead to a focus on static aspects and is therefore pertinent, especially when considering the complexities of automated driving. While safety goals define top-level safety objectives, ensuring these goals are met in all dynamic and varied real-world situations presents a significant challenge. Risks in automotive systems often derive from combinations in dynamic scenes, with behavior that changes over time.

Heterogeneous variances of external influencing parameters in the interaction with the real world and the behavior of other traffic participants make static safety argumentations nearly impossible. This complexity makes it difficult to anticipate and statically define all potential hazardous scenarios that an automated vehicle might encounter. Scenario-based verification involves testing the system's behavior in representative scenarios. Therefore, a scenario-based approach to complement other verification and validation methods is needed.

Regarding the three-pillar approach from VMAD (Validation Methods for Automated Driving), further aspects need to be considered. The series of standards ISO 34501 to ISO 34505 established to address these challenges. These standards have emerged to specifically address the validation of Automated Driving Systems. They provide guidance on scenario-based testing, virtual testing, and other methodologies needed to ensure the safety of these complex systems in the face of the challenges that have been described. The development of such standards reflects the industry's recognition that traditional approaches, while foundational such as with ISO 26262, need to be augmented with methods that can handle the dynamic and highly variable nature of automated driving.

8.4.1 Overview of ISO 34501 to ISO 34505

The ISO 34500 series is dedicated to addressing the complex safety considerations associated with the development, validation, and deployment of Automated Driving Systems (ADS). These standards aim to establish a common framework and terminology for ensuring the safety of ADS-equipped vehicles (Figure 8.5).

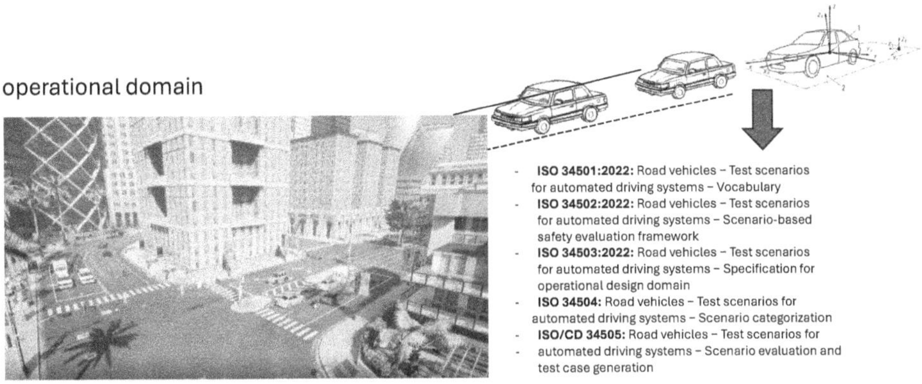

Figure 8.5 ISO 34500 series standards

A breakdown of the core standards within this series is listed in the following.

ISO 34501: Road Vehicles - Functional Safety - Development of Automated Driving Systems - Vocabulary

This standard establishes a unified vocabulary for automated driving systems. It defines key terms and concepts, promoting consistent communication and understanding among stakeholders. It is crucial for clarifying the meaning of terms used in the other standards of the ISO 34500 series. Having a clear and common vocabulary is a prerequisite for safety relevant topics.

The standard provides for example a scenario where the subject vehicle and the pedestrian are considered as actors in a dynamic entity (Figure 8.6). The standard describes scenarios which consist of static entities as a scenery and dynamic scenes as dynamic entities (Figure 8.7). These scenarios provide the context for situations where subject vehicle and actors are observed from different perspectives.

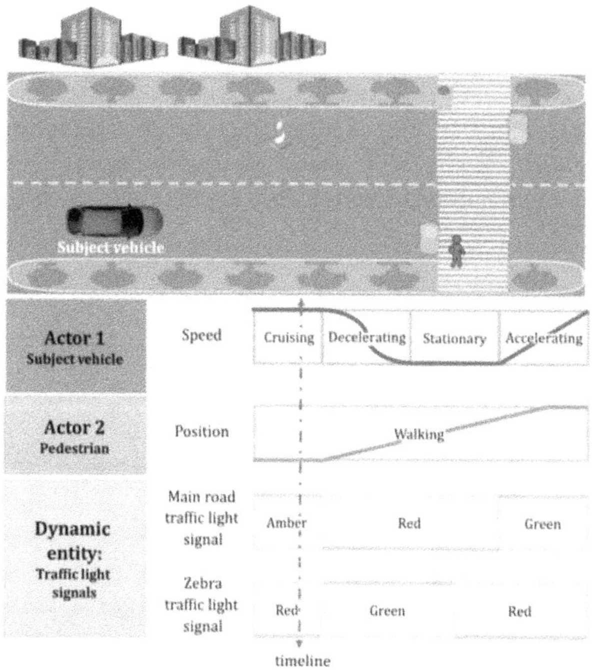

Figure 8.6 Actors in dynamic entity (source: ISO 34501:2022)

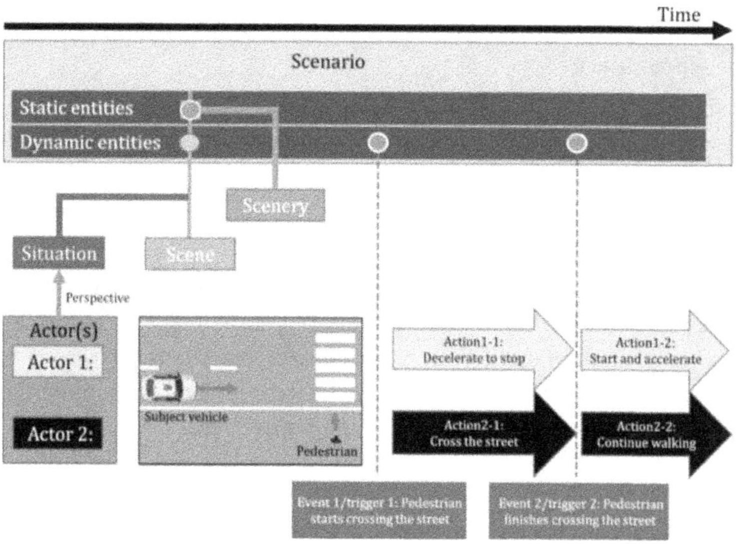

Figure 8.7 Scenery and scenes in static entities (source: ISO 34501:2022)

ISO 34502: Road Vehicles – Functional Safety – Development of Automated Driving Systems – Safety Requirements

This standard focuses on defining safety requirements for ADS. It outlines the necessary safety goals and objectives that ADS must achieve to mitigate potential risks. This includes the safety analysis and the definition of needed safety measurements for the system.

The standard provides an example for a control chain consisting of the following elements (Figure 8.8):

- **perception:** using different sensors for the environment
- **judgement:** planning of the driving strategy
- **control:** examples of how to control external impacts

You find more detailed approaches and further examples in previous chapters.

The description of the operational domain is very similar to the approach from the founding project "Pegasus," which also provides a fundamental approach about the methodology of "Operational Design Domain."

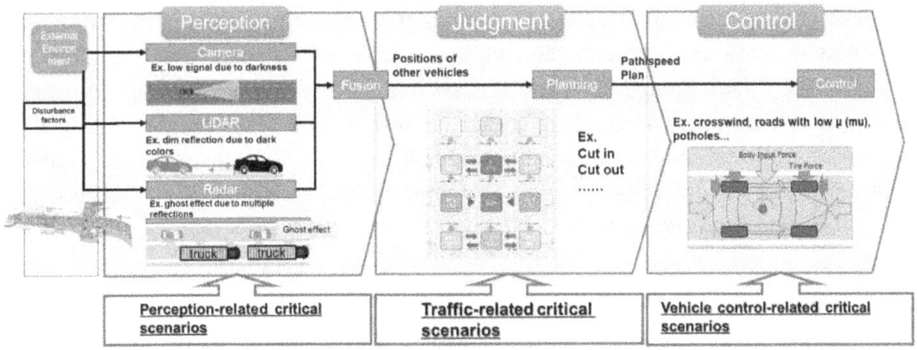

Figure 8.8 Test scenarios (source: ISO 34502:2022)

ISO 34503: Road Vehicles – Functional Safety – Development of Automated driving Systems – Operational Scenarios

This standard addresses the development and management of operational scenarios for ADS (Figure 8.9). It provides guidance on how to define, analyze, and use operational scenarios for testing and validation purposes. This is very important because ADS systems need to operate in a very wide variation of possible real world driving scenarios.

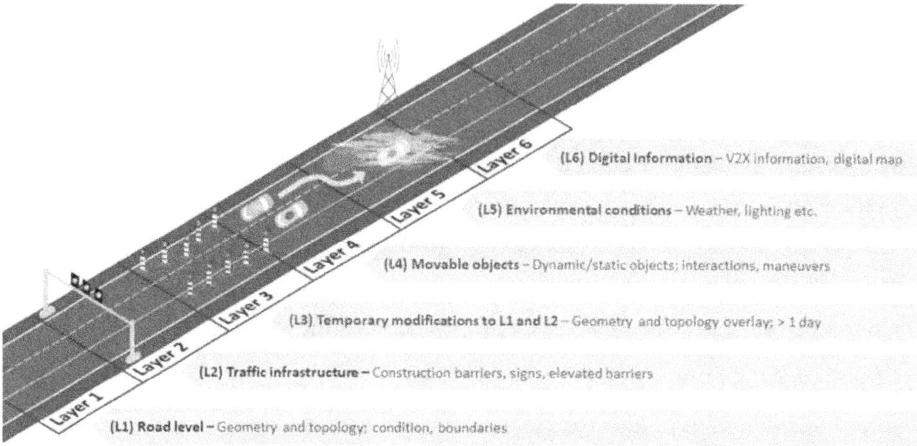

Figure 8.9 Layer of the operational domain (source: ISO 34503:2023)

ISO 34504: Road Vehicles – Functional Safety – Development of Automated Driving Systems – Performance and Validation

This standard focuses on the performance and validation of ADS. It outlines methods and procedures for evaluating the performance of ADS and ensuring that they meet the defined safety requirements. This standard deals with the question "How to prove that the ADS is safe enough?". It gives directions for test methods needed. ISO 34504 provides guidance on categorizing scenarios in the testing and verification of Automated Driving Systems (ADS). It emphasizes the use of "tags" as meta-attributes to categorize various driving scenarios, contributing to a harmonized approach to scenario classification (Figure 8.10).

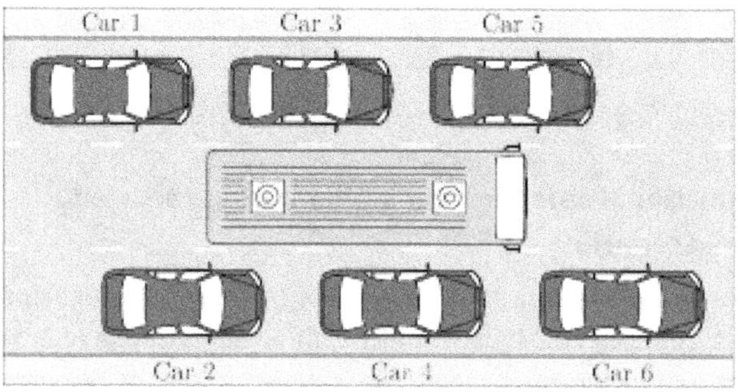

Figure 8.10 Tags and meta-attributes (source: ISO 34504:2024)

ISO/DIS 34505: Road Vehicles – Test Scenarios for Automated Driving Systems – Scenario Evaluation and Test Case Generation

ISO 34505 is not finally been published. It defines a methodology to evaluate scenarios and provides a procedure extending test scenarios to test cases for a given function in a traceable way based on the testability (Figure 8.11). This document also defines the necessary characteristics of a test case that include but are not limited to unified identifier, test objective, inputs, steps, platform, and expected results. It describes methods and criteria to evaluate the test case (e. g., frequency, criticality, complexity of a scenario), the coverage concerning functional and technical requirements, Operational Domain (OD) and test criteria, and also the optimizing of the set of prioritized test cases. It is applicable to Level 3 and higher ADS as defined in ISO/SAE PAS 22736 (taxonomy and definitions for terms related to driving automation systems for on-road motor vehicles). The focus of this document is on safety (functional safety and Safety of the Intended Functionality (SOTIF)). In general, the content is also applicable to non-safety relevant test scenarios.

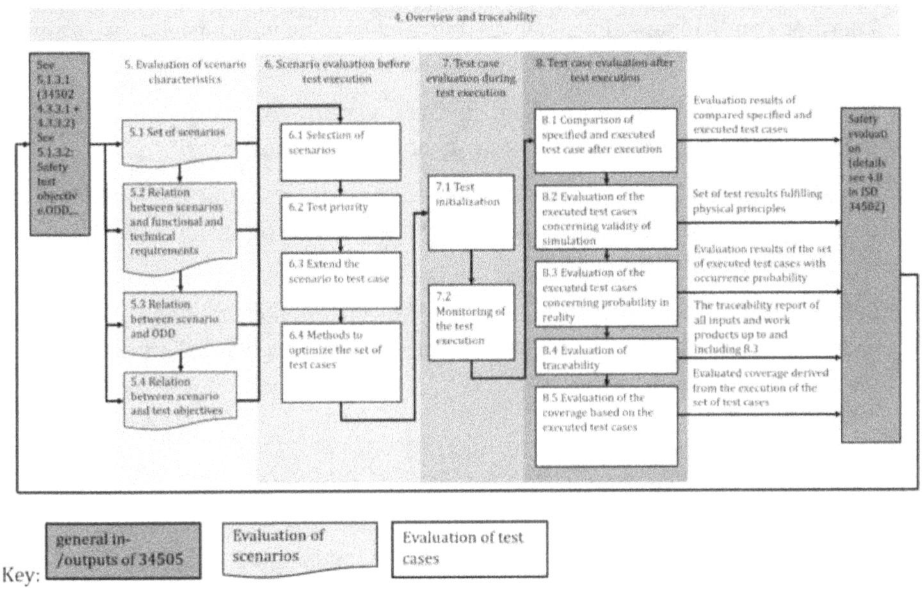

Figure 8.11 Workflow in test context (source: ISO/DIS 34505)

These standards are crucial for the safe development and deployment of Automated Driving Systems (ADS). There are many more standards under development that try to address the risk and how to provide evidence and measures on how to control the risks. Even for cybersecurity, several additional standards are under development and provide a significant basis to develop a safety argumentation for automated driving solutions in public road traffic environments. They provide a comprehensive framework that addresses various safety aspects, from vocabulary and requirements to validation and cybersecurity.

8.4.2 Complexity and Uncertainties

The contemporary automotive landscape, particularly with the advent of autonomous and highly automated driving systems, presents an unprecedented challenge in ensuring comprehensive vehicle safety. It is an axiom that no finite set of tests or simulations can definitively guarantee a vehicle's behavior across the infinite spectrum of real-world conditions. The sheer combinatorics of potential scenarios, encompassing myriad environmental factors, road conditions, and human interactions, render exhaustive validation an unattainable ideal.

Consider the multitude of external influence parameters that can affect vehicle behavior:

- **Weather conditions:** Rain, snow, ice, fog, and varying wind speeds can dramatically alter tire grip, sensor accuracy, and overall vehicle dynamics.
- **Road conditions:** Potholes, uneven surfaces, varying road gradients, and differing road materials introduce a vast array of potential challenges.
- **Traffic scenarios:** The behavior of other road users, including unpredictable actions by pedestrians, cyclists, and other vehicles, creates a constantly evolving and highly complex environment.
- **Sensor limitations:** Sensors can be affected by factors such as lighting conditions, obstructions, electromagnetic interference, leading to potential inaccuracies in perception.

8.4.3 Corridor Approach

The sheer variability of a vehicle's inherent behavior, coupled with the vast heterogeneity of external influences, renders an empirical, test-driven approach to safety assurance fundamentally inadequate. While the potential range of vehicle motion parameters is indeed finite, the combinatorial explosion of interaction between these parameters and external factors creates a virtually limitless landscape of scenarios.

The Corridor Concept: A Layered Approach to Safety

To address this complexity, we must adopt a layered "corridor concept," wherein a graduated series of safety zones surround the vehicle. These zones dynamically expand, or contract based on the vehicle's velocity, reflecting the increasing temporal and spatial demands of higher speeds.

Core Principles and Regulatory Alignment

This concept aligns with the established principles of the Geneva Convention on Road Traffic, which mandates that drivers must maintain control of their vehicles under all circumstances. For ADS, this translates to a proactive adaptation of speed and maneuverability in response to sensor limitations or environmental challenges. For instance, if sensors encounter difficulties due to adverse weather or challenging road geometry (e. g., an approaching hill), the vehicle must reduce its speed accordingly.

Lateral Movement and Urban Challenges

Similarly, lateral movement requires a dynamically adjusted safety corridor. At higher speeds, the vehicle must maintain a wider lateral buffer zone from road edges or Vulnerable Road Users (VRUs). However, this presents significant challenges in urban environments. For example, German traffic regulations stipulate a 1.5-meter clearance when passing cyclists. In dense urban settings, this often necessitates following cyclists at speeds as low as 15 km/h, highlighting the complexity of shared road usage.

Layered Corridor Structure

The corridor concept comprises three distinct layers (Figure 8.12):

- **Green Corridor (1 in Figure 8.12):** This innermost zone represents a "safe" operating space, within which the vehicle can tolerate minor errors and react to foreseeable external influences (e. g., weather variations, other road users' behavior).
- **Yellow Corridor (2 in Figure 8.12):** This intermediate zone signals an impending hazard. Upon entering this zone, the vehicle initiates a Minimal Risk Maneuver (MRM) aimed at returning to the green corridor. This active safety function relies on robust risk monitoring with high confidence levels.
- **Red Corridor (3 in Figure 8.12):** This outermost zone serves as a final safety net. If the yellow corridor's safety mechanisms fail to return the vehicle to the green zone, a more aggressive degradation of vehicle function is triggered, potentially culminating in an emergency stop.

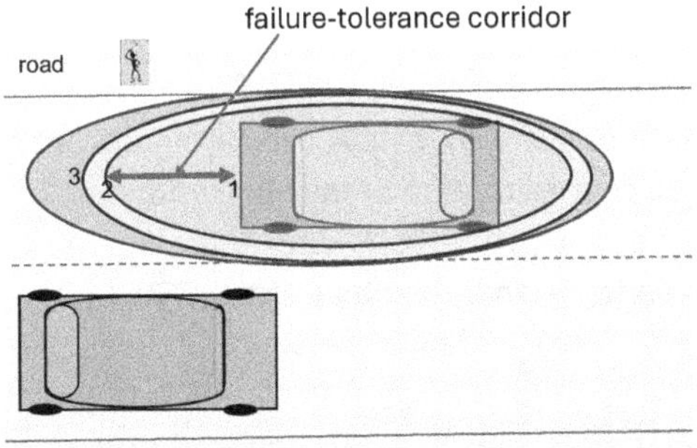

Figure 8.12
Failure tolerance corridor

Proactive Safety and Anticipatory Driving

The key to minimizing red corridor activations lies in proactive, anticipatory driving. By anticipating potential hazards, the vehicle can maintain a comfortable operating margin within the green corridor. However, in unavoidable scenarios, such as a child suddenly entering the vehicle's path, an immediate emergency stop is the only viable response. Even in cases of low confidence regarding such hazards, a significant speed reduction is imperative, achievable through anticipatory driving.

The complexity of automated driving necessitates a shift from purely empirical testing to a layered, corridor-based safety approach. Dynamic adjustment of safety corridors based on vehicle speed and environmental conditions is crucial. Proactive, anticipatory driving is essential for minimizing reliance on emergency interventions. The safety concept must be built in a way that the system can handle errors. The transition from one corridor to the next one must be defined by a clear set of rules.

8.4.4 Application of Safety Corridor Principles

Here is an example how to apply the safety corridor approach (Figure 8.13). We make the following assumptions:

Safety objective: adequate decelerating in case of objects within the driving corridor

We need to consider an upper range and a lower range of tolerance in that safety corridor:

- The upper range of tolerance means: Braking too abruptly can lead to a collision with following vehicles and cause injuries to the occupants.
- The lower range of tolerance means: Braking too low can lead to an accident with the object in the driving corridor.

The green corridor means allowable nominal range and could be considered as a safe corridor (medium light areas in Figure 8.13). In case of human-driven vehicles, the red corridor is limited by the following:

- driver capability to control the situation and perceive the possible objects that could enter the driving corridor,
- vehicle dynamics and the holding forces or grip of the road surface

The corridor approach, which defines a safe path for a vehicle, can be significantly enhanced by integrating it with a grid-based representation of the environment. This combination allows for a more detailed and dynamic understanding of the surroundings, particularly in complex scenarios involving Vulnerable Road Users (VRUs).

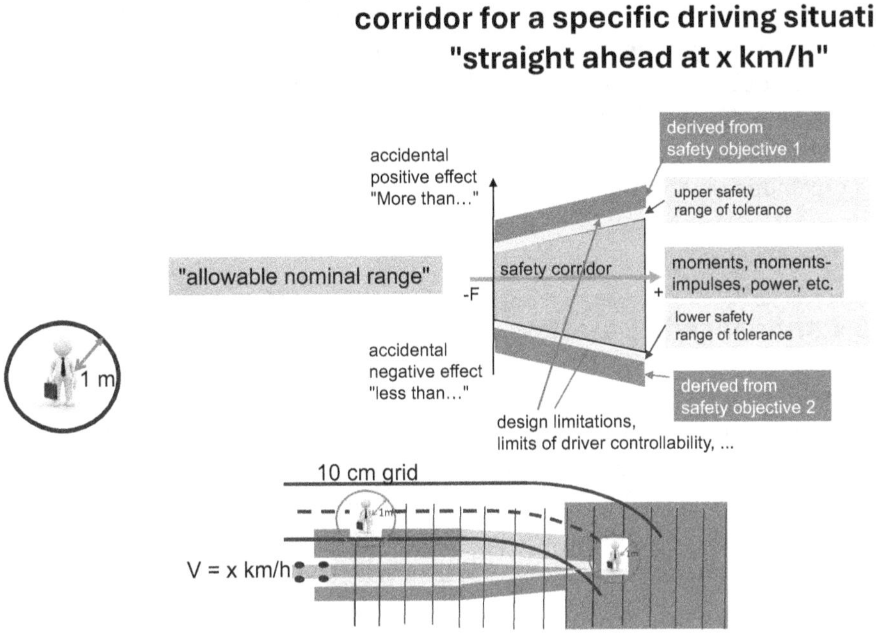

Figure 8.13 Safety Corridor Approach (green = medium light areas, red = dark areas)

Key Aspects of the Combined Approach

Detailed environmental representation:

- By overlaying a grid onto the corridor, free space and the motion profiles of both the vehicle and VRUs can be precisely tracked in both temporal and geometric coordinates.
- This granular level of detail enables the system to identify potential hazards and predict future trajectories with greater accuracy.

Rapid monitoring and assessment:

- The grid-based representation facilitates rapid monitoring of the environment using logically verifiable functions and system states.
- For example, a system state could be defined as: "The road surface is grippy, there is no water or debris on the road, and there are no objects within or approaching the driving corridor."

These states can be quickly evaluated, allowing for real-time risk assessment and decision-making.

By considering the motion profiles of VRUs within the grid, the system can anticipate potential conflicts and take proactive measures to avoid collisions. This is particularly crucial in situations where VRUs may exhibit unpredictable behavior. The combined approach enables the system to optimize the vehicle's trajectory within the corridor, taking into account the surrounding environment and the movement of other road users. This can lead to smoother and more efficient driving, as well as reduced fuel consumption due to accomplishment driving. The integration of corridor and grid approaches provides a robust framework for ensuring safety and efficiency in Automated Driving Systems. The detailed environmental representation and rapid monitoring capabilities enable the system to respond effectively to dynamic and complex situations.

8.4.5 Confidence in Dataflow

There are some aspects to be considered to achieve sufficient confidence in a dataflow in hard real-time capable systems. Here are some core principles and architecture of a communication protocol designed to ensure reliable and predictable communication within time-critical systems, with a particular focus on safety-relevant applications such as autonomous driving. The protocol emphasizes deterministic timing, redundancy, and state awareness to mitigate potential failures and ensure system integrity.

In the realm of safety-critical systems, particularly those employed in autonomous vehicles, the imperative for robust and deterministic communication protocols cannot be overstated. These systems necessitate real-time data exchange with stringent timing constraints to guarantee operational safety and reliability. This paper introduces a protocol structure conceived to address these exigencies through a synthesis of deterministic timing, redundancy, and state awareness.

The protocol's architecture is predicated on the following fundamental principles:

- **Deterministic timing:** All message transmissions and system actions must adhere to strictly defined timeframes. This ensures predictable system behavior and facilitates precise synchronization.
- **Redundancy:** The protocol mandates multiple transmissions of critical data to enable error detection and correction. This redundancy enhances the system's resilience against communication failures.
- **State awareness:** Each message encapsulates information regarding the system's current and intended states. This facilitates accurate state tracking and enables proactive error mitigation.

Protocol Architecture

The protocol's structure comprises the following key elements.

These are the basic message elements:

- **Start and stop bits:** These delimiters delineate message boundaries within the data stream.
- **Sequence or message counter:** This ensures message integrity by preventing message loss and maintaining message order.
- **Timestamp:** This enables precise temporal correlation and synchronization across system components.
- **Nominal value:** This represents the actual data being transmitted (e.g., target speed, steering angle).
- **CRC or hash:** These checksums facilitate error detection and correction.

Dynamic state flag

In addition to the basic message elements, a dynamic state flag conveys information regarding the actuator's intended state transitions. Examples of state transitions include the following:

- state maintenance (e.g., constant speed, steady steering angle)
- acceleration or deceleration adjustments (e.g., incremental changes)
- abrupt state changes (e.g., emergency braking, sudden steering manoeuvres)

Redundant Data Transmission

To augment reliability, critical data is transmitted redundantly, ideally across independent communication channels. The actuator compares received redundant messages to detect errors and validate data plausibility. Cross-comparisons between redundancies are recommended for further enhancement of safety diagnostics, predictive health monitoring, and diagnostics of the degree of synchronicity. In the event of inconsistencies among redundant messages, the actuator initiates a conservative response (e.g., reduced braking force, gradual steering adjustment) until subsequent messages resolve the discrepancy.

Operational Mechanism

The is how the message generation works:

- The sender (e.g., an electronic control unit) generates a message incorporating the basic message elements and the dynamic state flag.
- The message is transmitted redundantly.

This is how the message reception works:

- The receiver (e. g., an actuator) receives the redundant messages.
- It verifies message integrity (CRC/Hash) and compares the redundant data.

State transition implies the following:

- Based on the dynamic state flag and the received data, the actuator executes the corresponding state transition.
- Inconsistencies in redundant data trigger a conservative response.

There are additional alternative protocols or communication principles available, that could support the proposed principles:

- **Time-Triggered Protocol (TTP):** an established protocol for hard real-time systems based on a global time schedule
- **Time-Sensitive Networking (TSN):** a set of standards enabling deterministic communication over Ethernet
- **FlexRay:** a protocol employed in the automotive industry for time-critical applications

This protocol structure is pivotal for the development of safe and reliable autonomous systems. It enables the real-time execution of complex AI algorithms and rapid control loops, which are indispensable for autonomous driving.

8.4.6 Asymmetric Multiprocessing (AMP) for Robust and Reliable Real-Time Systems

In the pursuit of developing dependable and deterministic systems, particularly within safety-critical domains such as V2X (Vehicle-to-Everything) communication and automated driving, the choice of architectural implementation is paramount. Asymmetric Multiprocessing (AMP) presents a compelling alternative to traditional synchronized redundancy approaches, offering significant advantages in terms of performance, fault tolerance, and flexibility.

There are several advantages of AMP versus homogeneous redundancy:

- **Elimination of pre-emptive scheduling:** Unlike synchronized redundancy systems that often necessitate pre-emptive scheduling to maintain temporal alignment, AMP facilitates the dedicated allocation of tasks to individual processor cores. This eliminates the need for pre-emptive interruptions and the associated time overhead incurred by maintaining synchronized time slots. This is a critical factor in hard real-time systems where timing predictability is essential.

- **Enhanced fault tolerance:** AMP systems exhibit inherent tolerance to runtime variances and stochastic errors. Because each processor core operates independently, a delay or error on one core does not necessarily compromise the functionality of other cores. This robustness is vital in dynamic environments where unpredictable events can occur.
- **Flexible comparator decision-making:** AMP affords considerable flexibility in comparator decision-making. The comparator can be configured to utilize the first valid message to trigger an actuator, thereby minimizing latency. Alternatively, in a cross-comparison scenario, the first valid message in the data flow can be forwarded to the next processing level, enhancing system efficiency. This adaptability enables rapid response to events and efficient utilization of available computational resources.
- **Increased performance and reliability:** By the ability to use the first valid message, or in cross comparisons to forward the first valid message, the performance and reliability is increased.

AMP offers an efficient means of implementing redundancy without the drawbacks associated with synchronized redundancy. The ability to leverage the first valid message or, in cross-comparison scenarios, forward the first valid message, enhances both performance and reliability. By harnessing the capabilities of AMP, we can develop robust and reliable systems that meet the stringent demands of hard real-time and deterministic applications. This is of paramount importance for the safety and dependability of future V2X and automated driving solutions. The presented protocol, with its focus on determinism, redundancy, and state awareness, addresses the critical communication requirements of safety-critical systems. Its implementation can significantly contribute to the advancement of autonomous technologies.

8.4.7 Example for Asymmetric Data Processing in a Layered Basic Software

In the realm of embedded systems, particularly those in safety-critical applications like automotive, a layered architecture is essential for managing complexity and ensuring timely execution. AUTOSAR (AUTomotive Open System ARchitecture) provides a standardized framework for this.

The Purpose of Layered Architecture

Abstraction and modularity implies the following:

- A layered architecture separates concerns, allowing developers to focus on specific functionalities without being overwhelmed by the entire system.
- This modularity enhances maintainability, testability, and reusability.

Time management implies the following:

- Different layers handle tasks with varying time constraints.
- This separation allows for optimized scheduling and resource allocation.

Time Constraints in Different Layers

These are the characteristics of the hardware layer (Microcontroller Abstraction Layer, MCAL):

- This layer interacts directly with the microcontroller's hardware.
- It must handle very fine-grained timing requirements, often in the range of microseconds.

Modern microcontrollers can reliably ensure time intervals of 10 microseconds for basic hardware operations.

Basic Software Layer (BSW)

This layer provides essential services like diagnostics, communication, and memory management. Diagnostics, for example, might need to be checked and ensured within a 10-millisecond interval. High-clocked embedded microcontrollers are able to ensure those timeframes.

Application Layer (ASW)

This layer implements the core functionalities of the vehicle, such as steering and braking. Steering systems typically require tasks to be executed within a 10-millisecond time frame. This necessitates a hard round-robin scheduler to guarantee timely execution. Modern brake systems require also hard real-time constraints. A 50-millisecond hard cyclic time is needed. This cannot be guaranteed by a dynamic scheduling.

Hard Real-Time Requirements

Safety-critical systems have hard real-time requirements, meaning that tasks must complete within their deadlines. Failure to meet these deadlines can have catastrophic consequences.

Limitations of Dynamic Scheduling

Dynamic scheduling algorithms, while flexible, may introduce unpredictable delays. For tasks with strict timing constraints, a static, deterministic scheduler like round-robin is often preferred. A layered architecture, particularly in an AUTOSAR context, provides a structured approach to managing time-critical systems. By separating concerns and allocating tasks to appropriate layers, developers can ensure that deadlines are met, and system reliability is maintained. The need for deterministic scheduling, especially in the application layer, is crucial for guaranteeing the safety and dependability of these systems.

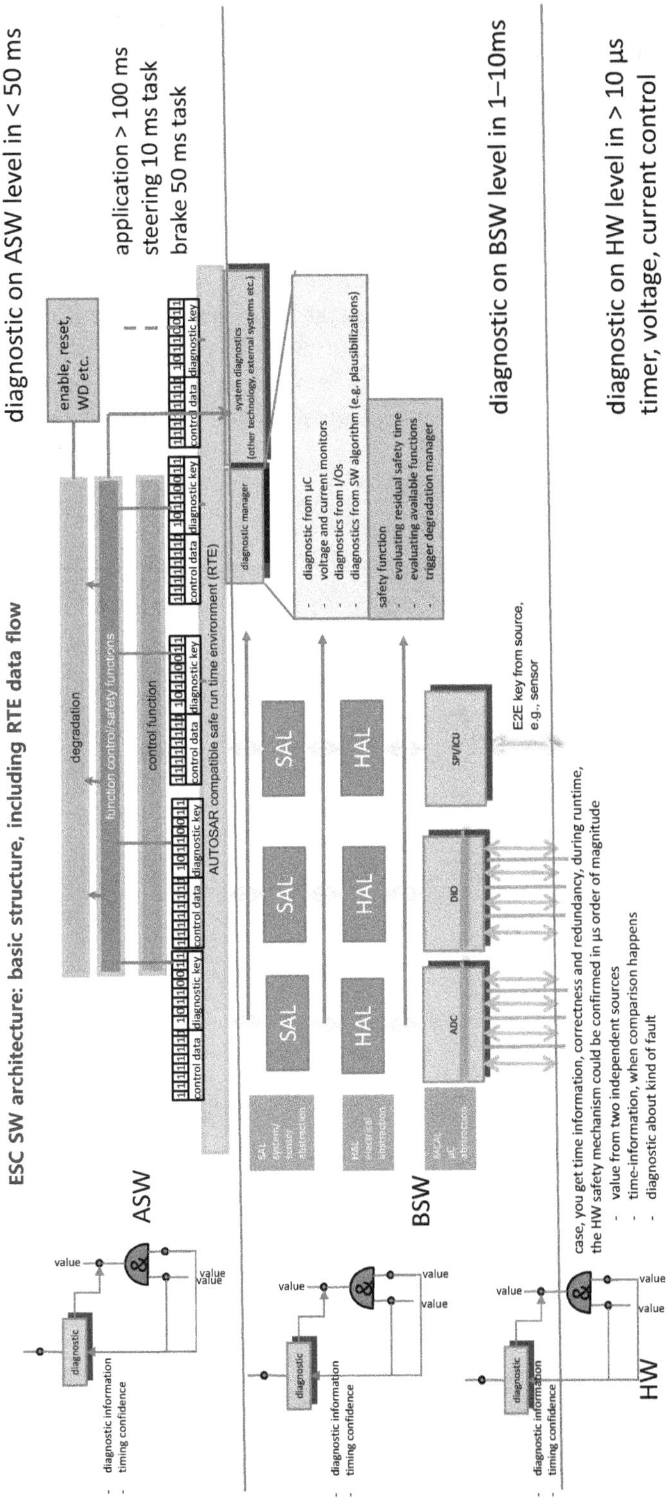

Figure 8.14 Example for a layered time critical application

In the context of safety-critical systems, particularly those adhering to ASIL D safety integrity according to ISO 26262, achieving deterministic real-time behavior is paramount. This necessitates a solution that transcends hardware dependencies and ensures consistent timing across various execution scenarios. The key lies in implementing continuous synchronization and runtime variance compensation within the BSW layer.

The Role of the BSW Layer

The BSW layer serves as an abstraction layer between the hardware (microcontroller) and the application software running on the RTE (Runtime Environment). It provides essential services, including time management, communication, and diagnostics. To achieve hardware independence, the BSW must compensate for variations in hardware performance and timing.

Separation of Intended Control and Control Measures

A separation of functional elements in a safety architecture for high time critical systems is necessary to avoid the possible common cause related to timing and functional impacts (Figure 8.15).

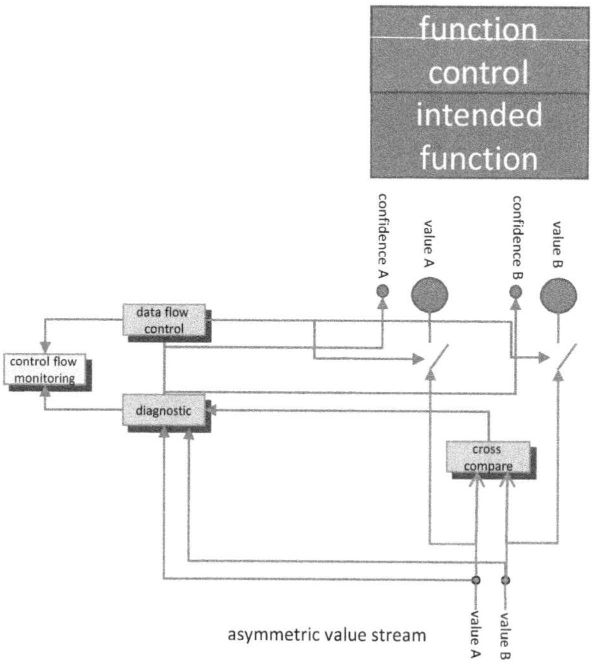

Figure 8.15 Separation in BSW

The following requirements should be considered:

- Data flow needs to be redundant, in case the source of information could not provide any safety confidence.
- The control of data flow needs to be separated due to common causes and effects.
- The monitoring for the correctness (fault monitoring) of the application requires an independent partition from the intended function.
- Deadline, timing or performance monitors are not the same as fault monitors.
- Monitoring of data and control flow shall be sufficiently independent.

Hardware Software Interface to RTE

The hardware and communication interface needs to support the redundant architecture, too. The "handler" which collects the data from the hardware registered needs to be controlled in an adequate time frame. The real-time diagnostic (dt) interval is the key for the performance of a hard real-time capable safety related system (Figure 8.16).

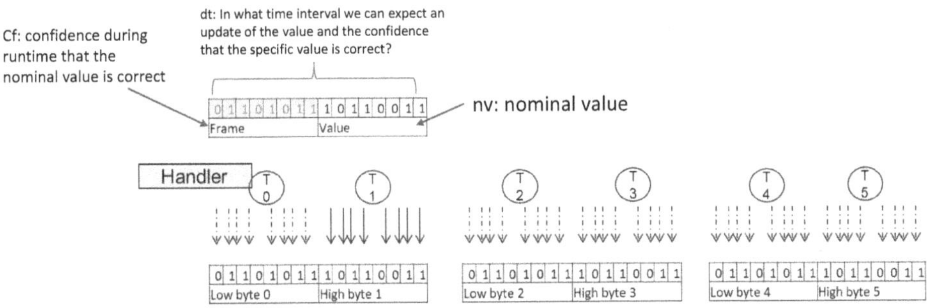

Figure 8.16 Hard real-time capability

The following aspects need to be considered:

- **dt:** This means in what time interval we can expect an update of the value and the confidence that the specific value is correct.
- **Cf:** This means the confidence during runtime that the nominal value is correct.

In embedded systems, diagnostic information and nominal data flow should never be mixed. Only when nominal data and their confirmed confidence are available simultaneously within a deterministic time window at the Real-Time Environment (RTE) can application software ensure safe and timely processing, resulting in correct output of the process.

Continuous synchronization and compensation go along with the following aspects:

- **Runtime variance:** Microcontrollers, even those within the same family, can exhibit variations in execution time due to factors like temperature, voltage fluctuations, and manufacturing tolerances.
- **Synchronization:** To counter these variations, the BSW must implement mechanisms for continuous synchronization. This typically involves using hardware timers and interrupt routines to maintain a consistent time base.
- **Compensation:** The BSW must also compensate for runtime variances by adjusting task execution times and scheduling parameters. This can be achieved through techniques like adaptive scheduling and jitter compensation.
- **RTE determinism:** By continuously synchronizing and compensating for runtime variances, the BSW ensures that the RTE provides a deterministic execution environment, independent of the underlying hardware.

ASIL D and Fault Tolerance

ASIL D systems, designed for the highest level of safety integrity, mandate multi-fault control, which leads to a significant need for failure tolerance. These requirements extend to the BSW, where redundancy and independence must be implemented to mitigate the risk of common-cause failures. From the hardware of the microcontroller to the RTE sufficient independence must be implemented.

Redundancy and Validation

To further enhance reliability, redundancy can be implemented at various levels within the BSW. This redundancy should be continuously monitored, including the validation of time budgets. By monitoring the time budgets on all layers, the first valid data packet can be made available to the RTE. This approach enables the system to detect and mitigate errors, ensuring that the RTE receives accurate and timely data.

Continuous synchronization and runtime variance compensation within the BSW are essential for achieving deterministic real-time capabilities on the RTE. By implementing these mechanisms, along with redundancy and continuous monitoring, safety-critical systems can ensure reliable and predictable operation, independent of hardware variations.

8.5 Methods to Develop a Safety Case

Dynamic systems, such as autonomous vehicles, robotic systems, or complex software applications, are characterized by their ability to adapt to changing environments in real-time. However, this adaptability poses inherent challenges for safety assessment:

- **Complexity and dependencies:** Dynamic systems are often highly complex and exhibit numerous dependencies between their components. These dependencies can lead to unexpected interactions that are not adequately accounted for in static safety cases.
- **Heterogeneous contextual influences:** The performance and safety of dynamic systems are strongly influenced by their context. This context can encompass a variety of factors, such as environmental conditions, user interactions, or other systems with which the dynamic system interacts. These influences are often heterogeneous and can change over time.
- **Non-deterministic behavior:** Dynamic systems can exhibit non-deterministic behavior due to factors such as sensor data noise, unpredictable user interactions, or complex algorithms. This non-deterministic behavior makes it difficult to predict and assess system safety.

The Necessity of Formal and Transparent Safety Argumentation

Given these challenges, it is essential that safety argumentation for dynamic systems follows formal structures and is transparent and comprehensible to the respective stakeholders. This means the following:

- **Formal methods:** The use of formal methods, such as mathematical models, logical argumentation, or formal verification techniques, can help to increase the precision and reliability of safety argumentation.
- **Transparency and comprehensibility:** Safety argumentation should be designed to be transparent and comprehensible to all relevant stakeholders, including developers, testers, regulatory authorities, and end-users. This requires a clear and precise presentation of the assumptions, arguments, and evidence underlying the safety assessment.
- **Dynamic safety cases:** Because static safety cases are not sufficient for dynamic systems, dynamic safety cases are needed. That means, that the safety case has to adapt to the system, and its context. This can for example be achieved by a constant monitoring of the system, and by changing the safety case, if something changes.

- **Continuous safety assessment:** The safety of dynamic systems should not be considered a one-time assessment, but a continuous process that takes place throughout the system's lifecycle. This makes it possible to react to changes in the system or its context and to ensure safety continuously.

The safety of dynamic systems is a complex challenge that requires formal, transparent, and dynamic safety argumentation. By using appropriate methods and tools, we can help ensure the safety of these systems and strengthen stakeholder confidence.

8.5.1 Basic Approach

Effective structuring of safety argumentation can help increase transparency and comprehensibility. One possible approach is to use argumentation patterns, such as Goal Structuring Notation (GSN). GSN makes it possible to structure safety argumentation hierarchically and to clearly represent the relationships between goals, strategies, arguments, and evidence.

Typical Aspects that Should Be Considered in a Safety Case Framework

Avoid long text documents that are outdated tomorrow but rather provide a framework to focus on the actual arguments and link the relevant documents that entail more information with descriptions, for those that require more details.

A safety case framework combines guidance from applicable laws and regulations, governmental organizations, best practices from safety-critical industries, voluntary industry standards and consortia, academic research, and what an organization has learned in its own work. In the automotive industry, it's an important tool for developing automated vehicles that will operate safely on public roads and deliver these services to customers, and the public. It evaluates the entire development lifecycle for both occupancies and vulnerable road users and should provide evidence for all stakeholders.

The safety case framework should support continuous learning and address testing during development as well as during operations until the end of the vehicle's lifecycle. UL 4600 provides such a framework as it requires claims supported by arguments and evidence. In general, the safety case must encompass all aspects of safety, including difficult-to-reproduce aspects of the item and include accepted risk.

UL 4600 defines "What is a safety case?" as follows: A safety case is a structured argument, supported by a body of evidence that provides a compelling comprehensible and valid case that a system is safe for a given application in a given environment (Koopman 2022).

Claims (C): statements regarding safety or related properties that are held to be true within a stated context (e. g., mid-level claim: "The Vay TD system does not collide with motorcycles"—if the claim is true, that portion of safety case likely supports a top-level claim such as "The Vay TD system is acceptably safe")

Strategy or process argument: If a claim is to be subdivided into several subclaims (e. g., because there are several steps), a process argument is required that proves which method or process is used for this subdivision. A strategy is also possible—but here you also need evidence.

Arguments (A): elements of the safety case that support the claims by presenting reasoning why the claim is thought to be true

Evidence (E): This is the information that supports an argument. Using such a structured safety case argumentation (also in the application of relevant standards and literature) supports also the legal safety argumentation and the general "defensibility" of potential product liability or connected claims.

Global argumentation categorization: For the classification of the argumentation, a global claim with subclaims is defined. This global claim and the subclaims are necessarily not final but are only intended to serve as a global classification for introduction to this specific argumentation.

8.5.2 Goal Structure Notation

Effective structuring of safety argumentation can help increase transparency and comprehensibility. Combining the approach from UL 4400 with formal notations should be recommended. Goal Structuring Notation (GSN) makes it possible to structure safety argumentation hierarchically and to clearly represent the relationships between goals, strategies, arguments, and evidence.

GSN is a graphical notation employed to articulate safety arguments. It serves to structure the argumentation for the safety of a system or product in a clear and comprehensible manner. The GSN is predicated on a hierarchical structure, comprising various constituent elements. The principal elements are as follows:

- **Goals:** These elements represent the safety objectives that are to be achieved. For example, within the automotive domain: "The braking system must function reliably at all speeds."
- **Strategies:** Strategies delineate the approaches by which the goals are intended to be realised. For instance: "The reliability of the braking system is ensured through redundant sensors and actuators."
- **Solutions:** Solutions are the concrete measures that are implemented to execute the strategies, for example, "implementation of a fault-tolerant algorithm for the processing of sensor data."

- **Contexts:** Contexts specify the framework within which the argumentation is valid, for instance, "The argumentation is valid for vehicles approved for use in Europe."
- **Assumptions:** Assumptions are premises that are made for the validity of the argumentation, for example, "The sensors of the braking system are correctly calibrated."
- **Justifications:** Justifications provide rationales for specific elements of the argumentation, for instance, "The utilisation of redundant sensors enhances reliability, as a failure of one sensor can be compensated for by the other sensors."

Example for GSN Within the Automotive Domain Considering Automated Driving

Goal: "The autonomous vehicle must be capable of navigating safely in complex traffic situations."

Strategy: "utilisation of multi-sensor fusion for the perception of the environment"

Solution: "implementation of an algorithm for object detection and tracking"

Example for the Electric Drive Systems

Goal: "The high-voltage system of the electric vehicle must be protected against short circuits."

Strategy: "utilisation of fuses and circuit breakers and independent redundancy."

Solution: "installation of two independent battery systems with high-performance fuses in the battery line"

Example for Advanced Driver-Assistance Systems (ADAS)

Goal: "The lane keeping assist system must maintain the vehicle safely within the lane."

Strategy: "Utilize cameras, radar for near field information and a map for long range vision. Determine continuously the confidence of the sensor data during runtime."

Solution: "Implement redundant independent algorithms for the detection of lane markings and the structure of the road edge and compare them with map and global positioning system (GPS)."

The GSN facilitates the representation of complex safety arguments in a structured and comprehensible manner. This is particularly crucial within the automotive domain, where the safety of vehicles is of paramount importance.

8.6 Summary and Conclusion

To introduce new technologies into public road traffic, a comprehensive understanding of the risks to the involved individuals is essential. People assume various roles, from stakeholders who seek to profit from the development and manufacture of the transport means, to those who wish to use, own, or offer services with it. This includes developers, those who aim to introduce the transport means into traffic, inspectors, and those responsible for corporate and regulatory approval, homologation, or certification. Additionally, there are the occupants, passengers, vulnerable road users, politicians, and interest groups who influence public opinion on the success of the transport means. Society, in particular, demands environmentally friendly solutions that do not introduce new hazards. This book could only provide examples and food for thought, presented as concisely as possible and from various perspectives. A consistent systems engineering approach forms the basis for considering societal aspects alongside technical ones, offering a tolerable level of expected safety.

References

ISO 26262:2011: Road vehicles – Functional safety. Scope

ISO 34501: Road vehicles – Test scenarios for automated driving systems – Vocabulary. 2022

ISO 34502: Road vehicles – Test scenarios for automated driving systems – Scenario based safety evaluation framework. 2022

ISO 34503: Road Vehicles – Test scenarios for automated driving systems – Specification for operational design domain. 2023

ISO 34504: Road vehicles – Test scenarios for automated driving systems – Scenario categorization. 2024

ISO/DIS 34505: Road vehicles – Test scenarios for automated driving systems – Scenario evaluation and test case generation. Final Draft

Koopman, P.: The UL 4600 Guidebook. What to Include in an Autonomous Vehicle Safety Case. 2022

Index

A

T

U

V

W

X

Y

www.ingramcontent.com/pod-product-compliance
Lightning Source LLC
LaVergne TN
LVHW060615300926
849162LV00008B/1627